Frontiers of Ferroelectricity:

A Special Issue of the Journal of Materials Science

Frontiers of Ferroelectricity:

A Special Issue of the Journal of Materials Science

SYDNEY B. LANG
Ben-Gurion University of the Negev
Department of Chemical Engineering
Beer Sheva, Israel

HELEN L.W. CHAN
The Hong Kong Polytechnic University
Department of Applied Physics
Hung Hom Kowloon, Hong Kong

Springer

Sidney B. Lang
Department of Chemical Engineering
Ben-Gurion University of the Negev
Ben-Gurion Boulevard
Beer Sheva, Israel

Helen L.W. Chan
Department of Applied Physics
The Hong Kong Polytechnic University
Hung Hom, Kowloon
Hong Kong

Frontiers of Ferroelectricity:
A Special Issue of the Journal of Materials Science

e-ISBN 0-387-38039-6
ISBN 978-1-4419-4254-8 e-ISBN 978-0-387-38039-1

Printed on acid-free paper.

9 8 7 6 5 4 3 2 1

springer.com

Contents

CONTENTS BY KEYWORD

Preface

The field of ferroelectricity is a very active one. Many hundreds of papers in this field are published each year and a large number of local and international conferences are held. We felt that it would be appropriate at this time to publish a set of papers in a single journal describing some of the most active areas in the field. The Journal of Materials Science agreed to publish a special issue on ferroelectricity. Accordingly, we sent requests for papers to a number of research groups around the world. It was difficult to select a small number of groups from among the many excellent ones in the field and we apologize to those not included. We received 24 manuscripts from groups in North America, Asia and Europe, each one of which was reviewed by two referees. The papers include reviews and current research, both experimental and theoretical. It was especially satisfying that the authors included not only established researchers but also many younger people who are destined to continue in the field in the future. The special issue entitled "Frontiers of Ferroelectricity" appeared as Volume 41, Issue 1 of the Journal of Materials Science in January 2006. Because we believed that many researchers and students would find great value in having the complete set of papers on their bookshelf, we suggested to the editors of Springer that *Frontiers of Ferroelectricity* should be published in book form. They concurred and this volume is a result of their efforts. We wish to thank the authors and the referees for their contributions to this work. We hope that *Frontiers of Ferroelectricity* will be of value to those in the field and will encourage others to join in this research effort.

Helen L.W. Chan
Department of Applied Physics
Hong Kong Polytechnic University
Hong Kong, China

Sidney B. Lang
Department of Chemical Engineering
Ben-Gurion University of the Negev
Beer Sheva, Israel

JOURNAL OF MATERIALS SCIENCE 41 (2006) 1–11

Self-assembled nanoscale ferroelectrics

MARIN ALEXE, DIETRICH HESSE
Max Planck Institute of Microstructure Physics, D-06120, Halle, Germany

Multifunctional ferroelectric materials offer a wide range of useful properties, from switchable polarization that can be applied in memory devices to piezoelectric and pyroelectric properties used in actuators, transducers and thermal sensors. At the nanometer scale, however, material properties are expected to be different from those in bulk. Fundamental problems such as the super-paraelectric limit, the influence of the free surface, and of interfacial and bulk defects on ferroelectric switching, etc., arise when scaling down ferroelectrics to nanometer sizes. In order to study these size effects, fabrication methods of high quality nanoscale ferroelectric crystals have to be developed. The present paper briefly reviews self-patterning and self-assembly fabrication methods, including chemical routes, morphological instability of ultrathin films, microemulsion, and self-assembly lift-off, employed up to the date to fabricate ferroelectric structures with lateral sizes in the range of few tens of nanometers.

1. Introduction

Ferroelectrics are, and will be, widely used in many applications that require sizes down to the nanometer range [1]. It is a challenge to fabricate structures in this range using both lithography ("top-down" approaches) and self-assembly methods ("bottom-up" techniques). Whereas conventional lithographic systems work usually with a resolution of about 100 nm, self-assembly approaches are mainstream methods for the fabrication of structures smaller than 50 nm.

All lithography-based patterning techniques are top-down approaches and represent the basis of nowadays microelectronics technology. They are very well suited to fabricate structures with very good spatial resolution and positioning precision, but are intrinsically limited to structures the sizes of which are well above the lowest nanometer-scale sizes. Thus there is a considerable interest in other fabrication methods which are not based on carving thin films, but rather on building structures from the bottom using atoms and molecules. These bottom-up methods will in principle allow the inexpensive fabrication of structures with sizes of 10 to 20 nm in a parallel approach. The primary disadvantage of the bottom-up methods is the random positioning of the obtained nanostructures that will make a precise interconnection of them virtually impossible.

Successful strategies and routes have been developed to synthesize nanoscale materials of numerous simple systems such as semiconductors (Si, CdS, InAs/GaAs) or metals. Complex systems such as ferroelectric oxides, or any muticomponent oxides belonging to the class of functional materials, are not systematically addressed so far, in spite of the general possibility to discover a new class of nanomaterials with unique properties. Recently, the first attempts to prepare ferroelectric nanostructures *via* bottom-up approaches have been published. The approaches can already be classified in two main routes, *viz.* physical and chemical routes, which are obviously based on different principles. The general characteristics of both routes are the following.

1.1. Self-patterning via physical routes

It was recognized that crystallization at all scales, but mostly at nano- and mesoscopic scales, is an important self-assembly strategy in the synthesis and fabrication of small structures. Well-known physical growth concepts, such as island growth, have already been successfully applied to grow nanosize dots of germanium on silicon, or of InAs on GaAs. An important role is played by the lattice mismatch between the substrate and the deposited film. Depositing an epitaxial thin film onto a single-crystal substrate with high lattice mismatch, the initial stages of the growth process are characterized by either the island (Volmer-Weber) or layer-than-island (Stranski-Krastanov) growth modes. In contrast to the layer-by-layer (Frank-van der Merwe) growth mode, which results in a smooth uniform film, the two former growth modes

0022-2461 © *2006 Springer Science + Business Media, Inc.*
DOI: 10.1007/s10853-005-5912-x

are suitable to grow crystalline nano-size dots. The effect is well known and has been carefully studied in the case of simple systems such as germanium on silicon [2], and has even found application in the growth of compound semiconductor lasers [3].

1.2. Self-patterning via chemical routes

A number of interesting fabrication approaches appear to be based on chemical routes. Using simple chemical routes it is possible to fabricate nanosize crystals or nanoparticles in colloidal suspensions, which can later be spread onto any substrate surface. After evaporating the solvent, the nanoparticles can be crystallized in two- or even three-dimensional arrays. Although obtaining a regular array of particles is not an easy task, self-assembly concepts based on the reduction of the interfacial energy at the fluid-solid or fluid-fluid interface—concepts that have proven valid for simple systems such as CdS [4]—can be used for more complicated systems such as ferroelectric oxides. One of the most promising routes is based on the microemulsion concept in which a water-in-oil emulsion is produced using a surfactant. Such an emulsion consists of nanometer-size water droplets uniformly dispersed in an oily solvent. This microemulsion can be used to hydrolyse a complex metalorganic precursor. The nanodroplets act as nano-reactors in which the hydrolysis of the precursors takes place and, if the optimum conditions are fulfilled, the final reaction product consists of monodisperse nanosize particles [5].

2. Self-patterning via physical routes

2.1. Self-patterned nano-electrodes

From the historical point of view the first paper that introduced the concept of ferroelectric nanocells was most probably the paper on nanosize bismuth oxide electrodes obtained by self-assembly [6]. Although not concerned with intentional fabrication, it yet triggered a first concerted effort to obtain and characterize nanoscale ferroelectrics. The authors had observed that during deposition of epitaxial $Bi_4Ti_3O_{12}$ (BiT) thin films onto silicon substrates, under certain deposition conditions rectangular planar arrays of bismuth-containing crystalline electrodes showing metallic conductivity—"nano-electrodes"—can be obtained on the surface of the epitaxial BiT thin films (Fig. 1).

The bismuth titanate film was deposited on top of an epitaxial layer of $La_{0.5}Sr_{0.5}CoO_3$ (LSCO) on Si(100), the LSCO layer serving both as an electrode and as an epitaxial template for the ferroelectric film. The nanoelectrodes were uniform and well separated from each other. Depending upon processing conditions the lateral size of the nanoelectrodes was typically 150 nm, yielding an equivalent density of 1 Gbit of ferroelectric cells per chip.

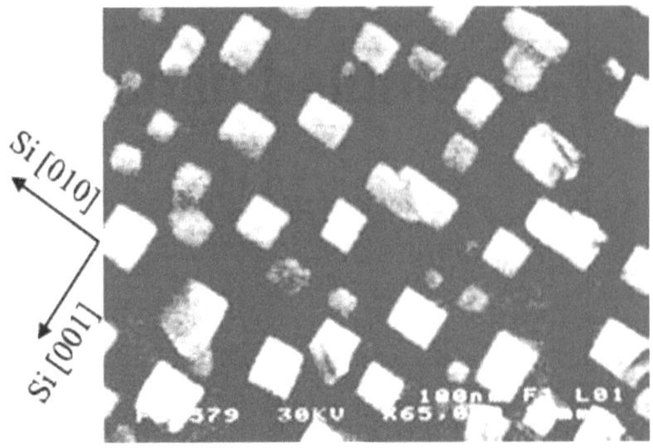

Figure 1 Plan view SEM micrograph of a bismuth-containing nanoelectrode array [6].

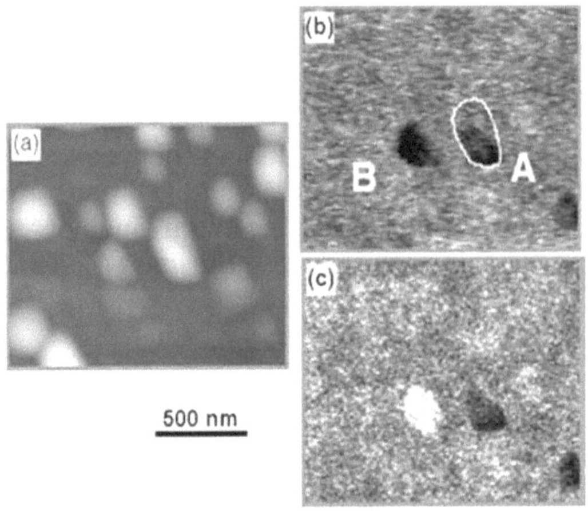

Figure 2 Topography image (a) and piezoelectric images (b, c) showing switching of a single memory cell. The contour of the cell marked "A" has been determined from the topography image. The cell "B" is shown before (b), and after (c) applying a single voltage pulse of +10 V for 100 ms.

The proposed growth mechanism is based on the fact that in bismuth-layered perovskite films any superficial bismuth segregates out as metallic, elemental bismuth at the surface. The deposition taking place in an oxygen atmosphere, the nanoelectrodes were mainly composed of crystalline cubic δ-Bi_2O_3, as determined from high-resolution transmission electron microscopy. Cubic δ-Bi_2O_3 has a defective fluorite structure and is the best ionic conductor known, with a resistivity of only 1 Ωcm at 1023 K [7]. Being ionic conductors, the bismuth-containing islands formed mesoscopic capacitors and offered the possibility of locally measuring the ferroelectric properties of the epitaxial c-oriented BiT film. Probing the nanocells *via* a conductive AFM tip and using piezoresponse force microscopy (PFM), local switching (see Fig. 2) and local hysteresis measurements were performed [8].

This work was significant, because it demonstrated the possibility of decreasing the size of ferroelectric memory cells into the sub-micron range, with a corresponding

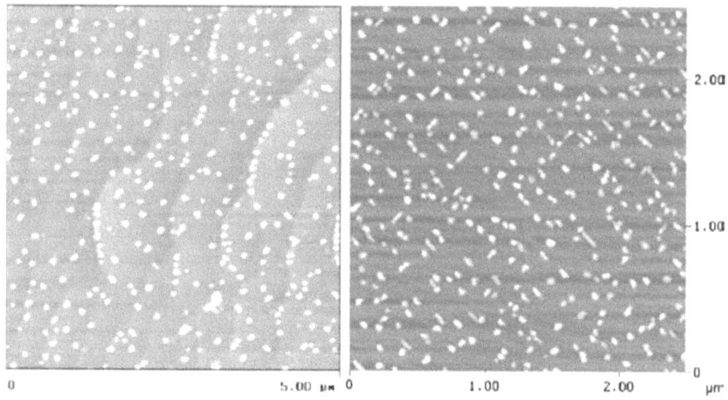

Figure 3 $BaTiO_3$ nanostructures deposited onto vicinal $SrTiO_3$ by pulsed laser deposition.

increase of the memory density into the Gbit range, by the time when the commercial and the R&D prototypes of ferroelectric random access memories (FRAMs) had only kBit densities and were using memory cells of the size of few tens of microns. The self-assembly concept was thus introduced into the fabrication of nanoscale ferroelectrics; however, lateron it was applied to fabricate ferroelectric nanostructures rather than nanoelectrodes.

2.2. Self-assembled nanostructures by physical vapor deposition

Very well known methods used for thin film deposition have been among the first methods used to fabricate nanosize structures. These are true self-assembly methods which make use of atoms as building blocks, and of physical growth concepts such as island growth, in order to grow nanosize structures. As it was already pointed out, the main role is played by the mismatch in the lattice constant during the initial stages of the growth process. This mismatch determines different growth modes such as the Stranski-Krastanov or the Volmer-Weber mode, in this way achieving the growth of crystalline nanosize structures instead of a continuous film. Assembling nanostructures is achieved *via* different mechanisms based on strain fields. This approach can be used to induce a *vertical* self-assembly, like in the case of germanium islands on silicon, where the germanium islands can "reproduce" themselves in a vertical direction *via* their strain fields [9]. Based on the same assembly concept, using strain fields generated by pre-patterned [10] or implanted [11] wafers, a *lateral* self-assembly can also be induced. This concept was successfully applied to simple systems (such as germanium on silicon) and even applied to the preparation of compound semiconductor lasers [3]. However, very little has been done so far—using the above concepts—to obtain nanostructures made from complex oxides. Only recently the growth of complex oxide islands on different substrates using several deposition methods has been achieved. Pulsed laser deposition (PLD) and metalorganic chemical vapor deposition (MOCVD) were the first deposition methods employed.

Pulsed laser deposition and ablation are potential methods to obtain nanoparticles either deposited on a substrate or simply in the form of a dispersed powder. Pulsed laser ablation has been used to produce monodisperse PZT nanoparticles with a fairly narrow size distribution. The experimental setup is rather complex and consists of a laser ablation chamber, a charger, and a furnace. Single crystalline nanoparticles with an average diameter of about 7 nm and a standard size deviation of 5.4% were obtained [12]. Using a simple PLD deposition process, Visinoiu *et al.* have grown $BaTiO_3$ nanostructures on $SrTiO_3$ (STO) substrates [13]. Furthermore, using vicinal STO substrates a certain degree of registration and self-organization can be induced as shown in Fig. 3.

In the case of systems that have a high lattice mismatch, the same concept of island growth leads to the formation of self-assembled structures during the initial growth stages of $SrRuO_3$ (SRO) films on $LaAlO_3$ (LAO) substrates [14]. In the initial growth stage (below a thickness of three monolayers) a periodic ripple forms as a result of stress release *via* the creation of an additional roughness. Further nucleation is confined to the ripples. As a consequence, at a film thickness of about 2.5 nm, three-dimensional islands assemble on the ripples in a row-like pattern. Due to an anisotropic grain coalescence, most probably originating from an anisotropic diffusion coefficient, a compact regular array of wires results [15]. The nanopatterned SRO films show metallic conductivity and are ferromagnetic (as are bulk films). They show a resistivity of about 280 $\mu\Omega$cm at room temperature, and a ferromagnetic Curie temperature of about 155 K, slightly lower than the bulk value, due to the presence of stress.

Concerning the growth of nanosize ferroelectric structures employing island growth, the best results so far were most probably obtained using MOCVD. It was found by Shimizu *et al.*, that during initial growth stages of $PbTiO_3$

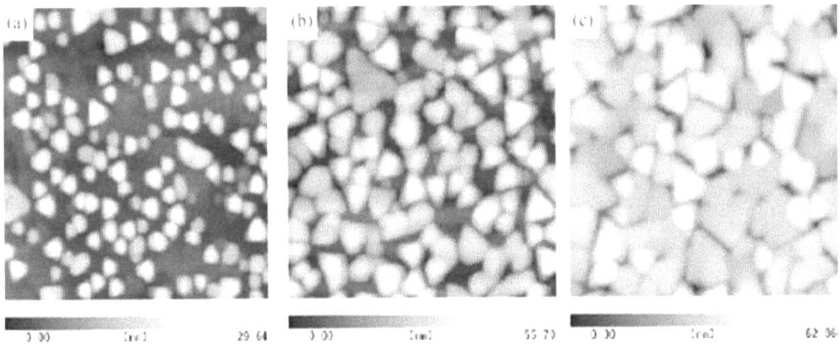

Figure 4 PZT (Zr/Ti = 24/76) island structures deposited for (a) 1 min, (b) 3 min and (c) 7 min. Scan area is 1×1 μm. [18].

(PTO) and PZT films on Pt substrates, islands were observed before a continuous film formed [16]. In the case of PZT films, triangular-shaped PZT islands were observed independently of the Zr/Ti ratio (Fig. 4). The triangular shape of the PZT islands directly reflects the (111)-orientation of the perovskite lattice. A very short deposition time of a few seconds resulted in triangular islands, the size of which increased from 50 nm to 160 nm as the deposition time increased from 30 s to 3 min. At the same time the height increased from 35 to 70 nm, while at longer deposition time the islands coalesced into a continuous film. For a certain deposition time the width and height of the islands decreased as the Zr content increased. This result corresponds to the fact that a higher growth temperature is required to crystallize Zr-rich PZT than Ti-rich PZT. Transmission electron microscopy (TEM) observations revealed that these PZT islands have tetragonal or rhombohedral structures, and that corresponding PTO islands have a twinned structure showing 90° domains originating in the tetragonal structure. This was taken as an initial hint to the nanosize island showing ferroelectricity [17].

PFM measurements indeed proved that the MOCVD-obtained PZT islands are ferroelectric, but the minimum width and height of the islands that exhibit ferroelectricity were relatively large, *viz.* 50 and 20 nm, 70 and 30 nm, and 70 and 8 nm for Zr/Ti ratios of 0/100, 24/76 and 74/26, respectively [18]. The poor ferroelectric behavior was attributed to a degradation of ferroelectricity due to internal stress that is retained in the isolated islands.

An interesting method to pre-define the position of the structures by inducing nucleation sites has been shown by Buhlmann and Muralt [19, 20]. A TiO_2 nucleation site is lithographically pre-defined using electron-beam lithography and etching of a TiO_2-coated substrate. The resulted TiO_2 dots are only locally increasing the nucleation probability of the PZT and are acting as nucleation sites for the subsequent sputter deposition of PZT. In such way, ferroelectric structures are formed only on the TiO_2 dots and not on the remaining free surface. Ferroelectricity in 120–150 nm lateral size structures has been subsequently shown by PFM.

2.3. Microstructural instability

One of the most convenient methods to obtain oxide nanoscale structures uses the effect of microstructural instability of ultrathin films. Although this method is frequently named a "chemical route", it is actually a physical self-patterning method. A chemical deposition method is used just to obtain an ultra-thin oxide layer on a substrate of choice, but the self-patterning process itself takes place in the later stage of high-temperature annealing. Briefly, using sol-gel or other metalorganic routes a precursor solution is prepared [21, 22]. This precursor solution is spun on a substrate in order to obtain a thin metalorganic gel film that subsequently is transformed into an amorphous oxide film by pyrolysis, i.e. by a thermal annealing process at a moderate temperature of 250 to 300°C in air or oxygen. During the subsequent crystallization anneal, depending on film parameters and annealing conditions, the system transforms either into a normal thin film, or it patterns "itself" into nanostructures of a large variety of shapes and sizes. Seifert *et al.* [23] were the first to observe microstructural instabilities in perovskite thin films, following previous work on yttria-stabilized zirconia (YSZ) [24]. For example, PTO films on $SrTiO_3$ (STO) substrates, with a nominal film thickness below 100 nm, develop holes during the crystallization anneal (performed in order to convert them into single crystal films). Upon further annealing these holes grow to a stable size, or even cause breaking up of the film into isolated, single crystal islands. In this way relatively thick films entirely cover the substrate, whereas thinner films develop faceted pits and/or holes and finally fall apart into single crystal islands. The above phenomenon is described as a microstructural instability based on the minimization of the global surface energy: During the crystallization anneal, initially isolated pores (that are always present due to the CSD process) grow into surface pits that eventually transform into holes and uncover the substrate. This process depends on the film thickness and pore spacing and implies four different stages [23]. The first stage involves the growth of pyramidal holes until they reach the film-substrate interface. In the second stage the holes continue to grow as truncated pyramids, until they

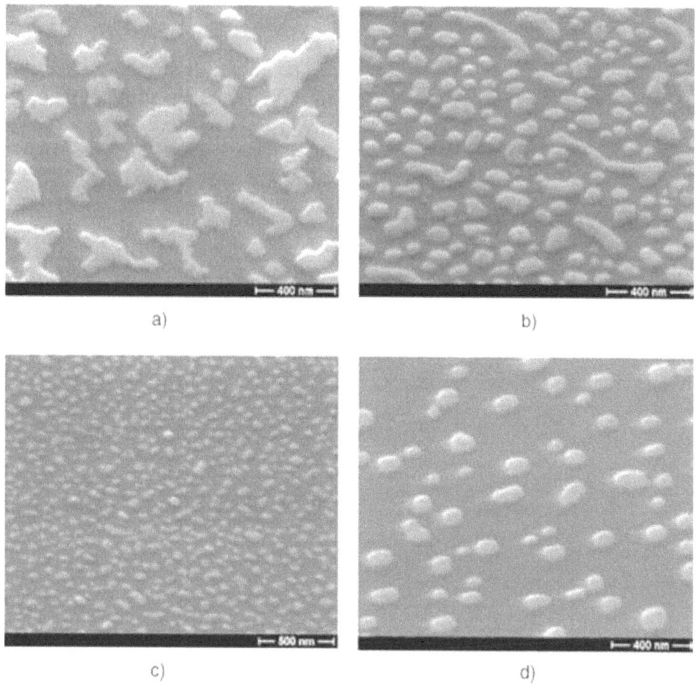

Figure 5 SEM images of PZT nanoislands obtained after deposition of (a) 1:20, (b) 1:25, (c) and (d) 1:40 diluted PZT precursor and after 800°C (a–c) and 1100°C (d) thermal treatment.

become connected. In the third stage the film is transformed into a collection of truncated pyramids still connected at the lowest edge, until finally the film becomes a collection of unconnected islands in the fourth stage.

The above well-described effect of instability of ultrathin films was used as a "bottom-up way" to fabricate nanoscale ferroelectric structures. Waser *et al.* obtained polycrystalline PTO nanoscale structures on Pt-coated silicon substrates using a highly diluted precursor [25]. In this way structures with lateral sizes down to 20 nm were successfully prepared. Using PFM, Roelofs *et al.* showed that such PTO grains with lateral sizes of about 20 nm do not exhibit the out-of-plane piezoelectric signal, suggesting that the size at which ferroelectricity vanishes in $PbTiO_3$ is about 20 nm [26].

If a single crystal substrate is used, the obtained structures are epitaxially oriented as was shown by Szafraniak *et al.* They obtained epitaxial single crystal ferroelectric structures using the same concept of microstructural instability [27]. Nanosize epitaxial PZT crystals with lateral dimensions of 40 to 90 nm and a thickness of 9 to 25 nm were obtained by conventional annealing of ultrathin amorphous oxide films at temperatures above 800°C (see Fig. 5). The size, shape and distribution of the nanocrystals could be tuned to some extent by modifying the initial film thickness and the crystallization temperature. The epitaxial nature of the crystals was revealed both by X-ray diffraction analysis and high resolution TEM investigations.

Dawber *et al.* [28] analyzed the shape and size distribution of both the self-patterned nanoelectrodes and the above PZT nanostructures. They discussed the results of the analysis in terms of island formation mediated by repulsive interactions between the islands and the substrate *via* strain fields, in a similar way as it had been described for Ge islands on Si(100) before [29, 30]. Dawber's theory predicts (i) three different kinds of structures such as pyramids, domes, and superdomes, (ii) the volume distribution and (iii) a shape map of the relative population of the structures, both as a function of coverage and crystallization temperature. The self-patterned nanoelectrodes showed a bimodal distribution (see Fig. 6), whereas the shape of PZT structures on STO - due to the very low thickness-transfers from the coexistence of superdomes and domes to the dominance of domes. Apparently this was the first time that the Shchukin-Williams theory was applied to a more complex system than a semiconductor. In addition this approach might offer a practical base for the registration of the islands using strain fields and engineered substrates.

The high quality of the PZT nanocrystals obtained by the above self-assembly method has enabled a detailed structural study performed by Chu *et al.* [31] on the impact of misfit dislocations on the polarization instability in nanoscale ferroelectrics. The main result of the high

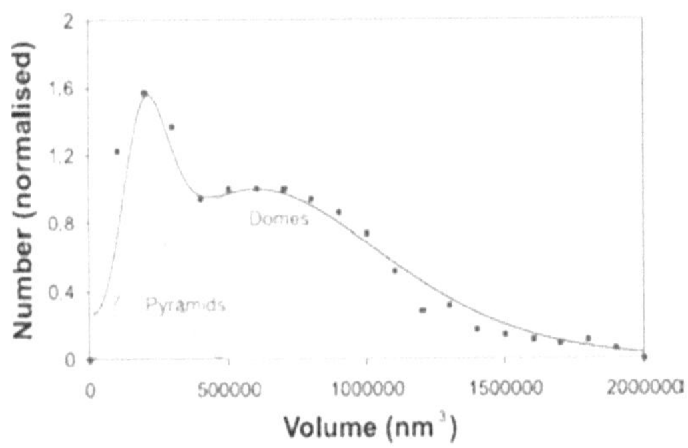

Figure 6 Bimodal distribution of bismuth oxide nanoelectrodes showing the coexistence of pyramids and domes and the excellent fit to the distribution function of Wiliams *et al.* [30].

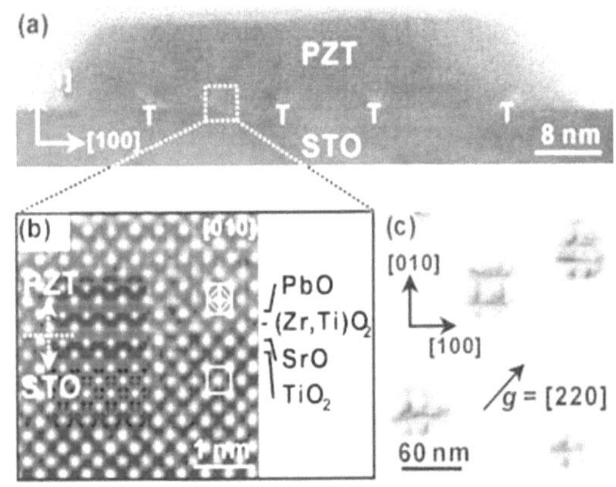

Figure 7 Cross-sectional and plan-view micrographs and calculated contrasts of the PZT (48/52) nanocrystals. (a) Cross-sectional HREM image of a (001)-oriented PZT nanoisland in [010] projection; misfit dislocations indicated by T. (b) An enlarged interface zone of a; the rectangle and square indicating respective PZT and STO lattices; inset, multislice contrast simulation at $t = 4$ nm and $\Delta f = -60$ nm with PbO-(Zr,Ti)O_2-SrO-TiO_2 stacking across the interface. At these imaging conditions, bright and dark contrasts represent the cation and anion columns, respectively. (c) Bright-field plan-view image recorded under $g = [220]$ on specimens annealed at 950°C for 1h, the dark contrasts showing the network of misfit dislocations [31].

resolution TEM investigations was the observation of a high-strain region formed around each misfit dislocation (Fig. 7). (Such misfit dislocations generally occur to relax the strain in epitaxially grown structures and films). For the studied case of PZT (48/52) nanoislands grown on STO substrates, the cross section of the highly distorted "tube" around each dislocation is about 8 nm in width and 4 nm in height. In this volume the tetragonal lattice is distorted due to triaxial strain fields, so that the ferroelectric polarization of an island with a height of about 9 nm and a lateral size of 20 nm becomes unstable. This was interpreted as a size effect which actually originates from an *extrinsic* effect (*viz.* the misfit dislocations) rather than from the super-paraelectric limit. It was also shown that avoiding the formation of misfit dislocations at the interface, the ferroelectricity could be retained in very small structures of this type.

The above studies have clearly shown that on the one hand high-quality structures approaching ideal systems are required to evidence true size effects, and that on the other hand single defects in these nanoscale structures can be most detrimental for the ferroelectric properties. In other words, the extrinsic size effects generated by interfaces, surfaces, strain, compositional inhomogeneities, etc. can have a higher negative impact on the polarization stability than intrinsic size effects.

3. Self-patterning via chemical routes

3.1. Hydrothermal growth

Hydrothermal synthesis is a unique technique for the low-temperature fabrication of powders or films. It involves the growth of oxides from an aqueous solution at moderate temperatures (about 100–400°C) and pressures (0.1–15 MPa). During the last decade this method was applied to produce various oxides at low temperatures [32]. Perovskite ferroelectrics can be easily synthesized by this technique at temperatures below 200°C. For instance, barium titanate was grown even at 90°C without any additional post-deposition thermal treatment [33–35]. Hydrothermal growth can be used as a bottom-up strategy in order to obtain complex oxide nanoscale structures. Usually, one needs to mix an equimolar amount of precursors in an aqueous solution with a very high pH value, and subject the solution for a certain time to a low-temperature annealing under pressure. For example, appropriate precursors for Ti, Pb, and Ba ions could be TiO_2, $Pb(NO_3)_2$, and barium acetate, respectively. The pH value of the solution plays a very important role to stabilize the desired phase. For the Ba–Ti–H_2O–CO_2 system, barium titanate is stable for a pH >12 [36].

Recently Szafraniak *et al.* obtained ferroelectric epitaxial nanocrystals using a hydrothermal route [37]. As expected, besides the initial stoichiometry and the pH value, temperature and time play an important role for the obtained structures. For $PbTiO_3$ grown on (111) STO the first visible crystals nucleate after 2 h while at 150°C the first pyramid-shaped structures appear after 4 h of treatment. For a longer time the nuclei grow and eventually coalesce in larger structures or films. As expected, the coalescence process takes longer time at lower temperatures. Interestingly, in few cases a self-assembled, fairly ordered arrangement was observed (Fig. 8). This fact is most probably due to the substrate preparation (e.g., the polishing process) which could induce a spatial arrangement of nucleation centers, for instance along steps of vicinal surfaces. This suggests that a certain degree of orientation and order can be induced, manipulating the nucleation process at the substrate surface. For instance, a pre-patterning or a deposition of monolayers of materials

Figure 8 Highly ordered epitaxial PZT nanostructures obtained by hydrothermal growth on Nb:STO (100) single crystal substrates [37].

with different wetting properties might induce a preferential nucleation and registration of the obtained structures.

3.2. Synthesis of nanoparticles

Provided that there will be appropriate methods to manipulate nanosize particles and to deposit them into monolayers on desired substrates, or even to arrange them in a registered way, the fabrication of nanoparticles via chemical routes might be of high interest. This interest should be due to the flexibility of the process of low-temperature processing by which very small nanosize particles made from complex materials can be obtained.

Vapor phase routes, such as vapor deposition (CVD) and pulsed laser deposition (PLD) are suitable to produce high-purity nanoparticles, but suffer from the high cost of equipment and from low yield. The solution-based chemical routes potentially offer better grain-size homogeneity and good stoichiometry control. A number of routes to produce fine powders have already been established. These include co-precipitation, sol-gel processing, hydrothermal synthesis, and other routes.

Two different approaches are considered in order to synthesize nanoparticles via chemical routes. One is the preparation of stabilized particles dispersed in liquids, and the second is an *in-situ* preparation using a surfactant assembly such as Langmuir-Blodgett films, micelles, microemulsions, or similar, all based on colloid chemistry. In all cases the desired output is represented by nanometer size particles dispersed in a solvent. The stability of the solution is given by the balance between the short range attractive forces among the particles and the repulsive forces [38]. The obtained particles can be transferred to a substrate using an appropriate technique such as the Langmuir-Blodgett method. Appropriate control of the particles' hydrophobicity, surface charge and charge density can result in monolayers of particles that later can be transferred to a solid substrate [39].

3.2.1. Sol-gel synthesis

Sol-gel synthesis is one of the most common routes to synthesize oxide particles, in particular particles of ferroelectric oxides. The synthesis follows a simple route, starting with the preparation of a complex alkoxide chosen according to the final formula of the desired oxide compound. The obtained alkoxyde is then hydrolyzed in a controlled way to obtain a precipitate. Finally the nano-particle precipitate is washed and dispersed in a certain solvent. To obtain a stable suspension from the nano-particles is the most difficult part of the process. A successful attempt to synthesize stable monodisperse nanosize particles using a simple sol-gel method was recently reported by Liu *et al.* [40]. They prepared a complex metalorganic compound from lead acetate, Ti ethoxide and Zr butoxide in pure acetic acid in order to avoid any reaction with water. The solution was slowly precipitated at 80°C by adding ammonia via an argon bubbler, and then kept at this temperature for several days to allow the particles to grow. The dried product of nanoparticles, which were initially amorphous, was crystallized at temperatures ranging from 400° to 800°C in argon atmosphere and then in oxygen at a lower temperature. A stable suspension can be formed after modifying the surface of the nanoparticles with tartrate ligands. Free standing nanoparticles of 10 nm to 30 nm diameter were prepared in this way.

Two key steps allow the preparation of free standing crystalline nanoparticles: (i) the annealing in argon which — unlike annealing in oxygen — prevents sintering of the nanoparticles, and (ii) a chemical treatment of the surface (in this case with a tartrate ligand) which allows the suspension in a solution. As it will be seen later, the last step is decisive to achieve a suspension of free-standing particles, because un-coated particles are usually quickly aggregating and precipitating.

A slightly different route was implemented by O'Brian *et al.* [41] to prepare nanosize $BaTiO_3$ particles. A single bimetallic alkoxide precursor (barium titanium ethylhexano-isopropoxide) was injected into a mixture of diphenylether and oleic acid as stabilizing agent at 140°C under an inert atmosphere. After distillation of 2-propanol the resulting solution was cooled to 100°C, and hydrogen peroxide was injected. The solution was maintained at this temperature for 48 h to promote hydrolysis and crystallization under the conditions of inverse micelles. Monodisperse $BaTiO_3$ particles of 4 to 8 nm diameter were obtained. The diameter could be tuned modifying the ratio between alkoxide precursor and oleic acid, which also passivated the surface of the particles, in this way preventing the agglomeration and enabling the transfer of the

Figure 9 Scanning electron microscopy image of (a) agglomerated $BaTiO_3$ nanoparticles and (b) $BaTiO_3$ nanoparticles deposited on a Nb:STO substrate.

particles into a nonpolar solvent. Two points are remarkable in this route. First, the crystalline phase occurs at temperatures as low as 100°C directly from the reaction, and second the reaction takes place in a so called inverse micelle, i.e. a colloidal-type solution. The colloidal or microemulsion route is a generic route to synthesize nanoparticles as discussed in the following.

3.2.2. Generation of nanoparticles in microemulsions

Among all the chemical routes, the microemulsion technique is the most promising method to prepare nanoscale perovskite particles. The basic characteristic of this route is the confinement of the involved reactions in so-called "nanoreactors". The particle formation in nanoreactors takes place simultaneously in a number of about 10^{18}-10^{20} nano-compartments separated from each other [42]. The microemulsion technique has already been used to prepare ultrafine powders from a variety of materials, including metals [43], metal oxides [44], and high-T_C $YBa_2Cu_3O_{7-x}$ superconductors [45]. Herrig and Hempelmann [5] were among the first to use the microemulsion approach for the preparation of ultrafine powders of simple perovskites. A comprehensive review on the preparation of nanosize oxides via microemulsion-mediated synthesis is given by Osseo-Asare [46].

It is not the goal of the present paper to review this route in detail. We would just like to point out the possibilities of the microemulsion route and to show a few examples. Briefly, a miniemulsion or microemulsion is a system where small droplets of high stability are created using high shear [42]. The system must have at least three components, *viz.* a continuous phase, a dispersed phase, and a surfactant. An unstable system is, e.g., given by a water-in-oil system which is simply obtained by mixing a dispersed phase (water) in the continuous phase (oil). The stability of such a system is obtained by adding an agent that dissolves a dispersed phase but is insoluble in the continuous phase. For a water-in-oil system, an oil, water, an emulsifier (surfactant) and a hydrophobic agent are mechanically homogenized (using, e.g., ultrasonic agitation) to obtain monodisperse droplets in the size range between 30 and 500 nm. A simple water-in-oil system is considered being unstable due to two droplet growth mechanisms, Ostwald ripening and coalescence. The surfactants control the coalescence, while the Ostwald ripening is controlled by the hydrophobic agent. Moreover, the type of surfactant used for the stabilization can be used to adjust the size of the droplets.

The nanodroplets of a water-in-oil system are an ideal medium for the preparation of very fine oxide particles, either by a sol-gel type hydrolysis of metal alkoxides in the microemulsion-provided nano-reactors, or by coprecipitation of an aqueous solution of mixed oxalates or nitrates from the corresponding microemulsion. According to Beck *et al.* [47], nanoscale $BaTiO_3$ particles can be fabricated using the sol-gel-type hydrolysis approach. First a microemulsion is prepared by mixing a surfactant (Tergitol TNP-35, TNP-10, or TNP-7) with cyclohexane and 1-octanol, adding water. The resulting microemulsion can then be used to hydrolyze a $BaTiO_3$ alkoxide precursor to yield fine $BaTiO_3$ particles with a crystallite size tailored in a broad range from 3 nm to about 60 nm, each size having a relatively narrow distribution. The particle diameter is nicely tuned by adjusting the droplet size of the microemulsion, knowing that the latter is determined by the length of the hydrophilic part of the surfactant. This effect was used by Beck *et al.* to tune the crystallite size of $BaTiO_3$ performing a hydrolysis of a complex alkoxide in different microemulsions prepared with different surfactants.

An important advantage of the microemulsion route is the low calcination temperature required for the formation of the perovskite phase, which is up to 200°C lower than in the conventional solid-state reaction methods [48]. Even

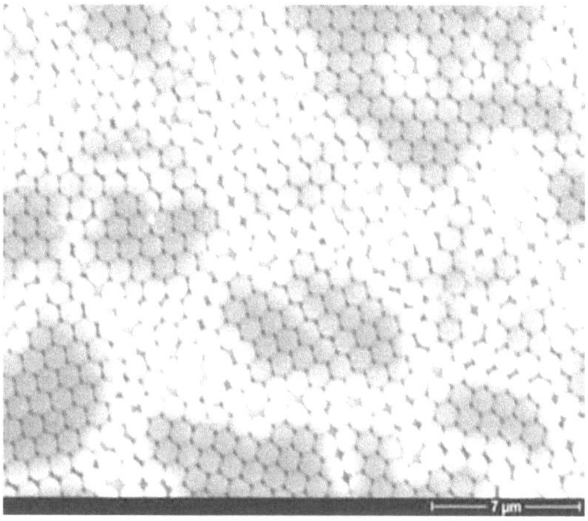

Figure 10 SEM image of a self-assembled monolayer of monodisperse latex spheres of 1 μm diameter.

the formation of a pure barium titanate perovskite was reported during the wet chemical precipitation followed by any calcination treatment [5].

Bhattacharyya *et al.* [49] have used a microemulsion route to produce monodisperse nanoparticles, i.e. particles dispersed in a liquid medium, and to obtain $BaTiO_3$ structures of 50–60 nm diameter in a random distribution on Nb:STO substrates, as shown in Fig. 9.

4. Self-assembled lithography

An economic, versatile means to fabricate arrays of ferroelectric nanostructures is given by the possibility to use a self-assembled lift-off deposition mask in combination with a physical or chemical vapour deposition technique. For example, a principal possibility is the combination of a self-assembled monolayer mask of hexagonally close-packed monodisperse latex spheres with the physical vapour deposition of an array of metal nanodots, called "Natural lithography", "Nanosphere lithography" or "Shadow nanosphere lithography" [50–52]. This technique can also be applied to the fabrication of ferroelectric nanostructure arrays, as has been demonstrated by Ma and Hesse [53–55].

A self-assembled monolayer of monodisperse latex spheres arranged in a close-packed hexagonal array was prepared according to [51, 52] and was used as a lift-off-type deposition mask for pulsed laser deposition (PLD) of $BaTiO_3$ (BTO) or $SrBi_2Ta_2O_9$ (SBT) nanostructures [53–55]. BTO and SBT were studied as prototypes for perovskite and bismuth-layered perovskite ferroeletric materials, respectively. The latex-sphere monolayer was prepared on a conductive (100)-oriented Nb-doped $SrTiO_3$ single crystal substrate (that later served as a bottom electrode) by a spin-coating process, using commercial monodisperse polystyrene latex sphere dispersions with spheres of 1 μm or 0.5 μm diameter. Fig. 10 is a scanning electron micrograph (SEM image) of such a monolayer. The bright regions correspond to stacking faults within the layer. Avoiding the formation of stacking faults is possible, e.g. on an area of 50 μm^2 [56]. This requires, however, some effort, skill and experience, cf. [56]. Stacking faults can, however, also made use of. They provide somewhat larger spacings for deposition, resulting in larger than usual lateral sizes of the prepared ferroelectric structures. These larger structures may be used to compare their physical properties with those of the regular, small nanostructures, as shown below.

Pulsed laser deposition of BTO and SBT followed by the lift-off of the polystyrene latex spheres in methylene chloride, and the subsequent crystallization anneal, yield a regular pattern of ferroelectric nanostructures as shown in Fig. 11.

The advantage of such a technique is that—using appropriate substrates—the nanostructures can be epitaxially crystallized [54, 55]. Moreover, using piezoresponse scanning force microscopy (PFM) the structures can be accurately characterized by measuring piezoelectric hysteresis loops (see Fig. 12). This enabled a detailed study of — and revealed a thickness-dependent imprint effect in

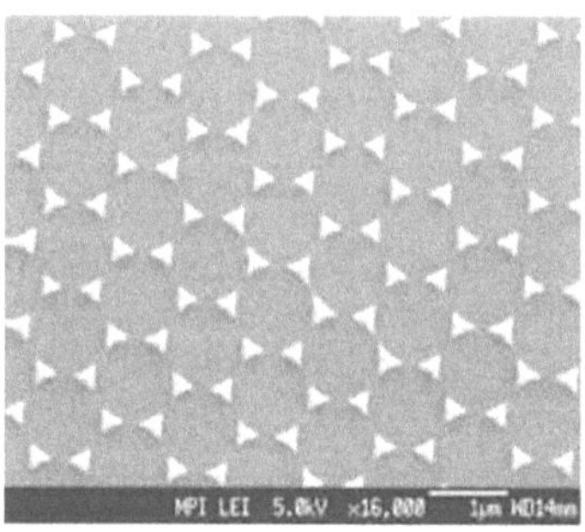

(a)

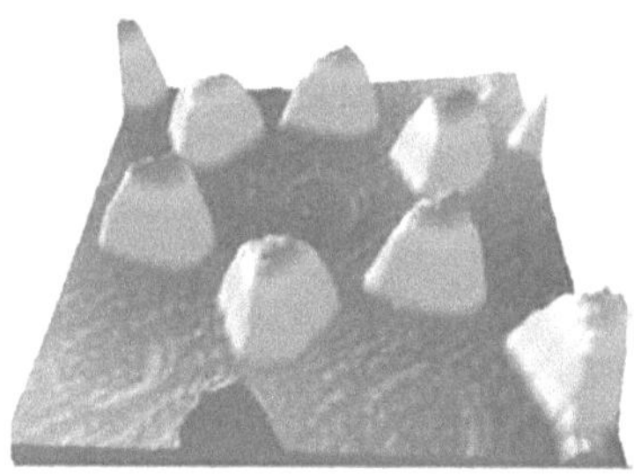

(b)

Figure 11 (a) SEM image and (b) Scanning force (AFM) topography image (2 × 2 μm) of part of a $SrBi_2Ta_2O_9$ array fabricated from a self-assembled monolayer of latex spheres of 1 μm diameter.

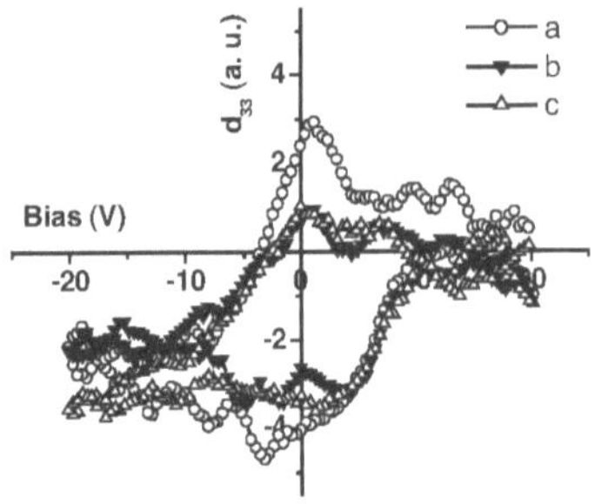

Figure 12 PFM hysteresis loops obtained from BTO nanostructures of different size. d_{33} is the effective piezoelectric coefficient along the vertical of the nanostructure. For curves a, b, and c, see the text.

- both the BTO and the SBT nanostructures. For example, Fig. 12 shows that the hysteresis curve (a) of a 44 nm high BTO nanostructure (fabricated from a 1 μm sphere mask) is more or less symmetric, whereas the loops (b) and (c) of two only 26 nm high nanostructures (fabricated using a 0.5 μm sphere mask) are shifted towards the negative d_{33} axis, i.e. are imprinted. This effect is independent of the lateral size, as is also shown in the figure: Curve (c) was recorded from a laterally extended structure, that had been formed at a stacking fault of the sphere mask, whereas curve (b) is from a laterally small structure.

In cases where three-dimensional nanostructures have clearly developed lateral surfaces, a similar *size*-dependent imprint effect may occur [57]. To explain this effect, as well as the *thickness*-dependent imprint effect observed in the laser-deposited arrays shown above, a phenomenological model has been developed. According to this model, the imprint effect in the flat $BaTiO_3$ nanostructures results from a domain locking near the ferroelectric-electrode interface, so that the nanostructures are composed of switchable regions and of non-switchable ones, the latter being close to the interface with the substrate. A similar effect occurs at the lateral surfaces of the three-dimensional nanostructures.

In summary, self-assembled lithography methods based on latex monolayers permit a rapid, cost-effective fabrication of high-quality ferroelectric nanostructures by physical or chemical vapor deposition enabling in-depth studies of size effects in nanoscale ferroelectrics.

5. Conclusions

The successful preparation of sub-micron and nanosize ferroelectric structures and the development of appropriate nanoscale measurement methods have opened a new emerging field, *viz. nanoscale ferroelectrics.* Bottom-up approaches offer the possibility of a low-cost fabrication of ferroelectric structures with lateral sizes of 10 to 50 nm, well below the sizes accessible by any of the top-down methods on relatively large substrates. Unfortunately, registration of the structures is still a major problem of all bottom-up approaches, and quite an effort will be necessary to solve this problem. Some of the self-assembly methods offer the possibility to fabricate high-quality epitaxial or single crystalline nanostructures, which represents a major advantage in addressing fundamental problems such as size effects, domain pinning at nanoscale, or determination of the super-paraelectric limit. In any case suitable fabrication methods to obtain high-quality nanoscale ferroelectrics are now available, which make us assert that *nanoscale ferroelectrics* will be part of our near future.

References

1. O. AUCIELLO, J. F. SCOTT and R. RAMESH, *Physics Today* **51** (1998) 22.
2. D. J. EAGLESHAM and M. CERULLO, *Phys. Rev. Lett.* **64** (1990) 1943.
3. R. L. SELLIN, C. RIBBAT, M. GRUNDMANN, N. N. LEDENTSOV and D. BIMBERG, *Appl. Phys. Lett.* **78** (2001) 1207.
4. Y. LIN, H. SKAFF, T. EMRICK, A. D. DINSMORE and T. P. RUSSELL, *Science* **299** (2003) 226.
5. H. HERRIG and R. HEMPELMANN, *Mater. Lett.* **27** (1996) 287.
6. M. ALEXE, J. F. SCOTT, C. CURRAN, N. D. ZAKHAROV, D. HESSE and A. PIGNOLET, *Appl. Phys. Lett.* **73** (1998) 1592.
7. T. TAKAHASHI and H. IWAHARA, *Mater. Res. Bull.* **13** (1978) 1447.
8. M. ALEXE, A. GRUVERMAN, C. HAMAGEA, N. D. ZAKHAROV, A. PIGNOLET, D. HESSE and J. F. SCOTT, *Appl. Phys. Lett.* **75** (1999) 1158.
9. O. G. SCHMIDT and K. EBERL, *Phys. Rev. B—Condensed Matt.* **61** (2000) 13721.
10. O. G. SCHMIDT, U. DENKER, K. EBERL, O. KIENZLE, F. ERNST and R. J. HAUG, *Appl. Phys. Lett.* **77** (2000) 4341.
11. H. OMI, D. J. BOTTOMLEY, Y. HOMMA and T. OGINO, *Phys. Rev. B—Condensed Matt.* **67** (2003) 115302.
12. K. S. SEOL, S. TOMITA, K. TAKEUCHI, T. MIYAGAWA, T. KATAGIRI and Y. OHKI, *Appl. Phys. Lett.* **81** (2002) 1893.
13. A. VISINOIU, *Private communication* (2001).
14. E. VASCO, R. DITTMANN, S. KARTHAUSER and R. WASER, *Appl. Phys. Lett.* **82** (2003) 2497.
15. S. KARTHAUSER, E. VASCO, R. DITTMANN and R. WASER, *Nanotechnology* **15** (2004) S122.
16. M. SHIMIZU, M. SUGIYAMA, H. FUJISAWA, T. HAMANO, T. SHIOSAKI and K. MATSUSHIGE, *J. Cryst. Gr.* **145** (1994) 226.
17. H. FUJISAWA, K. MORIMOTO, M. SHIMIZU, H. NIU, K. HONDA and S. OHTANI, *Japan Soc. Appl. Phys. Japanese J. Appl. Phys. Part 1—Regular Papers Short Notes & Review Papers* **39** (2000) 5446.
18. F. FUJISAWA, K. MORIMOTO, M. SHIMIZU, H. NIU, K. HONDA and S. OHTANI, *Ferroelectric Thin Films IX. Symposium (Materials Research Society Symposium Proceedings Vol. 655). Mater. Res. Soc.* (2001) CC10.
19. P. MURALT, S. BUHLMANN and S. VON ALLMEN, *Mater. Res. Soc. Symposium Proceedings* **784** (2004) 13.
20. S. BUHLMANN, P. MURALT and S. VON ALLMEN, *Appl. Phys. Lett.* **84** (2004) 2614.
21. K. D. BUDD, S. K. DEY and D. A. PAYNE, *Brit. Ceram. Proc.* (1985) 107.
22. T. SCHNELLER and R. WASER, *Ferroelectrics* **267** (2002) 293.

23. A. SEIFERT, A. VOJTA, J. S. SPECK and F. F. LANGE, *J. Mater. Res.* **11** (1996) 1470.
24. K. T. MILLER, F. F. LANGE and D. B. MARSHALL, *ibid.* **5** (1990) 151.
25. R. WASER, T. SCHNELLER, S. HOFFMANN-EIFERT and P. EHRHART, *Integ. Ferroelect.* **36** (2001) 3.
26. A. ROELOFS, T. SCHNELLER, K. SZOT and R. WASER, *IOP Publishing. Nanotechnology* **14** (2003) 250.
27. I. SZAFRANIAK, C. HARNAGEA, R. SCHOLZ, S. BHATTACHARYYA, D. HESSE and M. ALEXE, *Appl. Phys. Lett.* **83** (2003) 2211.
28. M. DAWBER, I. SZAFRANIAK, M. ALEXE and J. F. SCOTT, *J. Phys.—Condensed Matter* **15** (2003) L667.
29. V. A. SHCHUKIN, N. N. LEDENTSOV, P. S. KOPEV and D. BIMBERG, *Phys. Rev. Lett.* **75** (1995) 2968.
30. R. S. WILLIAMS, G. MEDEIROS-RIBEIRO, T. I. KAMINS and D. A. A. OHLBERG, *Ann. Rev. Phys. Chem.* **51** (2000).
31. M. W. CHU, I. SZAFRANIAK, R. SCHOLZ, C. HARNAGEA, D. HESSE, M. ALEXE and U. GOSELE, *Nat. Mater.* **3** (2004) 87.
32. F. F. LANGE, *Science* **273** (1996) 903.
33. A. T. CHIEN, J. S. SPECK and F. F. LANGE, *J. Mater. Res.* **12** (1997) 1176.
34. A. T. CHIEN, J. S. SPECK, F. F. LANGE, A. C. DAYKIN and C. G. LEVI, *ibid.* **10** (1995) 1784.
35. A. T. CHIEN, L. ZHAO, M. COLIC, J. S. SPECK and F. F. LANGE, *ibid.* **13** (1998) 649.
36. P. BENDALE, S. VENIGALLA, J. R. AMBROSE, E. D. VERINK, JR. and J. H. ADAIR, *J. Amer. Ceram. Soc.* **76** (1993) 2619.
37. I. SZAFRANIAK and M. ALEXE, *Ferroelectrics* **291** (2003) 19.
38. D.F. EVANS and H. WENNERSTROM, in "The Colloid Domain. Where Physics, Chemistry and Technology Meet" (Wiley-VCH, New York, 1994).
39. N. A. KOTOV, F. C. MELDRUN and J. H. FENDLER, *J. Phys. Chem.* **98** (1994) 8827.
40. C. LIU, B. ZOU, A. J. RONDINONE and J. Z. ZHANG, *J. Amer. Ceram. Soc.* **123** (2001) 4344.
41. S. O'BRIEN, L. BRUS and C. B. MURRAY, *J. Amer. Chem. Soc.* **123** (2001) 12085.
42. K. LANDFESTER, *Adv. Mater.* **13** (2001) 756.
43. F. JIYE, K. L. STOKES, J. WIEMANN and Z. WEILIE, *Mater. Lett.* **42** (2000) 113.
44. S. BANDOW, K. KIMURA, K. KON-NO and A. KITAHARA, *Japanese J. Appl. Phys. Part 1—Regular Papers Short Notes & Review Papers* **26** (1987) 713.
45. P. AYYUB, A. N. MAITRA and D. O. SHAH, *Physica C* **168** (1990) 571.
46. K. OSSEO-ASSARE, in "Handbook of Microemulsion Science and Technology", edited by P. Kumar and K. L. Mittal (Marcel Dekker, Inc., New York, Basel, 1999) p. 549.
47. C. BECK, W. HARTL and R. HEMPELMANN, *J. Mater. Res.* **13** (1998).
48. J. WANG, J. FANG, S.-C. NG, L.-M. GAN, C. H. CHEW, X. WANG and Z. SHEN, *J. Amer. Ceram. Soc.* **82** (1999) 873.
49. S. BHATTACHARYYA, S. CHATTOPADHYAY and M. ALEXE, *Nanomaterials for Structural Applications. Symposium (Mater. Res. Soc. Symposium Proceedings Vol. 740). Mater. Res. Soc.* (2003) 333.
50. A. KORIOSEK, W. KANDULSKI, P. CHUDZINSKI, K. KEMPA and M. GIERSIG, *Nano Lett.* **4** (2004) 1359.
51. H. W. DECKMAN and J. H. DUNSMUIR, *Appl. Phys. Lett.* **41** (1982) 377.
52. J. C. HULTEEN and R. P. VAN DUYNE, *J. Vac. Sci. Technol. A* **13** (1995).
53. W. MA and D. HESSE, *Appl. Phys. Lett.* **84** (2004) 2871.
54. W. MA, C. HARNAGEA, D. HESSE and U. GOSELE, *ibid.* **83** (2003) 3770.
55. W. MA and D. HESSE, *ibid.* **85** (2004) 3214.
56. J. RYBCZYNSKI, U. EBELS and M. GIERSIG, *Coll. Surf. A: Physicochem. Eng. Asp.* **219** (2003) 1.
57. M. ALEXE, C. HARNAGEA, D. HESSE and U. GOSELE, *Appl. Phys. Lett.* **79** (2001) 242.

JOURNAL OF MATERIALS SCIENCE 41 (2006) 13–25

Factors influencing the piezoelectric behaviour of PZT and other "morphotropic phase boundary" ferroelectrics

ANDREW J. BELL
Institute for Materials Research, University of Leeds, UK

Recent studies of the mechanism of piezoelectricity in PZT and related materials are reviewed and complemented by new analyses based on the Landau-Ginzburg-Devonshire theory of ferroelectricity. Particular attention is given to the nature of the morphotropic phase boundary between the tetragonal and rhombohedral perovskite phases and the accompanying peak in the piezoelectric coefficient. The importance of the changes in angular dependence of single crystal piezoelectric coefficients as a function of composition is highlighted together with the concept of field-induced rotation of the polarization in the (110) plane. It is shown that introducing the tendency to form monoclinic phases enhances this phenomenon.
The model that the monoclinic phase in PZT is due to the condensation of local disorder in the polar cation displacements from the macroscopic tetragonal and rhombohedral phases is examined in some detail using statistical analyses of the Zr/Ti conformation. Whilst the concept of monoclinic nano-domains is not inconsistent with statistically random distributions, it is argued that some ordering of the B-site cations may be required to enable the transformation to a macroscopically observable phase. The implications of this model on the contribution of polarization rotation to piezoelectricity in PZT are discussed.

1. Introduction

Lead zirconate titanate ($PbZr_{1-z}Ti_zO_3$ or PZT) has been the leading, high activity piezoelectric material for over 40 years and consequently is at the heart of the majority of piezoelectric actuators and sensors in production today [1]. However, in response to the introduction of environmental legislation, aimed at reducing the quantities of heavy metals entering the environment, research into alternative, lead-free piezoelectric materials is currently accelerating [2].

For the majority of applications of PZT, the optimum performance is found at the boundary between the tetragonal and rhombohedral perovskite phases (see Fig. 1), often known as the morphotropic phase boundary (MPB) [1]. Conventional wisdom suggests that for compositions close to the MPB, the piezoelectric coefficients maximise due to (i) a peak in the spontaneous polarization, to which the intrinsic piezoelectric coefficient is proportional and (ii) near degeneracy of the tetragonal and rhombohedral states, which allows for ease of reorientation of domains under applied fields and stresses, thereby maximising the extrinsic piezoelectric contributions. Consequently, much of the past research and development work on PZT has focused on the management of domain wall mobility through the control of defect chemistry. The discussions in this paper, however, focus mainly on the intrinsic contribution.

Since the morphotropic phase boundary in PZT ceramics is seen as central to their outstanding piezoelectric performance, the search for new or improved materials focuses on systems which possess a similar phase boundary. However, a number of relatively recent discoveries have driven a reassessment of our understanding of the MPB in PZT; these are:

(i) The unprecedented performance of complex-perovskite single crystals, such as [001]-oriented $Pb(Zn_{1/3}Nb_{2/3})O_3$-$PbTiO_3$ compositions on the rhombohedral side of their MPB; the mechanism appears to rely

0022-2461
DOI: 10.1007/s10853-005-5913-9

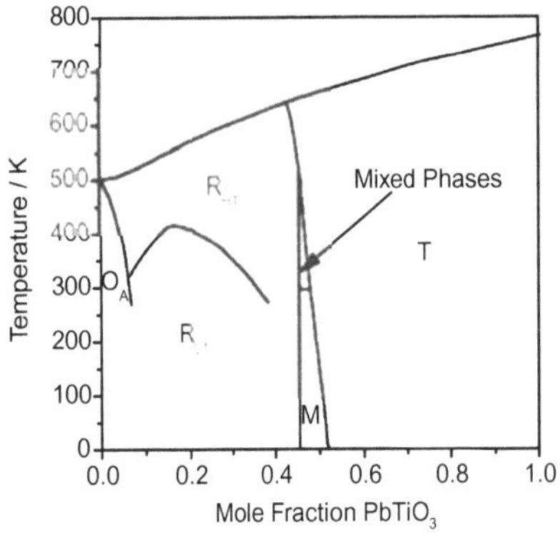

Figure 1 The phase diagram of $Pb(Zr_{1-z}Ti_z)O_3$ modified from Jaffe, Cook and Jaffe [1] to include the monoclinic phase [4].

on a field-induced rotation of the polarization vector in the (110) plane from the [111] to the [001] axis [3].

(ii) A previously undetected monoclinic phase at the morphotropic phase boundary in PZT [4] with polarization in the (110) plane and the presence of monoclinic [5] and orthorhombic [6] phases at the phase boundary in the complex-perovskites;

(iii) The local atomic displacements in PZT appear to differ from those consistent with the macroscopic symmetry [7]: for example, in the rhombohedral phase the lead cation displacements seem to be composed of local, short-range, random arrangements of tetragonal type displacements superimposed on the average, long range rhombohedral order.

This paper explores these findings, with an emphasis on using thermodynamic and statistical models to help understand how the anomalous piezoelectric properties close to the MPB arise.

2. Polarization rotation

A simplistic treatment of the intrinsic component of piezoelectricity might suggest that the piezoelectric effect is greatest parallel to the polar axis. That is, on applying an electric field parallel to the spontaneous polarization, P_s, it is augmented by an induced polarization, P_{ind}, which adds to the lattice strain, x:

$$\begin{aligned} x &= Q\,(P_s + P_{ind})^2 \\ &= QP_s^2 + 2QP_sP_{ind} + QP_{ind}^2 \approx x_s + 2QP_sP_{ind} \\ &\Rightarrow \Delta x \approx 2QP_s\varepsilon E \end{aligned} \tag{1}$$

where x_s is the spontaneous strain and Q is the relevant electrostriction coefficient. For small fields, the induced strain, Δx, is proportional to the electrostriction coefficient, the spontaneous polarisation and the permittivity, ε, where E is the applied electric field; the approximation for high permittivity materials, $\varepsilon \approx \frac{dP}{dE}$, is implicit. Hence the intrinsic piezoelectric coefficient, $d_{\rm int}$, is given by

$$d_{\rm int} = 2QP_s\varepsilon. \tag{2}$$

As the direction of P_s is normally taken to be the direction of maximum spontaneous polarisation, this textbook treatment can lead to the fallacy that the maximum intrinsic piezoelectric effect occurs parallel to the polar axis. Whilst this may be true for simple compounds well removed from phase transitions, it does not necessarily hold close to phase transitions at which the direction of the spontaneous polarization changes (for example at the morphotropic phase boundary in PZT or at the tetragonal-orthorhombic transition in barium titanate).

The misconception is highlighted by the large piezoelectric effect in "domain-engineered" complex perovskite single crystals [3]. It was noted that close to the MPB, in [001]-poled rhombohedral crystals, the piezoelectric coefficient is much greater parallel to [100] than it is along the polar [111] axis. It was proposed and confirmed [8] that on applying a field parallel to [001], the polarization rotated from the [111] axis continuously through the (110) plane to lie parallel, at a sufficiently high field, to the [001] axis. This "polarization rotation" was accompanied by a large induced strain. Support for this type of mechanism came in the form of calculations, from both the *ab initio* [9] and thermodynamic [10] schools, using $BaTiO_3$ as a prototype. In the latter case, the Landau-Ginzburg-Devonshire (LGD) approach [11] was used to show how the piezoelectric coefficients parallel to the non-polar low-index axes could exceed those parallel to the zero-field polar axes, again due to field induced rotation of the polarization vector. Furthermore, Damjanovic [12] calculated that in barium titanate the direction of maximum piezoelectric coefficient itself rotates as a function of temperature close to the tetragonal-orthorhombic transition. A number of *ab initio* calculations for PZT have also supported the rotation hypothesis [13, 14].

Whilst "polarization rotation" has been observed experimentally in complex perovskite crystals [8] and to a certain extent in barium titanate crystals [15], it is pertinent to ask how important is it in PZT ceramics and what is the role of the monoclinic phase?

3. LGD theory

The conventional Landau-Ginzburg-Devonshire model of ferroelectrics [11] as applied to perovskites is based on an elastic Gibbs' free energy expansion. In the stress-free case it is of the form:

$$\begin{aligned} \Delta G = {} & \alpha_{200}\left(P_1^2 + P_2^2 + P_3^2\right) + \alpha_{400}\left(P_1^4 + P_2^4 + P_3^4\right) \\ & + \alpha_{220}\left(P_1^2P_2^2 + P_2^2P_3^2 + P_3^2P_1^2\right) + \alpha_{600}\left(P_1^6\right. \\ & \left. + P_2^6 + P_3^6\right) + \alpha_{420}\left(P_1^4P_2^2 + P_1^2P_2^4 + P_2^4P_3^2\right. \end{aligned}$$

$$+ P_2^2 P_3^4 + P_3^4 P_1^2 + P_3^2 P_1^4) + \alpha_{222} P_1^2 P_2^2 P_3^2$$
$$- E_1 P_1 - E_2 P_2 - E_3 P_3 \qquad (3)$$

Here the series is terminated arbitrarily at the sixth power of polarization. P_i are the orthogonal components of polarization measured parallel to the pseudo-cubic axes of the perovskite unit cell. E_i are the corresponding components of the applied electric field. The coefficient α_{200} is necessarily temperature dependent. Hence, providing that the values of all the α_{lmn} coefficients are known, the values of the components of polarization can be determined for any combination of temperature and applied field by minimization of ΔG. Although zero-field analytical solutions exist, generally, and particularly for the case of non-zero fields, solutions are more easily found numerically.

The lattice strains are given by:

$$x_{ij} = Q_{ijkl} P_k P_l \qquad (4)$$

where $\mathbf{Q}$ is the electrostriction tensor.

For an arbitrary direction of electric field of amplitude E_0 and direction θ to the [001] axis and ϕ in the [110] plane, the polarization vector, $\mathbf{P}$, and strain tensor, $\mathbf{x}$, can be determined. Appropriate transformations can be applied to calculate the induced strain and piezoelectric coefficient $d_{\theta\phi}$ parallel to the direction of the field, using:

$$E_1 = E_0 \sin\theta \cos\phi$$
$$E_2 = E_0 \sin\theta \sin\phi, \qquad (5)$$
$$E_3 = E_0 \cos\theta$$

$$x_{\theta\phi} = x_3 \cos^2\theta - \cos\theta \sin\theta (x_6 \cos\phi + x_5 \sin\phi)$$
$$+ \sin^2\theta (x_1 \cos^2\phi - x_4 \cos\phi \sin\theta + x_2 \sin^2\phi) \qquad (6)$$

and

$$d_{\theta\phi} = \frac{x_{\theta\phi}(E_0) - x_{\theta\phi}(0)}{E_0}, \qquad (7)$$

where x_i represents elements of the strain tensor, according to the reduced notation convention [16] and $x_{\theta\phi}$ is the strain parallel to the applied field.

The set of coefficients, α_{lmn}, are dependent upon titanium concentration, z. The set employed throughout this paper, is adapted from the coefficients proposed by Haun [17, 18] and are shown in Table I. The value of α_{200} is for a temperature of 300 K. The present calculations are simplified by treating $PbZrO_3$ as a rhombohedral ferroelectric rather than an orthorhombic antiferroelectric. By extrapolation of the Haun coefficients in the rhombohedral phase of PZT to zero titanium content, a set of coefficients for $PbZrO_3$ were derived that satisfy this hypothetical scenario. To provide greater resolution close to the MPB, the coefficients for $0.4 < z < 0.6$ were fitted to fourth order polynomials in z, allowing calculations to be made for any value of titanium concentration in this range.

Fig. 2 shows results for the angular dependence of the piezoelectric coefficient, $d_{\theta\phi}$, from the numerical calculations under weak field conditions ($E_0 = 1000$ V m^{-1}). The data (excluding $PbZrO_3$) are identical to those of reference [19] obtained by analytical methods. It can be seen that for the end-members of the solid solution, $PbTiO_3$ and the fictitious, rhombohedral ferroelectric $PbZrO_3$, the maximum induced strain is indeed parallel to the polar axis, i.e. along [001] and [111] respectively. However, even for moderate levels of B-site substitution the direction of maximum strain shifts. In rhombohedral $Pb(Zr_{0.6}Ti_{0.4})O_3$, the direction of maximum strain is clearly along <100>, whilst in tetragonal $Pb(Zr_{0.4}Ti_{0.6})O_3$, the strain has become more isotropic than in $PbTiO_3$. The significance of such calculations is confirmed by experimental data for PZT thin-films: the piezoelectric coefficients of [001]-oriented, rhombohedral films is almost twice that of [111]-oriented films of the same composition [20] indicating that the non-polar [001] direction has the larger piezoelectric coefficient.

Although the LGD model suggests that on approaching the MPB the direction of maximum piezoelectric activity moves away from the polar axis, in this form it does not necessarily identify polarization rotation as being the dominant mechanism of piezoelectricity. For instance, calculations of the field-induced strain along [001] for rhombohedral compositions near the MPB [21] under high field, suggest a discontinuous polarisation switching between the [111] and [001] directions, that is a first-order field-induced transition. This is contrary to the experimental evidence for complex-perovskites which suggest a continuous polarization rotation mechanism, consistent with a second order transition [8].

4. Monoclinicity

The discovery that there exists a monoclinic phase in PZT, intermediate to the tetragonal and rhombohedral phases at the MPB [4], reveals that the 6th order Landau model used for the above calculations is flawed, in that it does not admit solutions for phases in which non-zero components of polarization can be unequal, as is the case in monoclinic phases. Subsequently it has been shown that thermodynamic functions allowing monoclinic and triclinic solutions require terms up to the 8th and 12th order respectively [22]. The inclusion of the higher order terms in the free energy expansion not only allows a monoclinic phase, but for tetragonal and rhombohedral ground states can encourage monoclinicity, that is rotation of the polarization vector in the (110) plane, under applied fields of the type identified in the complex perovskite single crystals. Revision of the Haun model to include higher or-

TABLE I 6th order free energy coefficients for $Pb(Zr_{1-z}Ti_z)O_3$

					Interpolation for $0.4 < z < 0.6$									
z	0.0	0.1	0.2	0.3	z^0	z^1	z^2	z^3	z^4	0.7	0.8	0.9	1.0	Multiplier
α_{200}	−4.582	−6.376	−7.47	−8.116	830.65	−7034.2	+21670	−29020	+14263	−12.47	−14.84	−16.17	−1.708	$\times 10^7$
α_{400}	52.35	41.25	31.29	22.3	−1161.4	+10056	−31486	+42802	−21443	0.6458	−3.05	−5.845	−7.9	$\times 10^7$
α_{220}	−16.71	−4.22	−0.0345	1.688	−366.50	+3070.4	−9447.1	+12716	−6313.8	5.109	6.32	7.063	7.5	$\times 10^8$
α_{600}	5.932	5.068	4.288	3.56	−287.77	+2452.1	−7617.7	+10307	−5141.2	2.348	2.475	2.518	2.61	$\times 10^8$
α_{420}	311.2	34.45	18.14	15.27	−1365.4	+11613	−36026	+48732	−24309	10.25	9.684	8.099	6.3	$\times 10^8$
α_{222}	−104.1	−8.797	−7.545	−7.052	647.96	−5506.9	+17075	−23090	+11514	−5.003	−4.901	−4.359	−3.66	$\times 10^9$

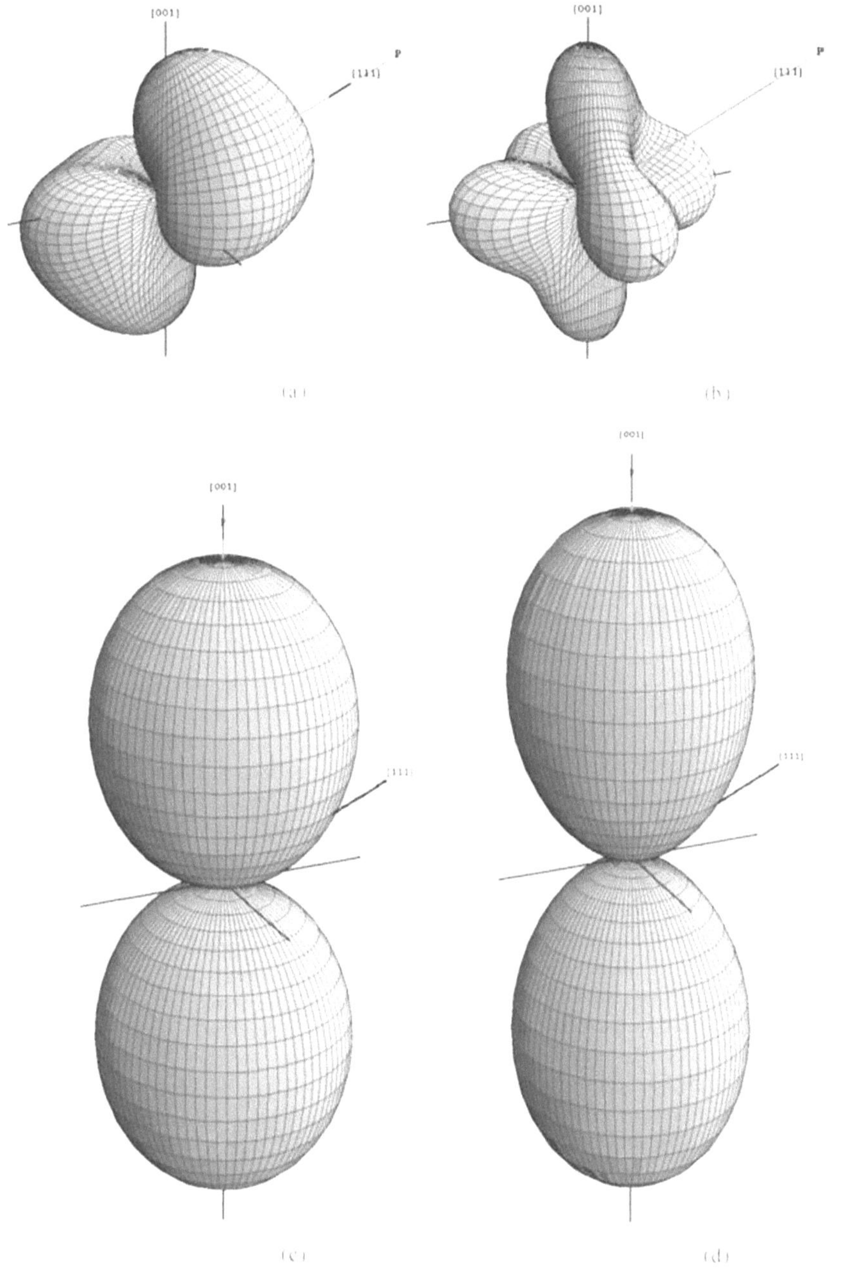

Figure 2 Single crystal piezoelectric coefficient as a function of angle, $d_{\theta\phi}$, for $Pb(Zr_{1-z}Ti_z)O_3$ from the 6th order LGD model: (a) $z = 0$, (b) $z = 0.4$, (c) $z = 0.6$ and (d) $z = 1$.

der terms can demonstrate how polarization rotation may be responsible for exceptional piezoelectric properties.

The additional terms for the expansion up to the eighth power of polarization are:

$$\Delta G = \ldots + \alpha_{800}\left(P_1^8 + P_2^8 + P_3^8\right) + \alpha_{620}\left(P_1^2\left(P_2^6 + P_3^6\right) + P_2^2\left(P_3^6 + P_1^6\right) + P_3^2\left(P_1^2 + P_2^2\right)\right)$$

$$+\alpha_{440}\left(P_1^4 P_2^4 + P_2^4 P_3^4 + P_3^4 P_1^4\right)$$

$$+\alpha_{422}\left(P_1^4 P_2^2 P_3^2 + P_1^2 P_2^4 P_3^2 + P_1^2 P_2^2 P_3^4\right)$$

Fig. 3 shows a contour plot of ΔG for $PbZr_{0.55}Ti_{0.45}O_3$ in the (110) plane for the 6th order set of coefficients of Table I. The x-axis represents the value of the polarization along [110] with $P_1 = P_2$, whilst the y-axis is P_{001} ($= P_3$).

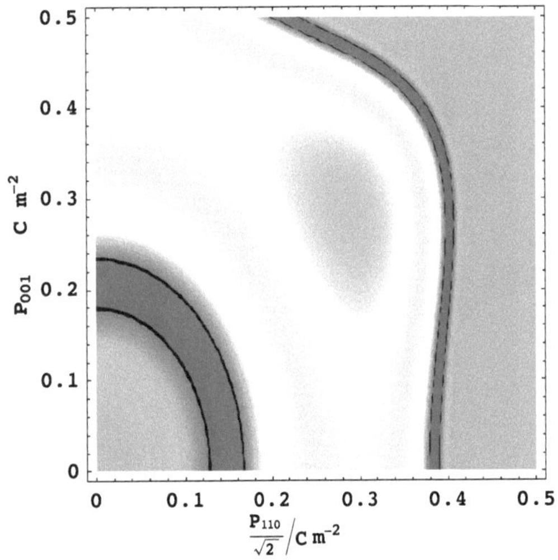

Figure 3 Contour plot of free energy as a function of polarization in the [110] plane for $Pb(Zr_{0.55}Ti_{0.45})O_3$ from the sixth-order LGD model. The energy at point {0,0} is 0 J m^{-3}; the minimum occurs at point {0.29, 0.29} with an energy of -8.06×10^6 J m^{-3}; contours occur at intervals of 8×10^4 J m^{-3}.

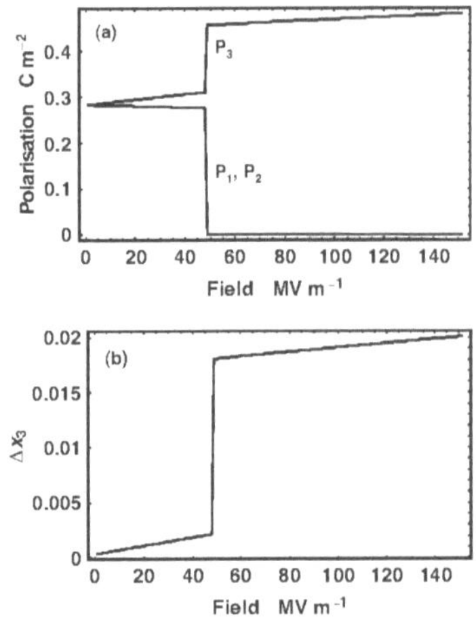

Figure 4 (a) Polarization and (b) strain as a function of applied field for $Pb(Zr_{0.55}Ti_{0.45})O_3$ from the sixth order LGD model.

In such plots a minimum lying on the y-axis represents tetragonal symmetry. A minimum on the x-axis indicates that an orthorhombic phase is stable, whilst a minimum on the line $P_{110} = P_{001}$ indicates a rhombohedral phase. Minima at any other points in the plane are of monoclinic symmetry. In this plot, the energy is zero at $P_{110} = P_{001} = 0$ and is positive for $P_{110} > 0.4$. All other values are negative. The absolute minimum can be seen to be at $P_{110} = P_{001} = 0.29$, indicating a rhombohedral phase is stable. A secondary minimum is seen at the point $\{P_{110}, P_{001}\} = \{0, 0.43\}$. However, there is a significant energy barrier between the 2 states, indicating that although a sufficient field parallel to [001] may result in the tetragonal phase becoming stable, the transition will occur via a discontinuity in the polarization as the barrier is traversed. This is demonstrated in Fig. 4a and b which show the components of polarization and strain calculated as a function of applied field along [001] for the same example. To demonstrate the concept of monoclinicity, a set of 8th order coefficients ('A' in Table II) have been derived.

In general the 8th order coefficients in the LGD expansion are expected to be relatively small. In principle they can be derived experimentally by fitting of the $P(E)$ characteristic to the LGD model for a single crystal sample. At present this is not possible for PZT, for which good examples of single crystals are notably rare. Alternatively, the 8th order coefficients can be estimated by fitting the temperature dependence of the lattice parameters to the LGD model. However, the accuracy of this method is expected to be poor unless the data encompasses a transition to the monoclinic state, which cannot be modelled without the 8th order terms. Hence the 8th order terms could be derived for that range of compositions exhibiting a monoclinic phase and for which could crystallographic is available [4]. The approach taken here is to employ 8th order coefficient values which are the minimum necessary to illustrate the effects in question.

Although coefficient set 'A' does not stabilize the monoclinic state at zero field, the tendency towards monoclinic symmetry is signalled by the removal of the energy barrier between the rhombohedral and tetragonal phases. Figs 5 and 6 illustrate this with free energy contour plots for the 8th order case. In Fig. 5, at zero field, the absolute minimum in energy is close to the point {0.26, 0.26}. However, there is a rather flat-bottomed valley connecting this point to the "tetragonal" point on the y-axis. This represents the easy path for rotation of the polarization. On applying a field along [001], the minimum moves continuously along this path. As an example Fig. 6 shows the plot for $E_3 = 5$ MV m^{-1} in which the minimum has shifted to the point {0.19, 0.37}. Again this is emphasised with plots of polarisation and strain as a function field (Fig. 7a and b). The path of the induced strain is similar to that seen in single crystal complex-perovskites [3]. Thus, under applied fields, the material passes through an extended region of monoclinic stability, corresponding

TABLE II Additional 8th order free energy coefficients for $Pb(Zr_{1-z}Ti_z)O_3$

	A	B
α_{222}	-2×10^9	as Table I
α_{800}	5×10^8	0
α_{620}	-2×10^9	-2.5×10^9
α_{440}	4×10^9	5×10^9
α_{422}	0	0

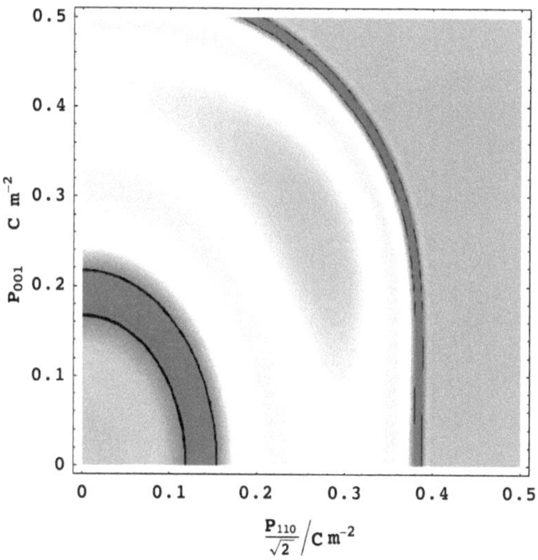

Figure 5 Contour plot of free energy as a function of polarization in the [110] plane for $Pb(Zr_{0.55}Ti_{0.45})O_3$ from the eighth-order LGD model. The energy at point {0,0} is 0 J m^{-3}; the minimum occurs at point {0.26, 0.26} with an energy of -7.05×10^6 J m^{-3}; contours occur at intervals of 7×10^4 J m^{-3}.

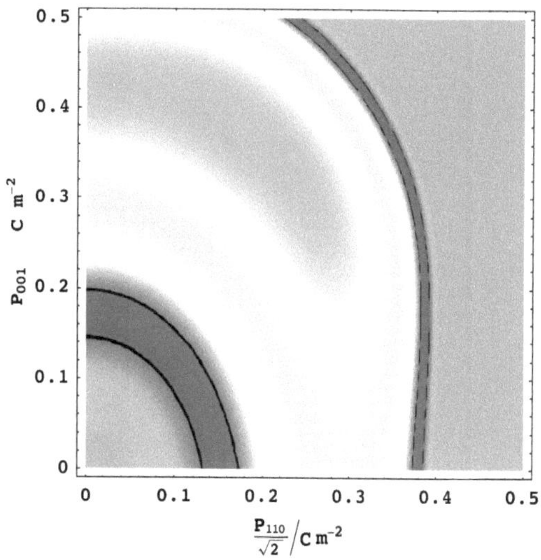

Figure 6 Contour plot of free energy as a function of polarization in the [110] plane for $Pb(Zr_{0.55}Ti_{0.45})O_3$ from the eighth-order LGD model for a field of 5×10^6 V m^{-1} applied parallel [001]. The energy at point {0,0} is 0 J m^{-3}; the minimum occurs at point {0.19, 0.37} with an energy of -8.6×10^6 J m^{-3}; contours occur at intervals of 9×10^4 J m^{-3}.

to rotation of the polarization vector in the (110) plane. These states are not accessible in the 6th order model. Although the high-field strain calculated under the 8th order model is slightly lower than the 6th order predictions, the weak field piezoelectric coefficient, i.e. the slope of the strain-field curves, is much larger when significant polarisation rotation is permitted. The fields employed for these simulations are consistent with those providing experimental for the complex perovskites.

It should be stressed that the coefficients listed in Table II do not necessarily represent a self-consistent set of free-energy coefficients, but have been selected purely to illustrate the concepts under discussion.

Fig. 8 shows $d_{\theta\phi}$ for $Pb(Zr_{0.4}Ti_{0.6})O_3$ at 300 K for the same set of additional 8th order coefficients. For these coefficients there is no stable monoclinic form in the zero-field phase diagram, but as can be seen by comparison with Fig. 2c, the set has a profound influence on $d_{\theta\phi}$. Although the response is still consistent with tetragonal symmetry, the maximum in piezoelectric coefficient lies away from the polar direction, its locus being in the form of a circle about the [001] axis and including the [111] direction.

The calculation of $d_{\theta\phi}$ can provide an estimate of the piezoelectric coefficient of a poled ceramic parallel to the poling direction, d_p, by averaging $d_{\theta\phi}$ over the appropriate range of angles. Fig. 9a compares the values of d_p close to the MPB, taking into account either tetragonal or rhombohedral phase poling, for the Haun model and for the 8th order coefficient set A. The greater freedom for polarization rotation provided by the 8th order terms results in a significant increase in piezoelectric coefficient. Fig. 9b

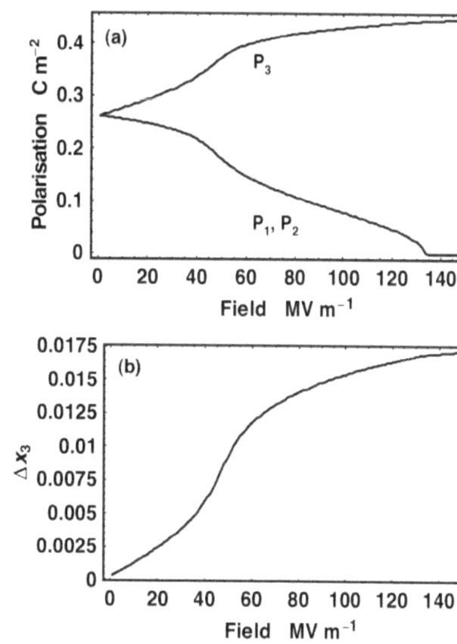

Figure 7 (a) Polarization and (b) strain as a function of applied field for $Pb(Zr_{0.55}Ti_{0.45})O_3$ from the eighth-order LGD model.

shows the results of the same calculation for coefficient set B. This set of 8th order coefficients has been selected to illustrate the stabilisation of a monoclinic phase at the MPB phase, over a range similar to that observed experimentally [4]. The magnitude of piezoelectric coefficient is greater than experiment suggests, but the data may account better for the sharp divergence of the piezoelectric properties at the MPB [1] than do other models. For com-

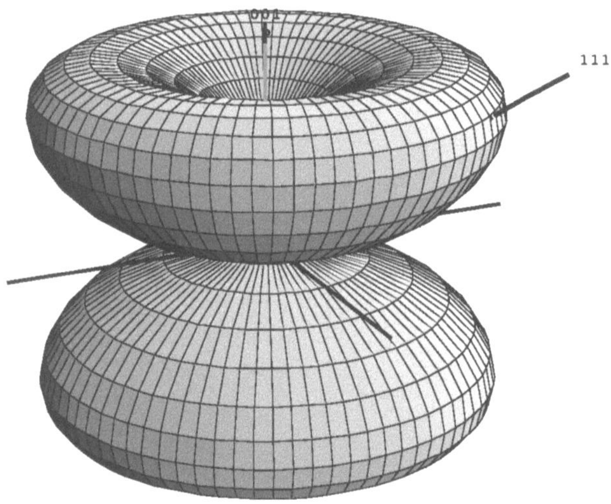

Figure 8 Single crystal piezoelectric coefficient as a function of angle, $d_{\theta\phi}$, for $Pb(Zr_{0.4}Ti_{0.6})O_3$, calculated from the 8th order LGD model using coefficient set A.

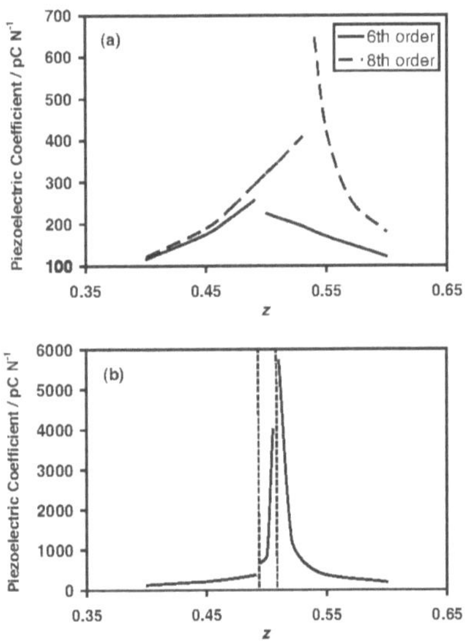

Figure 9 Piezoelectric coefficients as a function of titanium concentration calculated for poled ceramics from the LGD model employing (a) the 6th order and 8th order model (coefficient set A) and (b) the 8th order model (coefficient set B).

parison, $d_{\theta\phi}$ calculated for monoclinic $Pb(Zr_{0.5}Ti_{0.5})O_3$ from coefficient set B, is shown in Fig. 10.

The above examples give weight to the argument that monoclinicity, the capacity of a crystal to be stable in monoclinic forms over significant regions of parameter space, (although not necessarily at zero-field), is important in increasing piezoelectric activity by allowing the rotation of the polarisation vector in the (110) plane.

5. Monoclinic nanodomains

There is strong evidence that the monoclinic motif is present in PZT over a wide range of titanium concentration, not only in the region of the MPB. From structure refinements of neutron diffraction data [9] it appears that the "local symmetry," as indicated by the shift of the Pb-ions from the cubic setting, might best be described as monoclinic in both the rhombohedral and tetragonal phases. These suggestions were prompted by the observation that in the structure refinement process, if the cation displacements are constrained along directions which are consistent with the assumed macroscopic symmetry, the temperature factors associated with these displacements appear to be unreasonably large in directions perpendicular to the displacement. Temperature factors are normally expected to be spherically symmetric about the lattice position of the ion with which they are associated. To eliminate the anisotropy, without compromising the quality of the refinement, "off-axis" cation displacements are introduced, i.e. Pb ions may displace in directions other than [111] in the rhombohedral phase or [001] in the tetragonal phase. Thus, a static model structure which fit the data was identified: in the rhombohedral phase [9], the displacement of the Pb-ions from their cubic position comprises a long-range ordered, homogeneous [111] displacement, superimposed with lo-

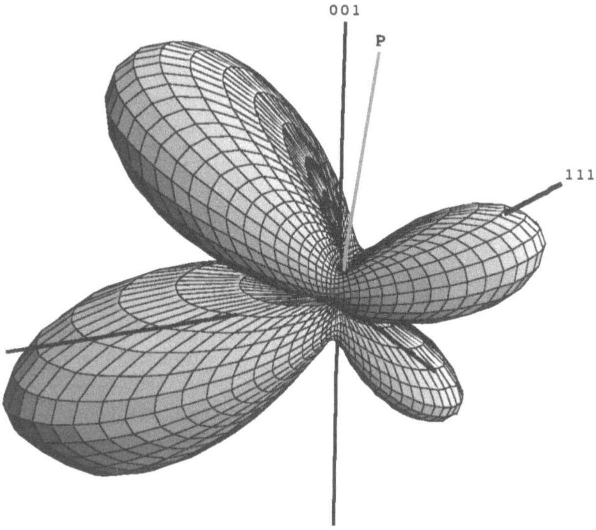

Figure 10 Single crystal piezoelectric coefficient as a function of angle, $d_{\theta\phi}$, for $Pb(Zr_{0.5}Ti_{0.5})O_3$, calculated from the 8th order LGD model using coefficient set B.

cal, randomly oriented, minor <001> shifts. Frantti [23] applied thc same arguments to data for the tetragonal phase, proposing that homogeneous [001] cation displacements are augmented with minor <110> (or possibly <111>) randomly oriented shifts. In both cases, the random shifts average to zero so that the macroscopic symmetries remain rhombohedral and tetragonal, respectively.

Bell and Furman postulated [24] that the macroscopic monoclinic phase may be due to condensation of the short-range random displacements into longer range ordered structures and proposed a mixed-crystal LGD model to describe this concept [25]. This concept has been more thoroughly explored independently by Glazer [26]. Working from the hypothesis that the monoclinic phase at the MPB was due to a change in coherence length of the monoclinic nano-domains that constitute the "rhombohedral" and "tetragonal" phases, it was concluded that for structure probes with sufficiently fine coherence length the morphotropic phase boundary does not exist.

First principles calculations [27, 28] tend to support the model of local monoclinic symmetry. DFT calculations carried out on 8 and 6 unit supercells showed that the Pb-ion displacements can be perturbed from the macroscopic polar direction by the particular conformation of nearest neighbour Ti and Zr ions, with resultant local polar displacements that need not conform to [001] or [111]. The implication is that for random placements of Zr and Ti on the B-site sub-lattice, there is disorder in the Pb displacements similar to those in the Glazer model [26].

The A-site cation response to the B-site disorder in PZT can be traced to the geometry of the Pb-O bonding of the end-members [29]. The fact that $PbTiO_3$ and $PbZrO_3$ tend to adopt different structures, may be attributed to the fact that PbO itself is polymorphic with an α-form exhibiting pyramidal Pb-O motifs of high geometric stability. In this conformation, the Pb lone-pair takes up a position opposite to the square oxygen base of the pyramid. An almost identical Pb-O arrangement is also present in $PbTiO_3$. However, β-PbO contains a trigonal bi-pyramidal Pb-O arrangement with less intrinsic stability, vestiges of which can be found in the structure of $PbZrO_3$. Local disorder in PZT may therefore be interpreted as the strong tendency of $PbTiO_3$ to form square pyramid Pb-O groups being frustrated by the destabilising influence of the Zr additions. Hence ambivalence in the oxygen coordination of the polar ion, combined with the presence of a lone pair can facilitate polar displacement disorder.

It is tempting to extend these arguments to cover known monoclinic type phases in the complex perovskite MPB systems. However, for these materials, an almost diametrically opposite model has been proposed [30], in which it is suggested that for domains with low wall energies, nanodomain structures of local tetragonal symmetry occur, which appear monoclinic when examined with long correlation length probes.

6. Monoclinic fluctuations

The main experimental evidence for the Glazer model comes from the relaxation of constraints on cation displacements employed to avoid the appearance of highly anisotropic temperature factors in structure refinements. The possibility of anisotropic dynamic fluctuations was suggested to be unphysical [7], but with little justification. It is therefore relevant to examine this suggestion in more detail.

The permittivity of ferroelectrics is generally highly anisotropic; perpendicular to the polar axis it is much larger than it is parallel to it. The permittivity is a measure of the ease of displacement of charge by an electric field imposed on the system and in the case of PZT appears to be dominated by the polarizability of the Pb ion. Any anisotropy observed in permittivity is hence a consequence of anisotropy in the free energy and a reflection of the anisotropy of the potential well for the Pb ion. Given the dominant role of Pb in PZT, it can be argued that the free energy landscape, as portrayed in Fig. 5, is an approximation of the potential seen by the Pb ions. Hence, thermal fluctuations within such a potential would be highly anisotropic and, given the broad flat nature of the well in the arc between [110] and [001], would be expected to be much larger than parallel to [111]. Hence it may not be so unusual for there to be significant, or even anomalous, dynamic fluctuations of the polar cations approximately perpendicular to the polar direction, producing the type of temperature factors seen in the constrained structure refinements. Indeed, extrapolating from Fig. 5 into three dimensions would suggest that any dynamic fluctuations observed would be constrained to a dish-shaped region with its symmetry axis lying along the [111] axis, similar in form to those described by Corker [7].

In this alternative model, the local symmetry at any instant in time would still appear monoclinic, but with temporal rather than spatial fluctuations of the Pb ion about the polar axis. Hence, the macroscopic monoclinic phase observed at the MPB would be interpreted as an ordering and freezing-in of these fluctuations, rather than a growth in the coherence length of static displacements.

Given the evidence from the *ab initio* calculations that the Pb displacements are strongly dependent upon the local Zr/Ti conformation, it would appear that the static interpretation of local random displacements is the more likely, although dynamic fluctuations may play a role.

Independent of whether the static or dynamic model offers the more accurate description of the structure, the common feature is a unit cell which is essentially monoclinic across the phase diagram. It is the scale of its correlation in space and time which determines the macroscopically observed phase. It is therefore pertinent to ask why the coherence of the local monoclinic structures should become macroscopic near the MPB?

7. Statistics of the Zr/Ti distribution

To understand this aspect of the problem, an exploration of the statistics of the Zr/Ti conformation is useful, especially in view of the rather small sample sizes considered in the *ab initio* studies [28]. The Zr/Ti distribution in PZT is, to date, believed to be totally random. A simple Monte Carlo type model has been constructed to determine the average Ti (or Zr) cluster size as a function of the Ti concentration under the assumption of random distributions.

A rigorous determination of cluster size for a 3-dimensional array of random B-site distributions is not a trivial exercise. Here an approximate approach is taken based on the construction of a list of cluster diameters from a 1-dimensional array of random B-site distributions. To construct a representative list of cluster volumes, each member of the list of cluster diameters is multiplied by all members of the list twice. Thus, for a Ti concentration, z, each element in an n-element linear array is populated with a random number r_i which controls the contents of a second array, $\mathbf{u}$: if $r_i < z$, then $u_i = 1$, otherwise $u_i = 0$, representing a Ti unit cell and a Zr unit cell respectively. Examining $\mathbf{u}$ for the number of contiguous occurrences of '1's provides an array of Ti cluster sizes, $\mathbf{s}$. To ensure that the analysis is valid in three dimensions, a list of cluster volumes is derived by flattening the cube of the array, $\mathbf{s} \times \mathbf{s} \times \mathbf{s}$ (i.e. the 3-dimensional array is re-listed as a 1-dimensional array). Although this method provides a distribution of cluster volumes, which approximates to a log normal distribution, to simplify the analysis only the mean of the distribution of cluster volume shall be considered here.

Using the above method, mean cluster volumes were calculated for $z = 0.1$ to 0.9 at intervals of 0.1. In each case, the calculation was carried out for an initial linear array of $n = 500$ unit cells. The resulting 3-dimensional analysis resulted in more than 1 million clusters for z in the range 0.4 to 0.6, and more than 100,000 outside this range. For each value of z, the average of 10 mean cluster volume calculations was determined. The mean cluster volume, v, expressed in unit cells and shown in Fig. 11, fits the surprisingly simple function

$$v = 1/(1 - z)^3. \tag{8}$$

Hence the mean titanium cluster diameter is equal to $1/(1-z)$ unit cells, whilst the mean zirconium cluster size is $1/z$ cells.

The data show that the mean number of contiguous titanium unit cells increases from 1 at vanishingly small concentrations, to just less than 10 at $z = 0.5$, to more than 1000 at $z = 0.9$. Hence, towards the end-members where one of the B-site elements is dominant and Zr/Ti conformation irrelevant, significant volumes of local monoclinic symmetry would seem unlikely. Even in the range $0.4 <$

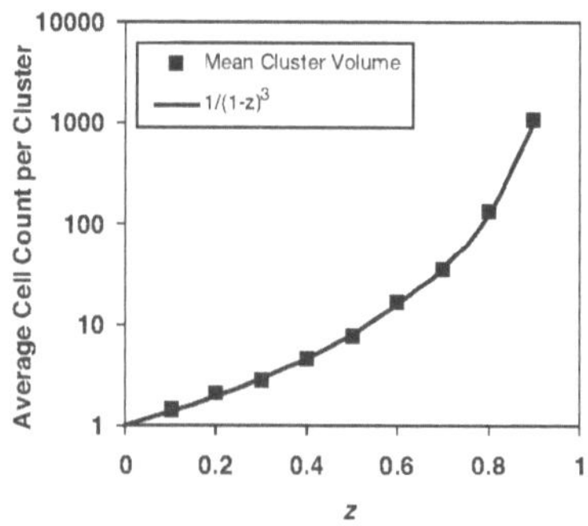

Figure 11 Mean Ti cluster volume calculated as a function of composition for randomly distributed Zr/Ti.

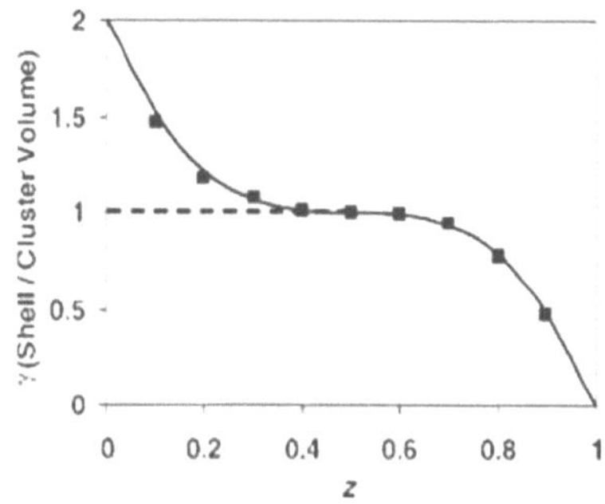

Figure 12 The fraction of the mean cluster volume constituted by surface unit cells as a function of composition.

$z < 0.6$, where the dominant B-site species is expected to have a cluster volume of between 10 and 20 unit cells the effect of B-site conformation might be thought to be minimal and might not conform to Grinberg's *ab initio* calculation [28] in which the assumed cluster sizes are somewhat smaller than 10 unit cells. However, consideration of cluster volumes alone can be misleading; analysis of the most likely nearest neighbour is also necessary. This can be estimated by comparing the number of unit cells at the "surface" of a cluster (i.e. the number of Ti unit cells with at least one Zr nearest neighbour) with the total volume of the cluster. For cuboid clusters of linear dimension l, the number of surface cells, or the volume of the "shell", equates to $l^3 - (l - 2)^3$. This is expressed as a shell volume to cluster volume ratio, γ, in Fig. 12, for both the Monte Carlo data and the assumption that $l = 1/(1-z)$. As a result of the assumption of cubic clusters, the relationship is only valid for $l \geq 2$ ($z \geq 0.5$); γ therefore saturates at unity for $z < 0.5$ and relates to the probability of a Ti-cell having a Zr-cell as a nearest neighbour. Representative 2-dimensional arrays of B-site occupancy are shown in Fig. 13 for values of z of 0.5, 07 and 0.9, however these can be misleading as the influence of the third dimension on nearest neighbour probability

Figure 13 2-dimensional maps of random B-site occupancy for values of z: (a) 0.5, (b) 0.7 and (c) 0.9; Zr occupied cells are shown in black.

is significant and only apparent in the interpretation of Fig. 12. For $z < 0.7$, virtually all Ti-cells have a Zr-cell as a nearest neighbour and it is only for $z > 0.9$ that more than 50% of Ti-cells do not have a Zr neighbour. This treatment therefore shows that for the majority of the phase diagram local Zr/Ti conformations are certainly able to influence the ion displacements in the majority of unit cells and lends further support to the existence of disorder in the local Pb displacements.

The gradual change of the orientation of the polarization with change in Ti concentration requires only that there be chemical disorder on the B-site coupled with the conflicting influences of the two B-site cations on Pb-O bonding. However, the stabilisation of a macroscopic monoclinic phase would seem to require a significant change in the coherence length of these two factors. For the monoclinic phase to appear macroscopic to neutrons and X-rays, the structure should be coherent over at least 0.1 μm [26]. Do the statistics of the Zr/Ti conformation around $z = 0.5$ provide such coherence ? The sum of the mean cluster size of the Zr and Ti clusters is actually a minimum at $z = 0.5$, with the most likely minimum volume in which an equal number of Zr and Ti ions is found being only 16 unit cells (i.e. 1 nm^3). Hence for a neutron-coherent monoclinic region to be observed we should expect some order in the Zr/Ti distribution that transcends the cluster sizes given by random statistics. Given their size difference, positional ordering of the Zr and Ti ions is clearly possible, but it has not yet been reported for PZT.

The two order-parameter LGD model of Bell and Furman [25] predicts a stable monoclinic phase in the region of the MPB, through a coupling of the free energy functions for $PbTiO_3$ and the fictional ferroelectric $PbZrO_3$ of Table I. Whilst this model was proposed as a means of introducing the influence of the disorder in the Zr/Ti distribution into the thermodynamic model, it does not address the issue of coherence length. There is no implication in the results that coherence in the direction of polarization in the monoclinic phase would be observable experimentally. The model was established in the context of the conventional rhombohedral and tetragonal phases being observed at all scale lengths and reflects this. However, it is consistent with the explanation given above for the existence of large anisotropic temperature factors when structure refinements are constrained to these conventional macroscopic symmetries.

8. Piezoelectricity and local structures

What are the implications of the local monoclinic distortions for piezoelectric properties? As demonstrated above, the inclusion of terms in the LGD polarization expansions which tend to stabilize monoclinic symmetry lead to large piezoelectric coefficients due to rotation of the polarization in the [110] plane. However, the calculations which produced Fig. 9, for example, are based on polynomial expansions which are invariant under cubic symmetry operations, hence there are 48 identical monoclinic solutions, corresponding to the 8 symmetry related directions of polarization in each of the 6 $\{110\}$ planes. That is, monoclinic perovskite ferroelectrics have 48 degenerate domain states. On this basis, poling of monoclinic ceramics, would result in orientation of the favourable [111]-[001] channel in the (110) plane within the quadrant containing the poling direction and allowing the maximum piezoelectric contribution of polarization rotation in this plane. However, the nano-domains postulated in the Glazer model [26] are not locally degenerate. The implication is that the local Zr/Ti conformation biases the free energy function towards a specific direction of polarization. Poling may be able to switch the major component of polarization (i.e. along <111> or <001>). However, as the potential topology for each of the 48 domains states will be different, it is not clear what the effect on the minor components of cation displacement will be. Due to the "internal bias" provided by random Zr/Ti conformations, the contribution of polarization rotation in the (110) plane would be less significant than the symmetrical LGD model suggests and may be one reason why the predictions of Fig. 9 overestimate the piezoelectric coefficients. However, in the case of Zr/Ti ordering close to $z = 0.5$, in which a greater degree of local symmetry and domain state degeneracy might be restored, the polarization rotation model would be expected to be more significant.

9. Conclusions

Recently proposed models of piezoelectricity in morphotropic phase boundary perovskites and for the structure of PZT have been examined employing Landau-Ginzburg-Devonshire theory and statistical models of Zr/Ti conformation. It is shown that the mechanism of polarization rotation in the (110) plane can enhance the piezoelectric effect in ceramics and that the tendency towards the formation of macroscopic monoclinic forms facilitates this by reducing the energy for realignment of the polarization vector. The proposed model of static, monoclinic nano-domains which are observed as rhombohedral or tetragonal macroscopic structures appears to be consistent with the statistics of random Zr/Ti conformations across a large region of the PZT phase diagram. However, it is argued that the increase in coherence length of the nano-domains, proposed as the mechanism for the appearance of the macroscopic monoclinic phase close to $z = 0.5$, may be a consequence of some undetected order in the B-site cations. Furthermore, the biasing of the nano-domain orientation by the local Zr/Ti conformation may impede the polarization rotation contribution to the piezoelectric effect.

It has been argued that significant anisotropic thermal fluctuations, of the type that are seen in constrained structure refinements, are not inconsistent with the dielectric anisotropy of PZT, particularly under the assumption of a tendency towards macroscopic monoclinicity. This proposition would also be more consistent with the concept of significant polarization rotation and hence should not yet be ruled out in favour of the nano-domain structure model.

Further studies toward deeper insight are: structure determinations under applied electric fields; modelling of temperature factors from dielectric data; modelling of the switching of locally biased monoclinic domains; high resolution studies of Zr/Ti conformations at the MPB; a more extensive study of the statistics of Zr/Ti conformations and larger scale *ab initio* simulations.

References

1. B. JAFFE, W. R. COOK and H. JAFFE, in "Piezoelectric Ceramics" (Academic Press Ltd, 1971).
2. Y. SAITO, H. TAKAO, T. TANI, T. NONOYAMA, K. TAKATORI, T. HOMMA, T. NAGAYA and M. NAKAMURA, *Nature* **432** (2004) 84.
3. S.-E. PARK and T. R. SHROUT, *J. Appl. Phys.* **82** (1997) 1804.
4. B. NOHEDA, D. E. COX, G. SHIRANE, J. A. GONZALO, L. E. CROSS and S. E. PARK, *Appl. Phys. Lett.* **74** (1999) 2059.
5. Z. G. YE, B. NOHEDA, M. DONG, D. COX and G. SHIRANE, *Phys. Rev. B* **64** (2001) 184114.
6. D. LA-ORAUTTAPONG, B. NOHEDA, Z. G. YE, P. M. GEHRING, J. TOULOUSE, D. E. COX and G. SHIRANE, *Phys. Rev. B* **65** (2002) 144101.
7. D. L. CORKER, A. M. GLAZER, R. W. WHATMORE, A. STALLARD and F. FAUTH, *J. Phys.- Cond. Matter,* **10** (1998) 6251.
8. D.-S. PAIK. S.-E. PARK, S. WADA, S.-F. LIU and T. R. SHROUT, *J. Appl. Phys.* **85** (1999) 1080.
9. H. FU and R. E. COHEN, *Nature* **403** (2000) 281.
10. A. J. BELL, *J. Appl. Phys.* **89** (2001) 3907.
11. A. F. DEVONSHIRE, *Adv. Phys.* **3** (1954) 85.
12. D. DAMJANOVIC, F. BREM and N. SETTER, *Appl. Phys. Lett.* **80** (2002) 652.
13. Z. WU and H. KRAKAUER, *Phys. Rev. B* **68** (2003) 014112.
14. N. HUANG, Z. LIU, Z. WU, J. WU, W. DUAN, B.-L. GU and X.-W. ZHANG, *Phys. Rev. Lett.* **91** (2003) 067602.
15. S. WADA, S. SUZUKI, T. NOMA, T. SUZUKI, M. OSADA, M. KAKIHANA, S.-E. PARK, L. E. CROSS and T. R. SHROUT, *Jpn. J. Appl. Phys.* **38** (1999) 5505.
16. J. F. NYE, in "Physical Properties of Crystals" (Oxford, 1985)
17. M. J. HAUN, E. FURMAN, S. J. JANG and L. E. CROSS, *Ferroelectrics* **99** (1989) 13.
18. M. J. HAUN, Z. Q. ZHUANG, E. FURMAN, S. J. JANG and L. E. CROSS, *ibid.* **99** (1989) 45.
19. X. DU, J. ZHENG, U. BELEGUNDU and K. UCHINO, *Appl. Phys. Lett.* **72** (1998) 2421.
20. D. V. TAYLOR and D. J. DAMJANOVIC, *ibid.* **76** (2000) 1615.
21. A. J. BELL, *ibid.* **76** (2000) 109.
22. D. VANDERBILT and M. H. COHEN, *Phys. Rev. B* **63** (2001) 4108.
23. J. FRANTTI, J. LAPPALAINEN, S. ERIKSSON, V. LANTTO, S. NISHIO, M. KAKIHANA, S. IVANOV and H. RUNDLOF, *Jpn. J. Appl. Phys.* **39** (2000) 5697.
24. A. J. BELL and E. FURMAN, *Ferroelectrics* **293** (2003) 29.
25. *Idem., Jpn. J. Appl. Phys. Pt. 1* **42** (2003) 7418.
26. A. M. GLAZER, P. A. THOMAS, K. Z. BABA-KISHI, G. K. H. PANG and C. W. TAI, *Phys. Rev. B* **70** (2004) 184123.
27. I. GRINBERG, V. R. COOPER and A. M. RAPPE, *Nature* **419** (2002) 909.
28. I. GRINBERG, V. R. COOPER and A. M. RAPPE, *Phys. Rev. B* **69** (2004) 144118.
39. A. J. BELL and P. GOUDOCHNIKOV, in "Proceedings of Material Technology and Design of Integrated Piezoelectric Devices", (2004) p. 29.
30. Y. M. JIN, Y. U. WANG, A. G. KHACHATURYAN, J. F. LI and D. VIEHLAND, *J. Appl. Phys.* **94** (2004) 3629.

JOURNAL OF MATERIALS SCIENCE 41 (2006) 27–30

Polar nanoclusters in relaxors

R. BLINC*
J. Stefan Institute, Jamova 39, 1000 Ljubljana, Slovenia
E-mail: robert.blinc@ijs.si

V. V. LAGUTA
J. Stefan Institute, Jamova 39, 1000 Ljubljana, Slovenia; Institute for Problems of Materials Science, Ukrainian Academy of Sciences, Krjijanovskogo 3, 03143 Kiev, Ukraine

B. ZALAR
J. Stefan Institute, Jamova 39, 1000 Ljubljana, Slovenia

J. BANYS
Faculty of Physics, Vilnius University, Sauletekio al. 9, 2040 Vilnius, Lithuania

The central problem in the physics of relaxors is the nature of the polar nanoclusters. Whereas relaxors are homogeneous at high enough temperatures, polar nanoregions immersed in a neutral matrix are formed below a certain temperature T_b. This should lead to a two component system. Here we present direct microscopic evidence for the two component nature of relaxors. We show that the chemical shift perturbed ^{207}Pb NMR spectra of these systems consist of an isotropic component corresponding to a spherical glassy matrix which does not respond to an applied electric field, and an anisotropic component, corresponding to frozen out polar nanoclusters which order in a strong enough electric field, forming a ferroelectric phase. This is as well reflected in the dynamic properties where the relaxation time distribution function starts to become asymmetric with decreasing temperature and a second maximum—which is never seen in dipolar glasses and is obviously due to polar clusters—appears on further cooling. We also show that the basic difference between dipolar glasses and relaxors is the fact that polar nanoclusters can be oriented in a strong enough electric field and a ferroelectric phase can be induced. This is not the case in dipolar glasses where the response is due to single dipoles which can not be ordered by applied electric fields.

1. Introduction

The central problem in the physics of relaxors [1, 2] is the nature of the polar nanoclusters which are believed to be responsible for the multi-scale dynamics, spatial inhomogeneity and many other physical properties of these materials such as giant piezo-electricity and electrostriction [3]. Whereas relaxors are homogenous at high enough temperatures, polar nanoregions immersed in a neutral matrix should be formed below a certain temperature T_B according to Burns and Dacol [4]. This would lead to a two component system. In spite of many investigations, direct physical evidence for the existence of polar nanoregions is still lacking. Here we present direct microscopic evidence for the two component nature of relaxors. We show that the ^{207}Pb NMR spectra of these systems consist of an isotropic component corresponding to a spherical glassy matrix which does not respond to an applied electric field, and an anisotropic component, corresponding to frozen out polar nanoclusters which order in a strong enough electric field, forming a ferroelectric phase. This is as well reflected in the dynamic properties where the relaxation time distribution function $f(\tau)$ starts to become asymmetric with decreasing temperature and a second maximum—which is never seen in dipolar

*Author to whom all correspondence should be addressed.

0022-2461 © *2006 Springer Science + Business Media, Inc.*
DOI: 10.1007/s10853-005-5914-8

glasses and is obviously due to polar clusters—appears on further cooling in the dielectric dispersion. Cross and Viehland [5] pointed out the similarity between relaxors and dipolar glasses and suggested that nanoclusters are dynamic entities with thermally fluctuating dipole moments which freeze out at low enough temperatures. The present results demonstrate that the basic difference between relaxors and dipolar glasses is their response to applied electric fields: Polar nanoclusters—corresponding to the anisotropic component in the NMR spectra—can be oriented in a strong enough applied electric field and a ferroelectric phase can be induced. This is not the case in dipolar glasses where the response is due to single dipoles which cannot be ordered by applied electric fields.

2. Results and discussion

We have investigated a single crystal of the prototype relaxor lead manganese niobate, $Pb(Mg_{1/3}Nb_{2/3})O_3$, abbreviated as PMN. It is a perovskite solid solution characterized by site and charge disorder. The dipolar glass studied for comparison was a $(ND_4)_{0.5}$ $Rb_{0.5}D_2PO_4$ single crystal abbreviated as DRADP-50. The techniques used for PMN were field cooled and zero field cooled ^{207}Pb magnetic resonance and broad band dielectric dispersion spectroscopy. The electron paramagnetic resonance (EPR) spectra of 0.1% $Mn^{2+}+$ doped PMN ceramics were investigated as well. DRADP-50 was studied by 2D and $1D^2H$ and ^{31}P NMR and dielectric spectroscopy for comparison. A broad band dielectric dispersion study was performed on a PSN-PZN-PMN mixed system.

The macroscopic symmetry of PMN is cubic between 100 K and 4 K. At high temperatures above $T_B \approx 617$ K PMN behaves like all other simple perovskites. The dynamics of the system is determined by the soft TO phonon which exhibits a normal dispersion and is underdamped at all wave vectors. Below T_B in addition to the soft mode – which becomes overdamped [6] – a new dielectric dispersion mechanism appears at lower frequencies which can be described by a correlation time distribution function $f(\tau)$.

As it can be seen from Fig. 1a the correlation time distribution function $f(\tau)$ for PSN-PZN-PMN (obtained from broad band dielectric dispersion data) becomes asymmetric with decreasing temperature and a second maximum at long correlation times appears. Such behaviour was also observed [7] for other relaxors, such as PMN, PLZT and SBN. This second maximum is not seen in dipolar glasses like DRADP-50 (Fig. 1b). It seems to be specific to relaxors and signals the formation of polar nanoclusters. This is supported by the fact that this maximum changes if an external electric field E larger than the critical field E_C is applied along the ferroelectric [111] direction at $T <$ 205 K. The first maximum of $f(\tau)$ corresponding to shorter

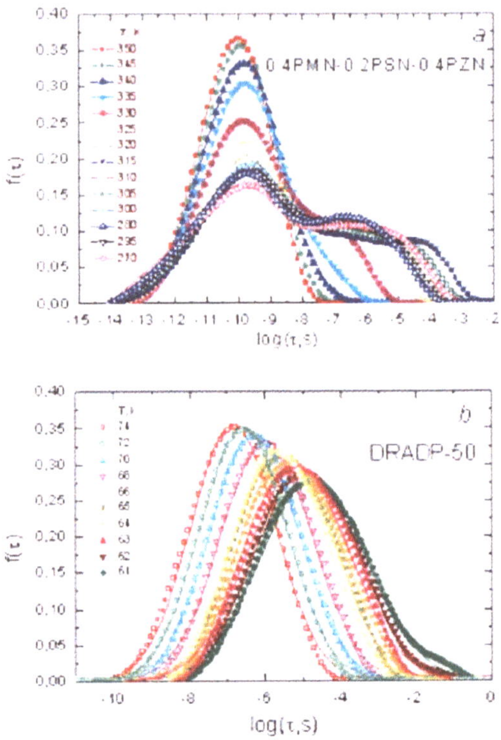

Figure 1 **(a) Relaxation time distribution function $f(\tau)$ describing the dielectric dispersion in relaxor 0.4PMN-0.2PSN-0.4PZN. The short time scale maximum describes the glassy type dynamics whereas the long time scale part refers to the polar cluster dynamics. The same features are obtained in PMN, PLZT, and SBN relaxors. (b) Relaxation time distribution function in the dipolar glass DRADP-50 where the long time scale maximum is absent.**

correlation times, on the other hand, is similar to the one in the dipolar glasses and is not affected by electric fields.

To investigate the nature of the polar nanoclusters and the surrounding glassy matrix on the microscopic level and to get spatially resolved evidence of the electric field response of PMN we performed field cooled (FC) and zero field cooled (ZFC) ^{207}Pb ($I = 1/2$) NMR experiments at different temperatures and orientations of the crystal in magnetic and electric fields. (Fig. 2a, b and c).

The ^{207}Pb spectrum at 290 K (Fig. 2a) is isotropic and of a Gaussian line shape. Two dimensional (2D) separation of interactions experiments show that the spectra are in fact frequency distributions and are composed of a large number of individual ^{207}Pb lines with different chemical shifts. This is incompatible with the assumption that the Pb ions sit at their high-symmetry cubic sites as in this case all Pb sites would be equivalent and only a single sharp line is expected. The fact that we see a Gaussian frequency distribution demonstrates that we deal with a spherical glass where all Pb nuclei are displaced but there is no preferential frozen out orientation or magnitude of displacement.

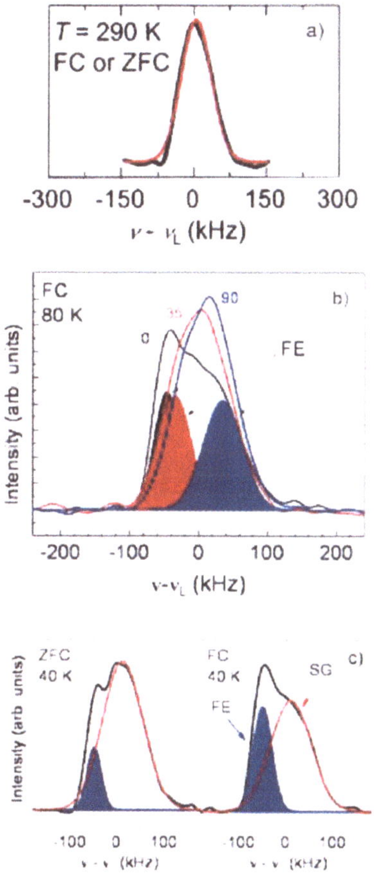

Figure 2 (a) FC and ZFC ^{207}Pb NMR spectra of PMN at 290 K. (b) FC ^{207}Pb NMR spectra of PMN at 80 K and different orientations of the crystal in the magnetic field. The spectra clearly consist of two components, an isotropic and an anisotropic one. (c) Field cooled and zero field cooled ^{207}Pb NMR spectra of PMN at 40 K showing the effect of the electric field on the concentration of the FE clusters.

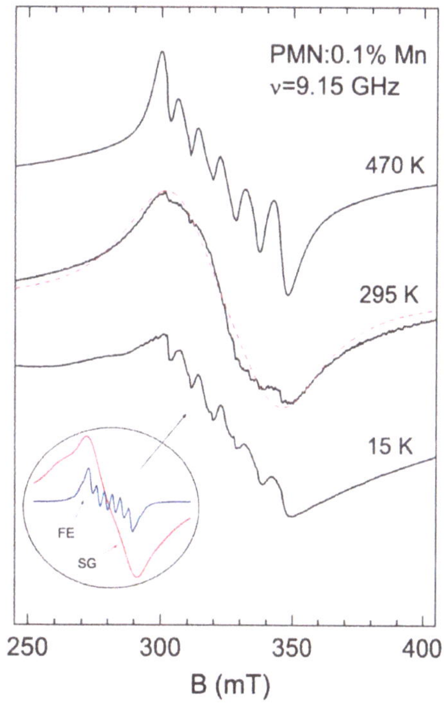

Figure 3 EPR spectra of 0.1% Mn^{2+}- doped PMN ceramics. The insert shows decomposition of the spectrum on two components, broad (SG) and narrow (FE) related, respectively, to glass matrix and ferroelectric clusters.

If the Pb ion shifts only varied in the orientation and not in magnitudes, a powder-like pattern rather than the observed Gaussian line shapes would be seen. Due to short-range order correlation among the displacements clusters are formed which fluctuate in time, orientation and magnitude of the dipole moment [4]. The orientational bias is here spherically symmetric. This agrees with the model proposed by Vakhrushev [8] which states that above the freezing temperature the displacements of the Pb nuclei lie in a spherical shell around the cubic position as well as with the mezoscopic spherical random bond–random field (SRBRF) model [9]. At lower temperatures an anisotropic component appears in the ^{207}Pb NMR spectrum in addition to the isotropic one (Fig. 2b). Its angular dependence in the external magnetic field follows the $(3\cos^2\vartheta - 1)$ law. The anisotropic component—which is itself a frequency distribution—corresponds to polar clusters frozen out on the NMR time scale and oriented along the ferroelectric [111] axis. Such a two component line shape is not seen in dipolar glasses [10].

As it can be seen from Fig. 2c, the anisotropic frozen polar cluster component increases in intensity if the crystal is cooled at low enough temperatures in an electric field larger than the critical field and applied along the [111] direction. A transition to the ferroelectric phase is induced for $E > E_C$. The difference between FC and ZFC Pb NMR spectra is striking and shows the existence of a two-component behavior—the spherical glassy matrix and the ferroelectric clusters—on a microscopic level. According to the intensities of the NMR lines, about 50% of the Pb nuclei still reside in the spherical glass matrix which does not respond to the electric field and 50% in the ferroelectric polar clusters which respond to electric fields. It should be noted that below $T_C = 210$ K a sudden increase in the intensity of the anisotropic component is also seen in the ZFC spectra, but the increase in the intensity of the anisotropic component is two times smaller then in the FC spectra at the same temperature. This shows that PMN is an incipient ferroelectric [11] and that in the absence of the electric field the concentration of the polar clusters is below the threshold for a percolation type glasses and is obviously due to polar clusters—appears on further cooling in the dielectric dispersion. Cross and Viehland [5] pointed out the similarity between relaxors and dipolar glasses and suggested that nanoclusters are dynamic entities with thermally fluctuating dipole moments which freeze out at low enough temperatures. The present results demonstrate that the basic difference between relaxors and dipolar glasses is their response to applied electric fields: Polar nanoclusters—corresponding to the anisotropic component in the NMR spectra—can be oriented in a strong enough applied electric field and a ferroelectric phase can be induced. This is not the case in dipolar glasses where the response is due to single dipoles which cannot be ordered by applied electric fields.

2. Results and discussion

We have investigated a single crystal of the prototype relaxor lead manganese niobate, $Pb(Mg_{1/3}Nb_{2/3})O_3$, abbreviated as PMN. It is a perovskite solid solution characterized by site and charge disorder. The dipolar glass studied for comparison was a $(ND_4)_{0.5}Rb_{0.5}D_2PO_4$ single crystal abbreviated as DRADP-50. The techniques used for PMN were field cooled and zero field cooled ^{207}Pb magnetic resonance and broad band dielectric dispersion spectroscopy. The electron paramagnetic resonance (EPR) spectra of 0.1% Mn^{2+}+ doped PMN ceramics were investigated as well. DRADP-50 was studied by 2D and 1D^{2}H and ^{31}P NMR and dielectric spectroscopy for comparison. A broad band dielectric dispersion study was performed on a PSN-PZN-PMN mixed system.

The macroscopic symmetry of PMN is cubic between 100 K and 4 K. At high temperatures above $T_B \approx 617$ K PMN behaves like all other simple perovskites. The dynamics of the system is determined by the soft TO phonon which exhibits a normal dispersion and is underdamped at all wave vectors. Below T_B in addition to the soft mode – which becomes overdamped [6]—a new dielectric dispersion mechanism appears at lower frequencies which can be described by a correlation time distribution function $f(\tau)$.

As it can be seen from Fig. 1a the correlation time distribution function $f(\tau)$ for PSN-PZN-PMN (obtained from broad band dielectric dispersion data) becomes asymmetric with decreasing temperature and a second maximum at long correlation times appears. Such behaviour was also observed [7] for other relaxors, such as PMN, PLZT and SBN. This second maximum is not seen in dipolar glasses like DRADP-50 (Fig. 1b). It seems to be specific to relaxors and signals the formation of polar nanoclusters. This is supported by the fact that this maximum changes if an external electric field E larger than the critical field E_C is applied along the ferroelectric [111] direction at $T <$ 205 K. The first maximum of $f(\tau)$ corresponding to shorter

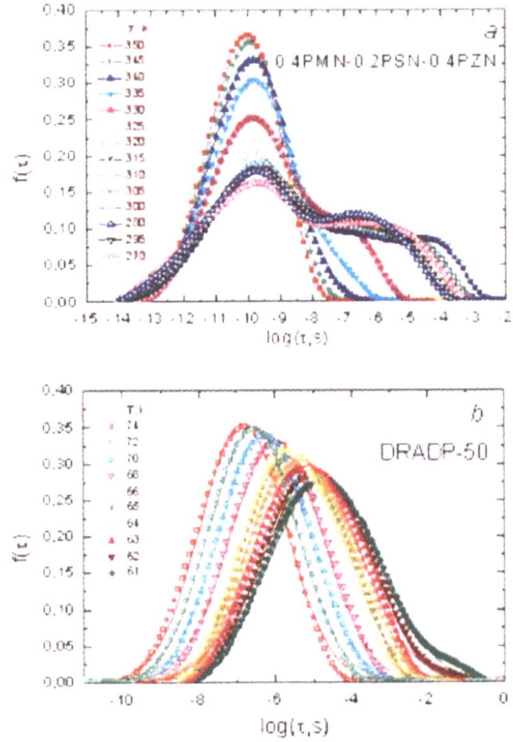

Figure 1 (a) Relaxation time distribution function $f(\tau)$ describing the dielectric dispersion in relaxor 0.4PMN-0.2PSN-0.4PZN. The short time scale maximum describes the glassy type dynamics whereas the long time scale part refers to the polar cluster dynamics. The same features are obtained in PMN, PLZT, and SBN relaxors. (b) Relaxation time distribution function in the dipolar glass DRADP-50 where the long time scale maximum is absent.

correlation times, on the other hand, is similar to the one in the dipolar glasses and is not affected by electric fields.

To investigate the nature of the polar nanoclusters and the surrounding glassy matrix on the microscopic level and to get spatially resolved evidence of the electric field response of PMN we performed field cooled (FC) and zero field cooled (ZFC) ^{207}Pb (I = 1/2) NMR experiments at different temperatures and orientations of the crystal in magnetic and electric fields. (Fig. 2a, b and c).

The ^{207}Pb spectrum at 290 K (Fig. 2a) is isotropic and of a Gaussian line shape. Two dimensional (2D) separation of interactions experiments show that the spectra are in fact frequency distributions and are composed of a large number of individual ^{207}Pb lines with different chemical shifts. This is incompatible with the assumption that the Pb ions sit at their high-symmetry cubic sites as in this case all Pb sites would be equivalent and only a single sharp line is expected. The fact that we see a Gaussian frequency distribution demonstrates that we deal with a spherical glass where all Pb nuclei are displaced but there is no preferential frozen out orientation or magnitude of displacement.

JOURNAL OF MATERIALS SCIENCE 41 (2006) 31–52

Recent progress in relaxor ferroelectrics with perovskite structure

A. A. BOKOV, Z.-G. YE
Department of Chemistry, Simon Fraser University, 8888 University Drive, Burnaby, BC, V5A 1S6, Canada

Relaxor ferroelectrics were discovered almost 50 years ago among the complex oxides with perovskite structure. In recent years this field of research has experienced a revival of interest. In this paper we review the progress achieved. We consider the crystal structure including quenched compositional disorder and polar nanoregions (PNR), the phase transitions including compositional order-disorder transition, transition to nonergodic (probably spherical cluster glass) state and to ferroelectric phase. We discuss the lattice dynamics and the peculiar (especially dielectric) relaxation in relaxors. Modern theoretical models for the mechanisms of PNR formation and freezing into nonergodic glassy state are also presented.

1. Introduction

Relaxor ferroelectrics or relaxors are a class of disordered crystals possessing peculiar structure and properties. At high temperature they exist in a non-polar paraelectric (PE) phase, which is similar in many respects to the PE phase of normal ferroelectrics. Upon cooling they transform into the ergodic relaxor (ER) state in which polar regions of nanometer scale with randomly distributed directions of dipole moments appear. This transformation which occurs at the so-called Burns temperature (T_B) cannot be considered a structural phase transition because it is not accompanied by any change of crystal structure on the macroscopic or mesoscopic scale. Nevertheless, the polar nanoregions (PNRs) affect the behaviour of the crystal dramatically, giving rise to unique physical properties. For this reason the state of crystal at $T < T_B$ is often considered as the new phase different from the PE one.

At temperatures close to T_B the PNRs are mobile and their behaviour is ergodic. On cooling, their dynamics slows down enormously and at a low enough temperature, T_f (typically hundreds degrees below T_B), the PNRs in the canonical relaxors become frozen into a nonergodic state, while the average symmetry of the crystal still remains cubic. Similar kind of nonergodicity is characteristic of a dipole glass (or spin glass) phase. The existence in relaxors of an equilibrium phase transition into a low-temperature glassy phase is one of the most interesting hypotheses which has been intensively discussed. Freezing of the dipole dynamics is associated with a large and wide peak in the temperature dependence of the dielectric constant (ε) with characteristic dispersion observed at all frequencies practically available for dielectric measurements (Fig. 1). This peak is of the same order of magnitude as the peaks at the Curie point in the ordinary ferroelectric (FE) perovskites, but in contrast to ordinary ferroelectrics it is highly diffuse and its temperature T_m ($>T_f$) shifts with frequency due to the dielectric dispersion. Because of the diffuseness of the dielectric anomaly and the anomalies in the temperature dependences of some other properties, relaxors are often called (especially in early literature) the "ferroelectrics with diffuse phase transition," even though no transition into FE phase really occurs.

The nonergodic relaxor (NR) state existing below T_f can be irreversibly transformed into a FE state by a strong enough external electric field. This is an important characteristic of relaxors which distinguishes them from typical dipole glasses. Upon heating the FE phase transforms to the ER one at the temperature T_C which is very close to T_f. In many relaxors the spontaneous (i.e. without the application of an electric field) phase transition from the ER into a low-temperature FE phase still occurs at T_C and thus the NR state does not exist.

Compositional disorder, i.e. the disorder in the arrangement of different ions on the crystallographically equivalent sites, is the common feature of relaxors. The relaxor behaviour was first observed in the perovskites with disorder of non-isovalent ions, including the stoichiometric complex perovskite compounds, e.g.

0022-2461 © *2006 Springer Science + Business Media, Inc.*
DOI: 10.1007/s10853-005-5915-7

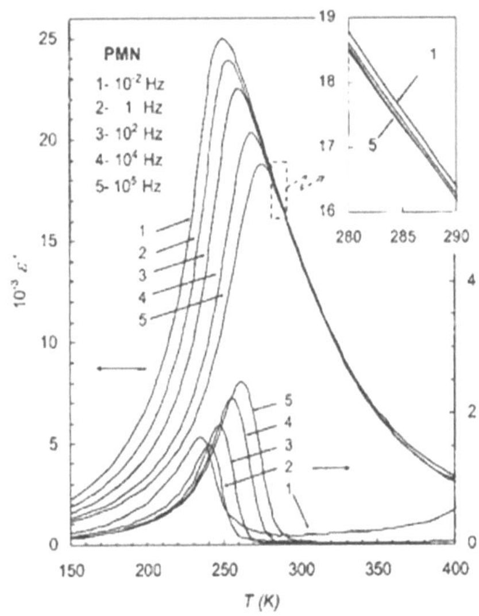

Figure 1 Temperature dependences of the real and imaginary parts of the relative dielectric permittivity measured at different frequencies in a crystal of the prototypical relaxor $Pb(Mg_{1/3}Nb_{2/3})O_3$. Enlarged view in the insert shows the universal relaxor dispersion (URD) (after Bokov and Ye, unpublished).

$Pb(Mg_{1/3}Nb_{2/3})O_3$ (PMN) [1] or $Pb(Sc_{1/2}Ta_{1/2})O_3$ (PST) [2] (in which Mg^{2+}, Sc^{3+}, Ta^{5+} and Nb^{5+} ions are fully or partially disordered in the B-sublattice of the perovskite ABO_3 structure) and nonstoichiometric solid solutions, e.g. $Pb_{1-x}La_x(Zr_{1-y}Ti_y)_{1-x/4}O_3$ (PLZT) [3, 4] where the substitution of La^{3+} for Pb^{2+} ions necessarily leads to vacancies on the A-sites. Recently an increasing amount of data reported has shown that many homovalent solid solutions, e.g. $Ba(Ti_{1-x}Zr_x)O_3$ (BTZ) [5, 6] and $Ba(Ti_{1-x}Sn_x)O_3$ [7] can also exhibit relaxor behaviour. Other examples of relaxor ferroelectrics are complex perovskites $Pb(Zn_{1/3}Nb_{2/3})O_3$ (PZN) $Pb(Mg_{1/3}Ta_{2/3})O_3$ (PMT), $Pb(Sc_{1/2}Nb_{1/2})O_3$ (PSN), $Pb(In_{1/2}Nb_{1/2})O_3$ (PIN), $Pb(Fe_{1/2}Nb_{1/2})O_3$ (PFN), $Pb(Fe_{2/3}W_{1/3})O_3$ (PFW) and the solid solutions: $(1-x)Pb(Mg_{1/3}Nb_{2/3})O_3-xPbTiO_3$ (PMN-PT) and $(1-x)Pb(Zn_{1/3}Nb_{2/3})O_3-xPbTiO_3$ (PZN-PT).

Although relaxor ferroelectrics were first reported nearly half a century ago, this field of research has experienced a revival of interest in recent years. In this paper, we try to provide an overview of the current understanding of the various issues of relaxors. Emphasis is put on the latest developments. For a review of the earlier studies, readers can refer to Refs. [8–11].

2. Compositional order-disorder phase transitions and quenched disorder in complex perovskites

As mentioned above, the disordered distribution of different ions on the equivalent lattice sites (i.e. compositional disorder, also called chemical, ionic or substitutional disorder) is the essential structural characteristic of relaxors. The ground state of the complex perovskites should be compositionally ordered, e.g. in the $A(B'_{1/2}B''_{1/2})O_3$ compounds each type of the cations, B′ or B″, should be located in its own sublattice, creating a superstructure with complete translational symmetry. This is because the electrostatic and elastic energies of the structure are minimized in the ordered state due to the difference in both the charge and the size of B′ and B″ ions. Thermal motion is capable of destroying the order at a certain nonzero temperature (T_t). This occurs in the form of structural phase transition, the order parameter (compositional long-range order, s) of which can be measured by the X-ray or other diffraction methods. Such kind of phase transitions had been known long ago (e.g. in many metallic alloys) and was also discovered comparatively recently at $T_t \sim 1500$ K in PST, PSN [12] and several other complex perovskites. Ordering implies the site exchange between B′ and B″ cations via diffusion. It is a relaxation process with a nearly infinite characteristic time at low temperatures, but at 1500 K it can be quite fast. As a result, in some perovskites (e.g. in PST, PSN, PIN), by annealing at temperatures around T_t and subsequent quenching, one can obtain the metastable states with different s at low temperatures. In some other materials (e.g. in PFN and PMN) the compositional disorder cannot be changed by any heat treatment because the relaxation time of ordering is too long. However in all known relaxors, at T_B and below, the compositional order is frozen (quenched), i.e. cannot vary during practically reasonable time.

In the real complex perovskite crystals and ceramics the quenched compositional disorder is often inhomogeneous, e.g. small regions of the ordered state are embedded in a disordered matrix. These regions can be regarded as a result of incomplete compositional order-disorder phase transformation or as quenched phase fluctuations. In the prototypical relaxor PMN this kind of inhomogeneous structure always exists and cannot be changed by any heat treatment.

In $Pb(B'_{1/2}B''_{1/2})O_3$ perovskites the ordering of B-site ions converts the disordered PE $Pm\bar{3}m$ structure into the ordered $Fm\bar{3}m$ structure in which B′ ions alternate with B″ ions along the ⟨100⟩ directions (1:1 ordering). In the ordered phase of many non-ferroelectric $A(B^{2+}{}_{1/3}B^{5+}{}_{2/3})O_3$ perovskites, B^{2+} ions alternate with two B^{5+} ions along the ⟨100⟩ directions (1:2 ordering). The type of ordering in lead-containing relaxor perovskites, $Pb(B^{2+}{}_{1/3}B^{5+}{}_{2/3})O_3$, has been the subject of debates. In the early works only inhomogeneous ordering (ordered regions within disordered surroundings) was found in the samples studied. High-resolution electron microscopy of PMN revealed nano-size (~2–5 nm) regions in which the ordering of 1:1 type ($Fm\bar{3}m$) was observed (see e.g. Refs. [13, 14]). These chemical nanoregions (CNR) give rise to weak superlattice reflections (the so-called F-spots). The results of anomalous X-ray scattering measurements [15] showed that the CNRs in PMN exhibit an isotropic shape and a temperature-independent (as expected for the quenched order) size in the temperature interval of 15–800 K. Alternating Mg^{2+} and Nb^{5+} ions, i.e the same type of ordering as in the ordered $Pb(B'_{1/2}B''_{1/2})O_3$ perovskites, were initially supposed to exist in these regions. This structural model was called "space charge model" because it implies the existence of the negatively charged compositionally ordered non-stoichiometric nanoregions, and the positively charged disordered non-stoichiometric matrix. Later, by means of appropriate high temperature treatments, Davies and Akbas [16] were able to increase the size of CNRs and obtained highly 1:1 ordered samples without the disordered matrix in the PMT and modified PMN ceramics. The existence of such ordering in overall stoichiometric samples is obviously inconsistent with the space charge model. The results of X-ray energy dispersive spectroscopy with a nanometer probing size revealed that the Mg/Nb ratio is the same in the CNRs as in the disordered regions of PMN [17], which also disagrees with the space charge model. A charge-balanced "random-site" model has been suggested in which one of the B-sublattices is occupied exclusively by B^{5+} ions while the other one contains a random distribution of B^{2+} and B^{5+} ions in a 2:1 ratio so that the local stoichiometry is preserved [16]. The inhomogeneous compositional disorder, characteristic of the $Pb(B^{2+}_{1/3}B^{5+}_{2/3})O_3$ perovskites, are shown schematically in Fig. 2.

The degree of compositional disorder can greatly influence the FE properties. For example, the disordered PIN crystals are relaxor ferroelectrics, but in the ordered state, they are antiferroelectrics with a sharp phase transition [18, 19], confirming the general rule that the relaxor

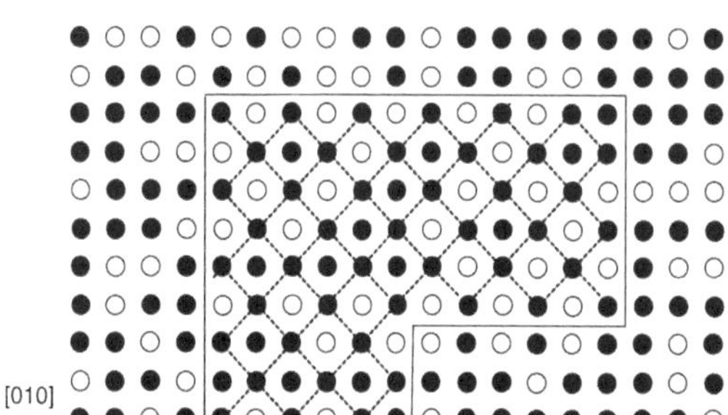

Figure 2 Schematic representation of the ordered chemical nanoregion, CNR (the area delimited by the solid line) within the disordered matrix in $Pb(B^{2+}_{1/3}B^{5+}_{2/3})O_3$ perovskites according to the random-site model. One of the two sublattices inside CNR (shown by dashed lines) is formed by B^{5+} ions only. Pb and O ions are not shown.

behaviour can only be observed in disordered crystals. The possibility for real perovskite samples to have different states of compositional disorder, depending on crystal growth or ceramic sintering conditions, should be taken into account in research work. More detailed discussion on the compositional ordering and its impact on FE properties can be found in Refs. [20–22], and the literature therein.

3. Relaxors in the ergodic state

3.1. Paraelectric structure

The PE phase of all perovskite ferroelectrics has the cubic $m\bar{3}m$ average symmetry, but locally the ion configuration can be distorted, i.e. the ions are not located in the special crystallographic sites of the ideal perovskite structure. For example, in the classical ferroelectrics $BaTiO_3$, the random displacements of Ti cations along the $\langle 111 \rangle$ directions caused by the multiple-well structure of potential surface were found [23, 24]. Such kind of displacements is due to the hybridization between electronic states of cations and the $2p$ states of oxygen (and should not exist in the case of purely ionic bonds). This effect is an important factor in the FE instability [25] and is also expected to occur in perovskite relaxors. Moreover, owing to the different sizes of the compositionally disordered cations and the random electric fields created because of the different charges of these cations in relaxors, all ions are expected to be displaced from special positions. These shifts should exist in the PE phase and also at lower temperatures.

Permanent uncorrelated displacements of ions from the high-symmetry positions of the (fully or partially compositionally disordered) cubic perovskite-type structure were indeed found in relaxors at temperatures much higher, as well as lower, than T_B. They are shown schematically in Fig. 3. The displacements of Pb^{2+} were detected by X-ray and neutron diffraction in PMN [26–28], PZN, PSN, PST, PIN, PFN, PZN-PT and PMN-PT with small x (see Refs. [29–31], and references therein). To describe the Pb distribution, a spherical layer model has been proposed [28] according to which the shifts of ions are random both in length and direction so that they are distributed isotropically within the spherical layer centred on the special Pb site.[1] The typical radius of the sphere is ~0.3 Å. It decreases slightly with increasing temperature. The off-symmetry displacements of Pb ions in PMN were found to vanish at $T > 925$ K [26] (for other relaxors no data up to so high temperatures are available). The spherical layer model for Pb displacements in PMN was confirmed by the NMR investigations [32] and by the pulsed neutron atomic pair-distribution function (PDF) analysis [33]. Note that the significant random off-centring of Pb ions in perovskites is not the result of compositional

[1] For PZN the shifts of Pb from the ideal positions along the eight equivalent $\langle 111 \rangle$ directions were reported, instead of spherical layer distribution [29].

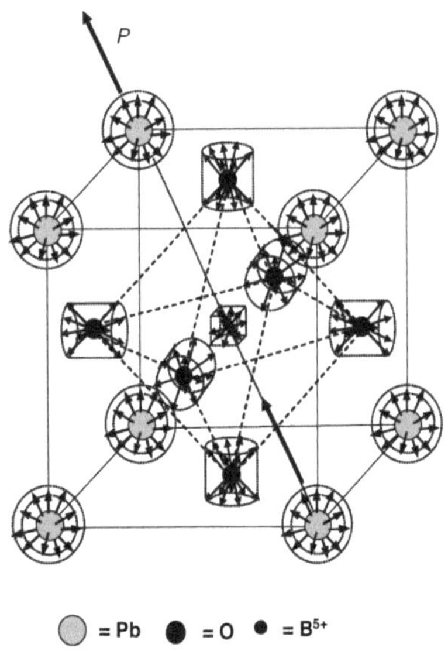

Figure 3 Typical uncorrelated ion displacements (shown by small arrows) in the unit cell of the lead-containing complex perovskite relaxor. Thick arrows show the direction of the local spontaneous polarisation P caused by the correlated displacements of ions inside PNRs.

disorder. It was also found in the PE phase of the ordinary perovskite $PbZrO_3$ [34]. On the other hand, in the PMN-PT solid solution with $x = 0.4$ which is still compositionally disordered, the Pb displacements from the special perovskite positions were not observed at $T > T_C$ [31].

According to neutron diffraction data [27, 29, 35] the shifts of oxygen ions in the planes parallel to the corresponding faces of the perovskite cubic cell are isotropic (in PMN the shifts are close to 0.2 Å). The oxygen ions are also shifted (by about 0.06 Å in PMN) in the perpendicular direction so that the distribution of shifts forms two rings parallel to the face of the cube. The displacements of B-site ions (Nb^{5+}, Mg^{2+}, Zn^{2+} etc.) from the ideal positions were not noticed in diffraction experiments [29, 35] (some authors found small seemingly isotropic displacements of about 0.1 Å in PMN [27]). Nevertheless the investigations of the extended X-ray absorption fine structure (EXAFS) and the pre-edge regions of absorption spectra revealed the off-centre random displacements of Nb in the direction close to ⟨110⟩ in PMN, PZN, PSN and PIN [36]. These displacements are not sensitive to the change of temperature (in the range of 290–570 K), nor to the degree of compositional disorder (in PSN and PIN). The pulsed neutron PDF studies confirm [33] that the Nb displacements (in PMN at room temperature) are comparatively small (much smaller than in $KNbO_3$).

In the canonical relaxors such as PMN, the average crystal symmetry seems to remain cubic with decreasing temperature (see, however, the discussion in Sections 7.1.1 and 7.2), but the local structure changes. In addition to the uncorrelated local distortions described above, the clusters of FE order (i.e. PNRs) appear at $T < T_B$ ($T_B \approx 620$ K in PMN). Due to their extremely small (nanometric) size, these clusters cannot be detected from the profiles of the X-ray and neutron Bragg diffraction peaks. Other experiments are needed to validate their existence.

3.2. Experimental evidence for PNRs

The first experimental (although indirect) evidence for the PNRs came from the temperature dependences of the optic index of refraction (n) which appear to be linear at $T > T_B$, as shown in Fig. 4 [37]. At lower temperatures a deviation from linearity was observed which was attributed to the variation of n induced (via quadratic electrooptic effect) by local spontaneous polarization inside the PNRs. The existence of PNRs was later confirmed by elastic diffuse neutron and X-ray scattering around the reciprocal lattice points [38–41]. In the PMN crystals, significant diffuse scattering appears at $T < T_B$ with the intensity increasing with decreasing temperature. This effect resembles the scattering caused by FE critical fluctuations, but an important difference (found in synchrotron X-ray experiments [41]) is that the shape of wavevector dependence of scattering intensity at large distances from the reciprocal lattice point deviates from the Lorentzian. This means [41] that the PNRs are more compact than the usual FE critical fluctuations and have better defined borders. The correlation length (ξ) of the atomic displacements contributing to the diffuse scattering, which is a measure of the size of PNR, can be derived from the experiment: it is inversely proportional to the width of the

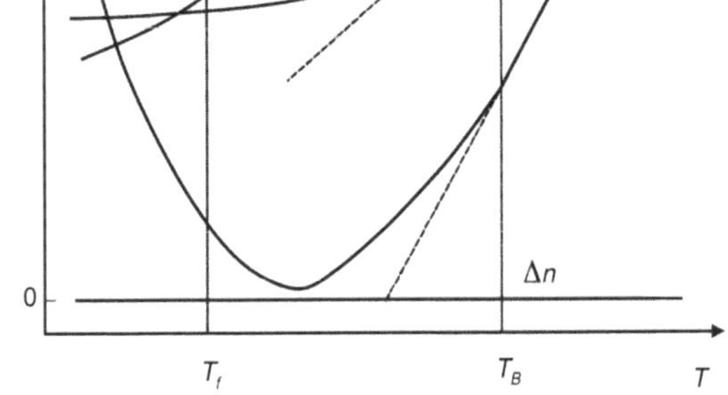

Figure 4 Schematical typical temperature dependences of the refractive index, n, unit cell volume, V, reciprocal dielectric permittivity, $1/\varepsilon$, and birefringence, Δn, in the canonical relaxor.

diffuse (Lorentzian) peak. According to the recent high-resolution neutron elastic diffuse scattering study of PMN [42], the size of the emerging PNRs is very small (ξ is around 1.5 nm) and practically temperature independent at high temperatures (Fig. 5). The perovskite unit cell parameter being ~0.4 nm, each PNR is composed of only a few unit cells. Below about 300 K, ξ begins to grow on cooling, reaching ~7 nm at 10 K. The most significant growth is found around T_f. Qualitatively the same behaviour was observed in the bulk of PZN crystals (the structure of PZN surface layers is different, see Section 7.2) but the size of PNRs is larger: they grow from ~7 nm at high temperatures to ~18 nm at 300 K [43, 44]. From the analysis of the relation between ξ and the integrated intensity of scattering, it was concluded [42] that the number of PNRs also increases on cooling, but in contrast to the temperature evolution of ξ, the increase begins right from T_B and at $T \approx T_f$ a sharp decrease of this number occurs (presumably due to the merging of smaller PNRs into larger ones). Below T_f the number of PNRs remains practically the same at any temperature.

Emergence of PNRs below T_B was also observed in the PMN crystal by means of transmission electron microscopy (TEM) [14], but their size was an order of magnitude larger than that determined from the neutron diffuse scattering, probably because of the influence of electron beam irradiation.

The directions of ionic displacements responsible for the spontaneous dipole moment of PNRs were investigated in several works. By means of dynamic structural analysis of diffuse neutron scattering in PMN crystals it was found that the *B*-site cations (Nb and Mg) and the O anions are displaced with respect to the Pb cations in the opposite directions along the body diagonal (i.e. the [111] direction) of the perovskite unit cell, forming a rhombohedral polar structure [45]. The rhombohedral *R*3*m* symmetry was also derived from the analysis of

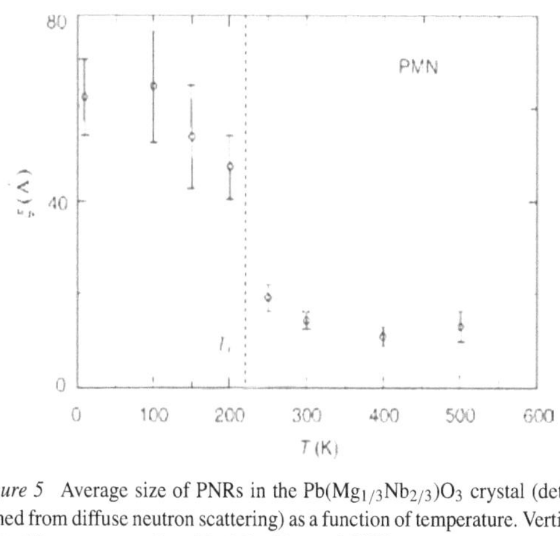

Figure 5 Average size of PNRs in the $Pb(Mg_{1/3}Nb_{2/3})O_3$ crystal (determined from diffuse neutron scattering) as a function of temperature. Vertical dashed line corresponds to T_f. (after Xu *et al.* [42]).

ion-pair displacement correlations obtained by an X-ray diffuse scattering technique [46], but according to this study, O displacements deviate from the body diagonal and remain parallel to the ⟨110⟩ direction. The shape of PNR was found to be ellipsoidal [46]. The same shape was revealed by TEM [14].

Besides the structural features, many properties of relaxors can be adequately explained on the basis of the idea of PNRs. For example, in contrast to ordinary ferroelectrics, where a sharp anomaly of specific heat is known to appear at phase transition, in relaxors such anomaly is smeared over a wide temperature range and thus is hardly distinguishable from the background of the lattice contribution. The excess specific heat (total minus lattice contribution) has been determined in PMN and PMT crystals using precise adiabatic and thermal relaxation techniques [47]. It appears as a diffuse symmetric maximum located within the same temperature interval where PNRs nucleate and grow (between 150 and 500 K in PMN). Therefore the anomaly is likely to be caused by the formation of PNRs and/or by dipolar interactions among them.

Brillouin spectra of PMN-PT at $T \gg T_C$ revealed significant relaxation mode (central peak) which was attributed to the thermally activated fast (10–100 GHz) relaxation of PNRs [48]. The intensity and the width of the peak increase with decreasing temperature, indicating an increase of the number of PNRs and a slowing-down of their dynamics, respectively. The hypersonic damping was also observed. It increases upon cooling, and is attributed to the scattering of acoustic mode by PNRs [48].

PNRs can be thought as unusually large dipoles whose direction and/or magnitude are dependent on an external electric field. Therefore the related properties are expected to be unusual. Indeed, at those temperatures where PNRs exist, relaxors are characterized by giant electrostriction [49–51], remarkable electrooptic effect [50] and extremely large dielectric constant (see Fig. 1 and Section 6).

Even though no unambiguous structural confirmations for the phase transition at T_B are known, the anomalies of properties at this temperature were reported. The frequency-independent maximum of the dielectric loss tangent was found at this temperature in PMN [52]. In the course of thermal cycling of PMN and PMN-PT crystals unannealed after growth, a narrow maximum of the acoustic emission activity is observed (and decreases with the increase of number of cycles) in the vicinity of T_B [53]. Not only the temperature dependence of the index of refraction deviates from linearity at $T < T_B$ (as discussed above in this section), but the temperature dependences of the reciprocal dielectric constant, lattice parameter [51] (see Fig. 4 and (consequently) thermal strain [10, 53] also do the same.

Little is known about the relation between the CNRs and the PNRs in relaxors, although such relation can *a priori* be expected. Based on the TEM data, it was concluded that the PNRs in PMN may contain CNRs inside and in this case the regions in which PNRs and CNRs overlap remain non-polar [14]. In the framework of the theoretical models discussed in Sections 5 and 7.3, the CNRs can be considered as one of the factors influencing the formation and behaviour of PNRs, but not necessarily the determining factor.

4. Lattice dynamics in relaxors

Phase transitions in displacive ferroelectrics (including perovskites) are known to be caused by softening and condensation of transverse optic (TO) phonon mode at the Brillouin zone centre. Since the frequency of this mode (ω_0) is connected to the static lattice dielectric constant (through the Lyddane-Sachs-Teller relation, $1/\varepsilon \propto \omega_0{}^2$), the divergence of ε in ferroelectrics at T_{CW} according to the Curie-Weiss law ($1/\varepsilon \propto T - T_{CW}$) implies that the mode condenses (its frequency tends to zero) at T_{CW}, too. In relaxors, the Curie-Weiss law also holds in the PE phase, i.e. at $T > T_B$, (see Fig. 4 and Section 6), but the corresponding softening of phonon modes had not been detected until recently. During the last few years neutron inelastic scattering technique was applied to investigate the lattice dynamics in relaxor crystals. In the PE phase of PMN far above T_B, the dispersion of the transverse acoustic (TA) and low-energy TO phonons were found to be very similar to that existing in the PE phase of classical displacive ferroelectrics $PbTiO_3$ [39, 54] (see the curves for 1100 K in Fig. 6). On cooling down to $T \approx T_B$, the optic branch softens in the same manner as in displacive ferroelectrics, i.e., the frequency of the mode at wave vector $\vec{q} = 0$ (zone centre mode) follows the Cochran law [54, 55]:

$$\omega_0^2 = A(T - T_0), \qquad (1)$$

with $A > 0$, as shown in Fig. 7. At $T < T_f$, the well-defined TO modes are also observed [55] (see the curve for 150 K in Fig. 6). Once again, the temperature evolution is consistent with the typical behaviour of a FE soft mode below the Curie temperature (Fig. 7), i.e. Equation 1 holds with $A < 0$ (note that T_0 is close to T_B in this case). The same dependence is observed at low temperatures in PZN crystal [44]. But in the temperature range between T_f and T_B, the lattice dynamics is different. The propagating TO modes are observed here only for the wave vectors larger than q_{wf}. For $q < q_{wf}$, the modes are overdamped. The TO phonon branch drops sharply into the TA branch at q_{wf}, resembling a waterfall (as shown in Fig. 6 for 500 K), and for this reason, the phenomenon is called "waterfall." It has been observed not only in PMN but also in other relaxor materials for which neutron inelastic experiments were performed, i.e. in PZN [56] and PZNT [57]. In the same temperature range where the TO mode is overdamped (between T_f and T_B), the damping of TA phonons is also enhanced (e.g. neutron inelastic experiments in PMN revealed a large maximum of the TA-phonon linewidth at ~400 K [55]).

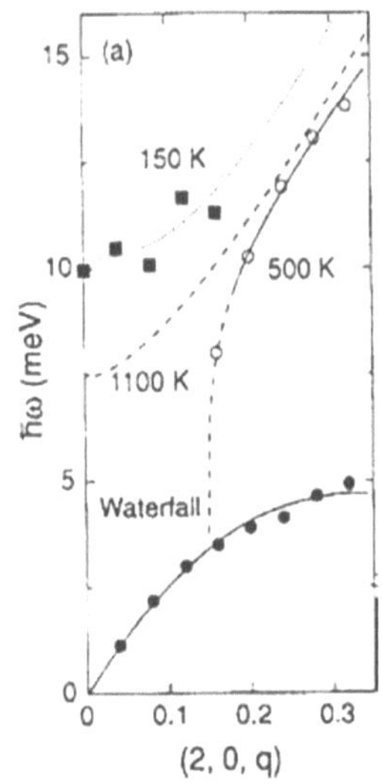

Figure 6 Phonon dispersion curves in PMN crystal for the TA branch at 1100 K (solid circles) and the lowest-energy TO branch at three different temperatures: $T = 1100$ K $> T_B$ (dashed line), $T = 500$ K $< T_B$ (open circles), and $T = 150$ K $< T_f$ (squares). (after Wakimoto *et al.* [55]).

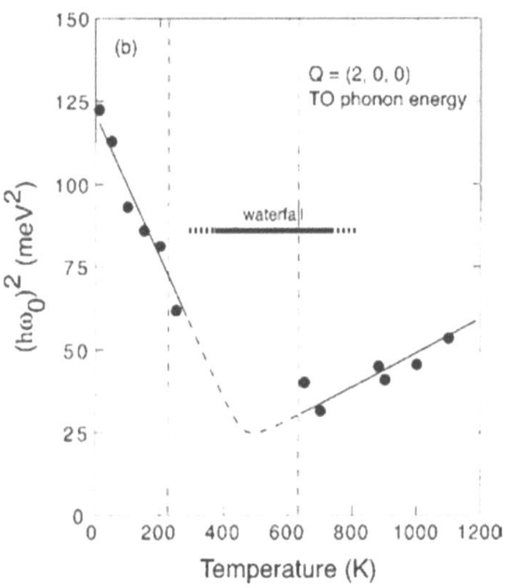

Figure 7 Temperature dependence of TO-phonon energy squared measured at (200) in PMN crystal. Vertical dashed lines correspond to T_f and T_B. The temperature range in which the waterfall feature appears is indicated by the thick horizontal line. The other dashed and solid lines are guides to the eye. (after Wakimoto *et al.* [55]).

It was initially proposed that the large damping of TO modes is due to the presence of PNRs which prevent the propagation of phonons with wavelength larger than the size of PNR, and thus q_{wf} is the measure of the average size of the PNRs [54, 57]. Later it was shown [58] that q_{wf} depends on the choice of the Brillouin zone and the relation of q_{wf} to the size of PNRs is improbable. The waterfall effect was then explained by the interactions of acoustic and optic branches. It was also noticed that damping of the soft mode near phase transition is not the unique feature of relaxors; similar effect can be found in some ordinary ferroelectrics [58].

An important question arises as to whether the low-energy TO mode found in relaxors is really the FE soft mode, i.e. whether the frozen mode displacements are responsible for the spontaneous dipole moment of PNRs and for the spontaneous polarization of the low-temperature FE phase. Initially the answer to the question was negative. Nabereznov *et al.* [39] compared the correlated atomic displacements in PNRs (found by elastic diffuse scattering experiments) and the TO-mode atomic displacements and concluded that they are incompatible. Besides, Vakhrushev and Shapiro [59] noticed that the value of T_0 derived from Equation 1 does not coincide with T_{CW} found in dielectric measurements and therefore this mode could not be the FE soft mode. They also identified an additional "quasioptic" branch lying significantly lower in energy than the TO branch and having the temperature variation consistent with the Curie-Weiss behaviour of the dielectric constant. The quasioptic mode was proposed to be the true soft mode. This opinion was later disputed. The apparent conflict between diffuse and soft mode scattering experiments was reconciled with the help of the "phase-shifted condensed soft mode" model proposed by Hirota *et al.* [40]. According to this model the total displacement of atoms inside a PNR consists of two components. The first component is created by the TO soft-mode condensation and gives rise to the spontaneous polarisation of PNR. The second one results from a uniform displacement of all atoms leading to the shift of the PNR along their polar direction relative to the surrounding non-polar matrix. Wakimoto *et al.* [60] described the lattice dynamics by a coupling between TA and TO modes without the need for considering any additional quasioptic modes. The concept of soft coupled optic mode was introduced. Being condensed, this mode has the optic component responsible for the dipole moment of PNR and the acoustic component giving rise to the uniform displacement of PNR as a whole. The microscopic origin of the uniform component is not yet clear.

Infrared (IR) spectroscopy of PMN between 20 and 300 K [61] and of PLZT between 10 and 530 K [62] revealed three main zone-centre TO modes as typical of cubic perovskites. High-energy TO modes do not show any pronounced temperature dependence both in PMN and PLZT. The lowest-energy mode (which was also resolved by time-domain THz spectrometry) was found to soften following Equation 1 with $A < 0$ and $T_0 \approx T_B$, i.e. in agreement with the above-discussed neutron scattering data. Nevertheless, in contrast to the neutron data, the mode remains underdamped in the whole temperature range. This discrepancy is related to the fact that phonons with different wave-vectors are probed in IR ($q \approx 10^{-5}$ Å^{-1}) and neutron ($q \geq 10^{-2}$ Å^{-1}) experiments.

Raman scattering, which is known to be an effective tool for studying the optic soft modes in the crystals with normal phase transitions, was also widely applied to relaxors (see Refs. [63–65], for a review), but no soft mode has been found here. Light scattering spectra in relaxors appear to be quite complex and their interpretation is not straightforward. In particular, it is not clear what kind of disorder gives rise to the observed first-order lines that are forbidden by the $Pm\bar{3}m$ (O_h^1) average symmetry of relaxors.

Recently, vibration spectra in PMN were determined theoretically from the first principles [66]. Although the computations were performed for the case of compositionally ordered structure (the real structure is disordered with the inclusions of ordered CNRs, see Section 2), it appeared to be possible to assign the calculated phonons to the main peculiarities of Raman and IR spectra.

5. Origin and evolution of PNRs: Models and theories

Although the very existence of PNRs in relaxors seems to be doubtless, the cause and mechanisms of their formation are not conclusively understood. At temperatures higher than T_B the structure and properties of relaxors closely resemble those of normal displacive ferroelectrics. When a relaxor becomes compositionally ordered after high-temperature annealing (without changing the chemical composition), a sharp ferro- or antiferroelectric (AFE) phase transition is observed (see Section 2). These facts seemingly suggest that the relaxor crystal tends to be ferro- or antiferroelectric at low temperatures, but the quenched compositional disorder somehow prevents the normal transition into the phase with macroscopic FE or AFE order from happening. Instead, the PNRs appear. There exist different approaches to explain the formation of PNRs. All of them can be schematically subdivided into two categories. The models of the first category [8, 67–70] consider the PNRs as a result of local "phase transitions" or phase fluctuations so that the crystal consists of nanosize polar islands embedded into a cubic matrix in which the symmetry remains unchanged (as shown in Fig. 8a). The models of the second category assume the transition to occur in all regions of the crystal and the crystal consists of low-symmetry nanodomains separated by the domain walls but not by the regions of cubic symmetry [71, 72] (the example is shown in Fig. 8b). Note that these two situations can hardly be distinguished experimentally by structural examinations [73] because the *local* symmetry of cubic matrix is not expected to be cubic

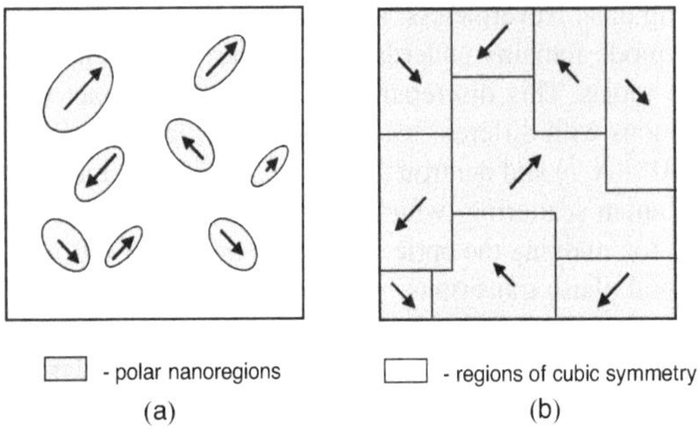

Figure 8 Schematic representation of PNRs in relaxors according to the different models.

and the thickness of domain walls (i.e. the regions where polarization is not well-defined) is comparable with the size of nanodomains.

The second category is represented by the random-field model proposed by Westphal, Kleemann and Glinchuk (WKG model) [71, 74], who applied the results of a theoretical work by Imry and Ma [75] to the relaxors. It was shown in Ref. [75] that in the systems with a continuous symmetry of order parameter, a second-order phase transition should be destroyed by quenched random local fields conjugate to the order parameter. Below the Curie temperature the system becomes broken into small-size domains (analogy of PNRs) instead of forming a long-range ordered state. It should be emphasized that this model does not consider the trivial case of the local spontaneous polarization which is directed parallel to the quenched field when the field is strong enough. Instead the situation is determined by the interplay of the surface energy of domain walls and the bulk energy of domains in the presence of *arbitrary weak* random fields [75].

For displacive transitions, continuous symmetry means that the spontaneous deformation is incommensurate with the PE lattice. However, this is not the case for the perovskite ferroelectrics in which the spontaneous deformation and the polarization (order parameter) are aligned along definite crystallographic directions (e.g. the $\langle 111 \rangle$ directions for the rhombohedral phase). Nevertheless, when the number of allowed directions is large (e.g. eight for the rhombohedral phase), the symmetry of order parameter can be considered quasi-continuous and the approach appears to be applicable. The disordered distribution of the heterovalent ions inherent to the compositionally disordered structure (e.g. Nb^{5+} and Mg^{2+} ions in PMN) provides the source for quenched random electric fields.

Ishchuk [72] analysed the thermodynamic potential in the framework of Landau phenomenological theory for the systems in which the energies of the FE and AFE phases are close to each other. It was shown that the state with coexisting FE and AFE domains may have lower thermodynamic potential than the homogeneous (FE or AFE) state. This effect is due to the interactions (electrostatic and elastic) between the FE and AFE domains. It was suggested that relaxors are just the crystals in which this effect occurs. In other words, the nonpolar regions, coexisting with PNRs (FE domains), are the domains of AFE structure.

The best-known model of the first category was developed in the early works by Isupov and Smolenskii [8, 22]. Due to the compositional disorder the concentrations of different ions (e.g. Mg^{2+} and Nb^{5+} in PMN) are subject to quenched spatial fluctuations. As the FE Curie temperature (T_C) depends on the concentration, spatial fluctuations of local T_C are expected. It was suggested that upon cooling, local FE phase transitions occur first in those regions where T_C is higher, whereas the other parts of the crystal remain in the PE phase. Therefore, PNRs are simply the regions with elevated Curie temperature.

Several other models use the microscopic approach and consider the structural evolution and formation of PNRs in terms of interatomic interactions. The FE lattice distortion in the ordinary perovskites is known to be determined by a delicate balance between the electrostatic (dipole-dipole) interactions and the short-range repulsions. Hybridization between the oxygen $2p$ states and electronic states of cations (covalent bonding) is able to change this balance, influencing thereby the phase transition temperature [25]. In the compositionally ordered (translationally symmetric) crystals, exactly the same forces affect all the atoms of a certain type because they have the same coordination neighbourhood. In the case of compositional disorder, the ions of different types may be found in the neighbouring unit cells on the same crystallographic positions (e.g. in the B-sublattice of PMN, both Mg and Nb ions are the nearest neighbours of Nb ions). The interatomic interactions which would cause ferro- or antiferroelectric order in the compositionally ordered state become random in this case, and as a result, the long-range polar order is disturbed. The models described below emphasize the importance of different interactions: the interactions under random local electric fields only (including dipole-dipole interactions) [67], and the dipole-dipole interactions together with random short-range repulsions [9, 68] or random covalent bonding [33].

In the random field theory developed for relaxors by Glinchuk and Farhi (GF model) [67] (see also Ref. [76] in which the related papers are reviewed), the transition is regarded as an order-disorder one, i.e. at high temperature the crystal is represented by a system of reorientable dipoles (dipoles caused by the shifts of ferroactive ions from their ideal perovskite positions, see Section 3.1). These random-site dipoles are embedded in highly polarizable "host lattice" (the high polarizability is due to the transverse optic soft mode existing in relaxors, see Section 4). The dipole-dipole interactions are indirect (they occur via the host lattice) and random. Nevertheless, according to the theory, they should lead to uniformly directed local fields and thus to FE ordering at low temperature (in contrast to *direct* dipole-dipole interactions which can lead to a dipole glass state). Thus to explain the absence of macroscopic FE order in relaxors, additional sources of random local electric fields are considered. These additional fields can be static (coming from quenched compositional disorder, lattice vacancies, impurities and other imperfections) or dynamic (associated with shifts of non-ferroactive ions from the special positions). In contras to the fields considered in the WKG model, these fields should be rather large (larger than critical value) to destroy the long-range FE order. The FE order parameter, phase transition temperature T_C, linear and nonlinear dielectric susceptibilities are calculated within the framework of statistical theory using the distribution function for local fields. It is found that depending on the model parameters (concentration of dipoles, other field sources and the host lattice correlation length), the low-temperature phase can be FE, dipole glass or mixed ferroglass. In the temperature interval between T_C and T_B, the short-range clusters may appear, in which the reorientable dipoles are ferroelectrically correlated (i.e. PNRs). In the ferroglass state these clusters coexist with the macroscopic regions in which the dipoles are coherently ordered.

Note that the GF model for relaxors is the extension of the analogous theory for incipient ferroelectrics with off-centre impurities (e.g. $KTaO_3$:Li, Nb, or Na). In the later case the off-centre impurities are the interacting dipoles. Due to their small concentration the crystal can be considered as a system of *identical* dipoles with random long-range interactions. In the case of complex perovskite relaxors, the dipole concentration cannot be considered small. The random interactions of different (short-range) nature are also involved and thus the dipoles are not identical. It was first recognized in the model proposed by one of the co-authors of the present review [9, 68]. In this model the PNRs are the result of local condensation of the soft phonon mode (which exists in relaxors as discussed in Section 4). The consideration is based on the model of coupled anharmonic oscillators which is often applied to ordinary ferroelectrics. The effective Hamiltonian is given by the sum of Hamiltonians of the individual unit cells:

$$H = \sum_{l}\left[0.5\Pi_l^2 + A_l\xi_l^2 + B_l\xi_l^4 - \sum_{l'} \upsilon_{ll'}\xi_l\langle\xi_{l'}\rangle\right],$$

where Π_l and ξ_l are the generalized momentum and coordinate of the soft mode displacements, A_l and B_l are parameters of one-particle potential, which are determined by the interactions (mainly short-range repulsive) between ions of the lth unit cell, and υ_{ll}' are parameters characterizing the interactions (long-range dipole-dipole) between the different cells. In the translationally invariant crystal, all the parameters, A_l, B_l, and υ_{ll}', would be the same. In the case of compositional disorder they are different. The distribution function for these parameters is introduced in the model. This distribution gives rise to the spatial distribution of local "Curie temperature" T_C. PNRs appear in the regions with enhanced local T_C. The model parameters are linked to the parameters of real structure (in particular, the size of ions). Based on the crystal composition, this model is able to predict quantitatively the degree of "diffusion" of the transition, i.e. the extent of temperature interval in which the PNRs develop before the crystal transform into the low-temperature nonergodic phase. In particular, the degree of diffusion increases with increasing difference in the radii of ions in the ferroactive sublattice (A or B perovskite sublattice) or with increasing compositional disorder in this sublattice.[2] On the other hand, the diffusion is much less sensitive to the disorder in the non-ferroactive sublattice. The influence of the degree of compositional disorder on T_C is also explained. Based on the arguments similar to those used in the original model [68] it was recently suggested [77] that, because of the randomness of microscopic forces responsible for the onset of spontaneous polarization, each PNR can consist of unit cells polarized in different directions. This model of "soft nanoregions" also implies that, due to thermally activated reorientations of some unit cells inside PNR, not only the direction (as believed before), but also the magnitude of the spontaneous dipole moment of individual PNR can strongly change with time (due to fluctuations or under the external field), while the size of PNR remains the same.

The Hamiltonian considered in the model by Egami [33] consists of two terms,

$$H = H_1 + H_2. \tag{2}$$

The first term is written in a standard form

$$H_1 = -\sum_{ij} J_{ij}\vec{S}_i \cdot \vec{S}_j, \tag{3}$$

where $\vec{S}_i$ is the local polarization caused by the displacement of i-th Pb ion from its special position (as discussed in Section 3.1), J_{ij} describes the random interaction between local polarizations mediated by oxygen and B-site ions. It is explained that in PMN the Pb ions cannot form the covalent bonds with those O ions which are bonded to Nb. On the other hand, Mg ions create purely ionic bonds and do not prevent the Pb–O bonding. Consequently the direction towards Mg is an "easy" direction for Pb displacement. This directional dependence of the energy of Pb displacements resembles the crystalline anisotropy in

[2] In lead-containing complex perovskites the Nb and Ta cations are supposed to be ferroactive.

magnetic systems. It is random in compositionally disordered crystal and can be described by model Hamiltonian H_2. This model was established to account for the relaxor properties in ER as wall as in NR phases, but the appearance of PNRs was not derived.

Timonin [69] suggested that the ergodic phase in relaxors is an antilog of Griffiths phase theoretically predicted long ago (but not yet experimentally found) for dilute ferromagnetics. Ferroelectric clusters of various sizes (i.e. PNRs) appear in this model at $T < T_C$ (where T_C is the Curie temperature for non-dilute crystal) and specific non-exponential relaxation is predicted.

Specific temperature evolution of PNRs can be explained in terms of the phenomenological kinetic theory of phase transitions in compositionally disordered crystals [70]. The emergence of PNR, i.e. the region of polar crystal symmetry within the cubic surrounding, should be accompanied by the creation of electric and elastic fields around PNR, which increase the total energy of the system. Due to the similar effects in the compositionally ordered crystals undergoing a first-order phase transition, the regions of the new phase (nuclei) are not stable. They tend to grow if their size is larger than the critical one or disappear otherwise. As follows from the theory [70], in disordered crystals the nuclei of the new phase can be stable and the equilibrium size of newly formed nuclei can be arbitrary small. The PNRs in relaxors which are really small (contain several unit cells) and stable can be regarded as such kind of nuclei. The theory predicts that PNRs begin to appear in the PE phase at T_B as a result of local "phase transitions" (e.g. condensation of phonon soft mode). Upon cooling, the number of PNRs increases but the equilibrium size of each PNR remains unchanged within a certain temperature interval just below the temperature at which it appears. Upon further cooling, the PNR grows slowly with decreasing temperature while remaining in a stable equilibrium, and finally at $T = T_C$, becomes metastable so that the size of PNR increases steeply due to phase instability. In other words, the behaviour predicted by this model is the same as experimentally observed in PMN (see Fig. 5). But this theory is unable to describe quantitatively the real behaviour at $T < T_C$, because it does not take into account the interactions between different PNRs, which are obviously significant at low temperatures. It was further explained [70, 78] that depending on the model parameters (in particular, the mean T_C and the width of the distribution of local transition temperatures), a sharp phase transition can occur, resulting in large FE domains at $T < T_C$ (in the case of a small width and a comparatively high T_C) or the transition is diffuse and the low-temperature polar regions are of nanometer size. The dipole-dipole interactions between them can lead to the formation of a glass-type phase at a certain temperature T_f. The intermediate situations are also possible with moderately diffuse transition and mesoscopic polar regions (domains). Note that these different types of behaviour have indeed been observed experimentally in different perovskite materials (see Section 7.2).

6. Dielectric response in relaxors

Small-signal dielectric response has been intensively studied in a large number of relaxor materials, but most investigations were restricted within the frequency range of 10–10^9 Hz or narrower. In the past few years, modern measurement facilities with enlarged frequency range have been applied to relaxors. It has been found that significant dielectric dispersion exists in the whole spectrum starting from the frequency of lattice vibrations down to the lowest practically measurable frequency of $f \sim 10^{-5}$ Hz. The present section will focus on these works.

The field-induced polarization in relaxors can be divided into several qualitatively different parts so that the total relative permittivity in the temperature range of permittivity maximum can be written as

$$\varepsilon = 1 + \chi_e + \chi_{Ph} + \chi_R + \chi_U + \chi_{LF}, \quad (4)$$

where χ_e, χ_{Ph}, χ_R, χ_U and χ_{LF} are the susceptibilities (complex numbers) describing the electronic, phonon, "conventional relaxor" (CR), "universal relaxor" (UR) and "low-frequency" contributions, respectively. All the contributions are frequency dependent [in Equation 4 the susceptibilities are ranked in the order of increasing typical characteristic time]. As in any materials, electronic contribution persists in relaxors at all temperatures and at frequencies up to 10^{15}–10^{17} Hz, but at lower frequencies, the value of $\chi_e' = (n^2 - 1) \sim 10$ is small as compared with other susceptibilities.

The phonon (lattice) susceptibility (caused by the mutual displacements of cation and anion sublattices) is active up to the frequencies of 10^{12}–10^{14} Hz. To separate χ_{Ph} from other contributions, measurements at these frequencies are necessary. In relaxors χ_{Ph} has been determined from IR reflectivity spectra at temperatures lower than T_B only. In PMN crystals, χ_{Ph} increases from ~40 at 20 K to ~100 at 300 K [61]. In other words, it constitutes less than 1% of the total low-frequency permittivity measured at T_m (see Fig. 1). This is an important difference of relaxors from ordinary displacive ferroelectrics in which phonon polarization totally accounts for the permittivity peak at the phase transition.

The susceptibilities χ_R and χ_U related to the relaxation-type polarizations are the main contributions giving rise to the peculiar relaxor peaks in the temperature dependences of permittivity (shown in Figs 1 and 9). The real part of χ_R is constant at low enough frequencies and decreases to zero when the frequency reaches the (temperature-dependent) characteristic value. This decrease is accompanied by the peak in the frequency dependence of the imaginary part.[3] Both real and imaginary parts of the UR susceptibility continuously (without any loss peak) decrease in the whole frequency range practically available for measurements according to the power law

$$\chi'_U = \tan(n\pi/2)\chi''_U \propto f^{n-1}, \qquad (5)$$

where n is close to but smaller than unity. Note that the same empiric classification (i.e. the monotonic frequency variation versus the variation with loss maximum) applies not only to relaxors, but also to the relaxation processes found in many other solids [79]. Nevertheless, the values of χ_R and χ_U in relaxors are extraordinary large as compared to other dielectrics. The CR dispersion (CRD) is observed at the low-temperature slope of permittivity peak giving rise to the frequency shift of T_m, while the UR dispersion (URD) exists at temperatures lower, as well as higher, than T_m (as shown in Figs 1 and 9).

The last term χ_{LF} in Equation 4 combines all possible relaxation contributions not related to the relaxor ferroelectricity, which may include the polarization of hopping charge carriers [79, 80], Maxwell-Wagner-type polarisation, etc. Typically, these contributions become significant in good-quality samples of relaxor perovskites at comparatively high temperatures and/or low frequencies. In the PMN crystal presented in Fig. 1 the contribution of χ_{LF} at lowest frequency $f = 10^{-2}$ leads to the noticeable increase of ε'' at temperatures above ~300 K.

In the radio- and audio-frequency ranges, where the dielectric properties of relaxors are most often studied, the value of χ'_R is much larger than χ'_U. That is probably the reason why the UR contribution has been discovered only recently with the help of the frequency response analyser that is able to work at ultra-low frequencies [as one can see from Equation 5, χ'_U increases with decreasing f]. The χ_U component was separated from the χ_R one by means of the analysis of dielectric spectra at $T > T_C$ (or $T > T_f$) first in PMN-PT [80–82] and then in PMN [77], PSN [83] and BTZ [84]. Since CR is the dominant contribution giving rise to the diffuse $\varepsilon'(T)$ peak, $\chi'_R \approx \varepsilon'$ in the vicinity of T_m (at least for the frequencies that are not very low or very high). Therefore, most of the dielectric investigations of relaxors dealt in fact with the CR contribution, even though it was not identified by the authors.

As shown recently for many relaxors [85, 86], the high-temperature slope of the diffuse $\varepsilon'(T) \approx \chi'_R(T)$ peak can be scaled with the empirical Lorenz-type relation,

$$\frac{\varepsilon_A}{\varepsilon'} - 1 = \frac{(T - T_A)^2}{2\delta^2}, \qquad (6)$$

[3]The well-known Debye relaxation is an example of such kind of the behaviour.

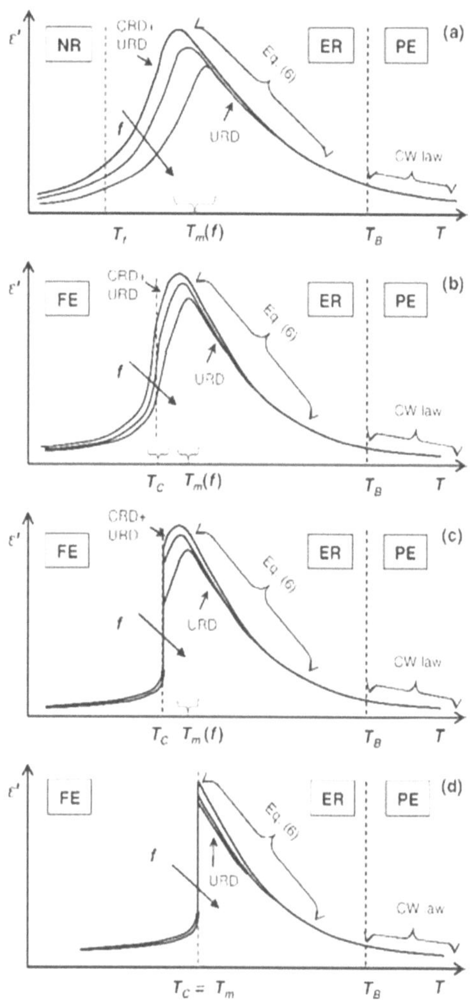

Figure 9 Different possibilities for the temperature evolution of structure and dielectric properties in compositionally disordered perovskites: (a) canonical relaxor; (b) crystal with a diffuse relaxor-to-ferroelectric phase transition at $T_C < T_m$; (c) crystal with a sharp relaxor-to-ferroelectric phase transition at $T_C < T_m$; (d) crystal with a sharp relaxor-to-ferroelectric phase transition at $T_C = T_m$. The temperature dependences of the dielectric constant at different frequencies f are schematically shown. The temperature intervals in which the Lorenz-type Equation 6 and the Curie-Weiss law hold, the regions of conventional relaxor dispersion (CRD) and universal relaxor dispersion (UR) and the types of structure [paraelectric (PE), nonergodic relaxor (NR), ergodic relaxor (ER), ferroelectric (FE)] are identified. Note the similar behaviour at high temperatures in all cases.

where T_A ($< T_m$) and ε_A ($> \varepsilon_m$) are the fitting parameters defining the temperature and magnitude of the Lorenz peak (6), and δ is as a measure of the degree of diffuseness of the peak. This formula gives a more adequate description of the experimental data than the previously used relation, $\varepsilon_m/\varepsilon'' - 1 \propto (T - T_m)^\gamma$ (where $1 < \gamma < 2$, and ε_m is the value of ε' at T_m). Equation 6 holds from temperature T_1, which is typically several degrees higher than T_m, to temperature T_2, which is a few dozens of degrees lower than T_B (see Fig. 9). The diffuse peak of $\varepsilon''(T) \approx \chi''_R(T)$ can be scaled with more complex relations (see Refs. [19, 87] for details).

At $T > T_B$, the dielectric constant is described by the Curie-Weiss law,

$$\varepsilon' = C/(T - T_{CW}),$$

where the Curie constant, C, has the same order of magnitude ($\sim 10^5$ K) as in ordinary displacive ferroelectrics and T_{CW} is typically higher than the low-frequency value of T_m (as shown in Fig. 4) but at high frequencies (e.g. in PMN at $f > 20$ GHz) T_m can become larger [52].

The CR contribution can consist of several components in itself involving different polarization mechanisms. Each of the mechanisms gives rise to the corresponding dispersion and can be seen in the $\varepsilon''(f)$ curve as an individual maximum (or an anomaly if neighbouring maxima overlap each other). For example, in PMN crystals three components (dispersion regions) have been found at $T < T_B$ [52, 61, 88], which were resolved simultaneously between 210 K ($\approx T_f$) and 290 K. The first component appears at the (temperature-independent) frequency of ~ 1 THz and gives rise to a comparatively small input to the static dielectric constant (about 130). The two other dispersion regions become broadened on cooling and their mean relaxation time increases [i.e. the frequency of the corresponding $\varepsilon''(f)$ peak decreases], so that at $T < T_f$ the low-frequency component shifts out of the measurement frequency range and the higher-frequency component develops into a constant (frequency-independent in the range of 10^2–10^{11} Hz) loss. This effect of constant loss is a noticeable property of the low-temperature nonergodic phase in relaxors. It can also be observed in other relaxors, e.g. in the PLZT ceramics [89] and the compositionally disordered PIN crystals [19]. The magnitude of constant loss decreases exponentially on cooling, but still remains measurable at liquid-helium temperature. At extremely low frequencies ε'' in NR phase is no longer constant and slightly increases with decreasing frequency (see Fig. 1).

To describe the CR dielectric spectra, the same empirical expressions as used for other dielectrics were applied. The Kohlrausch-Williams-Watts [80, 90], the Havriliak-Negami [91] and the simpler Cole-Cole [19, 89] formulae have been employed by different authors to fit the experimental $\varepsilon(f)$ data. The alternative way to analyse the dispersion is to find the appropriate function for the distribution of relaxation times. For example, Rychetsky *et al.* [89] fitted the relaxation in PLZT to a uniform distribution that broadens upon cooling.

A remarkable feature which was observed first in PMN [92] and then in many other relaxor ferroelectrics is the Vogel-Fulcher (VF) law connecting the temperature and the frequency of the $\varepsilon'(T)$ peak:

$$f = (2\pi\tau_0)^{-1} \exp\left[-E_a/(T_m - T_{VF})\right], \qquad (7)$$

where f is the measurement frequency, τ_0, E_a and T_{VF} are the fitting parameters. The same relation but with slightly different parameters has also been reported for the peak temperature in the $\varepsilon''(T)$ dependences (T_{mi}). Investigations of PIN crystals showed that the parameters of Equation 7 can be different in different frequency intervals [19]. The VF law was known in structural and spin glasses. When revealed in relaxors, it became one of the main reasons to postulate the existence of a dipole glass phase at $T < T_{VF}$. Equation 7 might (but not necessarily, see below) signify the similar VF relation for the characteristic relaxation time τ of the corresponding relaxation process:

$$\tau = \tau_\infty \exp\lfloor E_b/(T - T_f)\rfloor, \qquad (8)$$

where τ_∞, E_b are the parameters and T_f is the freezing temperature (i.e. the temperature below which the relaxation time becomes infinite). This divergence of τ indicates that the thermally activated reorientations of dipoles responsible for polarization slow down with decreasing temperature and become impossible (consequently dipoles cannot respond to the electric field) at $T = T_f$, but not at $T = 0$ as prescribed by the Arrhenius law for the dynamics of independent dipoles. In dipole glasses the interactions among the dipoles are the cause for such kind of freezing. These interactions (bonds) are frustrated (i.e. can be either FE or AFE but cannot be satisfied simultaneously) and thus favour the configurations with random directions of dipoles, in contrast to the ferroelectrics and antiferroelectrics in which the dipole directions are parallel and antiparallel, respectively.

The relations between Equations 7 and 8 in relaxors were studied by several authors. In the case of Debye relaxation (which can be expected in the system of identical non-interacting dipoles) this relation should be simple: τ follows the Arrhenius law [which is the same as the VF law (8), but with $T_f = 0\rfloor$ and this automatically means that law (7) also holds for T_{mi} with the same parameters, i.e. $\tau_0 = \tau_\infty$, $E_a = E_b$ and $T_{VF} = 0$. The relaxation in relaxors is much more complex and can be characterized by a wide spectrum for the distribution of the relaxation times. Simple relations between the parameters of Equations 7 and 8 are not evident. Furthermore, different relaxation times from the spectrum may have different freezing temperatures. It was also shown theoretically that the situations are possible in which Equation 7 holds with $T_{VF} \neq 0$, but no freezing at a non-zero temperature really takes place [93], i.e. the VF relationships for T_m and T_{mi} do not necessarily imply glass-type dipole dynamics. The problem seems to be solved in some relaxors by means of a special analysis of the frequency-temperature dependences of the real part of permittivity. It was shown that the longest relaxation time in the spectra of PMN [94, 95], PST [94] and PLZT [96] diverges according to the relationship (8) with $T_f = T_{VF}$ ($\approx$220 K in PMN), while the bulk of the distribution of relaxation times remains finite even below T_f [95, 96]. The divergence of the longest relaxation time means that, at least empirically, the behaviour of relaxors in small-signal electric field is similar to the behaviour of dipole glasses. However, a microscopic interpretation of this fact is not so clear. In contrast to ordinary dipole (or spin) glasses in which the susceptibility can by unambiguously attributed to the reorientation of certain permanent dipoles (spins), the structure of relaxors is more complex and the polarization mechanisms responsible for the large and diffuse $\varepsilon(T)$ peak have not been definitely identified.

Most of the existing explanations relate the dielectric relaxation in relaxors to the PNRs. The PNRs are very small and can be considered as individual thermally activated dipoles giving rise to the orientational polarization. Thus, the dominant contribution to the measured $\varepsilon(T)$ relaxor peak (i.e. the CR contribution according to the classification described above) may be attributed to the thermally activated reorientation of dipole moments of PNRs (local spontaneous polarization vectors). Many authors proceeded upon this assumption when analysing the dielectric data (see e.g. Refs. [97–99]). The dipole moments of PNRs are considered in many models as interacting (directly or via surrounding matrix) entities constituting a glassy system (see Sections 5 and 7.3). The reorientations may be affected by the random anisotropy and (in contrast to magnetic spin glasses) by an environment of random electric and elastic fields.

The second possible mechanism associated with PNRs is the side-way motion of their boundaries without the change of the orientation. In the course of such motion, the volume (and thereby the dipole moment) of the polar region changes, giving rise to the characteristic polarization response. This looks like breathing of PNRs and therefore, the corresponding model developed by Glazounov and Tagantsev is called "breathing" model [100]. The model considers the vibrating PNR boundaries in terms of the theory of randomly pinned interface, which was developed earlier for magnetic materials. In the case of relaxors the internal random local fields induced by charge disorder act as the pinning centres. Another approach was used by Rychetsky *et al.* who proposed a thermodynamic model for the polarization reversal near the PNR boundary, which is equivalent to the displacement of the boundary [89]. In particular, this model describes well the constant loss effect at low temperatures. From the analysis of the behaviour of PMN crystals in large dc and ac electric fields, it was suggested [101] that the dielectric response in the ergodic phase [i.e. in the vicinity of the $\varepsilon(T)$ maximum] is controlled by the vibration of PNR boundaries, rather than by the thermally activated reorientations of PNRs.

Note that in PMN and some other relaxors two main components determine the CR dielectric response in the ergodic phase (see above in this section). The low-frequency component may result from the reversal of the spontaneous dipole moments of PNRs and the high-frequency one may originate from the PNR boundary motion [88].

The value of the universal susceptibility [i.e. the susceptibility whose dispersion is described by Equation 5 at all frequencies] in relaxors is several orders of magnitude larger than in non-relaxor materials with the same n [80]. Thus it is reasonable to suggest that the UR polarization mechanism is also connected with PNRs which are inherent only in relaxors. Within the scope of the soft nanoregions model (see Section 5), the UR response has been attributed to the thermally activated reorientations of dipole moments of individual unit cells inside PNRs [77].

The Curie-Weiss law in relaxors can be treated in two different ways, depending on the polarization mechanism which is supposed to be valid in the temperature range of the law (i.e. at $T > T_B$). The first way (see e.g. Refs. [52, 59, 60, 88, 102]) implies that, as in the case of normal perovskite (displacive) ferroelectrics, the field-induced polarization is due to the phonon contribution. The second approach suggests that the polarization mechanism in the temperature range of the Curie-Weiss law is qualitatively the same as at T_m (by analogy to order-disorder ferroelectrics and spin glasses) and involves the relaxation of individual dipoles (see e.g. Ref. [103] in which the Sherrington-Kirkpatrick model was used to analyze the susceptibility in PMN). To determine which way is adequate, the experimental investigation of high-frequency (IR) dispersion at $T > T_B$ is needed.

7. Relaxors at low temperatures: A glassy state or a ferroelectric phase

In the previous sections we have mainly considered the relaxors at comparatively high temperatures, i.e. in the PE and ER phases. We have also discussed some basic aspects of the low-temperature behaviour of canonical relaxors, i.e. those in which the structure remains macroscopically cubic at all temperatures and the FE phase can be achieved only by poling (e.g. by applying an external electric field). In the canonical relaxors (e.g. PMN, PMT, PLZT with large x), a nonergodic (glassy) state appears at low temperatures. In many other materials (e.g. PSN, PST, PLZT with small x, and PMN-PT with large x) that exhibit relaxor properties and related structural features (e.g. PNRs) at high temperatures, a spontaneous (i.e. without poling) structural phase transition into the FE phase occurs. These two different paths of temperature evolution are shown schematically in Fig. 9. In this Section, we describe the low-temperature behaviour of relaxors in more detail.

7.1. Glassy nonergodic relaxor phase

7.1.1. Structure

As mentioned in Section 4 the soft mode in the prototypical relaxor PMN recovers below T_f so that the temperature dependence of the mode frequency shows the behaviour characteristic of a normal ferroelectric phase [i.e. follows Equation 1 with $A < 0$]. A sharp peak of hypersonic dumping was observed at T_f [104]. However, no other evidence of the structural phase transition at T_f has been detected. The average cubic symmetry of PMN at low temperatures was confirmed in many structural studies by the absence of any splitting of X-ray and neutron Bragg reflections (which means that the shape of unit cell is cubic) as well as by the analysis of the intensities of the reflections (which are sensitive to the positions of atoms in the cell). For instance, in Refs. [27, 73], the unit cell was determined to be cubic by X-ray and neutron powder diffraction experiments performed down to 5 K, but due to the limited number of reflections analysed, the positions of atoms and the thermal parameters could not be refined simultaneously. In Refs. [105, 106], the analysis of a large number of reflections obtained from X-ray diffraction of PMN single crystals confirmed the $Pm\bar{3}m$ space group in the range of 100–300 K. The cubic structure is also confirmed by the absence of birefringence [107, 108].[4]

Even though the structural phase transition in PMN is not definitely observed, some important structural changes not affecting the average symmetry are still found. With decreasing temperature, the average size of PNRs increases significantly around T_f (Fig. 5). The synchrotron X-ray scattering revealed the emergence of very weak and wide 1/2(hk0) superlattice reflections (α spots) in the vicinity of T_f [110]. These reflections were attributed to the antiferroelectric nanoregions (AFNR) formed by the correlated anti-parallel (static or dynamic) displacements of Pb ions along the $\langle 110 \rangle$ directions with a magnitude of ~0.2 Å. Significant enhancement of the intensity of α spots below T_f is believed to arise from an increase in the total number of the AFNRs, whose average size of ~30 Å (determined from the width of reflections) remains constant down to the lowest measured temperature of 10 K [110]. AFNRs appear to be different from PNRs and CNRs, and unrelated to either of them [110].

7.1.2. Broken ergodicity in relaxors

Relaxors show nonergodic behaviour resembling the behaviour of spin (or dipole) glasses. In the high temperature (ergodic) phase of glasses, the spins (or dipoles, which can be considered as pseudospins) are weakly correlated and free to rotate, so that after any excitation (e.g. after application and removal of an external field) the system quickly comes back to the state with the lowest free energy, i.e. the state with zero total magnetization. It is always the same state regardless of the initial conditions (i.e. the strength and direction of the field in our example). At lower temperatures, due to the correlations between spins, the free energy surface has very many minima of almost the same depth separated by energy barriers of different heights (each minimum corresponds to a specific configuration of spins). In the glass phase, some of these barriers are so high that the time needed to overcome them is larger than any practically reasonable observation time. Therefore, during this time the system cannot reach all the configuration states, and consequently, the usual thermodynamic averaging and the time averaging give different results, i.e. the system is in a nonergodic state. On its way to a new state of minimum free energy required by the changed external conditions, the system should pass many barriers of different heights. This leads to a process with a wide distribution of relaxation times. The maximum relaxation time from this distribution may be so large (infinite for an infinite crystal) that the system cannot effectively reach the equilibrium. As a result, the state and the physical properties of the material depend on the history (i.e. the external field applied, the temperature variations, the observation time, etc.). In particular, substantial ageing effects should be observed, i.e. the change of properties with time spent by the sample at certain fixed external thermodynamic parameters (temperature, field, etc.).

All the main (mutually related) characteristics of nonergodic behaviour typical of spin glasses, i.e. anomalously wide relaxation spectrum, ageing, dependence of the thermodynamic state on the thermal and field history of a sample, are observed in relaxors at temperatures around and below T_f. The slowing-down of dipole dynamics was already discussed in terms of small-signal dielectric response in Section 6. Slow relaxation manifests itself also in other properties related to the local and/or macroscopic polarization. In particular, the relaxation of optical linear birefringence induced in PMN by a weak ($E < E_{cr}$) external electric field was studied [107] (E_{cr} is the critical field needed to induce the transition to the FE phase). The Kohlrausch-Williams-Watts-type and the Curie-von Schweidler-type relaxations were found in the temperature intervals of $180 < T < 210$ K and $210 < T < 230$ K, respectively. The results were successfully described in terms of Chamberlin's approach to dynamic heterogeneity [111], implying a broad relaxation spectrum. Application of a strong ($E > E_{cr}$) d.c. field to the PMN crystal at $T < T_f$ results in a nearly logarithmic decay of dielectric permittivity [112] and a slow evolution of X-ray Bragg

[4] While most researchers agree that the average structure of PMN is cubic, the rhombohedral structure was also reported [109]. The possible reason for this discrepancy will be discussed in Section 7.2.

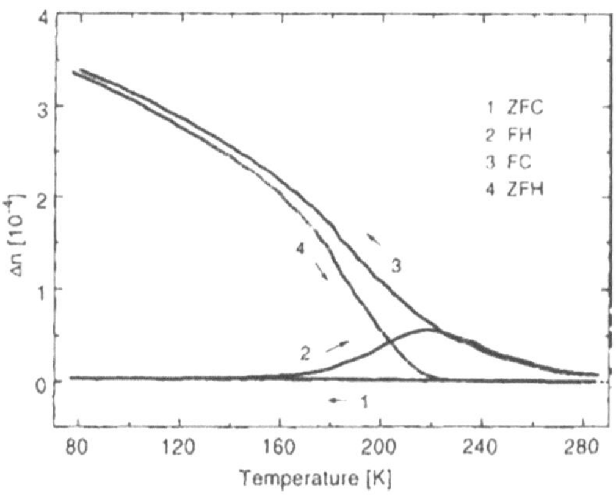

Figure 10 Linear birefringence measured subsequently as a function of temperature on zero-field cooling (ZFC), field heating with $E = 1.2$ kV/cm $< E_{cr}$ (FH), field cooling (FC) and zero-field heating (ZFH), illustrating the nonergodic behaviour of PMN crystal. (after Kleemann *et al.* [107]).

peaks reflecting the change of crystal symmetry [113].[5] The effects of ageing of susceptibility in the NR phase of PMN and in the typical spin glass phase were found to be very similar (and much stronger than in typical dipole glasses) [114]. The example of the dependence of properties on the thermal and electrical history of sample is shown in Fig. 10. The other examples are the splitting in the temperature dependences of the field-cooled and zero-field-cooled quasistatic dielectric constants in PMN and PLZT [95, 96] and the *P*(*E*) hysteresis loops (see Section 7.1.3),

The ergodicity is clearly broken in relaxors at low temperatures, but this does not necessarily mean that relaxors are really dipole glass systems. Many other systems may also be nonergodic [115]. In particular, an ordinary FE phase is also nonergodic, but its potential landscape contains only a few minima (which are symmetric and correspond to the different directions of spontaneous polarization). As a result, the properties are easily distinguishable from those of nonergodic spin glass (or relaxor) phase. Wide relaxation spectrum and ageing phenomena are absent in the ideal FE crystal. But in the compositionally disordered perovskite crystal the situation is very different and different explanations for the nonergodic behaviour are possible. For instance, the above-mentioned Kohlrausch-Williams-Watts-type relaxation of birefringence was explained by domain wall displacements, rather than by the reorientations of dipoles [107]. Furthermore, some peculiarities of the relaxor behaviour have never been observed in spin and dipole glasses. In particular, the Barkhausen jumps during poling process (detected optically in PMN) are not compatible with the glassy reorientation of dipoles, which takes place on a microscopic length scale and hence should be continuous and monotonic [71]. Field-induced FE phase and FE hysteresis loops have not been observed in typical dipole glasses. Thus, the nature of the nonergodic phase in relaxors remains the subject of intensive discussion. In particular, the WKG model suggests that the low-temperature phase of canonical relaxors is a ferroelectric state, but broken into nanodomains by quenched random fields. We will discuss the origin of nonergodic phase in more detail in Section 7.3.

Note also that in terms of compositional disorder, relaxors are frozen in a metastable state, as discussed in Section 2. The degree of compositional disorder can depend on thermal prehistory. This is also an effect of nonergodicity.[6] However, at temperatures around T_B and below, the compositional disorder remains unchanged on the experimental time scales (i.e. frozen), and at the same time, the motion of dipoles (at $T > T_f$) is fast. Thus, when considering the subsystem of dipoles at $T > T_f$, one can believe that the crystal reaches the equilibrium[7] and the phase is effectively ergodic. On the other hand, if the sample has been annealed during experiment at high temperatures ($\sim$700 K or higher) the possible effects of nonergodicity related to the compositional disorder should be taken into account.

7.1.3. Electric-field-induced ferroelectric phase in relaxors

An important feature of the NR state is that, it can be irreversibly transformed to the phase with the FE dipole order when poling by an electric field larger than the critical strength (in PMN the minimal E_{cr} is about 1.7 kV/cm at $T_C \simeq 210$ [108]). This feature points to the common nature of relaxor and normal ferroelectrics. The FE hysteresis loops, which are known to be the determinative characteristic of FE phase, are observed in relaxors with the values of remnant polarization and coercive field typical of normal ferroelectrics. Pyro- and piezoelectric effects are also observed after poling. X-ray diffraction [113, 117] and optical [108] investigations of poled PMN crystals showed that the field-induced phase has the rhombohedral 3m symmetry, i.e. the same symmetry as in several normal perovskite ferroelectrics. On the other hand, locally the structure is inhomogeneous, i.e. different from normal ferroelectric structure. The traces of cubic phase were observed at low temperature by X-ray diffraction experiments in poled PMN crystal [113]. The NMR investigations of PMN crystal poled by a field almost two

[5] After a long (several hours) waiting time the entire crystal suddenly transforms to the FE phase via a first-order transition.

[6] In contrast to spin (dipole) or FE state where the relaxation time is expected to become infinite in infinite crystal, the rate of compositional ordering does not depend on a crystal size. The corresponding relaxation time at nonzero temperature can be very large, but not infinite.

[7] According to Feynman [116], a system is in equilibrium if "all the fast things have happened and all the slow things have not".

times as large as E_{cr}, revealed that only about 50% of Pb ions are displaced parallel to the [111] poling direction in a FE manner, while the other 50% exhibit spherical layer-type displacements characteristic of PE phase [32]. The size and number of AFNRs found in PMN in the unpoled state (see Section 7.1.1) remain unchanged in the FE phase [110].

Upon heating, the FE phase transforms to the cubic (ER) phase at a well-defined temperature, T_C ($\approx$210 K in PMN). This first-order phase transition is accompanied by a step-like drop of spontaneous polarization (as determined from pyroelectric current), sudden vanishing of birefringence, and sharp peak of dielectric constant. A more detailed description of the field-induced transition and FE phase in relaxors is given in Ref. 11.

7.2. Spontaneous relaxor-to-ferroelectric phase transition

The transition from the ER to the FE phase typically takes place at temperature T_C, which is several degrees or several dozens of degrees lower than T_m, as schematically shown in Fig. 9b and c. Usually the transition is observed in those relaxors where the $\varepsilon'(T)$ peak is not very diffuse (i.e. the diffuseness parameter δ is relatively small). X-ray and neutron diffraction experiments unambiguously indicate the change of symmetry at T_C from the high-temperature cubic to a low-temperature tetragonal or rhombohedral (in most cases) one [118, 119]. The symmetry breaking is also confirmed by the appearance of Brillouin scattering peaks which are forbidden in cubic phase [48], the appearance of optical birefringence [120], and the formation of FE domains which are clearly observed by optical polarizing microscopy [121, 122], electron microscopy [123] and scanning force microscopy [124–126]. The FE phase in relaxors exhibits typical FE properties, namely the large dielectric constant, the FE hysteresis loops [118, 121, 127], the pyro- [128] and piezoelectric (see below) effects, etc. At temperatures slightly above T_C, double hysteresis loops can be observed [118], as is typical of normal ferroelectrics.

The spontaneous relaxor-to-ferroelectric phase transition can be accompanied by significant anomalies in the temperature dependences of structural parameters [118, 119, 128], dielectric [4, 82, 118], optical [4, 120, 128], thermal [118, 128] and other properties. The transition can be very sharp: e.g. in PFN crystals, the related jump of $\varepsilon(T)$ occurs in a temperature interval smaller than 0.1 K [130]. In many other cases, it is smeared for different reasons. The change of $\varepsilon(T)$ in ceramics is usually not as sharp as in single crystals, probably because of the inhomogeneity related to the existence of grains and boundaries [131]. Another type of inhomogeneity that can smear the transition in the relaxor-based solid solution crystals (e.g. PMN-PT, PZN-PT) is the macroscopic variation of x across the sample [120, 132]. Besides, some "intrinsic" causes for the smearing of phase transition also exist so that a clear boundary between the canonical relaxors (in which the anomalies of structure and properties are diffuse or absent) and the relaxors with a sharp FE transition cannot be defined. The intermediate behaviour can appear in different ways. X-ray diffraction studies of PZN crystals revealed the coexistence of the mesoscopic domains of FE phase and the regions of cubic (relaxor) phase in a temperature range of about 70 K around the mean T_C [78, 133]. The concentration of the cubic phase gradually decreases on cooling, i.e. the transition is highly diffuse. The size of FE domains in PZN (40–200 nm) [78, 133] and in disordered PST (25–75 nm) [123] is smaller than in a normal FE phase, but larger than the size of typical PNRs. The average domain size (at room temperature) in the $(1-x)$PMN$-x$PT solid solutions was found, by scanning force microscopy, to gradually increase from ~40 nm in the rhombohedral phase with $x = 0.1$, to ~2 μm (which is comparable to the domain size in ordinary ferroelectrics) in the tetragonal phase with $x = 0.4$ [126].

Because of the similarity between the FE phase in ordinary ferroelectrics and the FE phase in relaxor ferroelectrics, the transition at T_C was initially called "spontaneous relaxor-to-normal ferroelectric transition," but later investigations showed that the low-temperature phase in relaxors is not exactly a "normal" ferroelectric phase, even in those relaxors where the FE transition is relatively sharp and FE domains are large. In particular, in the PMN-PT crystals with $x = 0.35$, the central peak in the Brillouin spectra, which is related to the relaxation of PNRs, was observed not only at $T > T_C$ (see Section 3.2), but also at $T < T_C$, indicating that the PNRs persist in the FE phase [48]. Furthermore, macroscopic (1–2 μm) areas of average cubic symmetry were found alongside with the areas of FE phase [48]. In PMN-PT crystals with $x = 0.20$, the piezoresponse force microscopy revealed a continuous distribution of the sizes of polar regions starting from ~5 nm (resolution limit). The complex structure of the micron-size FE domains with the PNRs of the opposite polarity embedded in them was observed [125]. A mixture of the rhombohedral domains and the domains of a different low-symmetry (presumably monoclinic) phase was observed by synchrotron X-ray diffraction in the FE phase of PZN crystal [78]. Unlike the plane walls in ordinary ferroelectrics, the domain walls in relaxor ferroelectrics are usually diffuse and irregular [122, 124]. The IR spectroscopy of PZN-PT and PMN-PT crystals (with $x = 0.08$ and $x = 0.29$, respectively) did not reveal any phonon softening that was expected for normal FE phase transitions at T_C and below [134]. Accordingly, the phonon contribution to the dielectric constant at these temperatures is small (~100), i.e. much less than the low-frequency value that reaches ~5×10^4 at $T = T_C$ and ~5×10^3 at $T << T_c$. Therefore, similarly to the case of canonical relaxors (and in contrast to ordinary displacive ferroelectrics), the dielectric response is determined by the relaxation polarization at all temperatures around and below the dielectric peak. It was concluded that the transition into the FE phase consists in a stepwise increase in the size of PNRs which transform into FE domains [134].

Another phenomenon, which is unusual for ordinary ferroelectrics, is the specific macroscopic phase inhomogeneities discovered recently in good-quality crystals of some relaxor ferroelectrics. Diffraction experiments performed in PZN with X-ray of different energies (and thus different penetration lengths) revealed that the outer layer (an estimated thickness of $\sim$10–50 μm) undergoes a structural phase transition into the FE phase while the lattice inside the crystal maintains the cubic unit cell [135, 136].[8] Another interesting point is that at all temperatures the lattice parameter of the outer layers is slightly ($\sim$0.2%) smaller than that of the bulk (inside) [136]. The same feature, i.e. a FE "skin" (observed by low-energy X-ray diffraction) and a cubic phase in the bulk (observed by high-energy X-ray or neutron diffraction), was also found in PZN-PT [137, 138] and PMN-PT [139] crystals with small x. It was suggested that this cubic phase (named X-phase) is similar to the average cubic phase in pure PMN [140], i.e. it is a NR phase. As a typical NR phase, the X-phase can be irreversibly transformed into the FE phase by poling [137].

Using spatially resolved neutron diffraction technique, it was found that even in PMN crystals, where the rhombohedral phase has not been detected, the near-surface layer (of $\sim$100 μm thick) has the lattice constant noticeably smaller than the bulk structure [141]. It was supposed [140] that a very thin rhombohedral skin possibly exists in PMN also, with a thickness much smaller than the penetration length of X-rays, so that the skin could not be detected in the usual diffraction experiments. Note in this connection that, as reported in Ref. [109], the Rietveld refinement of neutron diffraction data collected on PMN powder revealed a rhombohedral macroscopic symmetry, namely $R3m$ at 300 K (i.e. above T_f) and $R3c$ at 10 K. Second harmonic generation signal was also detected, indicating a non-centric symmetry. In these experiments, very fine powder (of 4–5 μm particle size) synthesized by a special route was used, so the whole material can be considered a near-surface region. This is the possible reason why the rhombohedral phase was found instead of the cubic phase usually observed by other authors in crystals and large-grain powders of PMN.

The behaviour unusual for relaxor as well as for ordinary ferroelectrics was recently found in PMN-PT crystals with large x ($\sim$0.5) [132]. The dielectric properties typical of the ER phase are observed at $T > T_m$, namely the deviation from the Curie-Weiss law at $T < T_B$ (where $T_B \gg T_m$) according to Equation 6 and the behaviour permittivity, indicating the UR dispersion. Nevertheless, the CR dispersion is absent, and consequently, T_m coincides with the temperature of the FE phase transition, T_C. This is shown in Fig. 9d. Both the deviation from the Curie-Weiss law and the UR dispersion are believed to result from the existence of PNRs, so the high-temperature phase is the ER state. PMN-PT with high x seems to be the first examples of relaxor without the characteristic CR dispersion. To confirm this opinion, more direct identification of PNRs (e.g. by neutron scattering) is desirable.

While an external electric field transforms a NR state to a FE one, the hydrostatic pressure is able to induce the reverse transformation. In the crystals exhibiting the relaxor-to-FE phase transition, the FE phase does not appear if the sample is cooled under a high enough pressure and the behaviour typical of the canonical relaxors is observed. The pressure-induced crossover from a ferroelectric to a relaxor state was discussed in detail in the recent reviews by Samara [142, 143].

Excellent piezoelectric properties were found near the morphotropic phase boundary in PMN-PT, PZN-PT and some other solid solutions of complex perovskite relaxors with $PbTiO_3$. The transition from FE to ER phase is observed in these crystals at temperatures much higher than room temperature. Thus, they are considered and investigated as promising materials for practical applications. This field of research, currently very active, has been reviewed in a number of recent papers [144–148].

7.3. Theoretical description of nonergodic phase in relaxors

Early works on relaxors (e.g. the composition fluctuations model by Smolenskii and Isupov [8, 97] and the superparaelectric model by Cross [10]) considered the PNRs to be relatively independent noninteracting entities. It was later understood that the specific nonergodic behaviour of relaxors at low temperatures cannot be explained without taking into account the interactions among PNRs and/or quenched random local fields existing in the compositionally disordered structure. The interactions among PNRs may lead to anomalous slowing-down of their dynamics (nonergodicity effects) or, when becoming frustrated, even to the formation of the glass state in which the dipole moments of individual PNRs are randomly fixed in different directions. Note that these interactions are of dipole-dipole nature and can be considered as dynamic local fields. Additionally PNRs can be influenced (or probably even fixed) by quenched local random fields stemming from the compositional disorder or other types of lattice defects.

In Section 5 we have already discussed the modern theories explaining the formation of PNRs. Some of these

[8] A tetragonal unit cell was initially reported [135], but more elaborate investigations [136] later showed that it is indeed cubic.

theories can also explain the transition from the ergodic to nonergodic relaxor state. In particular, in the GF model, the PNRs naturally appear in the temperature interval between the PE and the low-temperature dipole glass or mixed ferroglass phase. In the WKG model, the formation of PNRs as well as the transition to the NR state is ascribed to the quenched random fields exclusively.

However, the mechanisms leading to the formation of PNRs at high temperatures are not necessarily responsible for their freezing and for the development of the low-temperature nonergodic state. The formation and freezing of PNRs are possibly two distinct phenomena requiring different approaches. The "semi-microscopic" models [98, 149, 150] of glass state in relaxors describe only the latter phenomenon, while PNRs are believed to be already-existing objects and the mechanisms of their formation are not examined.

In the spherical random-bond-random-field (SRBRF) model proposed by Pirc and Blinc [149, 150], the Hamiltonian is formally written with Equations 2 and 3, but the meanings of the parameters are different from those discussed in Section 5. Pseudospins $\vec{S}_i$ proportional to the dipole moments of PNRs are introduced so that the relation

$$\sum_i (\vec{S}_i)^2 = 3N \tag{9}$$

is satisfied (N is the number of pseudospins in the crystal). It is assumed that each component of $\vec{S}_i$ can fluctuate continuously and take any value,[9] i.e.

$$-\infty < S_{i\mu} < +\infty. \tag{10}$$

J_{ij} in Equation 3 are the random interactions (bonds) between PNRs which are assumed to be infinitely ranged. The second term in the Hamiltonian in Equation 2 describes the interaction of pseudospins with quenched random electric fields $\vec{h}_i$,

$$H_2 = -\sum_i \vec{h}_i \cdot \vec{S}_i.$$

Both random bonds J_{ij} and random fields $\vec{h}_i$ obey the (uncorrelated) Gaussian probability distributions with an rms variance of $J/\sqrt{N}$ and Δ, respectively. The mean value of the distributions equals J_0/N (for random bonds) and zero (for random fields). In the absence of random fields ($\Delta = 0$), if $J < J_0$, the theory predicts the transition from the PE phase (in the model this phase is equivalent to the ER phase) into an inhomogeneous FE phase with a nonzero spontaneous polarization; if $J > J_0$, the system transforms, at a well-defined temperature $T = J$, from the PE to a spherical glass phase without long range order, and the glass order parameter (which is equivalent in this model to the well-known Edwards-Anderson order parameter, q_{EA}) decreases linearly from 1 at $T = 0$ to zero at $T = J$. The presence of random fields ($\Delta \neq 0$) destroys the phase transition so that q_{EA} remains nonzero at $T = J$, and approaches zero when the temperature further increases. Fig. 11a shows the temperature dependence of q_{EA} determined experimentally from the NMR data of PMN (q_{EA} is shown to be proportional to the second moment, M_2, of the frequency distribution corresponding to the narrow ^{93}Nb NMR line) [150]. The solid line represents the fit with the parameters $J/k = 20$ K and $\Delta/J^2 = 0.002$, confirming the applicability of the model. The local polarization distribution function $W(\vec{p})$ (where $\vec{p} = <\vec{S}>$) predicted by the model and determined experimentally from the NMR lineshape also appears to be the same as shown in the inset of Fig. 11a [150]. The $W(\vec{p})$ shape observed in dipolar and quadrupolar glasses look very different, as shown in Fig. 11b. These results suggest that the NR phase in PMN cannot be described as a dipolar or quadrupolar glass. It is a new type of glass which can be called "spherical cluster glass" [151]. The SRBRF model is also able to explain the dielectric non-linearity in PMN. The dynamic version of SRBRF model describing the dispersion of liner and nonlinear dielectric susceptibility has been developed [152]. In the coupled SRBRF-phonon model [99], the coupling of PNRs with soft TO phonons leads to the modification of interactions among PNRs. The effect of pressure on the relative stability of different phases in relaxors are explained.

Vugmeister and Rabitz [98, 153] considered in their model the hopping of PNRs in multi-well potentials. The PNRs exist in a highly polarizable PE host lattice with a displacive dielectric response. The theory takes into account the broad distribution of the potential barriers controlling PNR dynamics and the effect of interactions between PNRs mediated by highly polarizable host. These two aspects are described in terms of the local field distribution function. In this model, the dipole glass freezing is believed to be accompanied by the critical FE slowing-down. It is shown that the true glass state in which all dipoles (PNRs) are frozen is not achieved in relaxors: the degree of the local freezing is rather small even at low temperatures.[10] The role of the critical slowing-down is shown to be significant in the dynamics of the system due to the closeness of FE instability. In other words, relaxors can be considered incipient ferroelectrics. This explains their very large dielectric constant. In the framework of this model, the shape of the frequency-dependent permit-

[9] Models in which the order parameter satisfies conditions (9), (10) are called "spherical" models. Due to these conditions the model is exactly solvable by the replica method.

[10] This is in agreement with the experimental finding that in relaxors only the longest relaxation time diverges at T_f, while the bulk of the relaxation spectrum remains active at low temperatures (see Section 6).

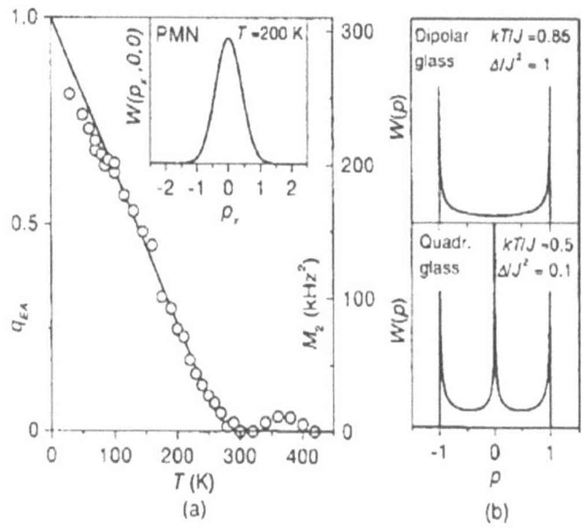

Figure 11 (a) Temperature dependence of the Edwards-Anderson glass order parameter q_{EA} in PMN. The solid line is the fit to the "spherical random bond random field" (SRBRF) model. The inset shows the local polarization distribution function $W(\vec{p})$ along the p_x axis according to the SRBRF model. (b) Examples of the $W(p)$ functions for dipolar and quadrupolar glasses are shown for comparison (after Blinc *et al.* [150]).

tivity as a function of temperature in typical relaxors is explained qualitatively. The glasslike freezing of the dynamics of PNRs is characterized by the non-equilibrium spin-glass order parameter, the temperature behaviour of which is consistent with the NMR experiments (shown in Fig. 11). The kinetics of the electric field induced transition from the NR to FE phase was also successfully reproduced [154] (while the glass models experience difficulties in explaining this transition).

The behaviour of PNRs can be influenced by the electronic subsystem. In particular, the thermo-localization of charge carriers on the defects in the temperature range of phase transition can change the relaxation dynamics. The direction of spontaneous polarization of PNRs can be pinned by localized charge carriers, preventing the alignment of PNRs in the external electric field. The related phenomena are studied in Refs. [155, 156].

As mentioned above, the models so far discussed in this Section consider PNRs (pseudospins) to be already-existing entities. In order to describe the process of their formation and development (which begins from $T_B \gg T_f$), other models are needed. Recently, it has been proposed that quenched random fields give rise to the formation of PNRs in the PE phase, as prescribed by the WKG model, and then, upon further cooling, the crystal undergoes a transition into the spherical cluster glass state due to random interactions between PNRs [151]. Alternatively, some other models can be used to describe the formation of SRBRF pseudospins, in particular, the soft nanoregions model [77] [which justifies the fulfilment of condition (10)] together with the kinetic model [70] (as discussed in Section 5).

Let us now discuss the mechanisms of the spontaneous relaxor-to-ferroelectric phase transition. There are two ways to explain the formation of FE phase at T_C from the system of disordered PNRs in relaxors. The first one suggests that the dipole-dipole interactions between PNRs (or individual ions) lead to their FE-type arrangement (as, e.g., in the SRBRF model discussed above). The second mechanism arises from the kinetic model of phase transitions in compositionally disordered crystals [70] (see Section 5) and suggests the thermally activated growth of PNRs at T_C. It is not easy to discriminate these two mechanisms from each other. In fact, it is also possible that both mechanisms contribute to the process of the formation of FE phase.

8. Conclusions

In this paper we have analyzed the peculiar behaviour of relaxor ferroelectrics that occurs in compositionally disordered perovskites. The quenched compositional disorder in these compounds gives rise to another type of disorder, i.e. the glassy nonergodic state that can be observed at low temperatures, instead of a FE or AFE ordering that exists in many simple perovskites. The research in this field has undergone such a tremendous growth that it was not possible to review all the important works in this short paper. Some subjects were discussed only briefly just to give the examples characterizing the peculiarities of the behaviour. Some other important topics have been left out, in particular, the materials technology of crystals, ceramics and thin films and applications of relaxor ferroelectrics.

Despite the remarkable progress achieved in the recent years, fundamental physics of the relaxors remains a fascinating puzzle. Some key questions, such as what the origin of relaxor behaviour is, still have no definite answers. Several theoretical models have been proposed; some of them contradict each other. Further experiments need to be performed in order to prove or reject these models, while new and more satisfactory theories are yet to be worked out. With their complex structures and intriguing properties, relaxors represent truly a frontier of research in ferroelectrics and related materials, offering great opportunities both for fundamental research and for technological applications.

Acknowledgements

This work was supported by the Natural Science and Engineering Research Council of Canada (NSERC) and the U.S. Office of Naval Research (Grant# N00014-99-1-0738).

References

1. G. A. SMOLENSKII, V. A. ISUPOV, A. I. AGRANOVSKAYA and S. N. POPOV, *Sov. Phys. Solid State* **2** (1961) 2584.

2. F. CHU, N. SETTER and A. K. TAGANTSEV, *J. Appl. Phys.* **74** (1993) 5129.
3. X. DAI, Z. XU and D. VIEHLAND, *Phil. Mag. B* **70** (1994) 33.
4. A. KRUMINS, T. SHIOSAKI and S. KOIZUMI, *Jpn. J. Appl. Phys. Pt. 1* **33** (1994) 4940.
5. PH. SCIAU, G. CALVARIN and J. RAVEZ, *Sol. Stat. Commun.* **113** (2000) 77.
6. A. SIMON, J. RAVEZ and M. MAGLIONE, *J. Phys.: Condens. Matter* **16** (2004) 963.
7. N. YASUDA, H. OHWA and S. ASANO, *Jpn. J. Appl. Phys. Pt. 1* **35** (1996) 5099.
8. G. A. SMOLENSKII, *J. Phys. Soc. Jpn.* **28** (Supl.) (1970) 26.
9. A. A. BOKOV, *Ferroelectrics* **131** (1992) 49.
10. L. E. CROSS, *ibid.* **76** (1987) 241; **151** (1994) 305.
11. Z.-G. YE, *Key Eng. Mater.* **155–156** (1998) 81.
12. C. G. F. STENGER and A. J. BURGGRAAF, *Phys. Stat. Sol. (a)* **60** (1980) 653.
13. C. BOULESTEIX, F. VARNIER, A. LLEBARIA and E. HUSSON, *J. Solid State Chem.* **108** (1994) 141.
14. M. YOSHIDA, S. MORI, N. YAMAMOTO, Y. UESU and J. M. KIAT, *Ferroelectrics* **217** (1998) 327.
15. A. TKACHUK and H. CHEN, *ibid.* **253** (2001) 1.
16. P. K. DAVIES and M. A. AKBAS, *J. Phys. Chem. Solids* **61** (2000) 159.
17. H. Z. JIN, J. ZHU, S. MIAO, X. W. ZHANG and Z. Y. CHENG, *J. Appl. Phys.* **89** (2001) 5048.
18. A. A. BOKOV, I. P. RAEVSKII and V. G. SMOTRAKOV, *Sov. Phys. Sol. Stat.* **26** (1984) 1708.
19. A. A. BOKOV, M. A. LESHCHENKO, M. A. MALITSKAYA and I. P. RAEVSKI, *J. Phys.: Condens. Matter* **11** (1999) 4899.
20. A. A. BOKOV, *Ferroelectrics* **183** (1996) 65.
21. P. K. DAVIES, *Curr. Opin. Sol. Stat. Mater. Sci.* **4** (1999) 467.
22. V. A. ISUPOV, *Ferroelectrics* **289** (2003) 131.
23. R. COMES, M. LAMBERT and A. GUINIER *Acta Crystallogr. B* **26** (1970) 244.
24. E. A. STERN *Phys. Rev. Lett.* **93** (2004) 037601.
25. R. E. COHEN, *Nature* **358** (1992) 136.
26. P. BONNEAU, P. GARNIER, E. HUSSON and A. MORELL, *Mat. Res. Bull.* **24** (1989) 201.
27. P. BONNEAU, P. GARNIER, G. CALVARIN, E. HUSSON, J. R. GAVARRI and A. MORELL, *J. Solid State Chem.* **91** (1991) 350.
28. S. VAKHRUSHEV, S. ZHUKOV, G. FETISOV and V. CHERNYSHOV, *J. Phys.: Condens. Matter* **6** (1994) 4021.
29. K. FUJISHIRO, T. IWASE, Y. UESU, Y. YAMADA, B. DKHIL, J.-M. KIAT, S. MORI and N. YAMAMOTO, *J. Phys. Soc. Jpn.* **69** (2000) 2331.
30. B. DKHIL, J. M. KIAT, G. CALVARIN, G. BALDINOZZI, S. B. VAKHRUSHEV and E. SUARD, *Phys. Rev. B* **65** (2001) 024104.
31. S. G. ZHUKOV, V. V. CHERNYSHEV and S. B. VAKHRUSHEV, *Ferroelectrics* **235** (1999) 143.
32. R. BLINC, V. LAGUTA and B. ZALAR, *Phys. Rev. Lett.* **91** (2003) 247601.
33. T. EGAMI, *Ferroelectrics* **267** (2002) 101.
34. S. AOYAGI, Y. KUROIWA, A. SAWADA, H. TANAKA, I. HARADA, E. NISHIBORI, M. TAKATA and M. SAKATA, *J. Phys. Soc. Jpn.* **71** (2002) 2353.
35. Y. UESU, Y. YAMADA, K. FUJISHIRO, H. TAZAWA, S. ENOKIDO, J.-M. KIAT and B. DKHIL, *Ferroelectrics* **217** (1998) 319.
36. V. A. SHUVAEVA, I. PIROG, Y. AZUMA, K. YAGI, K. SAKAUE, H. TERAUCHI, I. P. RAEVSKII, K. ZHUCHKOV and M. YU. ANTIPIN, *J. Phys.: Condens. Matter* **15** (2003) 2413.
37. G. BURNS and F. H. DACOL, *Sol. Stat. Commun.* **48** (1983) 853.
38. S. B. VAKHRUSHEV, B. E. KVYATKOVSKY, A. A. NABEREZNOV, N. M. OKUNEVA and B. P. TOPERVERG, *Ferroelectrics* **90** (1989) 173.
39. A. NABEREZNOV, S. VAKHRUSHEV, B. DORNER, D. STRAUCH and H. MOUDDEN, *Eur. Phys. J.* **11** (1999) 13.
40. K. HIROTA, Z.-G. YE, S. WAKIMOTO, P. M. GEHRING and G. SHIRANE, *Phys. Rev. B* **65** (2002) 104105.
41. S. VAKHRUSHEV, A. NABEREZNOV, S. K. SINHA, Y. P. FENG and T. EGAMI, *J. Phys. Chem. Solids* **57** (1996) 1517.
42. G. XU, G. SHIRANE, J. R. D. COPLEY and P. M. GEHRING, *Phys. Rev. B* **69** (2004) 064112.
43. D. LA-ORAUTTAPONG, J. TOULOUSE, J. L. ROBERTSON and Z.-G. YE, *ibid.* **64** (2001) 212101.
44. C. STOCK, R. J. BIRGENEAU, S. WAKIMOTO, J. S. GARDNER, W. CHEN, Z.-G. YE and G. SHIRANE, *ibid.* **69** (2004) 094104.
45. S. B. VAKHRUSHEV, A. A. NABEREZHNOV, N. M. OKUNEVA and B. N. SAVENKO, *Phys. Sol. Stat.* **37** (1995) 1993.
46. N. TAKESUE, Y. FUJII and H. YOU, *Phys. Rev. B* **64** (2001) 184112.
47. Y. MORIYA, H. KAWAJI, T. TOJO and T. ATAKE, *Phys. Rev. Lett.* **90** (2003) 205901.
48. F. M. JIANG and KOJIMA, *Phys. Rev. B* **62** (2000) 8572; *Appl. Phys. Lett.* **77** (2000) 1271.
49. V. V. LEMANOV, N. K. YUSHIN, E. P. SMIRNOVA, A. V. SOTNIKOV, E. A. TARAKANOV and A. YU. MAKSIMOV, *Ferroelectrics* **134** (1992) 139.
50. K. UCHINO, *ibid.* **151** (1994) 321.
51. J. ZHAO, A. E. GLAZOUNOV, Q. M. ZHANG and B. TOBY, *Appl. Phys. Lett.* **72** (1998) 1048.
52. YU. M. POPLAVKO, V. P. BOVTUN, N. N. KRAINIK and G. A. SMOLENSKII, *Sov. Phys. Sol. Stat.* **27** (1985) 1903.
53. E. DUL'KIN, I. P. RAEVSKII and S. M. EMEL'YANOV, *Phys. Sol. Stat.* **45** (2003) 158.
54. P. M. GEHRING, S. WAKIMOTO, Z.-G. YE and G. SHIRANE, *Phys. Rev. Lett.* **87** (2001) 277601.
55. S. WAKIMOTO, C. STOCK, R. J. BIRGENEAU, Z.-G. YE, W. CHEN, W. J. L. BUYERS, P. M. GEHRING and G. SHIRANE, *Phys. Rev. B* **65** (2002) 172105.
56. P. M. GEHRING, S.-E. PARK and G. SHIRANE, *ibid.* **63** (2001) 224109.
57. *Idem., Phys. Rev. Lett.* **84** (2000) 5216.
58. J. HLINKA, S. KAMBA, J. PETZELT, J. KULDA, C. A. RANDALL and S. J. ZHANG, *ibid.* **91** (2003) 107602.
59. S. B. VAKHRUSHEV and S. M. SHAPIRO, *Phys. Rev. B* **66** (2002) 214101.
60. S. WAKIMOTO, C. STOCK, Z.-G. YE, W. CHEN, P. M. GEHRING and G. SHIRANE, *Phys. Rev. B* **66** (2002) 224102.
61. V. BOVTUN, S. KAMBA, A. PASHKIN, M. SAVINOV, P. SAMOUKHINA, J. PETZELT, I. BYKOV and M. D. GLINCHUK, *Ferroelectrics* **298** (2004) 23.
62. S. KAMBA, V. BOVTUN, J. PETZELT, I. RYCHETSKY, R. MIZARAS, A. BRILINGAS, J. BANYS, J. GRIGAS and M. KOSEC, *J. Phys.: Condens. Matter* **12** (2000) 497.
63. I. G. SINY, S. G. LUSHNIKOV and R. S. KATIYAR, *Ferroelectrics* **231** (1999) 115.
64. I. G. SINY, S. G. LUSHNIKOV, R. S. KATIYAR and V. H. SCHMIDT, *ibid.* **226** (1999) 191.
65. O. SVITELSKIY, J. TOULOUSE, G. YONG and Z.-G. YE, *Phys. Rev. B* **68** (2003) 104107.
66. S. A. PROSANDEEV, E. COCKAYNE, B. P. BURTON, S. KAMBA, J. PETZELT, YU. YUZYUK, R. S. KATIYAR and S. B. VAKHRUSHEV, *ibid.* **70** (2004) 134110.

67. M. D. GLINCHUK and R. FARHI, *J. Phys.: Condens. Matter* **8** (1996) 6985.
68. A. A. BOKOV, *JEPT* **84** (1997) 994.
69. P. N. TIMONIN, *Ferroelectrics* **199** (1997) 69.
70. A. A. BOKOV, *Phys. Solid State* **36** (1994) 19; *Ferroelectrics* **190** (1997) 197.
71. V. WESTPHAL, W. KLEEMANN and M. D. GLINCHUK, *Phys. Rev. Lett.* **68** (1992) 847.
72. V. M. ISHCHUK, *Ferroelectrics* **255** (2001) 73.
73. N. DE MATHAN, E. HUSSON, G. CALVARIN, J. R. GAVARRI, A. W. HEWAT and A. MORELL, *J. Phys. Condens. Matter* **3** (1991) 8159.
74. W. KLEEMANN, *Int. J. Mod. Phys. B* **7** (1993) 2469.
75. Y. IMRY and S.-K. MA, *Phys. Rev. Lett.* 35 (1975) 1399.
76. M. D. GLINCHUK, *British Ceramic Trans.* **103** (2004) 76.
77. A. A. BOKOV and Z.-G. YE, *Phys. Rev. B* **66** (2002) 064103.
78. Y. BING, A. A. BOKOV, Z.-G. YE, B. NOHEDA and G. SHIRANE, *J. Phys. Condens. Matter* **17** (2005) 2493.
79. A. K. JONSCHER, *Nature* **267** (1977) 673; Universal relaxation law (Chelsea Dielectrics Press, London, 1996).
80. A. A. BOKOV and Z.-G. YE, *Phys. Rev. B* **65** (2002) 144112.
81. *Idem., J. Phys.: Condens. Matter.* **12** (2000) L541.
82. *Idem., Appl. Phys. Lett.* **77** (2000) 1888.
83. Y. H. BING, A. A. BOKOV and Z.-G. YE, unpublished.
84. A. A. BOKOV, M. MAGLIONE, A. SIMON and Z.-G. YE, *Ferroelectrics* (in press).
85. A. A. BOKOV and Z.-G. YE, *Sol. Stat. Commun.* **116** (2000) 105.
86. A. A. BOKOV, Y.-H. BING, W. CHEN, Z.-G. YE, S. A. BOGATINA, I. P. RAEVSKI, S. I. RAEVSKAYA and E. V. SAHKAR, *Phys. Rev. B* **68** (2003) 052102.
87. D. VIEHLAND, S. JANG, L. E. CROSS and M. WITTIG, *Phil. Mag. B* **64** (1991) 335.
88. V. BOVTUN, J. PETZELT, V. POROKHONSKYY, S. KAMBA and Y. YAKIMENKO, *J. Europ. Ceram. Soc.* **21** (2001) 1307.
89. I. RYCHETSKY, S. KAMBA, V. POROKHONSKYY, A. PASHKIN, M. SAVINOV, V. BOVTUN, J. PETZELT, M. KOSECand M. DRESSEL, *J. Phys. Condens. Matter* **15** (2003) 6017.
90. T. TSURUMI, K. SOEJIMA, T. KAMIYA and M. DAIMON, *Jpn. J. Appl. Phys., Part I* **33** (1994) 1959.
91. H. M. CHRISTEN, R. SOMMER, N. K. YUSHIN and J. J. VAN DER KLINK, *J. Phys. Condens. Matter* **6** (1994) 2631.
92. D. VIEHLAND, S. J. JANG, L. E. CROSS and M. WUTTIG, *J. Appl. Phys.* **68** (1990) 2916.
93. A. K. TAGANTSEV, *Phys. Rev. Lett.* **72** (1994) 1100.
94. A. E. GLAZOUNOV and A. K. TAGANTSEV, *Appl. Phys. Lett.* **73** (1998) 856.
95. A. LEVSTIK, Z. KUTNJAK, C. FILIPIC and R. PIRC, *Phys. Rev. B* **57** (1998) 11204.
96. Z. KUTNJAK, C. FILIPIC, R. PIRC, A. LEVSTIK, R. FARHI and M. MARSSI, *ibid.* **59** (1999) 294.
97. V. V. KIRILLOV and V. A. ISUPOV, *Ferroelectrics* **5** (1973) 3.
98. B. E. VUGMEISTER and H. RABITZ, *Phys. Rev. B* **57** (1998) 7581.
99. R. BLINC, V. BOBNAR and R. PIRC, *ibid.* **64** (2001) 132103.
100. A. E. GLAZOUNOV and A. K. TAGANTSEV, *Ferroelectrics* **221** (1999) 57.
101. A. K. TAGANTSEV and A. E. GLAZOUNOV, *Phys. Rev. B* **57** (1998) 18.
102. Z.-G. YE and A. A. BOKOV, *Ferroelectrics* **302** (2004) 227.
103. D. VIEHLAND, S. J. JANG, L. E. CROSS and M. WUTTIG, *Phys. Rev. B* **46** (1992) 8003.
104. C. S. TU, V. H. SCHMIDT and I. G. SINY *J. Appl. Phys.* **78** (1995) 5665.
105. A. VERBAERE, Y. PIFFARD, Z.-G. YE and E. HUSSON, *Mat. Res. Bull.* **27** (1992) 1227.
106. A. R. LEBEDINSKAYA and M. F. KUPRIYANOV, *Phase Trans.* **75** (2002) 289.
107. W. KLEEMANN and R. LINDNER, *Ferroelectrics* **199** (1997) 1.
108. Z.-G. YE and H. SCHMID, *ibid.* **145** (1993) 83.
109. N. W. THOMAS, S. A. IVANOV, S. ANANTA, R. TELLGREN and H. RUNDLOF, *J. Eur. Ceram. Soc.* **19** (1999) 2667.
110. A. TKACHUK and H. CHEN, Fundamental Physics of Ferroelectrics. (AIP Conference Proceedings No. 677, 2003) p. 55–64.
111. R. V. CHAMBERLIN, *Europhys. Lett.* **33** (1996) 545.
112. E. V. COLLA, E. YU. KOROLEVA, N. M. OKUNEVA and S. B. VAKHRUSHEV, *Phys. Rev. Lett.* **74** (1995) 1681.
113. S. B. VAKHRUSHEV, J.-M. KIAT and B. DKHIL, *Sol. Stat. Commun.* **103** (1997) 477.
114. E. V. COLLA, L. K. CHAO, M. B. WEISSMAN and D. VIEHLAND, *Phys. Rev. Lett.* **85** (2000) 3033.
115. R. G. PALMER, *Adv. Phys.* **31** (1982) 669.
116. R. P. FEYNMAN, Statistical mechanics (Benjamin, Reading, 1972).
117. G. CALVARIN, E. HUSSON and Z.-G. YE, *Ferroelectrics* **165** (1995) 349.
118. F. CHU, I. M. REANEY and N. SETTER, *J. Appl. Phys.* **77** (1995) 1671.
119. Z.-G. YE, Y. BING, J. GAO, A. A. BOKOV, P. STEPHENS, B. NOHEDA and G. SHIRANE, *Phys. Rev. B* **67** (2003) 104104.
120. Z.-G. YE and M. DONG, *J. Appl. Phys.* **87** (2000) 2312.
121. C.-S. TU, C.-L. TSAI, V. H. SCHMIDT, H. LUO and Z. YIN, *ibid.* **89** (2001) 7908.
122. A. A. BOKOV and Z.-G. YE, *ibid.* **95** (2004) 6347.
123. F. CHU, I. M. REANEY and N. SETTER, *J. Amer. Ceram. Soc.* **78** (1995) 1947.
124. M. ABPLANALP, D. BAROSOVA, P. BRIDENBAUGH, J. ERHART, J. FOUSEK, P. GUNTER, J. NOSEK and M. SULC, *J. Appl. Phys.* **91** (2002) 3797.
125. V. V. SHVARTSMAN and A. L. KHOLKIN, *Phys. Rev. B* **69** (2004) 014102.
126. F. BAI, J. LI and D. VIEHLAND, *Appl. Phys. Lett.* **85** (2004) 2313.
127. A. A. BOKOV and Z.-G. YE, *Phys. Rev. B* **66** (2002) 094112.
128. C. PERRIN, N. MENGUY, O. BIDAULT, C. Y. ZAHRA, A.-M. ZAHRA, C. CARANONI, B. HILCZER and A. STEPANOV, *J. Phys. Condens. Matter.* **13** (2001) 10231.
129. L. S. KAMZINA and N. N. KRAINIK, *Phys. Solid State* **42** (2000) 1712.
130. A. A. BOKOV and S. M. EMELIYANOV, *Phys Stat. Sol. (b)* **164** (1991) K109.
131. P. BAO, F. YAN, W. LI, Y. R. DAI, H. M. SHEN, J. S. ZHU, Y. N. WANG, H. L. W. CHAN and C.-L. CHOY, *Appl. Phys. Lett.* **81** (2002) 2059.
132. A. A. BOKOV, H. LUO and Z.-G. YE, *Mater. Sci. Eng. B* **120** (2005) 206.
133. A. LEBON, H. DAMMAK, G. GALVARIN and I. OULD AHMEDOU, *J. Phys. Condens. Matter* **14** (2002) 7035.
134. S. KAMBA, E. BUIXADERAS, J. PETZELT, J. FOUSEK, J. NOSEK and P. BRIDENBAUGH, *J. Appl. Phys* **93** (2003) 933.
135. G. XU, Z. ZHONG, Y. BING, Z.-G. YE, C. STOCK and G. SHIRANE, *Phys. Rev. B* **67** (2003) 104102.
136. *Idem., ibid.* **70** (2004) 064107.
137. K. OHWADA, K. HIROTA, P. REHRIG, FUJII and G. SHIRANE, *ibid.* **67** (2003) 094111.
138. G. XU, H. HIRAKA, G. SHIRANE and K. OHWADA, *Appl. Phys. Lett.* **84** (2004) 3975.
139. P. M. GEHRING, W. CHEN, Z. G. YE and G. SHIRANE, *J. Phys.: Condens. Matter* **16** (2004) 7113.

140. G. XU, D. VIEHLAND, J. F. LI, P. M. GEHRING and G. SHIRANE, *Phys. Rev. B* **68** (2003) 212410.
141. K. H. CONLON, H. LUO, D. VIEHLAND, J. F. LI, T. WHAN, J. H. FOX, C. STOCK and G. SHIRANE, *Phys. Rev. B* **70** (2004) 172204.
142. G. A. SAMARA, *Ferroelectrics* **274** (2002) 183.
143. *Idem., J. Phys.: Condens. Matter* **15** (2003) R367.
144. Y. YAMASHITA, Y. HOSONO, K. HARADA and N. YASUDA, *IEEE Transactions on Ultrasonics, Ferroelectrics, and Frequency Control* **49** (2002) 184.
145. P. W. REHRIG, W. S. HACKENBERGER, S.-E. PARK and T. R. SHROUT, in "Piezoelectric Materials in Devices,". edited by N. Setter (EPFL, Lausanne, 2002) p. 433.
146. Z.-G. YE, *Curr. Opin. Sol. Stat. Mater. Sci.* **6** (2002) 35.
147. B. NOHEDA, *ibid.* **6** (2002) 27.
148. A. A. BOKOV and Z.-G. YE, *Ceramic Transactions (Morphotropic Phase Boundary Perovskites, High Strain Piezoelectrics, and Dielectric Ceramics)* **136** (2003) 37.
149. R. PIRC and R. BLINC, *Phys. Rev. B* **60** (1999) 13470.
150. R. BLINC, J. DOLINSEK, A. GREGOROVIC, B. ZALAR, C. FILIPIC, Z. KUTNJAK, A. LEVSTIK and R. PIRC, *Phys. Rev. Lett.* **83** (1999) 424.
151. W. KLEEMANN, J. DEC, R. BLINC, B. ZALAR and R. PANKRATH, *Ferroelectrics* **267** (2002) 157.
152. R. PIRC, R. BLINC and V. BOBNAR, *Phys. Rev. B* **63** (2001) 0542203.
153. B. E. VUGMEISTER and H. RABITZ, *ibid.* **61** (2000) 14448.
154. *Idem., ibid.* **65** (2001) 024111.
155. R. F. MAMIN, *Phys. Sol. Stat.* **43** (2001) 1314.
156. R. F. MAMIN and R. BLINC, *ibid.* **45** (2003) 942.

JOURNAL OF MATERIALS SCIENCE 41 (2006) 53–63

Flexoelectric effects: Charge separation in insulating solids subjected to elastic strain gradients

L. ERIC CROSS
Evan Pugh Professor Emeritus of Electrical Engineering, The Materials Research Institute, Pennsylvania State University, University Park, PA 16802, USA

After a brief historical introduction this article will present a summary of experimental work carried through at Penn State to explore the flexoelectric coefficients μ_{ijkl} in ferroelectric, incipient ferroelectric and relaxor ferroelectric perovskites. The initial objective was to understand the magnitude of flexoelectricity in these systems to see whether it would be possible to develop a piezoelectric composite containing no piezoelectric element, which nonetheless could have practically useful properties. Recent discussions of the thermodynamic converse effect, ie. the generation of elastic strain by an electric field gradient, now suggest that such composites might be designed to have unique properties such as a direct but no converse effect, or vice-versa, and materials with this character could have important practical application. Present data already suggest that the direct effect may make an important contribution to the properties of epitaxial thin films where mismatch can give rise to very steep elastic strain gradients. Clearly, more work is needed to fully quantify the flexoelectric behavior. It will be important to measure single crystals in the ceramic systems which have been studied and to characterize the converse effect as a check of the measured values.

1. Introduction

Flexoelectricity describes the generation of electric polarization in an insulating solid by the application of an elastic strain gradient. The phenomenon is characterized by the tensor relationship

$$P_1 = \mu_{ijkl} \frac{\partial S_{ij}}{\partial x_k} \qquad (1)$$

Where

P_1 is the component of resultant polarization

μ_{ijkl} the flexoelectric coefficients, a fourth rank polar tensor

S_{ij} the component of the elastic strain

x_k the direction of the gradient in S.

The μ_{ijkl} have the same symmetry as the electrostriction constants Q_{ijkl} so that for a cubic crystal the non-zero components are μ_{1111}, μ_{1122}, μ_{1212} or in matrix notation μ_{11} μ_{12} and μ_{44}.

The first discussion of electric polarization induced in a centric crystal by inhomogeneous deformation is by Kogan [1]. The phenomenon was named flexoelectricity by Indenbom [2] by analogy with charge separation in non piezoelectric liquid crystals. The effect in soft elastomers was explored by Marvan [3] who still is active in the field. The first comprehensive theoretical discussions are by Tagantsev [4, 5], which underscore the complexity and give the first clear indication of the distinction between static and dynamic coefficients.

It is perhaps not surprising that practical measurement of flexoelectric coefficients languished as it was universally recognized that for simple homogeneous insulating solids the effect was very small with μ_{ij}~10^{-10} to 10^{-11} C/m. Theory [6] predicted four contributors to the effect:

(1) A bulk static flexoelectric effect ~10^{-10} C/m

(2) A bulk dynamic coefficient made up from two parts each ~10^{-10} C/m

(3) A surface flexoelectric coefficient also of order ~10^{-10} C/m

0022-2461 © *2006 Springer Science + Business Media, Inc.*
DOI: 10.1007/s10853-005-5916-6

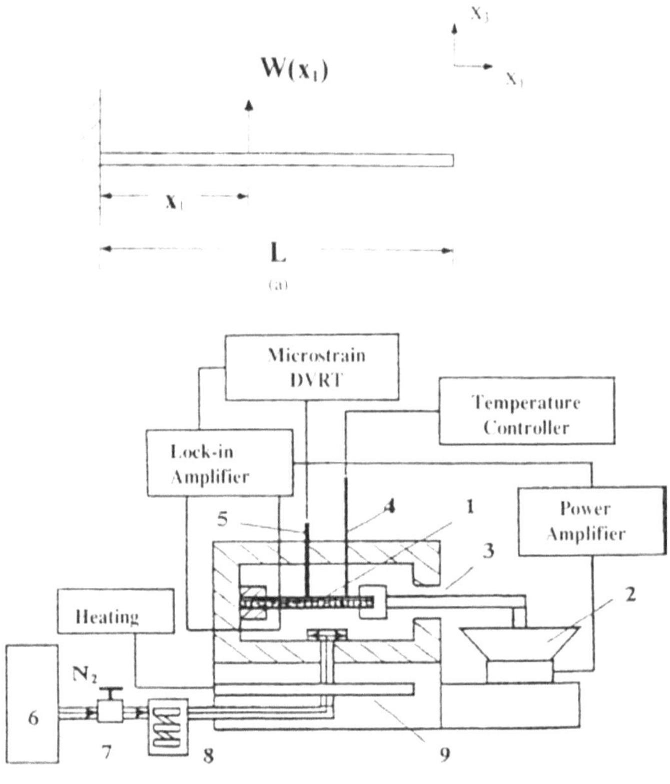

Figure 1 (a) Typical ceramic bar sample: ceramic bar 76.3×12.7×2.5 mm indicating measuring coordinate directions. (b) Experimental setup for the dynamic measurement of temperature dependence of flexoelectric μ_{12}. 1. PMN bar; 2. Loudspeaker; 3. Driving arm; 4. Thermocouple; 5. Microstrain transducer core; 6. High purity nitrogen; 7. Gas Flow meter; 8. Copper coils immersed in liquid nitrogen; 9. Heating elements.

(4) Surface piezoelectric effect whose magnitude depends markedly on the nature of the surface.

The situation is however more interesting for the paraelectric phase of the soft mode ferroelectric dielectric. Both static and dynamic bulk coefficients become larger in proportion to the enhanced dielectric susceptibility, the surface flexoelectric coefficient does not scale and so becomes unimportant, and the surface piezoelectric effect again depends on surface characteristics. It appeared natural that the place to start would be a high permittivity perovskite and since the laboratory had excellent experience with the relaxor lead magnesium niobate, this was an obvious choice upon which to start.

2. Experimental studies

2.1. Measuring techniques

Initial focus was on measuring the transverse flexoelectric coefficient μ_{12} in a range of cubic perovskite high-permittivity ceramics. Both quasi-static and low frequency dynamic techniques have been employed. For dynamic measurement a bar of ceramic of rectangular cross section, clamped at one end was driven in flexure at 1Hz by a small electro-magnetic actuator (Fig. 1). The sample with dimensions in the range 70×12×2 mm is equipped with a set of small circular electrodes equally spaced along the major surface. At each electrode the vertical AC displacement is monitored by a DVRT microstrain transducer and the local charge monitored by a charge amplifier phase locked to the strain generator. Taking the x_1 axis along the beam and the x_3 axis normal to the electroded surfaces (Fig. 1a), Equation 1 becomes

$$P_3 = \mu_{12}\frac{\partial S_{11}}{\partial x_3} \qquad (2)$$

P_3 is derived from the charge and electrode area $\frac{\partial S_{11}}{\partial x_3}$ from the local curvature.

The advantage of the flexure method is that there is a neutral axis down the middle of the beam so that any residual piezoelectricity cancels out. Asymmetry can be checked by inverting the beam which makes no difference to the flexoelectric signal but would be changed by piezoelectric asymmetry.

For static measurement the classic four point bending configuration was used (Fig. 2) and the resulting charge measured by electrometer. Here the objective was to

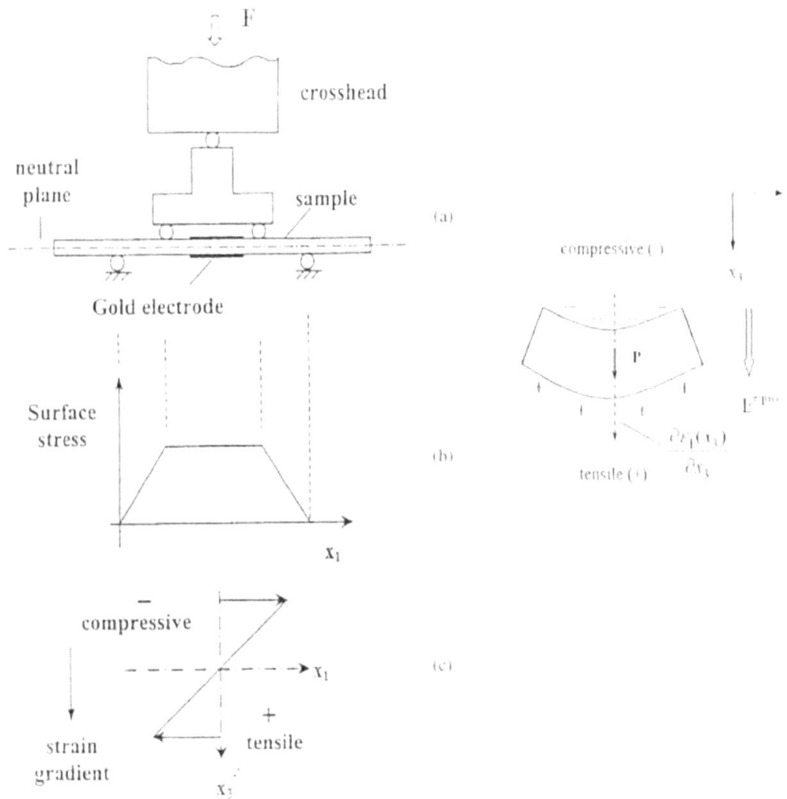

Figure 2 (a) Experimental arrangement for the static measurement of flexoelectric μ_{12} to high elastic stress levels. (b) Transverse stress profile for the four point bending experiment. (c) Illustration of the transverse strain gradient. Compressive in upper and tensile in the lower surface. (d) Sketch indicating the expected charge separation due to flexoelectric μ_{12}..

measure the effective μ_{12} for the centric $\infty\infty$ m symmetry of the unpoled ferroelectric ceramic and examine the possibility of a ferroelastic domain contribution to this response at high stress levels where obvious ferroelastic shape change could be observed.

To measure the longitudinal coefficient μ_{1111}, which is the most important for composite development, an AC Instron type 5866 was adapted to generate cyclic longitudinal stress at 0.5 Hz. The set up is shown schematically in plan and elevation in Fig. 3 and photographs of the holder, environment chamber and sample in Fig. 4. The sample was configured as a truncated triangle of constant width, to generate a uniform vertical stress gradient (Fig. 4c), and for ease of alignment the triangulated stressing jig (Fig. 3b) was developed. The charge release was measured by a charge amplifier phase locked to the AC drive of the Instron system.

2.2. Choice of materials

In order to permit a wide range of dielectric permittivities to be studied it was desirable to pick systems with a high weak-field dielectric permittivity maximum at a temperature T_m close to room temperature. For this reason the compound Lead magnesium niobate $Pb(Mg_{1/3}Nb_{2/3})O_3$ (PMN) and the solid solution $Ba_{0.67}Sr_{0.33}TiO_3$ (BST) were chosen for initial study. Above the dielectric maximum both are in cubic point group m$\bar{3}$m with μ_{11} μ_{12} and μ_{44} the only non-zero flexoelectric coefficients. PMN is the classic prototypic perovskite ferroelectric relaxor. BST is a near approximation to the soft-mode paraelectric, modified by order: disorder near T_c but largely free from lower frequency dielectric dispersion.

For the simple unpolarized ferroelectric perovskite ceramic in texture symmetry ∞ ∞ m, the choice is obviously very much broader. Because of the surprisingly large difference found between PMN and BST, it was decided to choose one lead containing and one non-lead ceramic. An excellent source for the soft lead zirconate titanate PZT 5H close at hand was TRS ceramics State College. For the lead-free ceramic barium titanate ($BaTiO_3$) is an obvious choice. Now however, since the ferroelectric 90° wall behavior was of interest a coarse grained (5–10 μm) microstructure was required and samples to this specification were fabricated in the laboratory. With the need for long bar-shaped samples and a severely

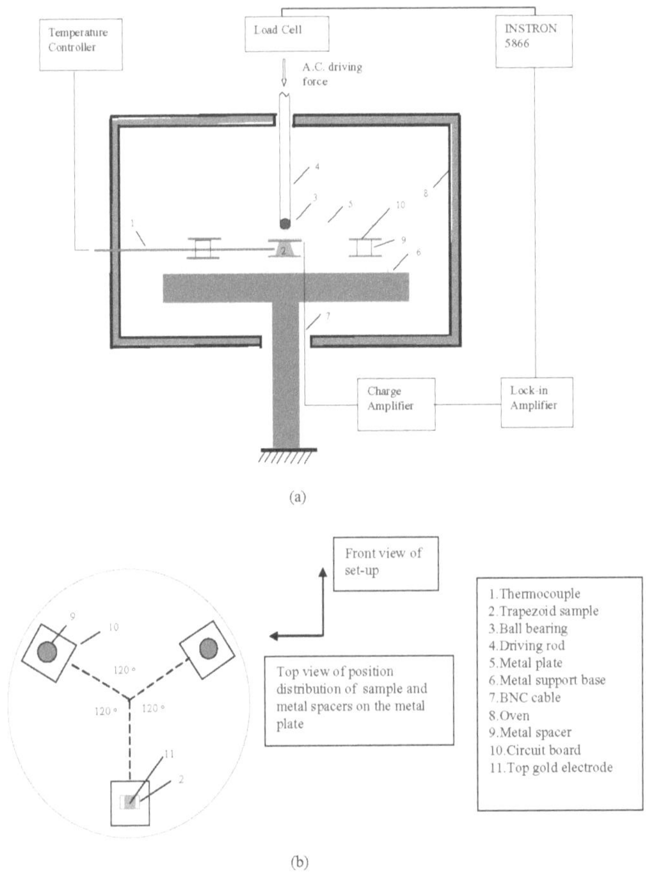

Figure 3 Schematic diagram of the arrangement used for measurement of flexoelectric μ_{11} as a function of temperature.

constrained budget, ceramic samples had to suffice, however at least for PMN and $BaTiO_3$ single crystals should be measured.

2.3. Measurements of flexoelectric μ_{12}

Detailed measurements for PMN, BST and PZT 5H have already been published [7–11], so only a brief synopsis of the key features will be reported.

In both PMN (Fig. 5) and in BST (Fig. 6) the enhancement of μ_{12} with increasing ϵ_ω is clearly evident. Over the intermediate range for ϵ_ω between 3000 to 11000 both materials follow the prediction by Tagantsev [5] that

$$\mu_{12} = \gamma \chi_{33} \left(\frac{e}{a}\right) \tag{3}$$

where χ_{33} is the weak field susceptibility. In both these high-permittivity dielectrics $\chi_{33} = \epsilon_\omega$

e is the electron charge

a the unit cell dimension

γ a constant.

For PMN γ=0.65 close to the unity value predicted, but in BST γ=9.3 much larger than expected.

The fit to Equation 3 in both materials is evidenced in Fig. 7. The enhancement above ϵ_ω~11000 in BST correlated closely with a departure in $\frac{1}{\varepsilon_\omega}$ from Curie–Weiss behavior and one suspects the introduction of a new polarizability associated with the occurrence of a few ferroelectric macro-domains. The reason for the drop in μ_{12} at permittivity values below 3000 is unclear as the sensitivity is quite adequate for much lower level signals.

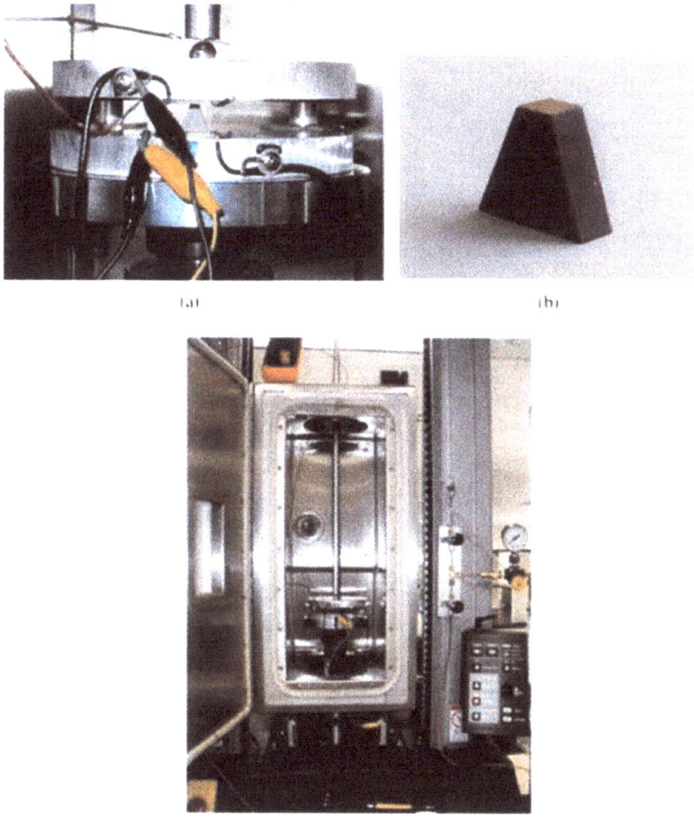

(a) (b) (c)

Figure 4 Photographs of the arrangement used for μ_{11} measurements. (a) Sample holder and stressing jig. (b) Typical electroded sample. (c) Environment chambers for holding the sample.

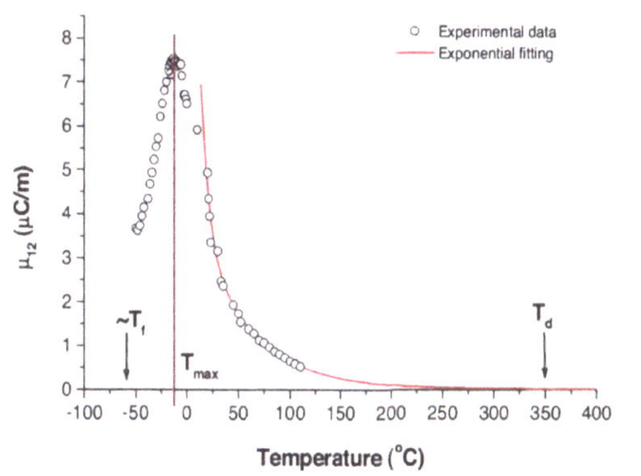

Figure 5 Flexoelectric μ_{12} in $Pb(Mg_{1/3}Nb_{2/3})O_3$ Ceramic as a function of temperature in the macro-cubic domain above the dielectric T_{max} measured at 1 Hz.

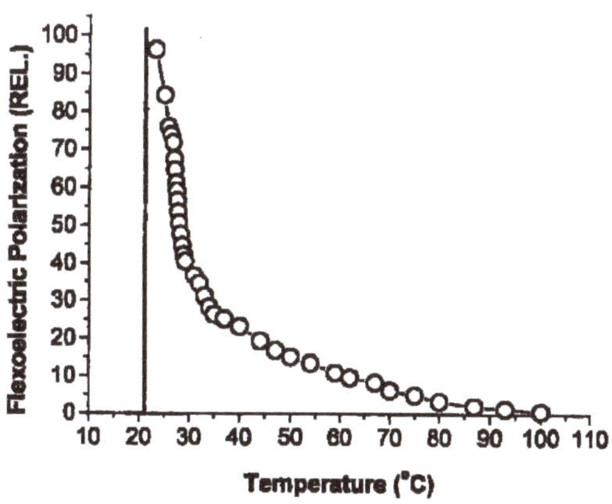

Figure 6 Flexoelectric polarization as a function of temperature in $(Ba_{0.67}Sr_{0.33})TiO_3$ ceramic, measured at 1 Hz. The vertical solid line indicates the Curie Temperature in this composition.

Initial objectives for the quasi-static flexure studies were to see whether the pattern of ferroelectric domains in an unpoled ferroelectric ceramic below T_c would markedly influence μ_{12}. At stress levels capable of driving ferroelastic domain walls, whether there would be an extrinsic domain contribution to response, and if so, whether the strain

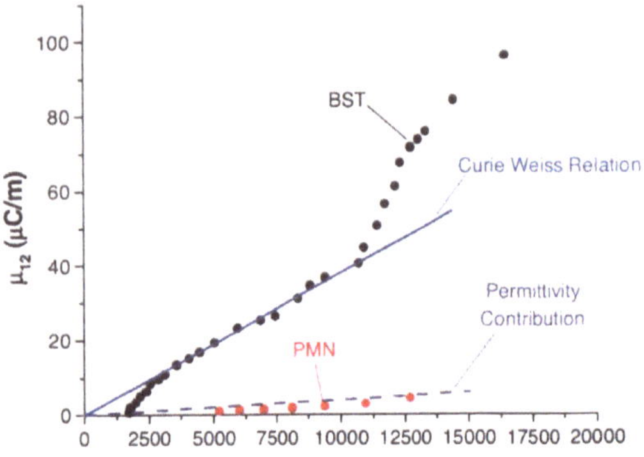

Figure 7 Comparison of Flexoelectric μ_{12} as a function of the dielectric permittivity ϵ_3 in $Pb(Mg_{1/3}Nb_{2/3})O_3$ and $Ba_{0.67}Sr_{0.33}TiO_3$.

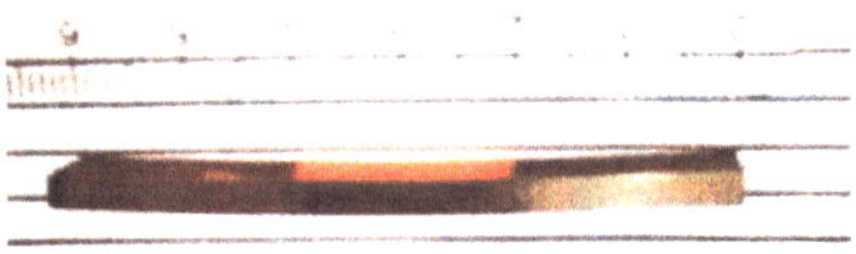

Figure 8 Illustration of the residual ferroelectric curvature in a typical PZT-5H bar after high stress four-point bend test. Clear indication that the sample has experienced ferroelastic domain wall motion.

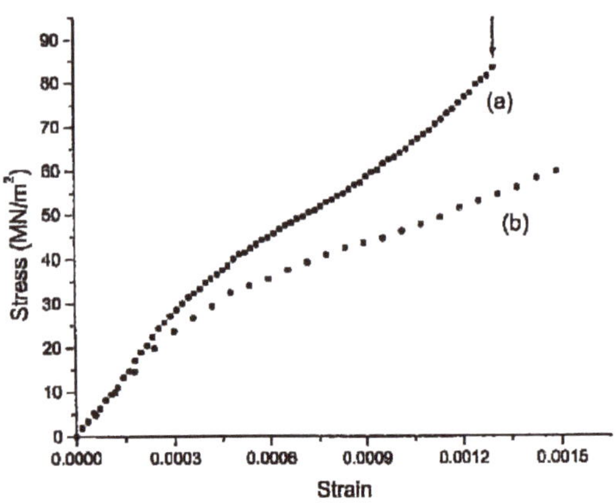

Figure 9 Surface strain as a function of surface stress during 4-point bending test. (a) Cross head speed 1 mm/min. (b) Crosshead speed 0.2 mm/min. Note the onset of time dependent softening due to ferroelastic:ferroelectric domain wall motion.

gradient could pole the ceramic to a piezoelectric state.

Residual curvature after high 4 point loading (Fig. 8) gives clear evidence of ferroelastic domain rearrangement, and the change in elastic stiffness at surface strain around $s_{11} \sim 0.0003$ (Fig. 9) suggests the onset of ferroelastic domain wall motion. This surface strain level corresponds to a strain gradient of 0.2 m^{-1} and it is ex-

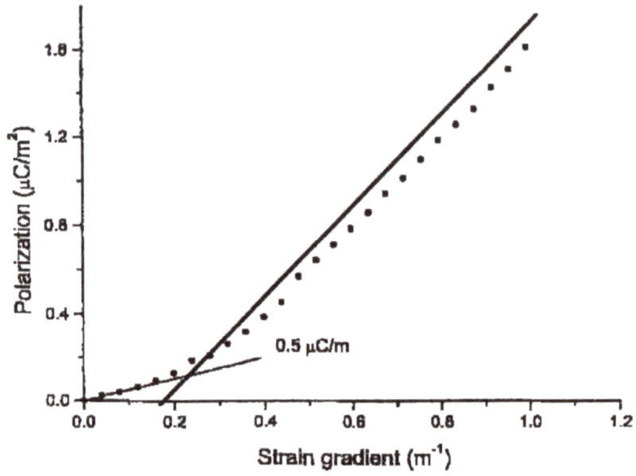

Figure 10 Flexoelectric polarization P_3 driven by the gradient in elastic strain S_1 Note the onset of enhanced μ_{12} above the strain level needed to drive ferroelectric domain wall motion.

citing to see in Fig. 10 the clear change of slope in the induced polarization corresponding to a change of μ_{12} from 0.5 to 2 μC/m clear evidence of an extrinsic domain contribution to the flexoelectric response. Careful testing by both phase sensitive d_{33} meter and IEEE resonance method however failed to show any evidence of an induced piezoelectric poling even after subjecting the samples to a maximum gradient of 1 m^{-1}.

In the high-temperature paraelectric phase where the response is just intrinsic it is interesting to see that for $BaTiO_3$ $\gamma = 11.4$.

2.4. Summary of results for μ_{12}

Lead magnesium niobate $PbMg_{1/3}Nb_{2/3}O_3$ $\gamma = 0.65$
Lead zirconate titanate PZT 5H Intrinsic $\gamma = 0.57$
Barium Strontium titanate $Ba_{0.67}Sr_{0.33}TiO3$ $\gamma = 9.3$
Pure barium titanate ceramic $BaTiO_3$ $\gamma = 11.4$
For PZT 5H in the ferroelectric phase
Intrinsic $\mu_{12} = 0.5$ μC/m Extrinsic 1.5 μC/m
For $BaTiO_3$ in the ferroelectric tetragonal phase Intrinsic $\mu_{12} = 5.5$ μC/m Extrinsic 30.2 μC/m

2.5. Measurements of flexoelectric μ_{11}

Initial studies have been focused on BST at the $(Ba_{0.67}Sr_{0.33})TiO_3$ composition and on lead strontium titanate (PST) at the $(Pb_{0.3}Sr_{0.7})TiO_3$ composition. As with BST, the PST composition is chosen to have T_m just below room temperature. It has classical Curie-Weiss behavior from high temperature to within 15°C of T_m with no low radio frequency dispersion and no evidence of relaxor ferroelectric character.

For the BST, again as expected, there is a good proportionality between the gradient and the induced polarization (Fig. 11) which leads to a well defined μ_{11} that

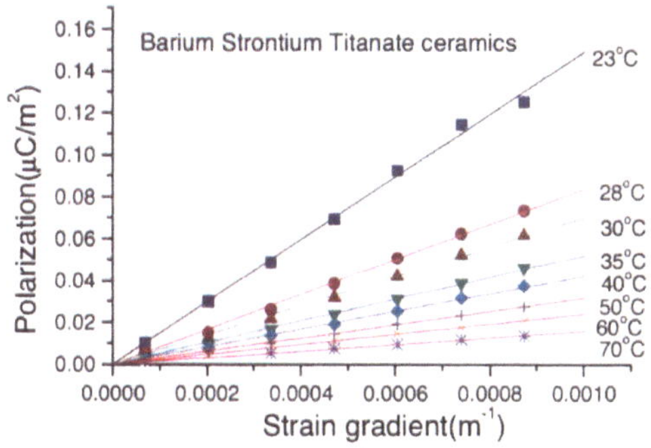

Fig. 11

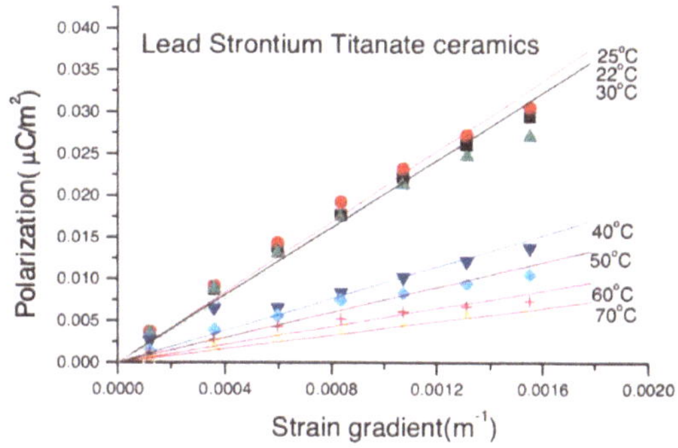

Fig. 13

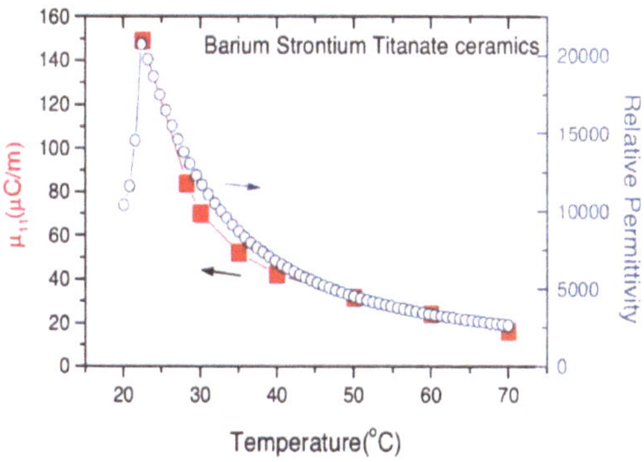

Fig. 12

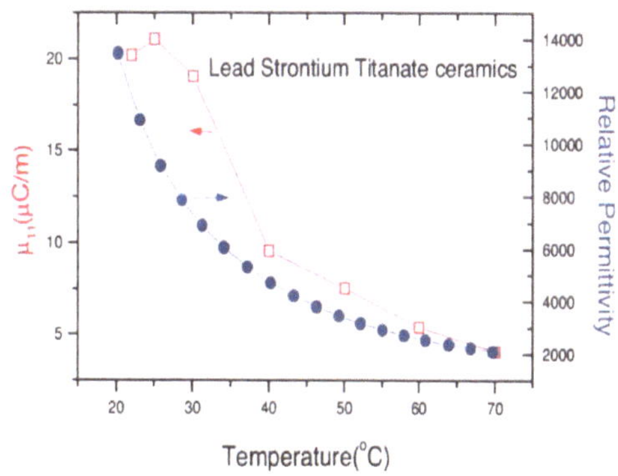

Fig. 14

Figures 11 and 12 AC Instron measurement of μ_{11} as a function of temperature $(Ba_{0.67}Sr_{0.33})TiO_3$ Ceramic. Frequency 0.5 Hz Gradient range 0 to 0.00088 m^{-1}.

Figures 13 and 14 AC Instron Measurement of μ_{11} as a function of temperature in $(Pb_{0.03}Sr_{0.07})$ TiO_3 ceramic frequency 0.5 Hz Gradient range 0 to 0.0015_5 m^{-1}.

also follows the proportionality to dielectric permittivity suggested in Equation 3 (Fig. 12) with a γ value of 18.3.

In PST, the general trend is the same, with good proportionality to the gradient level, (Fig. 13), but the tracking with dielectric permittivity is not so precise (Fig. 14). The mean γ value is 4.5.

Again it appears that the lead based compound has a significantly lower γ than the BST for μ_{11} even though it is not relaxor, but in both BST and PST it may be noted that the linear relation to ϵ breaks down as T approaches T_c with μ_{11} climbing more rapidly with decreasing temperature suggesting the intervention of a new polarization mechanism. Currently, although different samples of the same composition give similar values at the same temperature we must regard both the measured μ_{11} and γ values as preliminary as there is a spurious charge generation in the system which must be tracked down and corrected.

3. Thermodynamics of flexoelectricity

The constitutive equations for flexoelectricity take the form

$$P_i = \eta_{ij} E_j + \mu_{ijkl} \frac{\partial S_{kl}}{\partial x_j} \quad (4)$$

And

$$T_{k1} = c_{ijkl} S_{ij} - \mu_{ijkl} \frac{\partial E_i}{\partial x_j} \quad (5)$$

Where

P_i are the polarization components
E_i,E_j the electric field component
T_{kl} the elastic stress components
S_{ij} the elastic strain components
η_{ij} the dielectric susceptibility
c_{ijkl} the elastic constant
μ_{ijkl} the flexoelectric constants
$\partial S_{kl}/\partial \chi$ is the strain gradient in the jth direction
$\partial E_i/\partial \chi_j$ is the electric field gradient in the jth direction

For the electric polarization generated by an elastic strain gradient Equation 4 clearly reduces to

$$E_j = 0 \quad P_i = \mu_{ijkl} \frac{\partial S_{kl}}{\partial x_j}$$

which is the equation we have used for the direct flexoelectric effect.

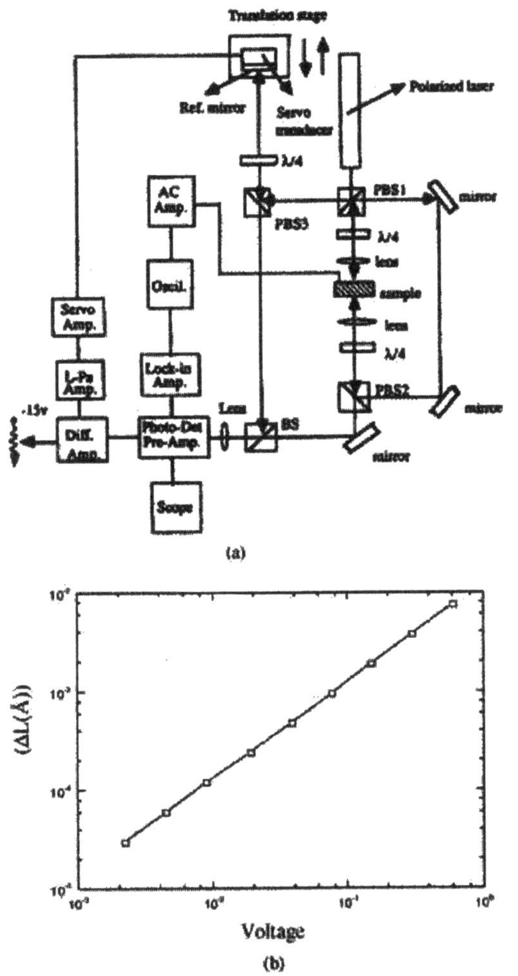

Figure 15 (a) Schematic of the Ultra-Sensitive laser double beam interferometer to be used for converse flexoelectric displacement measurements. (b) Calibration curve for the dilatometer using piezoelectric lithium niobate.

For the case (5) under zero applied stress Equation 5 reduces to

$$T_{kl} = 0 \quad c_{ijkl} S_{ij} = \mu_{ijkl} \frac{\partial E_i}{\partial x_j}$$

which clearly represents the converse effect i.e. the change of elastic strain at zero elastic stress generated by an electric field gradient.

3.1. Converse flexoelectric effect

So far we know of no attempts to measure the converse flexoelectric effect. However, interferometric techniques for measuring exceedingly small strains are very well developed and capable of measuring field induced displacements of order 10^{-4} Å[12]. The example we have is the system illustrated schematically in Fig. 15a. In this system, the photo-detector is phase locked to the low sample driving frequency. Drifts due to thermal changes and low frequency vibrations are taken out by the compensation mirror driven by a piezoelectric actuator. Stability and sensitivity of the system can be judged by the Calibration Curve in Fig. 15b which is linear down to 2×10^{-5} Å.

Unfortunately, initial studies were frustrated when it proved impossible to read the flexoelectric signal at frequency ω against the much larger signal at 2ω generated by electrostriction, one of the higher order effects not considered in Equation 5. To overcome this difficulty the sample is being miniaturized to steepen the gradient in E and the compensator illustrated in Fig. 16 is under construction.

TABLE I None zero components of piezoelectric d constant d_{ij} in matrix notation. In all groups the ∞ axis is taken as x_3; the axes x_1, x_2 are perpendicular to x_3 and to each other, but otherwise in arbitrary orientation

Matrices of d_{ijk} in Curie groups which are piezoelectric

∞					
0	0	0	d_{14}	d_{15}	0
0	0	0	d_{15}	$-d_{14}$	0
d_{31}	d_{31}	d_{33}	0	0	0
∞2					
0	0	0	d_{14}	0	0
0	0	0	0	$-d_{14}$	0
0	0	0	0	0	0
∞m					
0	0	0	0	d_{15}	0
0	0	0	d_{15}	0	0
d_{31}	d_{31}	d_{33}	0	0	0

4. Application to a piezoelectric composite

4.1. Introduction

The Penn State studies of flexoelectricity were initiated by discussions[1] on textures symmetries which pointed up the characteristic forms representing the Curie groups that are piezoelectric (Fig. 17a). If such forms are arranged in an orderly manner, as in Fig. 17b, to form a two phase composite, and both phases are insulators, even if neither phase is piezoelectric the composite ensemble must exhibit piezoelectricity. Further, if in all groups the ∞ axis is taken as x_3 the matrices of the non-zero piezoelectric constants will be as in Table I.

For the ∞m symmetry, discussing the possible charge separation mechanisms to account for the symmetry

[1] The Penn State flexoelectric program was initiated through discussion with Professor J. Fousek on possible mechanisms which could impart charge separation under elastic stress in textures symmetry ∞ m.

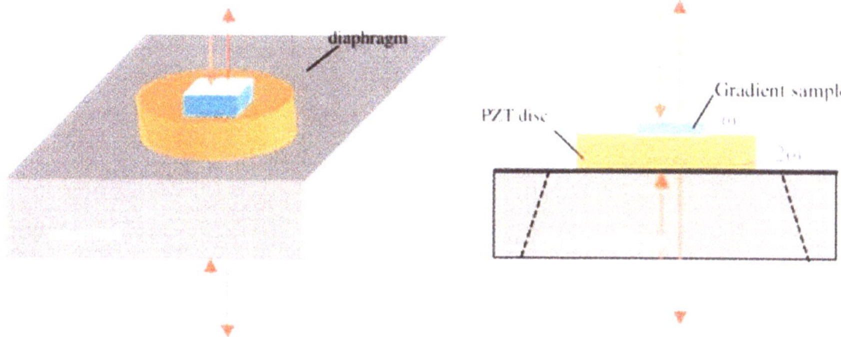

Figure 16 **Proposed sample arrangement for the converse measurement of μ_{11}. Sample driven at frequency ω', shaped to generate a strong field gradient in Ez Gradient generates a deformation at frequency ω', but electrostriction generates a much stronger deformation at $2\omega'$ PZT disc is driven at $2\omega'$ in anti-phase to the electrostriction deformation. Drive is varied to cancel out the total AC deformation at $2\omega'$ High sensitivity of the ultra-dilatometer can now be used to measure the deformation at frequency without perturbation by the much larger electrostriction signal.**

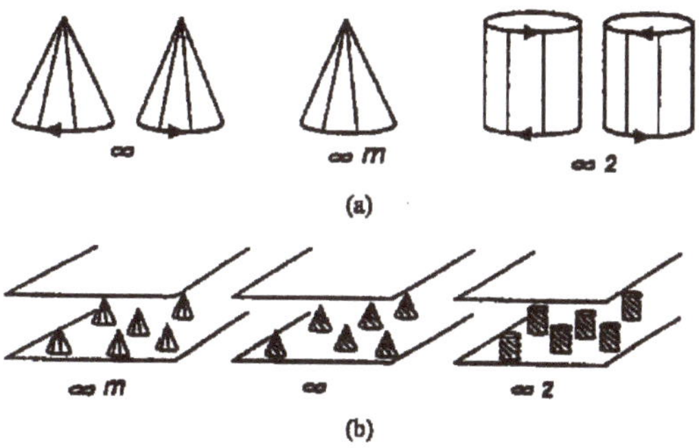

Figure 17 **(a) Characteristic forms representing symmetry of the Curie groups which allow for piezoelectricity. (b) Simple models of 0–3 composites allowing for piezoelectricity. There is an infinite number of shapes of the 0-constituent that could be tested for a maximum response.**

required behavior, it was evident that even if nonpiezoelectric the two phases differ in elastic properties so that the conical shape automatically gives rise to an axial stress gradient, even if the composite is subjected to a uniform stress. The gradient will then act through the flexoelectric effect to produce charge separation.

Remembering that symmetry only dictates what must be present (or absent), and nothing as to the magnitude, the natural question was whether flexoelectricity could lead to usable piezoelectric properties in a properly configured composite. Hence, the need to measure the μ_{ijkl} in materials likely to have large values of these constants.

To illustrate the potential, take for example the square truncated pyramid as in Fig. 18a as the element in a composite sheet (Fig. 18b). Suppose the upper square face has a side of length a_1, and the lower a length a_2 and the side wall is configured so that a_2 is a linearly increasing function of d the depth from a_1 to a_2. For a force F applied normal to upper and lower surfaces. Stress in the upper surface will be $T_{3(1)} = F/a_1^2$ and will give rise to a strain $S_{3(1)}$ given by $S_{3(1)} = F/a_1^2 c_{11}$ where c_{11} is the elastic constant of the truncated pyramid.

Similarly for the lower surface $T_{3(2)} = F/a_2^2$ and $S_{3(2)} = F/a_2^3 c_{11}$

Since the side walls are configured to make a^2 a linear function of d

$$\frac{\partial S_3}{\partial d} = \frac{S_{3(1)} - S_{3(2)}}{d} = \frac{F\left(\frac{1}{a_1^2} - \frac{1}{a_2^2}\right)}{dc_{11}}$$

If the pyramid material has a flexoelectric coefficient μ_{11} then

$$P_3 = \mu_{11}\frac{\partial S_3}{\partial d_3} = \mu_{11}\frac{F\left(\frac{1}{a_1^2} - \frac{1}{a_2^2}\right)}{dc_{11}} = \mu_{11}\frac{\frac{a_2^2 - a_1^2}{a_1^2}}{dc_{11}} \bullet \frac{F}{a_2^2}$$

i.e.

$$P_3 = \mu_{11}\frac{\left(\frac{a_2^2 - a_1^2}{a_1^2}\right)}{dc_{11}} T3$$

but for a piezoelectric sheet

$$P_3 = d_{33}T_3$$

So that

$$d_{33} = \mu_{11}\frac{\left(\frac{a_2^2 - a_1^2}{a_1^2}\right)}{dc_{11}}$$

For BST at room temperature $\mu_{11} \sim 100\ \mu\text{C/m}$

$$c_{11} = 1.66 \times 10^{11}\ \text{N}/m^2$$

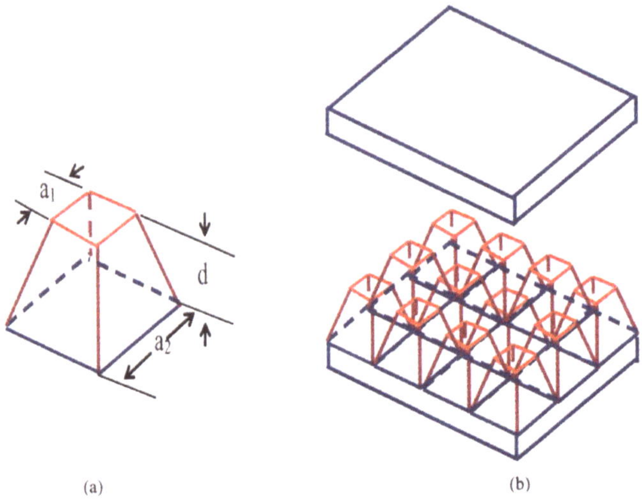

Figure 18 Proposed composite to generate piezoelectric response from flexoelectric polarization. (a) Individual truncated pyramidal building block from BST strongly flexoelectric stiff ceramic. (b) Composite arrangement of individual elements. Second phase in initial study will be air. Later a soft polymer. Since the vertical gradient of strain is uniform down the pyramids the whole upper surface may be electrode for charge collection.

for $a_1 = 50\ \mu\text{m}\ a_2 = 250\ \mu\text{m}\ d = 250\ \mu\text{m}\ d_{33} \approx 60$ pC/N
Scaling down $a_1 = 5\ \mu$ m, $a_2 = 25\ \mu\text{m}\ d = 25\ \mu\text{m}$ $d_{33} \approx 600$ pC/N.
A quite useful level at not impossibly small dimensions for the composite.

5. Future prospects

Immediate objectives of the current program are to uncover the spurious charge generation mechanisms in the μ_{1111} studies on BST and PST and to fabricate a coarse scale composite of the form described in 4 above to just demonstrate proof of principle. We believe that the problem of generating a completely uniform axial stress on a very stiff ceramic may be the source of the spurious charge generation that has plagued our μ_{11} measurements.

It is associated with the difficulty of generating an elastic equipotential surface, which was a major headache in earlier studies of the converse electrostrictive constants in simple solids obtained by measuring the uniaxial stress dependence of the dielectric susceptibility. Steps to be taken to explore the problem include elastic matching of piston and sample to avoid Poisson Ratio effects, soft elastic gaskets sealed at high stress levels, and if necessary, integrated with fired-in elastically matching electrodes.

For proof of principle a sub millimeter composite has already been fabricated and the assembly is in final stages. We hope that it will be possible to make the first measurements of the converse effect as the system is already fabricated, but pressure on the use of the ultra dilatometer is heavy and the current program may run out before measurements can be completed. We believe the topic bristles with intriguing possible longer-range opportunities.

Clearly, grain boundaries are a worrisome source of unneeded complexity, and the random crystallite arrangement can give rise to elastic inhomogeneity even in cubic systems. On the compounds already studied measurements should be carried out on single crystals to ascertain the validity of these concerns.

As was the case with electrostriction in simple low-permittivity solids, essential validation of the measured values is in close agreement between direct and converse measurements. Flexoelectric deformation associated with the strain generated by electric field gradient must be measured to confirm direct measurement.

If the piezoelectric composite demonstration validates the principle, it is possible to conceive of composites on a much finer scale which would have exciting properties. Further, it is not unreasonable to believe that systems could be developed where a strong strain gradient would develop under uniform stress but be unaccompanied by a strong polarization gradient under uniform electric field yielding composites with direct but no converse piezoelectricity and vice versa.

We believe that much remains to be done practically; also with current rapid progress in ab-initio theory for the perovskite it is clear that more refined theoretical approaches are certainly becoming accessible.

References

1. S. M. KOGAN, *Sov. Phys. Solid State* **5** (1964) 2069.

2. V. L. INDENBOM, E. B. LOGINOV and M. A. OSIPOV, *Kristalografija* **26** (1981) 1157.
3. M. MARVAN and A. HAVRANEK, *Science* **78** (1988) 33.
4. A. K. TAGANTSEV, *Phys. Rev. B* **34** (1986) 5883.
5. *Idem., Phase Trans.* **35** (1991) 119.
6. *Idem., Sov. Phys. JETP* **61**(6) (1985) 1246.
7. W. MA and L. E. CROSS, *Appl. Phys. Lett.* **78** (2001) 2920.
8. *Idem., ibid.* **79** (2001) 4420.
9. *Idem., ibid.* **81** (2002) 3440.
10. *Idem., ibid.* **82** (2003) 3293.
11. *Idem., ibid..* **86** (2005) 072905.
12. W. Y. PAN and L. E. CROSS, *Rev. Sci. Inst.* **60**(8) (1989) 2701.

JOURNAL OF MATERIALS SCIENCE 41 (2006) 65–76

Piezoelectric anisotropy: Enhanced piezoelectric response along nonpolar directions in perovskite crystals

D. DAMJANOVIC, M. BUDIMIR, M. DAVIS, N. SETTER
Ceramics Laboratory, Materials Institute, Ecole Polytechnique Fédérale de Lausanne—EPFL, 1015, Lausanne, Switzerland
E-mail: dragan.damjanovic@epfl.ch

This paper discusses the mechanisms that can contribute to the enhanced longitudinal piezoelectric effect along nonpolar directions in perovskite crystals, such as $BaTiO_3$, $PbTiO_3$, $KNbO_3$, $Pb(Mg_{1/3}Nb_{2/3})O_3$-$PbTiO_3$ and $Pb(Zn_{1/2}Nb_{2/3})O_3$-$PbTiO_3$. Piezoelectric anisotropy is discussed in relation to temperature induced phase transitions, compositional variation in solid solutions with morphotropic phase boundaries, applied electric fields, the domain wall structure and domain wall displacement.

1. Introduction

The anisotropy of elastic, dielectric and piezoelectric properties plays an important role in the application of piezoelectric materials. Several types of the anisotropy have been discussed in the literature, including the orientation dependence of the material coefficients and the difference in the values of the longitudinal, transverse and shear coefficients. The two best known examples are the temperature stabilized cuts of quartz [1] and $LiNbO_3$ crystals [2] where the temperature dependences of different elastic compliances compensate, leading to low temperature coefficients of resonant frequencies and acoustic velocities. Anisotropy of piezoelectric properties was intensively studied in the 1980s when unusually strong piezoelectric anisotropy was found in lead titanate ceramics with random grain orientation [3]. In these materials, for example, the longitudinal d_{33} coefficient behaves in the expected way whereas the transverse coefficient d_{31}, which in most perovskite materials is negative and 2–4 times smaller than d_{33}, can be close to zero or can even exhibit a positive sign [4]. Another example is the strong anisotropy of dielectric, elastic and piezoelectric properties in textured ceramics of bismuth based Aurvillius structures [5]. The anisotropy of the elastic and piezoelectric properties is exploited in piezoelectric–ceramic composites with 1–3 connectivity to reduce lateral coupling in transducers for underwater and medical imaging applications [6].

Interest in the electromechanical anisotropy of ferroelectric materials has been spectacularly renewed in the last several years after the (re)discovery [7, 8] of the large electromechanical coupling coefficient k_{33} (>90%) and longitudinal piezoelectric coefficient d_{33} (>2000 pC/N) measured along nonpolar directions in relaxor-ferroelectric solid solutions [e.g., $Pb(Mg_{1/3}Nb_{2/3})O_3$-$PbTiO_3$ or PMN-PT and $Pb(Zn_{1/3}Nb_{2/3})O_3$-$PbTiO_3$ or PZN-PT]. The reason why this result was surprising can probably be found in the fact that the most widely used piezoelectric material over the last 50 years has been ceramic $Pb(Zr,Ti)O_3$. Poled ceramics exhibit conical symmetry and their largest longitudinal, transverse and shear piezoelectric responses are measured along the axes of the orthogonal coordinate system whose x_3 axis lies parallel to the poling (polar) direction [9]. A maximum of the longitudinal piezoelectric response along nonpolar directions, as observed in relaxor-ferroelectrics, was therefore unexpected. However, subsequent studies have found that such behavior is common and has been reported in many perovskite crystals, both in those having simple compositions (e.g., $BaTiO_3$ [10] and $KNbO_3$ [11]) and in other complex solid solutions (e.g., $BiScO_3$-$PbTiO_3$ [12] and $Pb(Yb_{1/2}Nb_{1/2})O_3$ -$PbTiO_3$ [13]). Similar anomalies have been observed for the transverse and shear coefficients. In contrast to poled ceramics, the transverse coefficient is found to be largest in a plane that is not perpendicular to the polar direction. The shear

0022-2461 © *2006 Springer Science + Business Media, Inc.*
DOI: 10.1007/s10853-005-5925-5

effect behaves oppositely and is the highest when the field is applied and response measured along the corresponding axes of the crystallographic coordinate system [14, 15]. As we shall see later these results are nontrivial as some important perovskites, like $PbTiO_3$ [16], behave similarly to ceramics, exhibiting highest piezoelectric effects along the axes of the crystallographic coordinate system.

While the origin of the large piezoelectric activity in crystals of complex solid solutions (e.g., PMN_PT, PZN-PT) is still not clear, the related research has led to unprecedented activity in the field of ferroelectric materials, directly or indirectly leading to the discovery of the monoclinic phase in $Pb(Zr,Ti)O_3$ [17], to the development of textured ceramics with enhanced piezoelectric properties [18], and to the renewed interest in the role of engineered domain states on the electromechanical properties of ferroelectric crystals. Interpretation of experimental results obtained on single crystals has been further complicated by the fact that virtually all experimental data have been reported for multidomain samples, and the role of the intrinsic anisotropy and the presence of the engineered domain structure have not been separated.

In this article we shall discuss selected aspects of the piezoelectric anisotropy in perovskite crystals, focusing on mechanisms that may contribute to the enhancement of the longitudinal d_{33} piezoelectric coefficient when measured along nonpolar directions. The text is structured in the following way. First, basic relations defining the orientation dependence of piezoelectric coefficients in the tetragonal *4mm*, orthorhombic *mm2*, and rhombohedral *3m* point groups are given and briefly discussed in Section 2. The piezoelectric anisotropy is then discussed in Section 3 in relation to the proximity of ferroelectric-ferroelectric phase transitions, in Section 4 in terms of composition in materials exhibiting a morphotropic phase boundary, in Section 5 as a function of external electric fields, and in Section 6 in terms of the domain wall structure. Finally the effects of the extrinsic contributions on the electromechanical response are briefly discussed in Section 7.

2. Orientation dependence of the piezoelectric coefficients

The piezoelectric coefficient is a tensor of the third rank that can be transformed between two coordinate systems using the relation [19]:

$$d^*_{ijk}(\varphi, \theta, \psi) = a_{il} a_{jm} a_{kn} d_{lmn} \tag{1}$$

where φ, θ, ψ are the Euler angles and a_{ij} are the elements of the Euler matrix that describes the rotation defined by the Euler angles. In this text d is the tensor of piezoelectric coefficients in the crystallographic coordinate system with axis x_1, x_2 and x_3, while the asterisk denotes the tensor in the rotated system. A caution is necessary when comparing data from different sources since Euler angles are not uniquely defined in the literature. Here we follow the definition of Euler angles given in Refs. [2 , 20] where φ describes the first counterclockwise rotation around the x_3 axis, θ the second counterclockwise rotation around the new x_1 axis and ψ the third counterclockwise rotation around the new x_3 axis. The corresponding rotation matrix is:

$$a = \begin{pmatrix} \cos\psi\cos\varphi - \cos\theta\sin\varphi\sin\psi & \cos\psi\sin\varphi + \cos\theta\cos\varphi\sin\psi & \sin\psi\sin\theta \\ -\sin\psi\cos\varphi - \cos\theta\sin\varphi\cos\psi & -\sin\psi\sin\varphi + \cos\theta\cos\varphi\cos\psi & \cos\psi\sin\theta \\ \sin\theta\sin\varphi & -\sin\theta\cos\varphi & \cos\theta \end{pmatrix} \tag{2}$$

If another definition of the Euler angles is used, the terms in equations given below may have different signs [21, 22] and numerical coefficients. Furthermore, the piezoelectric tensor is usually given in the form of a matrix with reduced indices using Voigt notation [19]. A mistake sometimes made in the literature is to use for the piezoelectric d coefficients the reduced matrix of the third rank electro-optical tensors or the piezoelectric stress tensor e [23]. The reduced matrices for these tensors may differ from the matrix for the d tensor in the relationship between the coefficients. For example, in point group *3m* in the reduced notation $d_{26} = -2d_{11}$ but $e_{26} = -e_{11}$ [24].

In this article we shall discuss only the orientation dependence of the longitudinal d_{33} piezoelectric coefficient. This coefficient has the simplest form and is the easiest to analyze, yet allows a point to be made about the main features of anisotropy that will be discussed. In the most common symmetries and for the chosen definition of the Euler angles, the Equation 1 has the following forms:

in crystals belonging to the tetragonal *4 mm* group:

$$d^{t*}_{33}(\theta) = \cos\theta\left(d^t_{15}\sin^2\theta + d^t_{31}\sin^2\theta + d^t_{33}\cos^2\theta\right) \tag{3}$$

in the orthorhombic *mm2* group:

$$d^{o*}_{33}(\theta, \phi) = \cos\theta\left[\left(d^o_{15} + d^o_{31}\right)\sin^2\theta\sin^2\phi + \left(d^o_{24} + d^o_{32}\right)\sin^2\theta\cos^2\phi + d^o_{33}\cos^2\theta\right)\right] \tag{4}$$

and in the rhombohedral *3m* group:

$$d^{r*}_{33}(\theta, \phi) = d^r_{15}\cos\theta\sin^2\theta - d^r_{22}\sin^3\theta\cos 3\phi + d^r_{31}\sin^2\theta\cos\theta + d^r_{33}\cos^3\theta \tag{5}$$

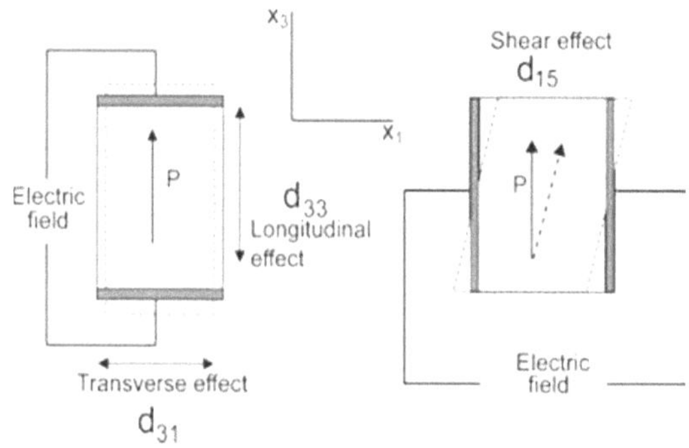

Figure 1 Illustration of longitudinal, transverse and shear piezoelectric effects in a tetragonal material and piezoelectric ceramic. The dark regions at the sample edges represent electrodes.

Indices *t, o* and *r* refer to the crystal symmetries. In the rest of the text terms 'tetragonal', 'orthorhombic' and 'rhombohedral' will refer to point groups *4mm*, *mm2*, and *3m*. Definitions of the longitudinal, transverse and shear coefficients are given in Fig. 1.

3. Piezoelectric anisotropy in the proximity of temperature induced ferroelectric-ferroelectric phase transitions

Let us consider $BaTiO_3$ mono domain single crystals. This material is of interest since it undergoes a series of phase transitions (cubic→tetragonal→orthorhombic→rhombohedral) as it is cooled from the paraelectric phase [25]. These crystal phases are typical of perovskite ferroelectrics and this fact, together with its chemical simplicity, makes $BaTiO_3$ convenient for modeling purposes. The temperature dependences of piezoelectric coefficients of $BaTiO_3$ are experimentally available only for the tetragonal phase [26], however, the coefficients of the Gibbs free energy expansion are known so that piezoelectric coefficients can be predicted as a function of temperature in all three phases, as shown in Fig. 2. The orientation dependence of piezoelectric coefficients, $d_{33}^{*}(\theta, \varphi)$, can be calculated at any temperature using Equations 3–5, as discussed in Ref. [16]. Behavior of $d_{33}^{*}(\theta, \varphi)$ in the tetragonal, Fig. 3, and the orthorhombic, Fig. 4, phases illustrates how orientation dependence changes with temperature and as the phase transition temperatures are approached.

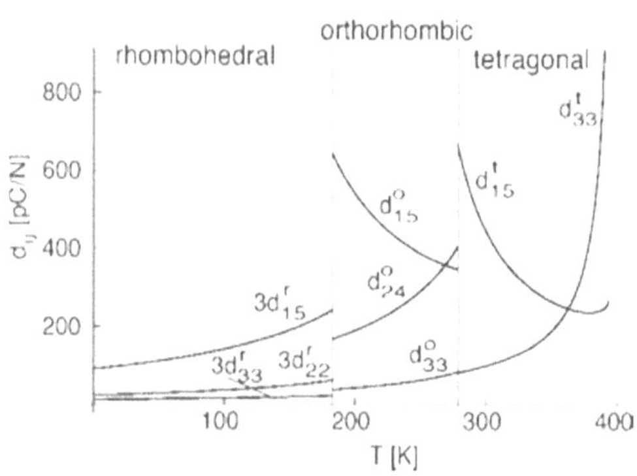

Figure 2 Temperature dependence of piezoelectric coefficients in all three ferroelectric phases of $BaTiO_3$ predicted by the LGD theory. Parameters of the LGD function were taken from Ref. [27].

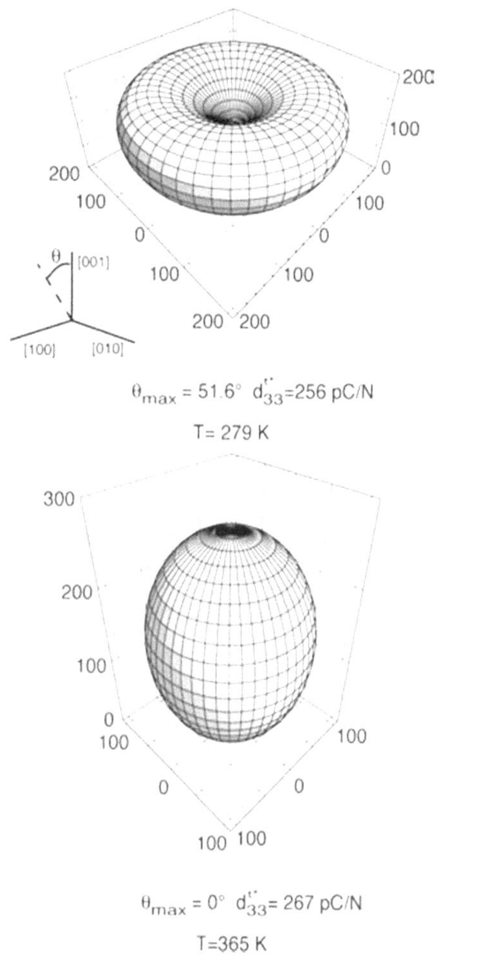

Figure 3 Orientation dependence of $d_{33}^{t*}(\theta)$ in the tetragonal phase of $BaTiO_3$ at two temperatures, close to the tetragonal-orthorhombic (upper figure) and the tetragonal-cubic (lower figure) phase transition temperature. Numbers on axes indicate values of $d_{33}^{t*}(\theta)$. The coordinate system indicates crystallographic axes.

The evolution of the $d_{33}^{t*}(\theta)$ surface near the tetragonal-orthorhombic and the orthorhombic-rhombohedral phase transitions temperatures can be analyzed and understood on several levels. The simplest approach is by inspecting the Equations 3–4 and Fig. 2 and by analyzing the competing influences of the shear (term with sinθ) and the longitudinal (term with cosθ) piezoelectric coefficients, while neglecting for simplicity the influence of the small transverse coefficient d_{31}^{t}. In the tetragonal phase the $d_{33}^{t*}(\theta)$

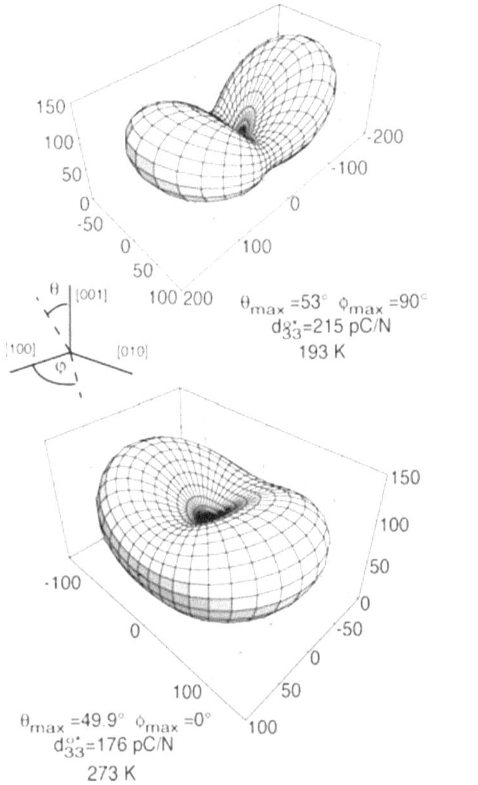

Figure 4 Orientation dependence of $d_{33}^{o*}(\theta)$ in the orthorhombic phase of $BaTiO_3$ at two temperatures, close to the rhombohedral-orthorhombic (upper figure) and the orthorhombic-tetragonal (lower figure) phase transition temperature. Numbers on axes indicate values of $d_{33}^{o*}(\theta)$. The coordinate system indicates crystallographic axes.

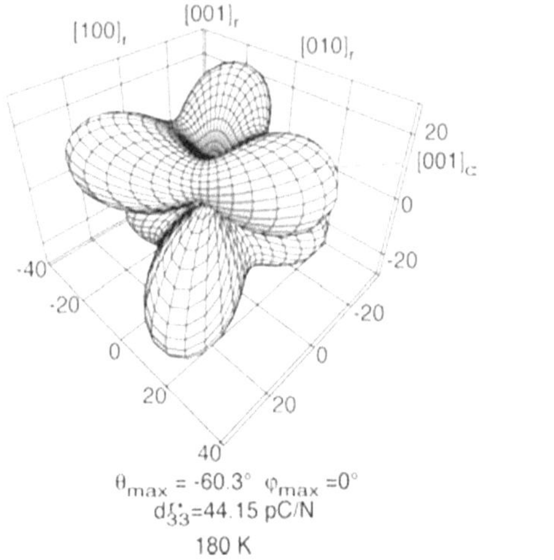

Figure 5 Orientation dependence of $d_{33}^{r*}(\theta)$ in the rhombohedral phase of $BaTiO_3$ at 180 K. The form of the surface does not change as the temperature is decreased. Numbers on axes indicate values of the $d_{33}^{r*}(\theta)$. The orientation of crystallographic axes is indicated by [*hkl*]. The maxiumum $d_{33}^{r*}(\theta)$ lies near the quasicubic $\langle 001 \rangle_c$ direction.

exhibits maximum value along the polar axis as long as the ratio of the shear and longitudinal coefficients, d_{15}^{t}/d_{33}^{t}, is below some critical value [28]. When this is no longer the case (i.e., when d_{15}^{t} is sufficiently high), $d_{33}^{t*}(\theta)$ will exhibit a maximum along a nonpolar direction. As can be expected from Fig. 3 the direction along which $d_{33}^{t*}(\theta)$ exhibits maximum is strongly temperature dependent [28].

More complex behavior is predicted for the orthorhombic phase, Fig. 4, which has two different shear coefficients, one dominating near the orthorhombic-rhombohedral and the other near the orthorhombic-tetragonal phase transition temperature, Fig. 2. In this phase, as is the case for the rhombohedral one, the longitudinal piezoelectric coefficient is always largest along a nonpolar direction. The rotation of the $d_{33}^{o*}(\theta)$ surface with temperature is directly caused by the change of the shear coefficient that dominates expression (4). For details see Ref. [16].

In the rhombohedral phase of $BaTiO_3$, the form of the $d_{33}^{r*}(\theta, \varphi)$ surface is typical of that found in other materials with rhombohedral *3m* symmetry, Fig. 5. The $d_{33}^{r*}(\theta, \varphi)$ surface has its maximum approximately along the [001] pseudocubic direction, just as in complex relaxor ferroelectrics. The form of $d_{33}^{r*}(\theta, \varphi)$ remains constant with temperature, and the anisotropy (the ratio between the maximum $d_{33}^{r*}(\theta, \varphi)$ and d_{33}^{r}along the polar direction) slightly decreases as the temperature is decreased [16].

It is useful at this point to comment on the very large piezoelectric response along off polar axes in relaxor ferroelectrics, such as PMN-PT and PZN-PT. The anisotropy in these materials is giant [14, 15, 23]. For example, the measurements of the properties of monodomain 0.67PMN-0.33 PT crystals show d_{33}^{r} of about 190 pC/N and $d_{15}^{r} > 4000$ pC/N in the crystallographic coordinate system, leading to maximum $d_{33}^{r*}(\theta, \varphi) > 2400$ pC/N along a nonpolar direction [15, 22]. Thus, in the simplest approach [in terms of Equation 3], it is the large d_{15}^{r} which is responsible for the large $d_{33}^{r*}(\theta, \varphi)$ in PMN-PT and PZN-PT along nonpolar directions. Any model of origins of the strongly enhanced longitudinal response along nonpolar directions in these materials should therefore aim to explain the large shear coefficients in the crystallographic coordinate system. From the point of view of the intrinsic behavior, we see that simple perovskites and complex relaxor ferroelectrics behave qualitatively in the same way. The main difference is that the shear coefficients are anomalously large in relaxor ferroelectrics.

To gain a deeper understanding of the origins of the maximum $d_{33}^{*}(\theta, \varphi)$ along nonpolar directions, it useful to look at the thermodynamic relations for the piezoelectric coefficients derived in the framework of the Landau-Ginzburg-Devonshire (LGD) theory [29, 30]. More detailed discussion for all crystal phases and piezoelectric coefficients can be found elsewhere [16]. As an example,

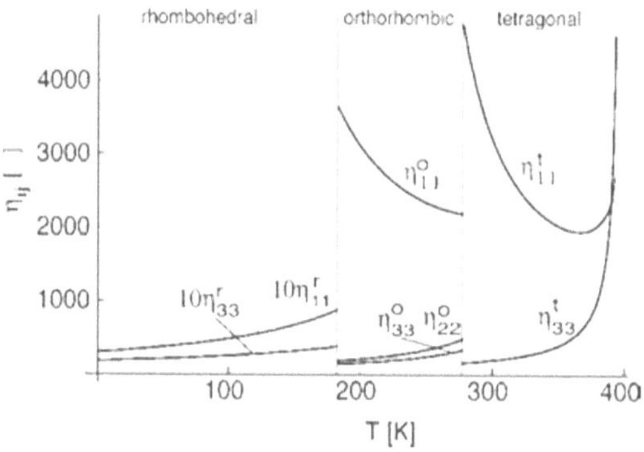

Figure 6 Predicted temperature dependence of dielectric susceptibilities in all three ferroelectric phases of $BaTiO_3$.

only the tetragonal phase is considered here. The longitudinal and the shear coefficients can be expressed as:

$$d^t_{15} = d^t_{24} = \varepsilon_0 \eta^t_{11} Q_{44} P^t_3 \qquad (6)$$

$$d^t_{33} = \varepsilon_0 \eta^t_{33} Q_{11} P^t_3 \qquad (7)$$

where Q are the electrostrictive coefficients, P is the spontaneous polarization and ε_0 the permittivity of vacuum. The temperature dependences of the susceptibilities parallel, η^t_{33}, and perpendicular, η^t_{11}, to the polarization are given in Fig. 6. The increase of the shear coefficient d^t_{15} on the tetragonal side of the tetragonal-orthorhombic phase transition temperature is clearly caused by the anomalous behavior of the η^t_{11}. In analogy to pretransitional behavior in the paraelectric phase where susceptibility follows the Curie Weiss law and diverges as the cubic-tetragonal phase transition temperature is approached on cooling, the η^t_{11} susceptibility in the tetragonal phase diverges as the crystal is cooled toward the tetragonal-orthorhombic phase transition temperature. Note that it is the susceptibility η^t_{11} and not η^t_{33} which is anomalous near the tetragonal-orthorhombic phase transition temperature, anticipating the change in the polarization direction from the [001] axis in the tetragonal phase to the pseudocubic ⟨011⟩ axes in the orthorhombic state. The large value of d^t_{15} and the large d^t_{15}/d^t_{33} ratio, essential for having a maximum of $d^{t*}_{33}(\theta)$ along a nonpolar direction, then follow immediately from Equations 6, 7 and 3. Similar arguments can be invoked for the orthorhombic and rhombohedral phases [16].

The association of the large dielectric susceptibility perpendicular to the polarization with the large longitudinal response along nonpolar directions is in agreement with the results of first principle calculations. These studies interpret the enhanced piezoelectric coefficients along off-polar directions in perovskite materials by the large polarization rotation induced by the strong external electric fields [31, 32]. The above discussion shows that the polarization rotation argument is valid in the weak field limit as well: the large η^t_{11} indicates enhanced polarization rotation (dielectrically soft material), and the large polarization rotation is indicative of the large shear piezoelectric coefficients [16] (see also Fig. 1). These results obtained by analyzing behavior of $BaTiO_3$ are directly relevant to relaxor ferroelectrics, which exhibit huge shear coefficients in the crystallographic coordinate system [14, 15 , 22], and to $Pb(Zr,Ti)O_3$ solid solution [33] (see Section 4).

In this context it is interesting to consider the behavior of ferroelectric materials in the absence of any ferroelectric-ferroelectric phase transitions. In $PbTiO_3$, which is believed to be tetragonal at all temperatures below the Curie temperature [34], the phenomenological calculations and the experimental results [35, 36] show that the anisotropy in η^t_{11} and η^t_{33} is much weaker than in the tetragonal $BaTiO_3$ and that $d^t_{15} < d^t_{33}$ at all temperatures. The maximum $d^*_{33}(\theta)$ in this material appears thus always along the polar direction [16 , 28]. Intriguingly, this is not the case for non-perovskite $LiNbO_3$ that possesses rhombohedral *3m* symmetry below and $\bar{3}m$ above the Curie temperature. At room temperature this crystal exhibits a $d^*_{33}(\theta, \varphi)$ surface characteristic of rhombohedral perovskites (similar to Fig. 5) with the maximum $d^*_{33}(\theta, \varphi)$ approximately along the [001] pseudocubic direction and not along the polar [111] pseudocubic axis. In contrast to $PbTiO_3$, the shear d_{15} coefficient in $LiNbO_3$ is much larger than d_{33} [2]. This difference could be related to the crystal structure of $LiNbO_3$ that can be considered as a modified perovskite structure [37], but which may be softer then the proper perovskite structure under shear fields. It would be interesting to see if there are any rhombohedral materials that exhibit the qualitative change in $d^*_{33}(\theta, \varphi)$ with temperature analogous to that predicted for tetragonal $BaTiO_3$.

A further step in understanding the origins of the piezoelectric anisotropy in the framework of the phenomenological thermodynamic theory is to analyze changes of the free energy function as the crystal is cooled through the successive phase transitions. The large longitudinal piezoelectric effect along nonpolar directions can be related to the flat Gibbs free energy G near the phase transition temperatures, as shown in Fig. 7. In tetragonal phase polarization is oriented along [001] axis, i.e., $P_3 \neq 0$, $P_1 = P_2 = 0$ while in the orthorhombic phase $P_3 = P_2 \neq 0$, $P_1 = 0$. In Fig. 7 the Gibbs energy is plotted as a function of P_2 for the same temperatures as in Fig. 3, and for P_3 values corresponding to the minimum of G at each temperature and giving stable tetragonal phase. The free energy becomes flatter as the tetragonal-orthorhombic phase transition is approached on cooling, indicating incipient orthorhombic phase and onset of minima in G at $P_2 \neq 0$ at lower temperatures. The flat G in the (100) plane indicates a large $1/\eta_{ii} = \partial^2 G/\partial P_i^2$ i.e., a large susceptibility η_{11} ($=\eta_{22}$) along [100] and [010] axes (compare

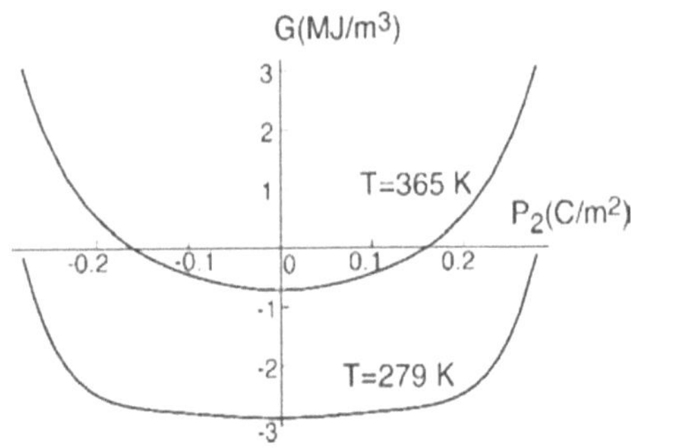

Figure 7 Gibbs free energy for tetragonal $BaTiO_3$ at $T = 365$ K and 279 K (compare with Fig. 3). The energy is presented as a function of polarization P_2 along [010] axis, with polarization P_3 along [001] axis fixed at $P_3 = 0.267$ C/m^2 at T=279 K, and fixed at $P_3 = 0.242$ C/m^2 at $T = 365$ K. The minimum of G at $P_2 = 0$ indicates stable tetragonal phase. The orthorhombic phase is obtained for $P_2 = P_3$. The flat energy near the orthorhombic-tetragonal phase transition temperature ($T = 279$ K) indicates large dielectric susceptibility η_{11}^t, (compare with Fig. 6) and easy polarization rotation in the P_2-P_3 plane.

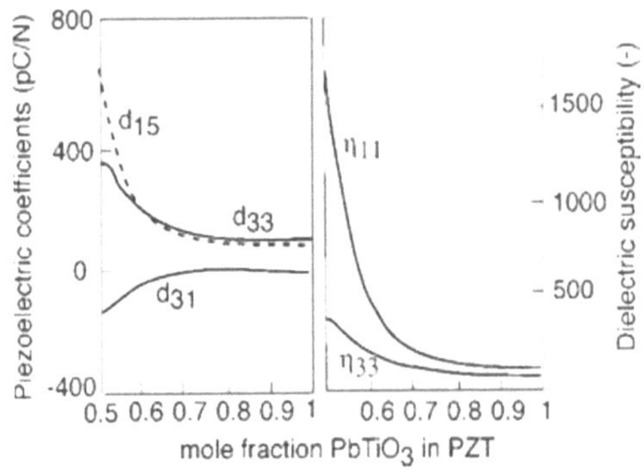

Figure 8 Piezoelectric coefficients and dielectric susceptibilities in $Pb(Zr,Ti)O_3$ crystals calculated using the LGD theory as a function of composition, for the tetragonal side of the morphotropic phase boundary. After [48].

with Fig. 6). As already discussed above, the large susceptibility perpendicular to the polar direction signifies enhanced polarization rotation and large shear piezoelectric coefficients (compare with Figs 2 and 6). A similar approach can be also applied for the free energy evolution as a function of the composition in solid solutions [38–42], and as a function of external fields [27]. Another approach is to use first principle calculations, and to attempt to explain the dielectric, elastic and piezoelectric properties of the perovskite structure in terms of ionic and electronic contributions to polarization [31, 43–47].

4. Piezoelectric anisotropy-composition relationships in materials exhibiting a morphotropic phase boundary

The previous section discussed enhancement of the piezoelectric anisotropy as a function of the proximity of the phase transition temperatures. In analogy, one can expect to observe similar effects in the vicinity of the compositionally driven phase changes in materials exhibiting a morphotropic phase boundary (MPB) [42]. Fig. 8 shows the evolution of d_{33}^t, d_{31}^t, and d_{15}^t piezoelectric coefficients and η_{11}^t and η_{33}^t dielectric susceptibilities in single crystals of $Pb(Zr,Ti)O_3$ (PZT) solid solution, as a function of composition, on the tetragonal side of the MPB. In this diagram, which is derived using the LGD thermodynamic phenomenological theory, the MPB was chosen to appear at a 50:50 Ti/Zr ratio [48]. At compositions close to the MPB, where polarization changes direction from the [001] axis (tetragonal side) to the [111] quasicubic direction (rhombohedral side), d_{15}^t and η_{11}^t strongly increase, just as in tetragonal $BaTiO_3$ near the tetragonal-orthorhombic phase transition temperature (compare Fig. 8 with Figs 2 and 6). Note that coefficients of the LGD function for PZT are determined experimentally, that is, they are independent on the recent discovery of the monoclinic distortion in PZT, except in the region very close to the MPB where rhombohedral and tetragonal symmetries are assumed. Values of coefficients will change if measured more precisely and if presence of the monoclinic phase was taken into account, however, trends indicated in Figs 7 to 9 should qualitatively remain the same, especially for compositions not lying too close to the MPB.

While in the case of tetragonal PZT the d_{15}^t fails to reach a high enough value [28] to bring the maximum $d_{33}^{t*}(\theta)$ away from the [001] direction, the general tendency is similar to that in the tetragonal $BaTiO_3$, as illustrated in Fig. 9: as the Zr content increases and the crystal approaches the MPB, the $d_{33}^{t*}(\theta)$ becomes flatter at the top, indicating tendency toward developing a local minimum along the [001] axis and a maximum away from this direction. As mentioned above, it is very likely that the coefficients of the Gibbs energy are not optimized in the case of PZT and for this reason the maximum in the calculated $d_{33}^{t*}(\theta)$ is not obtained along a nonpolar direction at compositions very close to the MPB.

A similar tendency is calculated for the rhombohedral side of the MPB in PZT. Fig. 10 shows the $d_{33}^{r*}(\theta, \varphi)$ surface and its cross section for two rhombohedal compositions with Zr/Ti ratio 90/10 and 60/40, at room temperature. The composition 90/10, which lies well into the rhombohedral side of the phase diagram, demonstrates a reduced piezoelectric anisotropy with respect to the composition 60/40 that is close to the MPB. Unfortunately, PZT crystals are not available and these theoretical predictions cannot be verified experimentally. However, reports on the piezoelectric properties of highly textured thin films do show a maximum in $d_{33}^{r*}(\theta, \varphi)$ approximately along the quasicubic [001] direction in 60/40 PZT composition and not along the polar quasicubic [111] axis, as shown in Fig. 11 [49]. Note that the former orientation also exhibits weaker nonlinearity with respect to the

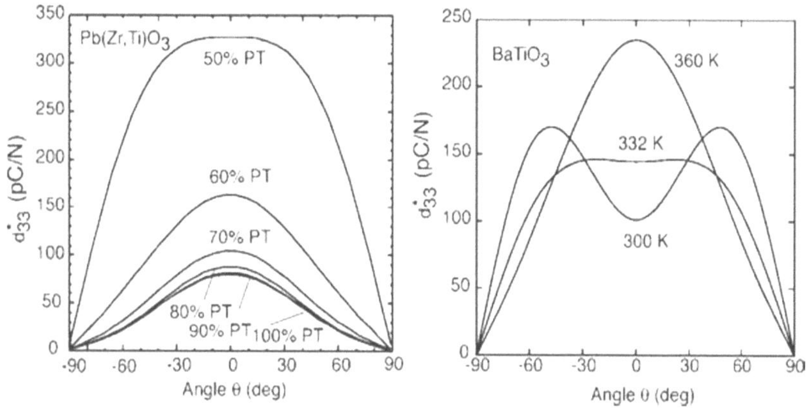

Figure 9 On the left: orientation dependence of $d_{33}^{l*}(\theta)$ for Pb(Zr,Ti)O_3 (PZT) at room temperature as a function of $PbTiO_3$ (PT) concentration, on the tetragonal side of the MPB. On the right: orientation dependence of $d_{33}^{l*}(\theta)$ for tetragonal $BaTiO_3$ at different temperatures. Note similarities between $d_{33}^{l*}(\theta)$ at 60% PT in PZT and at 360 K in $BaTiO_3$; and at 50% PT in PZT and at 332 K in $BaTiO_3$. Data in both figures were calculated for mono domain single crystals using the LGD approach. LGD parameters for PZT were taken from Ref. [48].

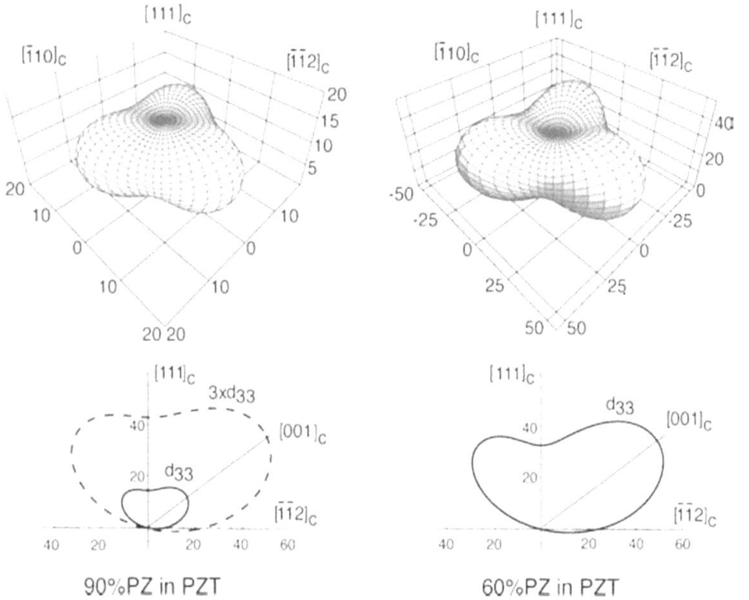

Figure 10 Orientation dependence of $d_{33}^{l*}(\theta, \varphi)$ for two rhombohedral compositions of PZT, one containing 90%$PbZrO_3$ (PZ) and the other 60%PZ. Cross section of the surfaces is shown in the bottom part of the figure. Note the reduced anisotropy in the composition with 90% PZ, which lies deep into rhombohedral part of the phase diagram. Dashed curve for this composition shows data magnified by factor three, for better comparison with the composition containing 60% PZ. Data were calculated at room temperature, for mono domain single crystals using the LGD approach. The LGD parameters were taken from Ref. [48].

driving field amplitude, which is consistent with the small contribution of domain walls to the piezoelectric effect. (Sections 6 and 7).

5. Effect of external electric fields on piezoelectric anisotropy

The effect of an electric field on the piezoelectric properties of perovskite crystals has been recently examined in $BaTiO_3$ by first principle calculations [31] and LGD theory [27], and in PZT by first principle calculations [32]. Results show that the application of the field along nonpolar directions enhances $d_{33}^{*}(\theta, \varphi)$ and other piezoelectric coefficients. In the case of very large fields, the field induced phase transitions are accompanied by huge shear piezoelectric coefficients, and thus a large $d_{33}^{*}(\theta, \varphi)$ along nonpolar directions [32].

These results and the discussion in previous sections suggest that the enhanced $d_{33}^{*}(\theta, \varphi)$ along nonpolar direc-

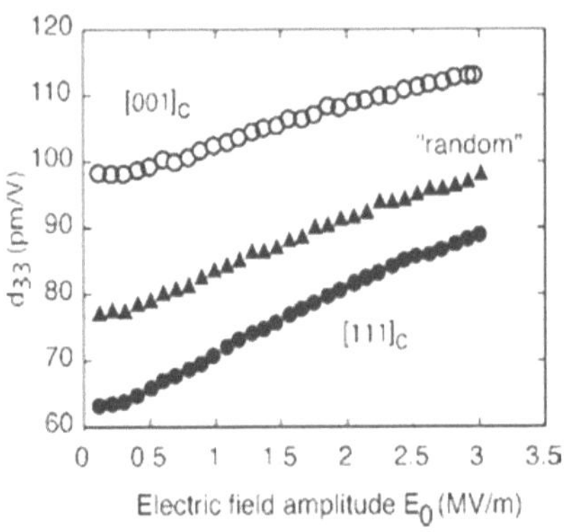

Figure 11 The longitudinal piezoelectric coefficient in textured 60PZ/40PT thin films as a function of the driving electric field amplitude. The largest response is observed in films with $[001]_c$ preferential orientation, in qualitative agreement with results predicted for single crystals. The relatively smaller piezoelectric nonlinearity of the sample with $[001]_c$ orientation is consistent with reduced contribution of domain walls for this orientation. For details see Ref. [49].

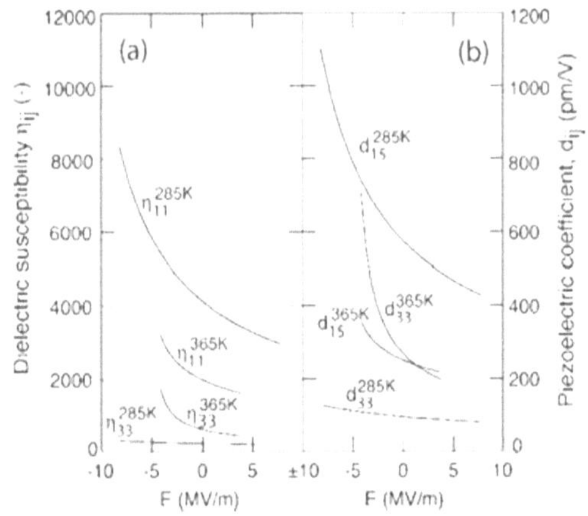

Figure 12 (a) Dielectric suscpetibilities and (b)piezoelectric coefficients of the tetragonal $BaTiO_3$ at two temperatures as a function of the electric field bias. The negative field is applied against the spontaneous polarization. The field dependence was calculated using the LGD theory. For details, see Ref. [50].

tions may be associated with instabilities related to incipient phase transitions, which are manifested by dielectric softening in directions perpendicular to the polarization axis. In this section we show that such instabilities may be also induced by application of the electric field bias applied *antiparallel* to polarization, leading to a huge enhancement of the $d^*_{33}(\theta, \varphi)$.

For simplicity, we again consider the tetragonal phase of a mono domain $BaTiO_3$ single crystal under electric bias field applied either along the polar axis [001] or antiparallel to it, i.e. along the $[00\bar{1}]$ direction. The effect of the field on the piezoelectric coefficients and susceptibilities is shown in Fig. 12. Calculation details can be found in Ref. [50]. The orientation dependence of $d^{t*}_{33}(\vartheta)$ can be calculated using Equation 3 and is shown in Fig. 13 for different bias fields at $T = 285$ K and 365 K. The $d^*_{33}(\vartheta)$ exhibits its maximum value for all bias fields at approximately $\vartheta_{max} \approx 50°$, i.e. close to the [111] axis. The value of $d^{t*}_{33}(\vartheta_{max})$ depends strongly on the bias field. For $E_3 = 0$, $d^{t*}_{33}(\vartheta_{max}) = 227\,\mathrm{pm/V}$. While positive bias fields decrease $d^{t*}_{33}(\vartheta_{max})$, the calculations predict that negative bias fields (anti-parallel to polarization) strongly *enhance* the piezoelectric coefficient. A field of −9 MV/m (applied along $[00\bar{1}]$) increases $d^*_{33}(\vartheta_{max})$ to 497 pm/V, which represents more than a five fold increases with respect to the value measured along the polar axis at zero bias field [$d^*_{33}(\vartheta = 0°) = d_{33} = 89\,\mathrm{pm/V}$] and more than a two fold increase with respect to the maximum value measured at $E_3 = 0$ [$d^*_{33}(\vartheta = 50°) = 227\,\mathrm{pm/V}$].

The above results can be again understood by analyzing Equations 6–7 and the field dependences of the susceptibilities and piezoelectric coefficients, shown in Fig. 12. The negative bias fields strongly increase η_{11} and, consequently, lead to a high d_{15} coefficient, which is then directly responsible for a maximum of $d^*_{33}(\vartheta)$ along a nonpolar direction. In contrast to earlier studies where strong bias fields were always applied along nonpolar directions, these results show that a weak-field polarization rotation may be facilitated by application of strong bias fields anti-parallel to polarization. It has been shown in Section 3 that the proximity of the tetragonal-orthorhombic phase transition temperature by itself leads to a maximum $d^{t*}_{33}(\vartheta_{max})$ approximately along the [111] axis, however, we now show that this effect is enhanced considerably by anti-parallel bias fields.

Interesting results are also observed at higher temperatures, closer to the tetragonal-cubic phase transition that occurs at 393 K. At these temperatures the condition for having a maximum in $d^{t*}_{33}(\vartheta_{max})$ along a non-polar direction is not fulfilled [28] and the piezoelectric response exhibits its maximum value along the polar axis. The positive bias field decreases the piezoelectric coefficient, similarly to what has been reported in ferroelectric thin films [51], while the negative field *enhances* the piezoelectric response, Fig. 13. At $E_3 = -4$ MV/m and at 365 K, the maximum $d^*_{33}(\vartheta = 0)$ is 700 pm/V, compared to 268 pm/V at $E_3 = 0$. Note that this field enhancement of the piezoelectric coefficient is not related to the polarization rotation.

It is interesting to note the similarities among the temperature, composition and field dependences of the piezoelectric coefficients in Figs 2, 8 and 12. The common denominator for each case is the increase of the susceptbility i.e., the dielectric softening of the crystal, perpendicular to the polarization. This in turn leads to an increase in the shear piezoelectric coefficients and enhanced $d^{t*}_{33}(\vartheta)$ along nonpolar directions. As discussed above, the dielectric softening is the direct consequence of the flattening of the Gibbs free energy.

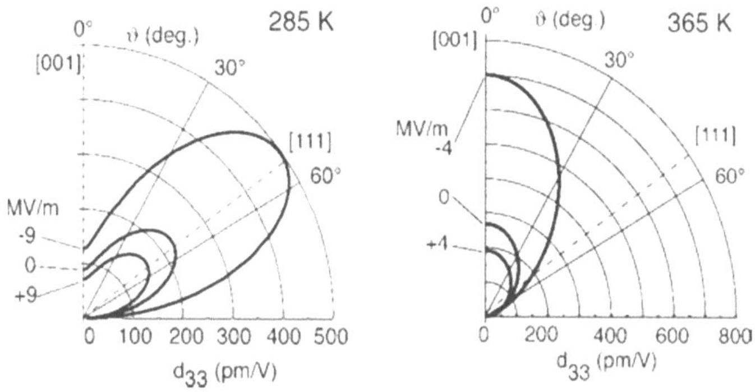

Figure 13 Orientation dependence of $d_{33}^{l*}(\theta)$ for tetragonal $BaTiO_3$ at two temperatures and for different electric fields applied along polarization (positive fields) and antiparellel to polarization (negative fields). The field dependence was calculated using the LGD theory. For details, see Ref. [50].

Finally, one obtains similar effects under applied external stresses [52]. For example, calculations on thin films show a large increase in the shear piezoelectric coefficients due to misfit strains [53] which reflects on the orientation dependence of piezoelectric coefficients.

6. Effect of domain wall structure on piezoelectric anisotropy

In previous sections we have discussed the orientation dependence of piezoelectric properties assuming that the crystal is in the mono domain state. Practically, the mono domain state is difficult to achieve in obliquely cut crystals, and in relaxor ferroelectrics the monodoman state appears to be unstable [14]. When a crystal is cut and poled along special directions so that the resulting domain states are equivalent and the piezoelectric properties are enhanced, the resulting domain structure is often called the "engineered domain state" [7]. Clearly there can be more than one "engineered domain state" for a given symmetry. In tetragonal crystals with polarization along the [001] axis, one of these states is obtained by poling the crystal along one of the ⟨111⟩ axes. In rhombohedral materials with polarization along the ⟨111⟩ quasicubic directions, an engineered domain state can be obtained by poling the crystal along the [001] quasicubic axis. In this case, four equivalent domain states are defined by polarization vectors oriented along the four equivalent quasicubic directions: [111], [$\bar{1}$11], [1$\bar{1}$1] and [$\bar{1}\bar{1}$1], and are shown in Fig. 14. With respect to an electric field applied along the poling direction, i.e. along the [001] quasicubic axis, these rhombohedral domain states are energetically equivalent and the associated domain walls are expected to be stable. Likewise, this rhombohedral domain-engineered state should be stable to stress: uniaxial pressure applied along the [001] psuedocubic direction will not favor any one of the four domain states.

The question is posed whether the high properties in poled relaxor ferroelectric crystals are a simple consequence of the material anisotropy, or whether the "engineered domain state" somehow further enhances the properties along nonpolar directions beyond what is expected from the intrinsic anisotropy. Considering the equivalence of the engineered domain states, the associated domain walls should be immobile and should not contribute to the piezoelectric properties. An indication that this is indeed so is found in the anhysteretic converse piezoelectric response of relaxor ferroelectric single crystals with engineered domain states (see [7] and Section 7) and in the reduced nonlinearity of piezoelectric properties in highly textured thin films, Fig. 11 [49].

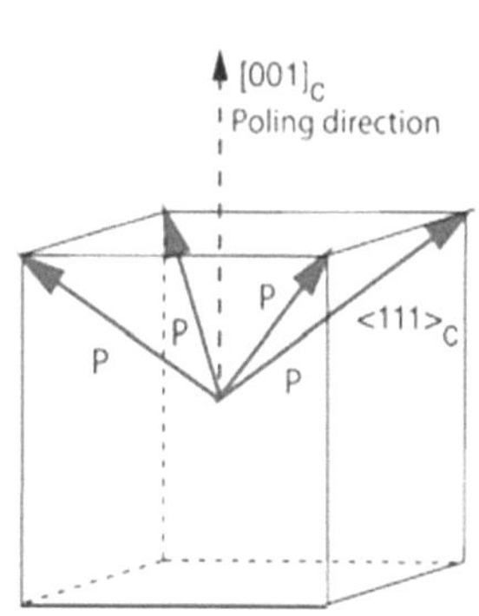

Figure 14 Schematic representation of the equivalent polarization directions in a rhombohedral *3m* crystal poled along quasicubic $[001]_c$ direction. For more examples and discussion see Ref. [54].

This question was first addressed by Nakamura et al. [11, 55]. They showed that in tetragonal $BaTiO_3$ and orthorhombic $KNbO_3$ crystals the piezoelectric response in crystals with an engineered domain state is approximately the same as in mono domain crystals cut and measured along the same direction. Later, the same was shown [15, 22] for $0.67Pb(Mg_{1/3}Nb_{2/3})O_3$-$0.33PbTiO_3$ crystals although this case is perhaps more controversial. Firstly, as a function of preparation and poling conditions, this latter composition can exhibit several different crystal structures [56] and secondly, in order to stabilize the mono domain state, the properties of the crystal had to be measured under an electric bias field [15]. It is interesting to note that in all cases mentioned above the response of monodomain samples along the nonpolar directions was calculated and not measured.

Only recently, Wada *et al.* [54, 57] have carried out a systematic study of effects of engineered domain structure on properties of perovskite crystals. They showed that in $BaTiO_3$ and $KNbO_3$ crystals the transverse piezoelectric response of crystals with an engineered domain structure depends on the density of domain walls, while the domain structure remains the same. The piezoelectric d_{31} coefficient of crystals with a coarse domain structure are similar to those of monodomain crystals measured along the same crystallographic direction. However, as the domain structure becomes finer, the response of the multidoman crystal becomes significantly higher than that of the monodomain sample. Considering the supposed equivalence of the domain states in crystals with engineered domain states this result, while significant, is unexpected and puzzling.

It is interesting to speculate about possible origins of the enhancement of piezoelectric response in crystals with dense, engineered domain walls if the contribution of domain wall displacement is, as usually assumed, excluded. One possibility is that as the number of domain walls increases, the material within the domains is no longer the same, i.e. its properties are different to those of the mono domain crystal or coarse-structured material. This can happen because with a denser domain wall structure the compensation of charges or stresses at the domain walls [58] may require more material to be involved in the compensation process; alternatively, internal fields associated with the compensation process may be stronger in crystal with finer domains. The previous sections have shown that antiparallel electric fields can enhance the intrinsic anisotropy. Additionally, the domain walls present, in a sense, a phase transition between two states; again, the transition related instabilities, discussed in previous sections, may lead to enhanced dielectric and piezoelectric properties.

7. Piezoelectric anisotropy and domain wall displacement

One of the most interesting features of crystals with "engineered domain states" is the absence of hysteresis in the piezoelectric strain-electric field relation, as illustrated in Fig. 15 for a rhombohedral crystal with nominal composition 0.67PMN-0.33PT. In ferroelectric materials the strain-electric field hysteresis is usually associated with the movement of domain walls. As mentioned in the previous section, in tetragonal and rhombohedral crystals with the special engineered domain structure the domain walls are not expected to move and their response should therefore be anhysteretic, as shown in Fig. 15.

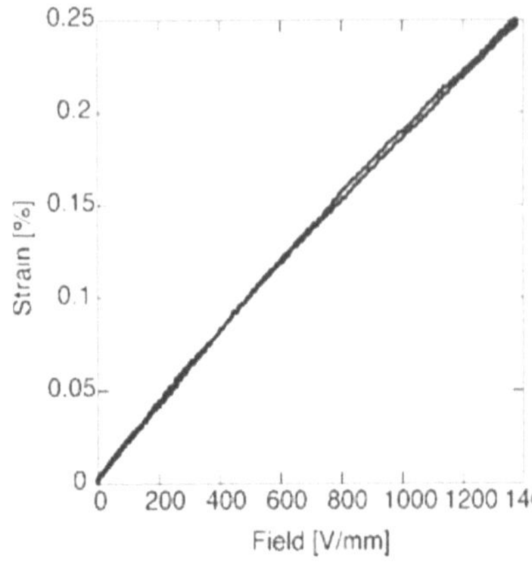

Figure 15 Anhysteretic strain-electric field relation in a 0.67PMN-0.33PT crystal poled and measured with unipolar field applied along the [001] pseudocubic direction.

It has, however, been reported that the strain-electric field relationship becomes hysteretic under uniaxial compressive stress [59], indicating a possible contribution from the moving domain walls. This fact does put into doubt the widely accepted picture, Fig. 14, of the engineered domain structure with equivalent sets of domain walls whose motion will not lead to sample size change. Note that, in the first approximation, the associated domain structure should be insensitive to both uniaxial pressure and electric field applied along the [001] quasicubic direction.

An investigation of the direct piezoelectric effect [60] has shown that in several compositions of PZN-PT and PMN-PT crystals poled along the [001] quasicubic orientation the longitudinal piezoelectric charge-force response is hysteretic and nonlinear, Fig. 16. Pressed crystals will even (partially) depole at sufficient uniaxial stresses. These experimental results may again be incompatible with the simple, usually assumed picture of the engineered domain structure. One can speculate that the absence of hysteresis in the converse effect may not necessarily be due to the presence of energetically equivalent engineered domain states, but rather due to pinned domain walls, which can be released by uniaxial compressive field, but not by unipolar electric field applied along the poling direction. If this is true then the simple picture of equivalent domain states no longer holds, and the actual domain wall structure in crystals with engineered domain states is more complex than is usually assumed.

Interestingly, in some cases (e.g., 0.68PMN-0.32PT and 0.045PZN-0.955PT) it was observed that the direct transverse piezoelectric response could be anhysteretic even when the longitudinal response is hysteretic, Fig. 17. The difference between the two cases is that in the transverse response the compressive static and dynamic pressures are applied perpendicular to the poling direction, whereas in

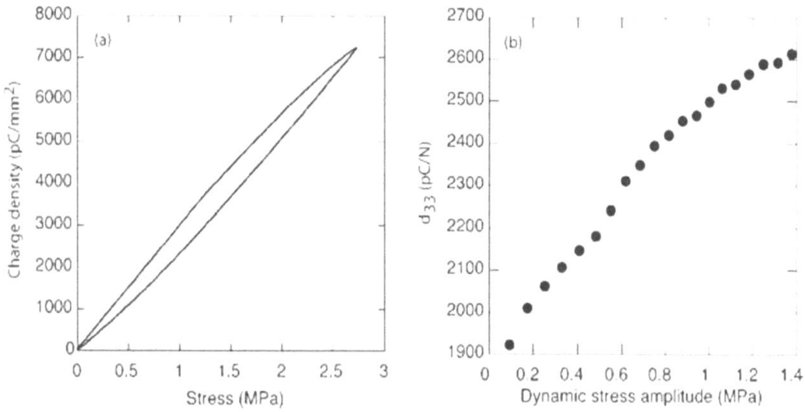

Figure 16 (a) Hysteretic charge density-stress relation and (b) field dependence of the piezoelectric coefficient in 0.67PMN-0.33PT single crystal poled and measured along the [001] quasicubic axis. For details see Ref. [60].

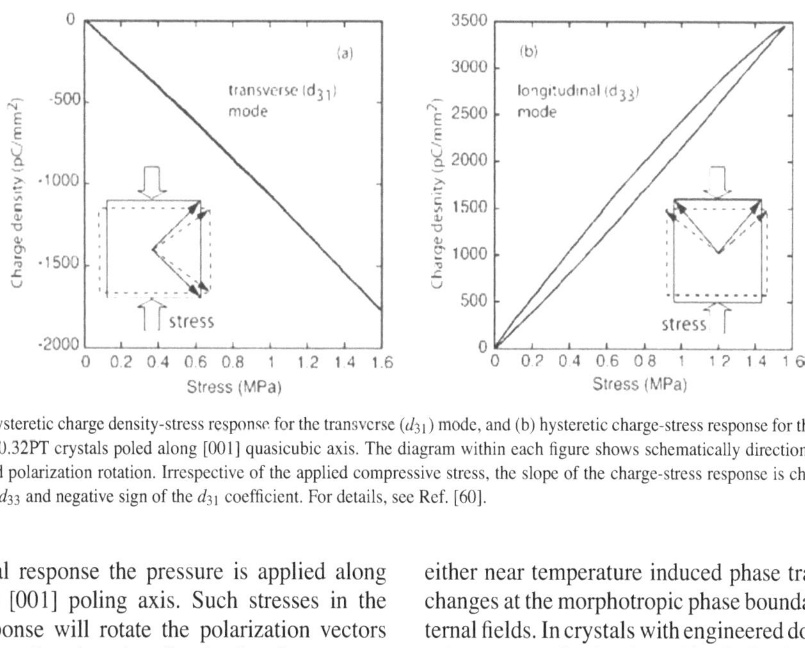

Figure 17 (a) Anhysteretic charge density-stress response for the transverse (d_{31}) mode, and (b) hysteretic charge-stress response for the longitudimal (d_{33}) mode in 0.68PMN-0.32PT crystals poled along [001] quasicubic axis. The diagram within each figure shows schematically direction of applied dynamic stress and presumed polarization rotation. Irrespective of the applied compressive stress, the slope of the charge-stress response is chosen to illustrate the positive sign of the d_{33} and negative sign of the d_{31} coefficient. For details, see Ref. [60].

the longitudinal response the pressure is applied along the quasicubic [001] poling axis. Such stresses in the transverse response will rotate the polarization vectors toward the poling direction, therefore having the same action as an electric field applied along the [001] axis. In the longitudinal case, however, compressive stresses will rotate polarization vectors away from the poling direction (see diagrams in Fig. 17). In either case the rotation direction, or indeed the rotation path, appears to be important. This apparent anisotropy in the extrinsic piezoelectric behavior should be verified for other perovskite materials.

8. Conclusions

Intrinsic piezoelectric anisotropy, manifest in the strong orientation dependence of the piezoelectric coefficients, is present in many simple and complex perovskites. The anisotropy can be associated to instabilities in crystals, either near temperature induced phase transitions, phase changes at the morphotropic phase boundary, or under external fields. In crystals with engineered domain states, the anisotropy can be dominated by the intrinsic anisotropy of the monodomain crystals if the domain structure is coarse. However, it appears that the anisotropy can be further enhanced when the poled crystal possesses a fine engineered domain structure.

The usual picture of the simple domain structure in crystals with engineered domain state and immobile domain walls appears to be questionable. Under external compressive pressure both the longitudinal converse and the direct piezoelectric effect become hysteretic suggesting that the domain walls are mobile. In contrast, the stress-free converse piezoelectric effect with a unipolar electric field applied along the poling axis, and, at least in some cases, the direct piezoelectric transverse effect where compressive pressure is applied in a direction perpendicular to the poling axis, are anhysteretic.

References

1. J. C. BRICE, *Rev. Mod. Phys.* **57** (1985) 105.
2. T. IKEDA, "Fundamentals of piezoelectricity" (Oxford Unversity Press, Oxford, 1990).
3. H. TAKEUCHI, S. JYOMURA, E. YAMAMOTO and Y. ITO, *J. Acoust. Soc. Am.* **72** (1982) 1114.
4. D. DAMJANOVIC, T. R. GURURAJA and L. E. CROSS, *Am. Ceram. Soc. Bull.* **66** (1987) 699.
5. T. TAKENAKA, K. SAKATA and K. TODA, *Jpn. J. Appl. Phys.* **24** (1985) 730.
6. R. NEWNHAM, D. P. SKINNER and L. E. CROSS, *Mat. Res. Bull.* **13** (1978) 52.
7. S. E. PARK and T. R. SHROUT, *J. Appl. Phys.* **82** (1997) 1804.
8. J. KUWATA, K. UCHINO and S. NOMURA, *Jpn. J. Appl. Phys.* **21** (1982) 1298.
9. M. KAHN, R. P. INGEL and D. LEWIS III, *Ferroelectrics* **102** (1990) 225.
10. S. WADA, S. SUZUKI, T. NOMA, T. SUZUKI, M. OSADA, M. KAKIHANA, S.-E. PARK, L. E. CROSS and T . R. SHROUT, *Jpn. J. Appl. Phys.* **38** (1999) 5505.
11. K. NAKAMURA, T. TOKIWA and Y. KAWAMURA, *J. Appl. Phys.* **91** (2002) 9272.
12. S. ZHANG, C. A. RANDALL and T. R. SHROUT, *ibid.* **95** (2004) 4291.
13. S. ZHANG, P. W. REHRIG, C. A. RANDALL and T. R. SHROUT, *J. Crystal Growth* **234** (2002) 415.
14. R. ZHANG, B. JIANGand W. CAO, *Appl. Phys. Lett.* **82** (2003) 787.
15. *Idem. ibid.* **82** (2003) 3737.
16. M. BUDIMIR, D. DAMJANOVIC and N. SETTER, *J. Appl.Phys.* **74** (2003) 6753.
17. B. NOHEDA, D. E. COX, G. SHIRANE, J. A. GONZALO, L. E. CROSS and S.-E. PARK, *Appl. Phys. Lett.* **74** (1999) 2059.
18. E. M. SABOLSKY, A. R. JAMES, S. KWON, S. TROLIER-MCKINSTRY and G. L. MESSING, *ibid.* **78** (2001) 2553.
19. J. F. NYE, "Physical Properties of Crystals," (Oxford University, Oxford, 1985).
20. H. GOLDSTEIN, "Classical Mechanics," (Addison-Wesley, Reading, MA, 1978).
21. D. DAMJANOVIC, M. BUDIMIR, M. DAVIS and N. SETTER, *Appl. Phys. Lett.* **83** (2003) 2490.
22. *Idem. ibid.* **83** (2003) 527.
23. S. ZHANG, L. LAURENT, S. LIU, S. RHEE, C. A. RANDALL and T. R. SHROUT, *Jpn. J. Appl. Phys.* **41** (2002) L1099.
24. Y. I. SIROTIN and M. P. SHASKOLSKAYA, "Fundamentals of crystal physics," (Mir, Moscow, 1982).
25. M. E. LINES and A. M. GLASS, "Principles and Applications of Ferroelectrics and Related Materials," (Clarendon, Oxford, 1979).
26. A. GARCIA and D. VANDERBILT, *Appl. Phys. Lett.* **72** (1998) 2981.
27. A. J. BELL, *J. Appl. Phys.* **89** (2001) 3907.
28. D. DAMJANOVIC, F. BREM and N. SETTER, *Appl. Phys. Lett.* **80** (2002) 652.
29. A. F. DEVONSHIRE, *Phil. Mag.* **40** (1949) 1040.
30. *Idem., ibid.* **42** (1951) 1065.
31. H. FU and R. E. COHEN, *Nature* **403** (2000) 281.
32. L. BELLAICHE, A. GARCIA and D. VANDERBILT, *Phys. Rev. B* **64** (2001) 060103.
33. R. GUO, L. E. CROSS, S.-E. PARK, B. NOHEDA, D. E. COX and G. SHIRANE, *Phys. Rev. Lett.* **84** (2000) 5423.
34. J. KOBAYASHI, Y. UESU and Y. SAKEMI, *Phys. Rev. B* **28** (1983) 3866.
35. M. J. HAUN, E. FURMAN, S. J. JANG, H. A. MCKINSTRY and L. E. CROSS, *J. Appl. Phys.* **62** (1987) 3331.
36. Z. LI, M. GRIMSDITCH, X. XU and S.-K. CHAN, *Ferroelectrics* **141** (1993) 313.
37. J. M. HERBERT, "Ferroelectric Transducers and Sensors," (Gordon and Breach, New York, 1982) Vol. 3.
38. Y. ISHIBASHI and M. IWATA, *Jpn. J. Appl. Phys.* **37** (1998) L985.
39. Y. ISHIBASHI and M. IWATA, *Japanese Journal of Applied Physics Part 1-Regular Papers Short Notes & Review Papers* **38** (1999) 800.
40. M. IWATA and Y. ISHIBASHI, *Jpn. J. Appl. Phys. Part 1* **38**(9B), (1999) 5670.
41. M. IWATA and Y. ISHIBASHI, *Japanese Journal of Applied Physics Part 1-Regular Papers Short Notes & Review Papers* **39** (2000) 5156.
42. M. IWATA, H. ORIHARA and Y. ISHIBASHI, *Ferroelectrics* **266** (2002) 57.
43. D. VANDERBILT, *ibid.* **301** (2004) 9.
44. L. BELLAICHE, A. GARCIA and D. VANDERBILT, *Phys. Rev. Lett.* **84** (2000) 5427.
45. L. BELLAICHE and D. VANDERBILT, *ibid.* **83** (1999) 1347.
46. G. SAGHI-SZABO and R. E. COHEN, *Ferroelectrics* **194** (1997) 287.
47. R. E. COHEN, *Nature* **358** (1992) 136.
48. M. J. HAUN, E. FURMAN, S. J. JANG and L. E. CROSS, *Ferroelectrics* **99** (1989) 63.
49. D. V. TAYLOR and D. DAMJANOVIC, *Appl. Phys. Lett.* **76** (2000) 1615.
50. M. BUDIMIR, D. DAMJANOVIC and N. SETTER, *ibid.* **85** (2004) 2890.
51. L. CHEN, V. NAGARAJAN, R. RAMESH and A. L. ROYTBURD, *J. Appl. Phys.* **94** (2003) 5147.
52. M. BUDIMIR, D. DAMJANOVIC and N. SETTER, (2005) Unpublished.
53. V. G. KOUKHAR, N. A. PERTSEV and R. WASER, *Phys. Rev. B* **64** (2001) 214103.
54. S. WADA, H. KAKEMOTO and T. TSURUMI , *Mater. Trans.* **45** (2004) 178.
55. K. NAKAMURA, T. TOKIWA and Y. KAWAMURA, in "Proceedings of the 12th IEEE International Symposium on the Applications of Ferroelectrics," (IEEE Service Center, Honolulu, 2000) Vol. II, pp. 717.
56. Y. LU, D.-Y. JEONG, Z.-Y. CHENG, Q. M. ZHANG, H.-S. LUO, Z.-W. YIN and D. VIEHLAND, *Appl. Phys. Lett.* **78** (2001) 3109.
57. S. WADA, K. MURAOKA, H. KAKEMOTO, T. TSURUMI and H. KUMAGAI, *Jpn. J. Appl. Phys.* **43** (2004) in print.
58. V. Y. TOPOLOV, *J. Phys.-Cond. Matter* **16** (2004) 2455.
59. D. VIEHLAND, L. EWART, J. POWERS and J. F. LI , *J. Appl. Phys.* **90** (2001) 2479.
60. M. DAVIS, D. DAMJANOVIC and N. SETTER, *J. Appl. Phys.* **95** (2004) 5679.

JOURNAL OF MATERIALS SCIENCE **41** (2006) 77–86

Voltage tunable epitaxial $Pb_xSr_{(1-x)}TiO_3$ films on sapphire by MOCVD: Nanostructure and microwave properties

S.K. DEY*, C.G. WANG, W. CAO, S. BHASKAR, J. LI
Department of Chemical and Materials Engineering & Electrical Engineering, Ira A. Fulton School of Engineering, Arizona State University, Tempe, AZ 85287-6006, USA
E-mail: Sandwip.dey@asu.edu

G. SUBRAMANYAM
Department of Electrical and Computer Engineering, University of Dayton, Dayton, OH 45469-0226, USA

Frequency and phase agile microwave components such as tunable filters and phase shifters will require ferroelectric thin films that exhibit a nonlinear dependence of dielectric permittivity (ε_r) with dc electric bias, as well as a high material ($\Delta\varepsilon_r/\tan\delta$) and device (or K-factor in phase shift/dB) figure of merits (FOM). Therefore, voltage tunable $(Pb_{0.3}Sr_{0.7})TiO_3$ (PST) thin films (90–150 nm) on (0001) sapphire were deposited by metalorganic chemical vapor deposition at rates of 10–15 nm/min. The as-deposited epitaxial PST films were characterized by Rutherford backscattering spectroscopy, X-ray methods, field emission scanning electron microscope, high resolution transmission electron microscopy, Raman spectroscopy, and electrical methods (7–17 GHz) using coplanar waveguide test structures. The epitaxial relationships were as follows: out-of-plane alignment of [111] PST//[0001] sapphire, and orthogonal in-plane alignments of $[1\bar{1}0]$ PST//$[10\bar{1}0]$ sapphire and $[\bar{1}\bar{1}2]$ PST//$[1\bar{2}10]$ sapphire. The material FOM and device FOM (or K-factor) at 12 GHz were determined to be 632 and ~13 degrees/dB, respectively. The results are discussed in light of the nanostructure and stress in epi-PST films. Finally, a rational basis for the selection of PST composition, substrate, and process parameters is provided for the fabrication of optimized coplanar waveguide (CPW) phase shifters with very high material and device FOMs. © *2006 Springer Science + Business Media, Inc.*

1. Introduction

Perovskite materials (e.g., $SrTiO_3$ and $Ba_xSr_{(1-x)}TiO_3$) with voltage tunable dielectric behavior (i.e., percent change in dielectric permittivity, ε_r, with dc bias) have been used for components such as phase-shifters (in phased-array antennas for radar) and preselect filters (in receivers for communication and radar) [1, 2]. Such applications require high ε_r for reducing device size, large dielectric tunability for frequency and phase agility, fast polarization response, and low dielectric loss tangent (or high Q) at microwave frequencies [3]. Frequency and phase agile microwave components such as tunable resonators, filters, local oscillators, and phase shifters have been already demonstrated at Ku- and K-band frequencies using ferroelectric thin films and microstrip transmission lines [4–6]. Although microstrip circuits are the most common transmission line component for microwave frequencies, when used in monolithic microwave integrated circuits (MMICs), the ground-plane is difficult to access for shunt connections necessary for active devices. Therefore, a coplanar waveguide (CPW) design is attractive since the ground conductor runs adjacent to the conductor strip, and the ease of monolithic integration. Findikoglu *et al.* demonstrated a tunable filter in CPW configuration with large tunability (>15%), and the electrical tuning resulted in improved filter characteristics [7]. Wilbur *et al.*

*Author to whom all correspondence should be addressed.

0022-2461
DOI: 10.1007/s10853-005-5926-4

[8] studied CPW-based phase shifter circuits in which the frequency and phase agility was achieved through the use of ferroelectric thin films that exhibit a nonlinear dependence of ε_r with dc electric bias. The material figure of merit (FOM), used for tunable microwave phase shifters, is an electrical parameter, which is directly proportional to the tunability and inversely proportional to the loss tangent (tan δ), whereas, the device FOM is characterized by the K-factor (phase shift/dB).

From a materials standpoint, $SrTiO_3$ has to be used at low temperatures, due to its lower incipient Curie transition temperature, T_c, and its structural compatibility with high temperature superconductor thin films. Therefore, to date, $Ba_xSr_{(1-x)}TiO_3$ or BST has been the most widely studied ferroelectric thin-film system for room-temperature microwave applications. However, due to the relatively high tan δ and difficulty in processing stoichiometric BST films, the search of alternative materials has been warranted. Recently, an alternative perovskite material system, $Pb_xSr_{(1-x)}TiO_3$ (or PST), has been explored for potential microwave applications at room temperature [9]. In this PST system, the T_c may be readily controlled by changing x; for $x \sim 0.3$, the $T_c \sim 20°C$ and room temperature operation can be in the Curie-Weiss region. Due to the disappearance of ferroelectric hysteresis at room temperature, bulk $(Pb_{0.3}Sr_{0.7})TiO_3$ is expected to show low dielectric loss in its paraelectric state. Indeed, the low frequency (1 MHz) tan δ and tunability for bulk PST (30/70) were measured to be <0.001 and 70%, respectively [9]. From a structure-property viewpoint, an understanding of the origin and subsequent engineering of the stress or strain is also important since previous studies have reported a strong correlation between the strain and microwave dielectric properties in BST films [10, 11]. In one such study, films deposited in oxygen pressures, varying from 2–100 mTorr, exhibited a range (0.996–1.003) in degree of tetragonal distortion, D (D = ratio of the in-plane to out-of-plane lattice parameters). Interestingly, at microwave frequencies (1–20 GHz), films with the highest material FOM of ~800 were observed in films that had the minimum strain or D = 1 [10].

In this current work, a reproducible metalorganic chemical vapor deposition (MOCVD) process for the deposition of epitaxial PST thin films (at high deposition rates) on sapphire (hexagonal Wurtzite structure) is demonstrated by using a direct liquid injection (DLI) system. Since sapphire has low ε_r (9.4 for c-cut) and tan δ (0.0003 for c-cut), and is available at a reasonable price, it was a substrate of choice. The PST films were first characterized by various techniques (RBS, XRD, FESEM, Raman, and HRTEM) to determine the composition and nanostructure. Next, CPW test structures (i.e., transmission line and resonator structures typical for microwave devices) were fabricated to evaluate the tunability, loss, and material and device FOMs between 7 to 17 GHz. Specifically, the conductor and dielectric losses were extracted using quasi-static conformal mapping techniques [12, 13]. Here, the ε_r, tunability, and tan δ are discussed in light of the nanostructure and stress in epi-PST films on sapphire. Finally, a rational basis for the selection of PST composition, substrate, and process parameters is provided for the fabrication of optimized CPW phase shifters with very high material and device FOMs.

TABLE I MOCVD process parameters for $Pb_xSr_{(1-x)}TiO_3$ thin films on sapphire

Precursors/solvent	0.1 M $Pb(thd)_2$ in C_4H_8O (THF)
	0.1 M $Sr(thd)_2$ in C_4H_8O
	0.3 M $Ti(^tOBu)_2(thd)_2$ in C_4H_8O
	thd = $O_2C_{11}H_{19}$; ^tBu = tert-C_4H_9
Deposition parameters	
Shimadzu DLI flow rate (Pb, Sr, Ti)	0.1–0.2, 0.1–0.3, 0.5–0.7 cc/min
Carrier gas (Ar) flow rate	300 sccm
Oxygen flow rate	1000–2000 sccm
Vaporizer temperature	250°C
Vaporizer pressure	3–8 Torr
Reactor pressure	~1 Torr
Substrate temperature	500–700°C
Deposition rates	10–15 nm/min

2. Experimental

2.1. Processing of epitaxial $Pb_xSr_{(1-x)}TiO_3$ films on sapphire by MOCVD

The precursors and MOCVD processing parameters are tabulated in Table I. For the deposition of epi-$Pb_xSr_{(1-x)}TiO_3$ thin films (90–150 nm) on 2-inch (0001) sapphire (Saint-Gobain Crystals & Detectors), $Pb(thd)_2$, $Ti(^tOBu)_2(thd)_2$, and $Sr(thd)_2$, dissolved in THF solvent and contained in three metal containers, were used as the liquid precursors. A Shimadzu direct liquid injection (DLI) system was used to deliver these precursors, at controlled flow rates, to a mixer and then into a vaporizer. A high purity Ar carrier-gas delivered the vaporized precursors into the reactor via a showerhead, while a separate line was used to introduce oxygen for reaction with the precursors near the substrate surface.

2.2. RBS, FESEM, Raman

The Rutherford backscattering spectroscopy (RBS) was performed for composition analysis using a 1.7 MV Tandetron Accelerator (General Ionex Corp.). The morphology of the films was analyzed using a Hitachi S4700 field emission scanning electron microscope (FESEM). For the Raman spectra of PST and $PbTiO_3$ films, the 514.5 nm line of an Ar^+ laser (2020, Spectra-Physics, Mountain View, CA) was used. The laser beam was focused onto the sample surface, using a 50X objective (Mitutoyo, Japan), to a spot size of ~3–4 μm. The spectra were recorded using a single-stage monochromator

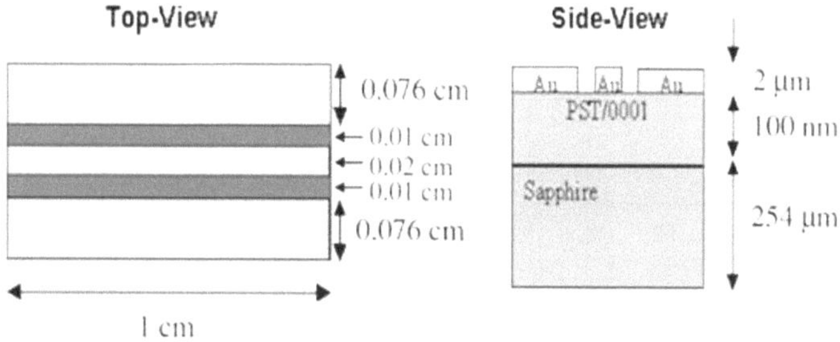

Figure 1 Coplanar waveguide (CPW) transmission line to determine the K-factor or figure of merit as phase shift/dB loss.

(HR640, ISA Instruments, Cedex, France) with a liquid-nitrogen-cooled CCD detector, and a filter (SuperNotch, Kaiser Optical Systems, Inc., Ann Arbor, MI) to eliminate plasma lines.

2.3. XRD, pole figure, and HRTEM

A Rigaku D/Max-IIB diffractometer was used for various X-ray diffraction (XRD) techniques to characterize the crystallinity and texture of the samples. A typical θ–2θ plot was generated to determine the composition, structure, and out-of-plane texture. The pole figures were obtained using the Rigaku pole figure attachment on a rotating anode generator. The data was recorded in reflection mode, with both alpha (15°–89°) and beta (0°–360°) steps of 2 degrees and collection time of 1 s at each step. For determination of nanostructure and for corroboration with X-ray analyses, high-resolution transmission electron microscopy (HRTEM; JEOL 4000EX, Japan) was also carried out on cross sectional samples. The relative in-plane and out-of-plane orientations between the PST film and sapphire substrate were determined by indexing the optical diffractograms obtained from HRTEM images. In order to confirm whether the PST film was under any in-plane stress, an off-axis X-ray scan was carried out. In this method, the effect of in-plane stress on the magnitude of the interplanar spacing of planes parallel (or inclined at angles <45°) to the film surface is determined. To accomplish this, the PST film was tilted so that the (110) planes of PST were perpendicular to the X-ray source-detector plane.

2.4. High frequency measurements

In the current approach, tunable CPW components were designed using the modified dielectric/ferroelectric/conductor configuration, and fabricated. The geometric layout consisted of a dielectric substrate (254 μm thick sapphire), a PST ferroelectric thin-film layer (100 nm thick), a 2 μm gold thin-film for the conductor (center line) and ground lines (adjacent parallel lines). The test structures, used to determine the material and device FOMs, included a CPW transmission line and a CPW resonator. The transmission lines (Fig. 1) were designed for 0.6 cm, and 1 cm length with a center conductor width of 0.02 cm, and a gap between center conductor and the ground line of 0.01 cm. The resonator (Fig. 2) was a transmission type resonator coupled to input and output feed lines. It was primarily used to determine the tan δ of the PST film. The CPW line and resonator electrode structures were deposited on the samples in an electron beam evaporation system using shadow masks. The exposed (open) areas of the shadow mask defined the electrode structures on the samples, and after Au electrode deposition, the samples were annealed at 550°C for approximately 30 min in flowing nitrogen.

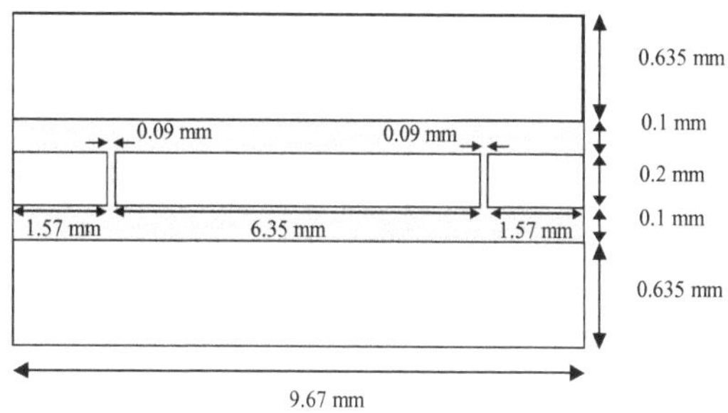

Figure 2 Transmission type resonator to determine the loss tangent of PST thin film.

The high frequency measurements were performed inside a closed cycle cryogenic system modified for on-wafer CPW measurements. A HP 8510C automatic network analyzer was used for swept frequency microwave measurements. The dc bias voltage to the signal conductor was applied using a high voltage bias tee. A Thru-Reflect-Load (TRL) calibration was performed before the actual measurements.

As a preliminary check for tunability and magnitude of tan δ for PST film, the frequency dependence of the insertion loss (magnitude of S_{21}) was plotted for applied dc bias of 0 V bias and 400 V between the center conductor and the ground lines. The phase of S_{21} (plotted as total phase angle versus frequency) was compared at a specific frequency to determine the phase shift between zero bias and 400 V bias. Note, a large phase shift corresponds to a high tunability of the sample.

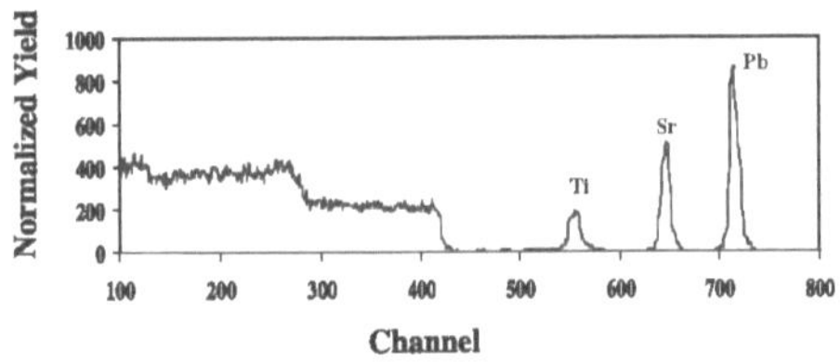

Figure 3 RBS of 100 nm $(Pb_{0.3}Sr_{0.7})TiO_3$ film on (0001) sapphire.

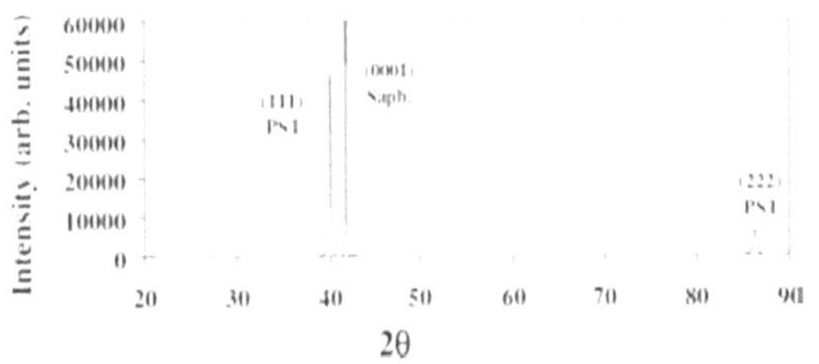

Figure 4 A θ–2θ XRD scan for 100 nm $(Pb_{0.3}Sr_{0.7})TiO_3$ film on (0001) sapphire.

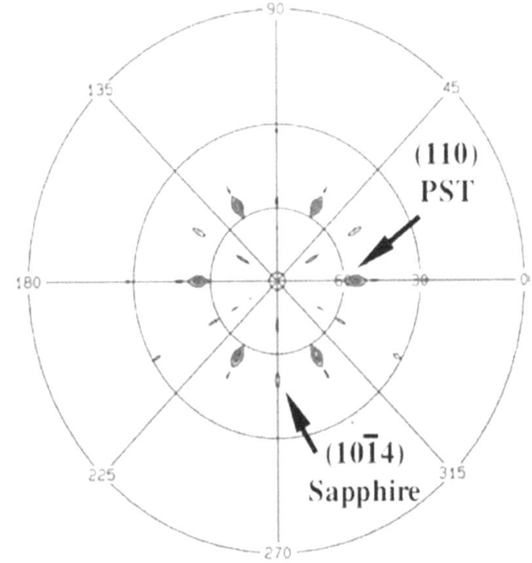

Figure 5 (110) Pole figure of (111) oriented $(Pb_{0.5}Sr_{0.7})TiO_3$ film (100 nm thick) on (0001) sapphire.

The swept frequency S-parameters were used to extract the ε_r (V) of the PST film by using the conformal mapping techniques [12, 13]. Also, from the resonance peak of the CPW resonator, the loaded Quality factor (i.e., Q_L = center frequency/3dB bandwidth) was determined, from which the unloaded Quality factor (Q_u) was calculated ($Q_u = Q_L \times S_{21}/(1-S_{21})$ and tan δ of the PST film is approximately $1/Q_u$). From the S-parameter measurements, the material FOM and the device FOM (i.e., K-factor in phase shift/dB) were determined. Please note that for calculation of the latter, the worst-case insertion loss was used. Once the ε_r and the tan δ of PST were obtained, the dielectric and conductor losses (in units of dB/cm) in the CPW line were extracted using a quasi-static TEM approximation of the CPW line, and conformal mapping techniques [12, 13].

3. Results and discussion

3.1. RBS, XRD, and pole figure

Fig. 3 shows the RBS spectrum of a 100 nm PST film deposited at 700°C by MOCVD on sapphire. The calculated composition of the PST film from a RUMP simulation was $(Pb_{0.3}Sr_{0.7})TiO_3$. The carbon contamination in the film was below the detection limit (0.5 atomic %). The XRD θ–2θ scan of the same film is shown in Fig. 4. Three major peaks in the regular spectrum are indexed as (111) PST, (0001) sapphire, and (222) PST. The high intensity of (111) PST peak indicates the general out-of-plane alignment in the [111] PST direction. The (110) pole figure, illustrated in Fig. 5, was carried out to demonstrate the high in-plane orientation of the film. The {110} PST poles with 6-fold symmetry were identified from the angular relationship (~35°) between the diffraction spots and the (111) plane. Also, the {10$\bar{1}$4} poles of sapphire with 3-fold symmetry are depicted in the same plot. The sharp contours of {110} poles of PST and {10$\bar{1}$4} poles of sapphire, along with their narrow intensity distribution, indicates clearly that PST grows epitaxially on sapphire. It is noted that all other low-density spots are related to the imperfection of the sapphire single crystal, which were verified by a pole figure (not shown here) obtained from a bare sapphire substrate under the same condition.

3.2. FESEM, HRTEM, Raman spectroscopy and origin of stress

A cross sectional FESEM image of a PST (150 nm)/sapphire interface is shown in Fig. 6. The PST film is dense and exhibits a single crystal like cleavage. A TEM image at low magnification (Fig. 7) shows threading dislocations; the number density of such dislocations which stem from the sapphire substrate was estimated to be $\sim 10^{10}$ cm^{-2}.

Fig. 8 illustrates a cross-sectional HRTEM image of the PST/sapphire interface, and optical diffractograms from regions within PST and sapphire. The observed interplanar spacing of sapphire: d_{006}(Sapphire) = 2.165 Å, d_{104}(Sapphire) = 2.551 Å, $d_{\bar{1}02}$(Sapphire) = 3.479 Å, and d_{300}(Sapphire) = 1.374 Åare as expected, whereas of PST: d_{110}(PST) = 2.780 Å, d_{111}(PST) = 2.276 Å, and d_{201}(PST) = 1.778 Åare consistently larger with respect to the bulk values [14]; similar observations for BST films on MgO substrate, in which the lattice parameters of the film were larger than their bulk values, have been

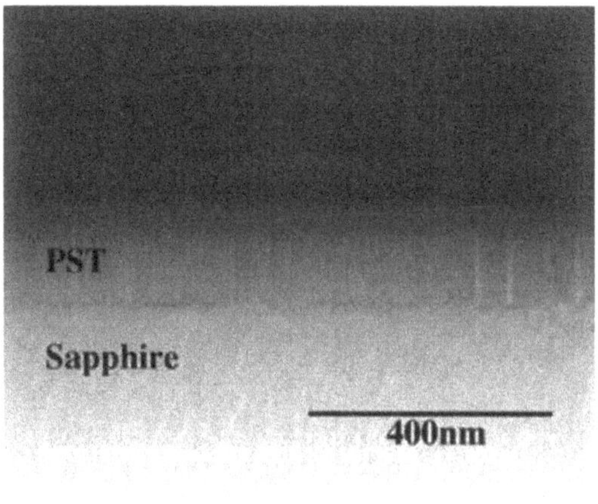

Figure 6 FESEM of $(Pb_{0.3}Sr_{0.7})TiO_3$ film on (0001) sapphire by MOCVD.

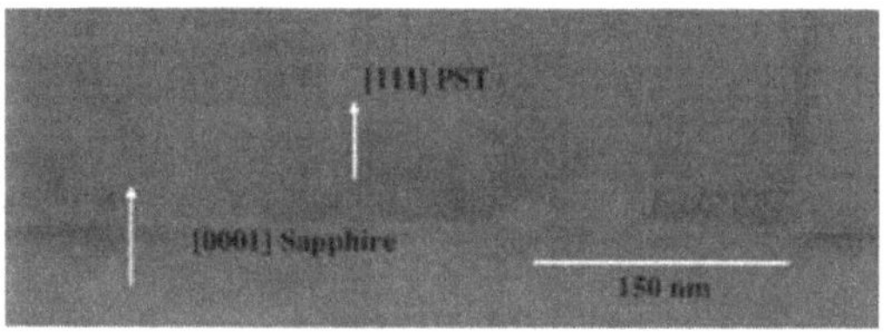

Figure 7 A low magnification TEM image showing defects in $(Pb_{0.3}Sr_{0.7})TiO_3$ film on sapphire.

reported elsewhere [15]. However, the as-deposited PST film is epitaxial and the orientation relationships, determined from the analyses of the diffractograms as well as all interplanar spacing, were as follows: out-of-plane alignment of [111] PST // [0001] sapphire, and orthogonal in-plane alignments of $[1\bar{1}0]$ PST//$[10\bar{1}0]$ sapphire and $[\bar{1}\bar{1}2]$ PST//$[1\bar{2}10]$ sapphire. Based on the HRTEM analyses, the in-plane atomic arrangements on (0001) sapphire and (111) PST planes, the overall orientation relationship of PST with respect to sapphire may be represented by the schematics of Fig. 9a and b, respectively; results which are consistent with the orientational relationships of PLZT on sapphire [16].

The observation of epitaxy in this large lattice-mismatched system (e.g., $d_{10\bar{1}0}$(sapphire) = 4.122 Åand $d_{1\bar{1}0}$(PST) = 2.780 Å) may be characterized by geometric epitaxy or domain matched epitaxy [17], i.e., although the misfit, f_1 (where, $f_1 = [d_{10\bar{1}0}\text{-}d_{1\bar{1}0}] \times 2/[d_{10\bar{1}0} + d_{1\bar{1}0}]$), is apparently large (38.9%), three interplanar spacing of $(1\bar{1}0)$ PST closely match with two interplanar spacing of $(10\bar{1}0)$ sapphire. This geometric matching gives a residual domain misfit, f_d (where, $f_d = [3d_{1\bar{1}0} - 2d_{10\bar{1}0}] \times 2/[3d_{1\bar{1}0} + 2d_{10\bar{1}0}]$), of only 1.16%, but the PST film in this direction still remains under compression. Similarly, the PST is also under a compressive stress in the orthogonal in-plane $[\bar{1}\bar{1}2]$ direction through geometric epitaxy; which means that if there exists bi-axial (in-plane) compressive stress, the out-of-plane [111] PST direction should be under tension [18]. Indeed, an off-axis (110) θ–2θ scan (Fig. 10) clearly shows a 0.49% extension or strain of the (110) lattice spacing of PST, which is also indicative of a residual tensile stress in the [111] PST direction.

Fig. 11 illustrates a Raman spectrum of 100 nm PST film on (0001) sapphire. For comparison, the Raman spectrum of iso-structural $PbTiO_3$ film (420 nm thick) is shown in the same figure. The sharp peaks at 417 cm^{-1},

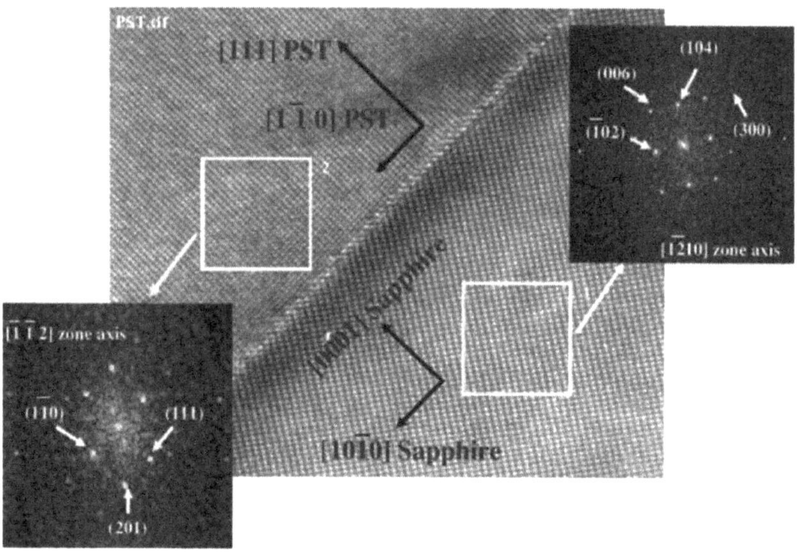

Figure 8 HRTEM of $(Pb_{0.3}Sr_{0.7})TiO_3$ film by MOCVD and optical diffractograms from regions within PST and sapphire.

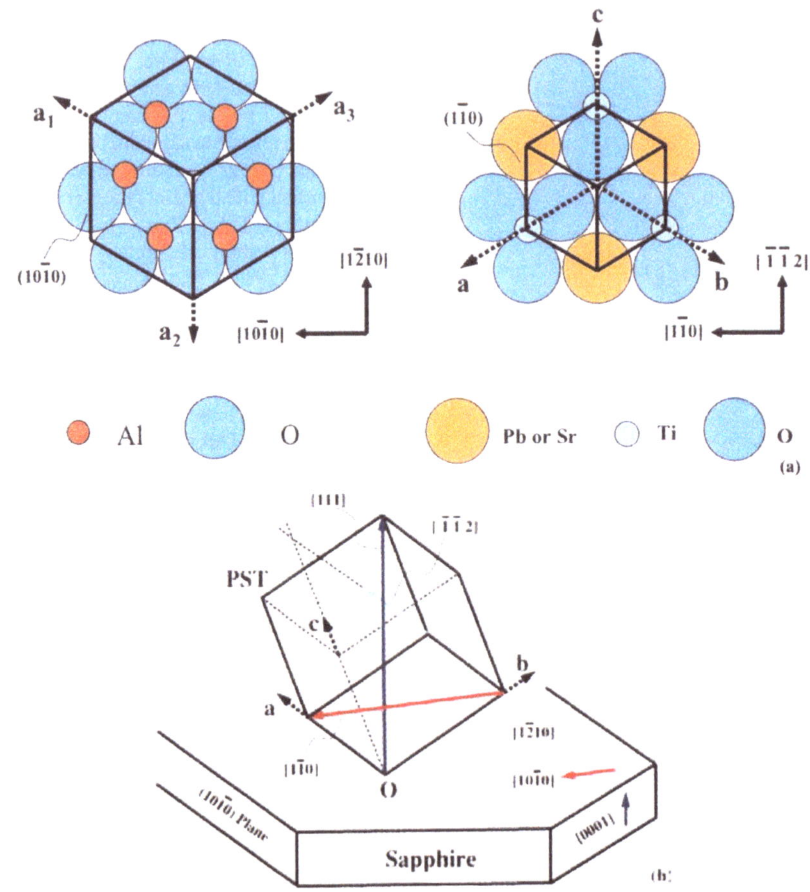

Figure 9 (a) Planar atomic configurations for sapphire basal plane and (111) PST plane; (b) Schematic of the orientation relationships for epitaxial PST on (0001) sapphire.

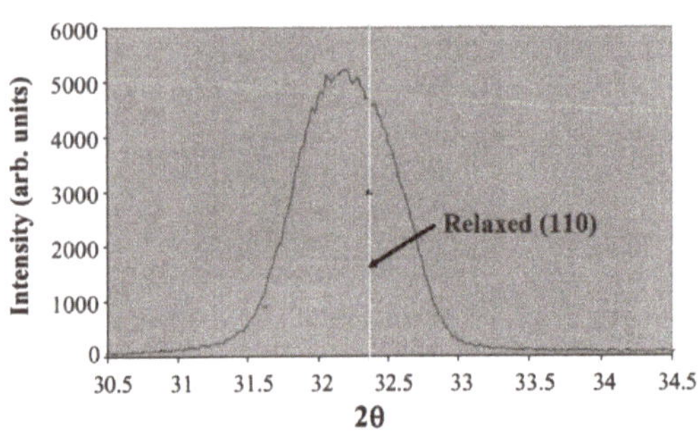

Figure 10 An off-axis θ–2θ scan for the (110) PST plane. Here, d_{110} (0.278 nm, $2\theta = 32.2°$) of PST film is greater than d_{110} (0.2766 nm, $2\theta = 32.36°$) of relaxed or bulk PST.

578 cm^{-1}, and 752 cm^{-1} are assigned to sapphire, and the PST spectrum has been cut out in the low frequency region (<100 cm^{-1}, where the "soft mode" exists in $PbTiO_3$ system) due to artifacts from the measurement system. The PST spectrum indicates that the incorporation of Sr into the $PbTiO_3$ lattice results in a structural disorder, and the difference between the wavenumbers for the E(TO) and A1(TO) modes are closer compared to the $PbTiO_3$ spectrum. These are indicative of lattice transformation from tetragonal ferroelectric (in pure $PbTiO_3$) to the cubic paraelectric state (in PST); as found in a previous study E(TO) and A1(TO) modes, these phonons are indistinguishable in the paraelectric state [19]. Additionally, a pseudo-cubic structure of the PST film may exhibit broad and diffuse modes due to residual scattering from A-site cation inhomogeneity (i.e., composition inhomogeneity in Pb and Sr sites) and/or stress.

The aforementioned anisotropy and sign of stress in the PST film: in-plane (bi-axial) compressive stress and out-of-plane tensile stress can stem from a number of factors: (a) thermal mismatch between PST and Al_2O_3, (b) presence of grains with different orientations in the film, (c) presence of oxygen vacancies, (d) lattice mismatch

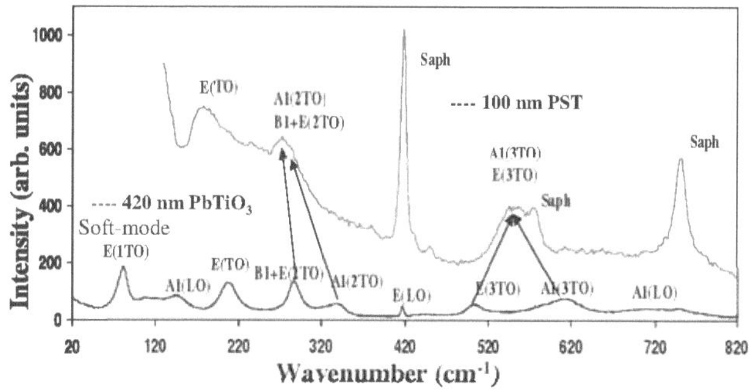

Figure 11 Raman spectra of 100 nm $(Pb_{0.3}Sr_{0.7})TiO_3$ film on (0001) sapphire and 420 nm $PbTiO_3$ film.

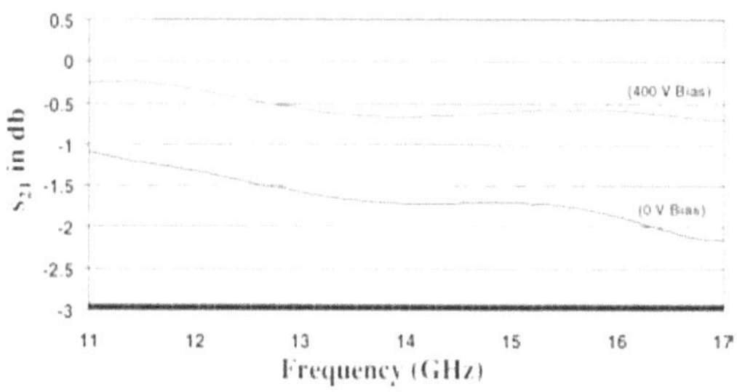

Figure 12 Insertion loss versus frequency for PST film at zero bias and 400 V bias between the center conductor and the ground lines.

between (111) PST and (0001) sapphire, and (e) A-site cation inhomogeneity. In the PST/sapphire heterostructure, considering the small difference between thermal coefficient of $SrTiO_3$ ($\alpha_{SrTiO_3} = 10 \times 10^{-6}$/K) and sapphire ($\alpha_{Al_2O_3} = 8.8 \times 10^{-6}$/K), a thermal mismatch contribution to stress may be neglected. The observation of epi-PST films eliminates the contribution of stress from grains with different orientations. However, due to the high deposition rates (10–15 nm/min) and as-deposited (and unannealed) condition of the PST film, oxygen vacancies may substantially contribute to this complex stress pattern [10]. Please note that the deposition conditions (such as oxygen partial pressure and deposition temperature) are yet to be optimized. Moreover, the effect of stress-inducing threading dislocations and composition inhomogeneity, as reflected in Figs 7 and 11, may also manifest as a source of stress.

3.3. High frequency (7 to 17 GHz) data

Fig. 12 shows the measured insertion loss (magnitude of S_{21}) for 0 V bias and 400 V bias between the center conductor and the ground lines. The insertion loss at zero bias is between 1 dB at 11 GHz and 2.2 dB at 17 GHz, whereas the insertion loss at 400 V bias improved to 0.25 dB at 11 GHz and 0.7 dB at 17 GHz. The improvement in insertion loss is an indication of tunability with applied dc bias, and the low insertion loss is also an indication of the relatively low tan δ in the PST film.

The phase shift of the PST film is illustrated in Fig. 13. It shows the swept frequency measurement of the phase of S_{21} in degrees for 0 V bias and 400 V bias (40 kVcm^{-1}) between the center conductor and the ground lines. The calculated ε_r at 12 GHz for the PST film is 200 (at zero bias) and 140 (at 400 V bias); i.e., a tunability of approximately 30%. The loss-tangent of the film at 12 GHz was estimated to be ~0.05 and 0.03 at 400 V bias. The material FOM and device FOM (or K-factor) were determined to be 632 and approximately 13 degrees/dB, respectively.

Fig. 14 shows the calculated dielectric loss in the PST thin film at zero bias and 400 V bias. As one may observe, the dielectric loss contribution due to the film is greatly reduced at 400 V; this due to the large tunability in both the ε_r and tan δ.

3.4. Microwave property-nanostructure-composition-stress relationships

Numerous past studies have shown that magnitude of ε_r in ferroelectric thin films, when compared to single crystals and/or bulk polycrystalline materials of the same composition, is significantly lower due to the effects of elastic and electrical boundary conditions [20], interfacial layer [21–24], and/or Schottky barriers at the film-electrode interfaces [25, 26]. The effects of elastic stress on the diminution of ε_r have been explained using a phenomenological approach [27–31]. In comparison to the ε_r (>10000) of bulk PST [9], the current study of epi-PST film on sapphire and in CPW structure gives a zero-bias ε_r of 200; a result that is consistent with the results of biaxially stressed $SrTiO_3$, where the permittivity was also significantly lower [32, 33]. As noted in the discussions of the observed nanostructure in Section 3.2, in-plane compressive stress is probably the primary factor for the low in-plane ε_r of epi-PST film. However, due to the non-uniform field distribution in a CPW structure, it may be

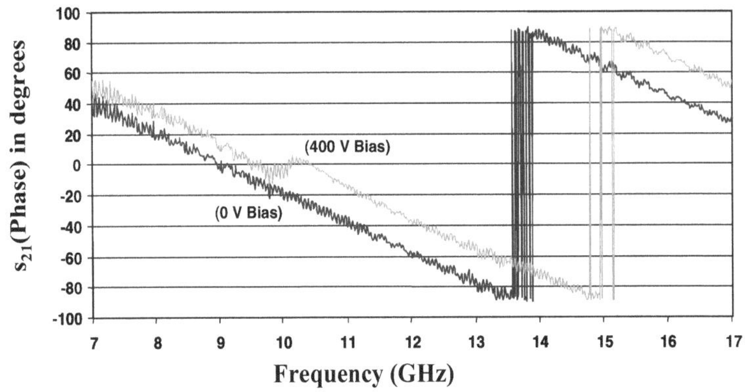

Figure 13 Comparison of the phase of S_{21} for PST film at zero bias and 400 V bias between the center conductor and the ground lines.

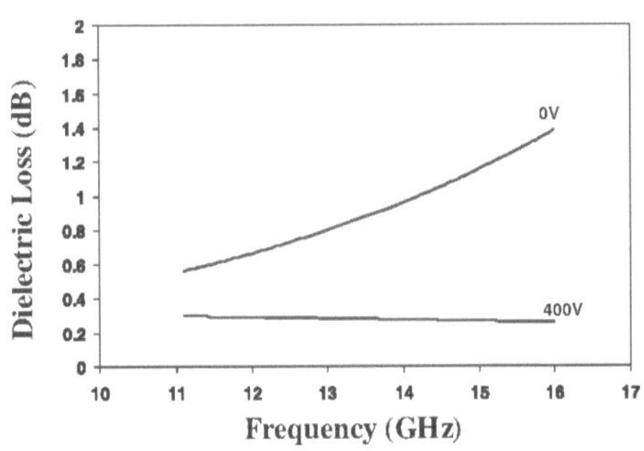

Figure 14 Frequency dependence of the dielectric loss contribution due to the PST thin film at zero bias and 400 V bias.

difficult to quantitatively address the influence of stress on ε_r and tunability.

In general, dielectric loss stems from intrinsic losses [34–36] due to multiple-phonon absorption (single crystal), losses related to charged defects (one-phonon absorption and phonon scattering), relaxation losses and losses due to domain walls [37–39]. Although, in this current study on PST films, it was intended to minimize the domain wall contribution to tan δ by controlling the $Pb_xSr_{(1-x)}TiO_3$ composition (i.e., $x = 0.3$), the potential existence of micro domains from composition inhomogeneity and/or stress are sources of loss. Additionally, stress-inducing threading dislocations, which are charged defects [40], could introduce one-phonon absorption and phonon scattering. However, the anisotropy and sign of stress in the PST film (discussed in Section 3.2) must be the predominant factor for extrinsic dielectric loss. Indeed, a previous study [10] has correlated the extrinsic dielectric loss with oxygen vacancy-mediated tetragonal distortion, D (D = ratio of the in-plane to out-of-plane lattice parameter), in BST films. At microwave frequencies (1–20 GHz), BST films with D = 1 exhibited the highest material FOM. From the aforementioned discussions, one may infer that a combination of different factors lead to the observed tan δ. The presence of charged and stress inducing threading dislocations, oxygen vacancies, lattice mismatches between PST and sapphire, and A-site composition inhomogeneity contribute to tan δ of epi-PST film.

3.5. Selection of composition, substrate, and process parameters

The current discussions on processing, nanostructure, and microwave properties of epi-PST thin films provide a rational basis for the selection of PST composition, substrate, and process parameters for the fabrication of optimized CPW phase shifters with very high figure of merit and operable at room temperature.

The composition of $Pb_xSr_{(1-x)}TiO_3$ must be in the Curie–Weiss range (i.e., $x < 0.3$) to minimize the domain wall contribution to dielectric loss, while maintaining the high tunability. The use of MOCVD in conjunction with the advanced DLI system (with its accurate control of the flow rates of individual cation precursors) can make this approach feasible.

To minimize the in-plane, biaxial stress and to achieve a tetragonal distortion (D) of unity in PST films, cubic substrates with close lattice match to PST or suitable buffer layers on substrates may be more appropriate. For example, Jia *et al.* [41] have studied the microwave properties of pulsed laser deposited $SrTiO_3$ on homoepitaxial buffer layers (2–25 nm) of $LaAlO_3$ on single crystal $LaAlO_3$ substrate. Cross-sectional TEM showed that the buffer layers had many defects, but aided in reducing the stress so that high quality and stress-free $SrTiO_3$ grew epitaxially. It was also found that the quality factor of the microwave device was improved by more than 50%. A similar approach may be implemented with the PST/sapphire system, by using thin buffer layers of crystalline oxides to achieve epitaxial cube-on-cube growth, as well as to minimize stress and stress-induced charged defects (threading dislocations) in the over layer of PST.

To ascertain the deposition of epi-PST films in this study, the MOCVD growth was carried out in the mass-transport controlled region, but at very high deposition rate. Given a small thermal and lattice mismatch with an appropriate substrate or buffered surface on sapphire, a high temperature (>500°C) and low deposition rate (<5 nm/min) will aid in the deposition of very high quality of epi-PST films with phase purity because of the ease of ad-atom migration to equilibrium positions. However, the control of oxygen stoichiometry and A-site composition homogeneity, to achieve defect-free and stress-free epi-PST films having D = 1, must be achieved through designed CVD experiments, coupled with doping and appropriate post annealing treatments in oxygen [42].

4. Summary

Voltage tunable $Pb_{0.3}Sr_{0.7}TiO_3$ (PST) thin films (90–150 nm) on (0001) sapphire were deposited by metalorganic chemical vapor deposition at rates of 10–15 nm/min. The as-deposited epitaxial PST films were characterized by RBS, X-ray methods, FESEM, HRTEM, Raman spectroscopy, and electrical methods (7–17 GHz) using coplanar waveguide test structures. The epitaxial relationships were as follows: out-of-plane alignment of [111] PST//[0001] sapphire, and orthogonal in-plane alignments of $[1\bar{1}0]$ PST//$[10\bar{1}0]$ sapphire and $[\bar{1}\bar{1}2]$ PST//$[1\bar{2}10]$ sapphire. The microwave properties, characterized using CPW transmission lines and resonators, indicated that the insertion loss at zero bias is between 1 dB at 11 GHz and 2.2 dB at 17 GHz, whereas the insertion loss at 400 V bias improved to 0.25 dB at 11 GHz and 0.7 dB at 17 GHz. The calculated ε_r at 12 GHz is 200 (at zero bias) and 140 (at 400 V bias); i.e., a tunability of approximately 30%. The loss-tangent of the film at 12 GHz was estimated to be ~0.05 and 0.03 at 400 V bias. The in-plane compressive stress is probably the primary factor for the low in-plane ε_r of epi-PST film, and the presence of charged and stress inducing threading dislocations, oxygen vacancies, in-plane (bi-axial) compressive stress and out-of-plane tensile stress, and A-site composition inhomogeneity in PST contribute to tan δ. It is noted that the material FOM and device FOM (or K-factor) were determined to be 632 and approximately 13 degrees/dB, respectively.

The guidelines for selection of PST composition, substrate, and process parameters for the fabrication of optimized CPW phase shifters with very high FOM and operable at room temperature are as follows: (1) a PST composition with $x<0.3$, (2) the use of cubic substrates with close lattice match to PST or suitable buffer layers on sapphire substrates, and (3) MOCVD of PST at low deposition rate (<5 nm/min), coupled with doping and appropriate post annealing treatments in oxygen.

Acknowledgements

SKD would like to acknowledge support of Dr Deborah Van Vechten of the Office of Naval Research, Dr. Stuart A. Wolf of the Defense Science Office at Defense Advanced Research Projects Agency (DARPA), and the Frequency Agile Materials for Electronics (FAME) program at DARPA (contract number N00014-00-1-0471).

References

1. M. J. DALBERTH, R. E. STAUBER J. C. PRICE, C. T. ROGERS, and D. GALT, *Appl. Phys. Lett.* **72** (1998) 507.
2. H.-D. WU and F. S. BARNES, *Integr. Ferroelectr.* **22** (1998) 291.
3. D. S. KORN and H.-D. WU, *ibid.* **24** (1999) 215.
4. G. SUBRAMANYAM, F. W. VAN KEULS, and F. A. MIRANDA, *IEEE Microwave Guided Wave Lett.* **8** (1998) 78.
5. F. W. VAN KEULS, R.R. ROMANOFSKY, D. Y. BOHMAN, M. D. WINTERS, F. A. MIRANDA, C. H. MUELLER, R. E. TREECE, T. V. RIVKIN, and D. GALT, *Appl. Phys. Lett.* **71** (1997) 3075.
6. S. S. GEVORGIAN, D. I. KAPARKOV and O. G. VENDIK, *IEEE Proc. Microwave, Antennas and Propagation* **141** (1994) 501.
7. A. T. FINDIKOGLU, Q. X. JIA, X. D. WU, G. J. CHEN, T. VENKATESAN and D. W. REAGOR, *Appl. Phys. Lett.* **68** (1996) 1651.
8. W. WILBUR *et al., Integr. Ferroelectr.* **19** (1998) 149.
9. Y. SOMIYA, A. S. BHALLA and L. E. CROSS, *Int. J. of Inorg. Mater.* **3** (2001) 709.
10. W. J. KIM, W. CHANG, S. B. QADRI, J. M. POND, S. W. KIRCHOEFER, D. B. CHRISEY and J. S. HORWITZ, *Appl. Phys. Lett.* **76** (2000) 1185.
11. R. A. YORK, A. S. NAGRA, P. PERIASWAMY, O. AUCIELLO, S. K. STREIFFER and J. IM, *Integr. Ferroelectr.* **34** (2001) 177.
12. E. CARLSSON and S. GEVORGIAN, *IEEE T. Micro. Theory* **47** (1999) 1544.
13. S. GEVORGIAN, T. MARTINSSON, A. DELENIV, E. KOLLBERG and I. VENDIK, *IEEE Proc. Microwave, Antennas and Propagation* **144** (1997) 145.
14. S. NOMURA and S. SAVADA, *J. Phys. Soc. Jpn.* **10** (1955) 108.
15. W. J. KIM, H. D. WU, W. CHANG, S. B. QADRI, J. M. POND, S. W. KIRCHOEFER, D. B. CHRISEY and J. S. HORWITZ, *J. Appl. Phys.* **88** (2000) 5448.
16. K. WASA and S. HAYAKAWA, in "Handbook of Sputter Deposition Technology" (Noyes Publications, Park Ridge, New Jersey, 1992) p. 175.
17. T. ZHELEVA, K. JAGANNADHAM and J. NARAYAN, *J. Appl. Phys.* **75** (1994) 860.
18. K. ABE, N. YANASE, K. SANO and T. KAWAKUBO, *Integr. Ferroelectr.* **21** (1998) 197.
19. J. MENG, G. ZOU, Y. MA, X. WANG and M. ZHAO, *J. Phys.-Condens. Mat.* **6** (1994) 6549.
20. O. G. VENDIK and L. T. TER-MARTIROSYAN, *Sov. Phys.-Solid State* **36** (1994) 1778.
21 K. ABE and S. KOMATSU, *Jap. J. Appl. Phys.* **32** (1993) L1157.
22. C. ZHOU and D.M. NEWNS, *J. Appl. Phys.* **82** (1997) 3081.
23. C. BASCERI, S.K. STREIFFER, A.I. KINGON and R. WASER, *J. Appl. Phys.* **82** (1997) 2497.

24. H.C. LI, W. SI, A.D. WEST and X.X. XI, *Appl. Phys. Lett.* **73** (1998) 464.
25. G.W. DIETZ, W. ANTPOHLER, M. KLEE and R. WASER, *J. Appl. Phys.* **78** (1995) 6113.
26. G. W. DIETZ and R. WASER, *Thin Solid Films* **299** (1997) 53.
27. S. STREIFFER, C. BASCERI, C.B. PARKER, S.E. LASH, and A.I. KINGON, *J. Appl. Phys.* **86** (1999) 4565.
28. N. A. PERTSEV, A. G. ZEMBILGOTOV and A. K. TAGANTSEV, *Phys. Rev. Lett.* **80** (1998) 1988.
29. N. A. PERTSEV, A. G. ZEMBILGOTOV, S. HOFFMANN, R. WASER and A.K. TAGANTSEV, *J. Appl. Phys.* **85** (1999) 1698.
30. N. A. PERTSEV, A. K. TAGANTSEV and N. SETTER, *Phys. Rev. B* **61** (2000) R825.
31. B. DESU, V. P. DUDKEVICH, P. V. DUDKEVICH, I. N. ZAKHARCHENKO and G.L. KUSHLYAN, *MRS Symp. Proc.* **401** (1996) 195.
32. E. HEGENBARTH and C. FRENZEL, *Cryogenics* **7** (1967) 331.
33. T. M. SHAW, Z. SUO, M. HUANG, E. LINIGER, R. B. LAIBOWITZ and J. D. BANIECKI, *Appl. Phys. Lett.* **75** (1999) 2129.
34. V. L. GUREVICH and A. TAGANTSEV, *Adv. Phys.* **40** (1991) 719.
35. A. K. TAGANTSEV, in "Ferroelectric Ceramics: Tutorial Reviews, Theory, Processing, and Applications," edited by N. Setter and E. L. Colla (Springer-Verlag, New York, LLC, 1992) p. 127.
36. O. VENDIK, L. TER-MARTIROSYAN and S. ZUBKO, *J. Appl. Phys.* **84** (1998) 993.
37. J. O. GENTNER, P. GERTHSEN, N. A. SCHMIDT and R.E. SEND, *ibid.*. **49** (1978) 4595.
38. K.H. HÄRDTL, *Ceram. Int.* **8** (1982) 121.
39. G. ARLT, U. BOTTGER, and S. WITTE, *Appl. Phys. Lett.* **63** (1993) 602.
40. S. MAHAJAN, *Prog. Mater. Sci.* **42** (1997) 341.
41. Q. X. JIA, A.T. FINDIKOGLU, D. REAGOR and P. Lu, *Appl. Phys. Lett.* **73** (1998) 897.
42. C. K. BARLINGAY and S. K. DEY, *Thin Solid Films*, **272** (1996) 112.

JOURNAL OF MATERIALS SCIENCE 41 (2006) 87–96

Studies on the relaxor behavior of sol-gel derived $Ba(Zr_xTi_{1-x})O_3$ ($0.30 \leq x \leq 0.70$) thin films

A. DIXIT, S. B. MAJUMDER, R. S. KATIYAR*
Department of Physics, University of Puerto Rico, San Juan, PR 00931-3343
E-mail: rkatiyar@rrpac.upr.clu.edu

A. S. BHALLA
Materials Research Laboratory, Pennsylvania State University, University Park, PA 16802

We have studied the relaxor behavior of sol-gel derived $Ba(Zr_xTi_{1-x})O_3$ ($0.30 \leq x \leq 0.70$) thin films. The plausible mechanism of the relaxor behavior has been analyzed from the dielectric data and micro-Raman spectra. Substitution of Zr^{+4} for Ti^{+4} in $BaTiO_3$ lattice reduces its long-range polarization order yielding a diffused paraelectric to ferroelectric phase transition. The solid solution system is visualized as a mixture of Ti^{+4} rich polar region and Zr^{+4} rich regions and with the increase in Zr contents the volume fraction of the polar regions are progressively reduced. At about 25.0 at% Zr contents the polar regions exhibit typical relaxor behavior. The degree of relaxation increases with Zr content and maximizes at 40.0 at% Zr doped film. The frequency dependence of the polar regions follows Vogel-Fulcher relation with a characteristic cooperative freezing at freezing temperature (T_f). Below T_f, a long range polarization ordering was ascertained from the polarization hysteresis measurement.

1. Introduction

Relaxors exhibit unusually higher dielectric constant and piezoelectric constants, which are very attractive for high energy density capacitors and actuators. Lead based perovskite ceramics has been extensively studied in last couple of decades. Studies on relaxors in thin film form are scanty and the available reports mostly are based on lead based perovskite thin films. These relaxor, irrespective in bulk ceramics or thin film forms have the following general characteristics: (i) all of them exhibit a diffuse phase transition (DPT) like behavior, where the temperature corresponding to the dielectric maxima (T_m) does not correspond any paraelectric-ferroelectric phase transition like normal ferroelectrics, (ii) T_m shifts to higher temperature with the increase in measurement frequency, (iii) a frequency dispersive behavior is also observed in the dielectric loss tangent, however, the peak temperatures (T'_m) for dielectric loss do not coincide with the corresponding T_m (iv) relaxors follow the Curie-Weiss law far above T_m, (iv) if the temperature is significantly lowered as compared to T_m, some of these relaxors exhibit stable polarization hysteresis.

The classic ferroelectric relaxors are $A(B'_xB''_{1-x})O_3$ (e.g. $Pb(Mg_{1/3}Nb_{2/3})O_3$) type complex compounds where the nano-size cation ordered polar regions, distributed in a disordered matrix, are known to be responsible for their relaxor behavior. Several model has been proposed to explain the relaxor behavior in PMN ceramics which includes a super-paraelectric model [1], glass-like freezing of the nano-polar regions [2], random field model [3] involving the occurrence of dynamic nanopolar regions in the paraelectric phase and their transformation into frozen domain upon cooling below the freezing temperature etc. Relaxor behavior is also reported in tetragonal tungsten bronze type compounds, doped quantum paraelectrics, *A* and *B* site doped perovskite oxides [4, 5] etc.

Recently, the attention has moved towards barium based relaxors as these are environmentally friendly lead free oxides. From the studies on the bulk ceramics and single

*Author to whom all correspondence should be addressed.

0022-2461 © *2006 Springer Science + Business Media, Inc.*
DOI: 10.1007/s10853-005-5929-1

crystalline barium zirconate titanate it is apparent that the Zr contents strongly affect the ferroelectric, piezoelectric, dielectric as well as relaxor characteristics of BZT ceramics [6–8]. We have already reported the relaxor characteristics in $BaZr_{0.40}Ti_{0.60}O_3$ (BZT-40) thin films, which has also been confirmed later by others, however, the mechanism lead to the observed relaxor behavior so far remain poorly understood even in case of bulk BZT ceramics [6, 9, 10]. The available literature reports are contradictory in nature as apparent from the following review. Sciau *et al.* reported that unlike $PbMg_{1/3}Nb_{2/3}O_3$ (PMN), $BaZr_{0.35}Ti_{0.65}O_3$ (BZT-35) does not have any cation ordered regions and the loss of the polar ordering of $BaTiO_3$ is due to the dilution of Ti^{+4} ions by Zr^{+4} substitution. The same group has also reported that, as observed in case of PMN ceramics, the application of external electric field does not induce any long-range polar order in BZT ceramics due to the difficulty of polarizing the Zr rich regions [11, 12]. In contrary to these reports, from the measurement of pyroelectric current, Farhi *et al.* [13] claimed that a field induced ferroelectric phase can be stabilized below the freezing temperature of $BaZr_{0.40}Ti_{0.60}O_3$ (BZT-40) ceramics. The freezing temperature was estimated to be in the range of 100–140 K from dielectric, Raman and pyro-current measurements and it represent to ergodic to non-ergodic phase transition. Apparently, in a solid solution system like BZT, no cationic ordering is expected between Zr and Ti. In several relaxor compounds (viz. PMN, PST etc.), cationic order of B site ions has been confirmed to be related to the dielectric relaxation. Artificial control of *B* site cation ordering has been demonstrated to be possible by constructing superlattices, however, from this report the relaxor characteristics of $BaZr_{0.20}Ti_{0.80}O_3$ (BZT-20) thin films with different degree of cation ordering are not properly elucidated [14].

In the present work we have prepared BZT thin films with Zr content varies from 30.0–70.0 at%. To the best of our knowledge no attempt has so far been made to study the dielectric behavior of BZT ceramics or thin films with Zr contents more than 40.0 at%. The relaxor behavior of BZT thin films were confirmed in the whole range of Zr contents. The dielectric relaxations in these films found to follow the Vogel-Fulcher type behavior originally reported for the spin-glass as well as complex perovskite systems and below a characteristics freezing temperature relaxor to ferroelectric transition has been demonstrated. Based on the dielectric, ferroelectric and micro-Raman scattering analyses, the observed relaxor behavior has been hypothesized due to the breakdown of the long range polar ordering in $BaTiO_3$ (BTO) by Zr^{+4} substitution in Ti^{+4} sites yielding a nano-size polar regions in a Zr rich matrix. Beyond an optimal dopant content ($\geq$25.0 at%) probably a critical size of the polar regions are reached to yield the relaxor behavior of BZT thin films.

2. Experimental

In the present work, we present a comprehensive study to support the relaxor like behavior in $BaZr_xTi_{1-x}O_3$ ($0.30\leq x\leq 0.70$) thin films prepared on platinum substrates by the sol-gel technique. Details of the precursor sol preparation have been reported elsewhere [15]. As deposited films were fired at 600°C for 5 min. for organic removal and initial crystallization. The coating and firing sequence was repeated for 20 times to attain a film thickness of about 600 nm. The films were finally annealed at 1100°C for 2 h in air for complete perovskite phase formation and better crystallinity. Polycrystalline perovskite nature of films was investigated using the X-ray diffraction method.

To measure the electrical properties circular Pt electrodes of 0.2 mm diameters were deposited on BZT films by dc magnetron sputtering. Electrical measurements and temperature dependent dielectric properties were studied in metal-insulator-metal configuration and the films were characterized using an impedance analyzer (HP 4294A, from Agilent Technology Inc.) in conjunction with a temperature controlled probe station (MMR Technology). The Polarization hysteresis nature of the film was analyzed using a ferroelectric tester system RT-6000 HVS from Radiant Technologies Inc. Raman spectra of these films were recorded using an ISA T64000 Raman microprobe. An optical microscope with 80 X objective was used to focus the 514.5 nm radiation from a Coherent Innova 99 Ar^+ laser on the sample. The same microscope objective was used to collect the backscattered radiation. The scattered light collected by the microscope and dispersed by the spectrometer was detected by a charge-coupled device (CCD) detection system. With 1″ CCD and 1800 grooves/mm grating, the spectral resolution was typically less than 1 cm^{-1}.

3. Results and discussions

3.1. Dielectric behavior

The temperature dependent capacitance and loss tangent of $Ba(Zr_xTi_{1-x})O_3$ ($0.30 \leq x \leq 0.70$) thin films were measured in a frequency range of 10 kHz-1 MHz using an impedance bridge and computer controlled thermal stage probe station. Fig. 1 shows the temperature dependent (a) dielectric constant (K) and (b) loss tangent (tanδ) of $Ba(Zr_{0.40}Ti_{0.60})O_3$, measured in a frequency range from 1 KHz to 1 MHz. A broad dielectric anomaly is apparent both in dielectric constant and loss tangent plots; however, the temperatures correspond to the dielectric constant maxima (T_m) do not coincide to that of the corresponding loss tangent maxima (T'_m). Both T_m as well as T'_m shift to higher temperature with the increase in frequency. For the other thin film compositions, similar features were observed in the temperature dependent dielectric constant and loss tangent behaviors as a function of frequencies;

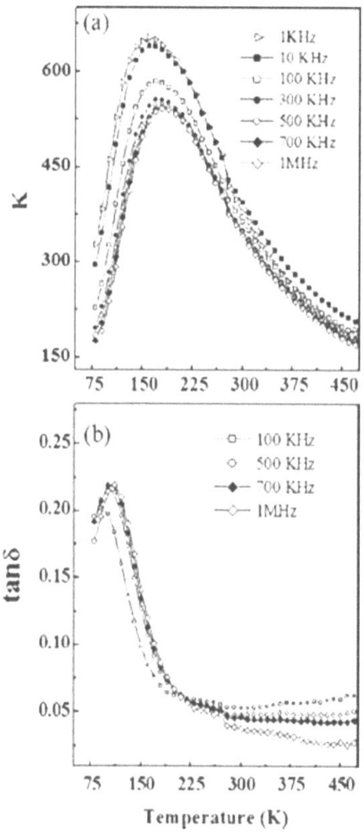

Figure 1 Temperature variation of (a) dielectric constant and (b) loss tangent of $Ba(Zr_{0.40}Ti_{0.60})O_3$ thin film measured in a frequency range of 1 kHz to 1 MHz.

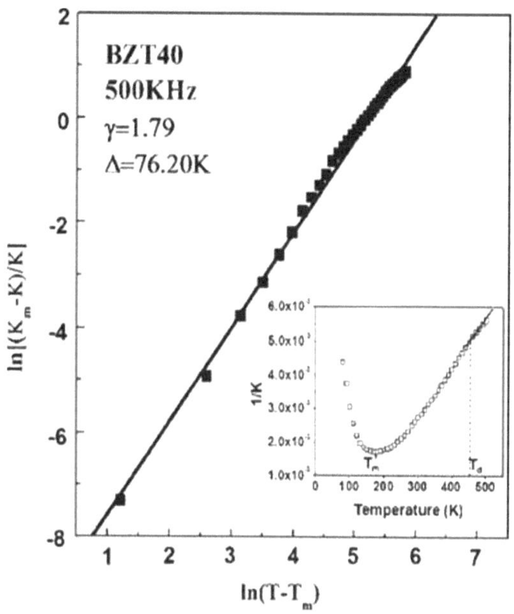

Figure 2 Plot of $\ln[(K_m - K)/K]$ vs. $\ln(T - T_m)$ for $Ba(Zr_{0.40}Ti_{0.60})O_3$ thin films. The symbols represents experimental data and the solid line fitted according to Equation 1. The inset shows the plot of 1/K vs. T for $Ba(Zr_{0.40}Ti_{0.60})O_3$ thin film. The solid line, fitted according to the Curie-Weiss law ($1/K = (T-\theta)/C$, C is the Curie constant, θ is the Curie-Weiss temperature) is valid far above T_m (see text).

however, a systematic decrease of T_m was observed with the increase in Zr contents varied from 30.0 to 70.0 at%. From the observed frequency dependence of the dielectric response it is imperative that in the composition range $0.30 \leq x \leq 0.70$, $Ba(Zr_xTi_{1-x})O_3$ thin films exhibit typical relaxor behavior with the estimated dielectric constant decreases and T_m shifts to higher temperature with the increase in measurement frequencies.

One of the main features of the phase transition in relaxor ferroelectrics in general, is the broadening of the transition termed as diffuse phase transition (DPT). In the vicinity of the broadening of dielectric maxima, the temperature variation of dielectric constant (K) is known to follow a power relation [16]:

$$1/K(f, T) = 1/K_m(f)\{1 + [T - T_m(f)]^{\gamma}/2\Delta^2\} \quad (1)$$

where K is the dielectric constant, K_m is the peak dielectric constant, T is the temperature, T_m is the peak temperature corresponding to the dielectric constant maxima, γ is the degree of relaxation, and Δ is the broadening parameter. From the linear fit between $\ln[(K_m/K)-1]$ vs $\ln(T-T_m)$ we have estimated the values of γ (from slope) and Δ (from intercept) as a function of frequencies for BZT thin films with Zr contents in the range of 30.0 to 70.0 at%. Fig. 2 shows such a typical linear fit for $Ba(Zr_{0.40}Ti_{0.60})O_3$ thin films measured at 500 kHz and the estimated values of γ and Δ for various Zr contents are 1.79 and 76°K respectively. From reported literature, the value of γ is equal to 2 for a perfect relaxor material, whereas for a perfect ferroelectric $\gamma = 1$ and Equation 1 reduced to Curie-Weiss law [6]. As shown in the inset of Fig. 2 the dielectric constant of $Ba(Zr_{0.40}Ti_{0.60})O_3$ film follows the Curie–Weiss behavior at temperatures much higher than T_m. Although a certain degree of subjectivity arises from the choice of the region where the linear fitting is applied, "far" from the temperature T_m, the Curie-Weiss law is considered valid. Accordingly, the temperature (T_d) where K^{-1} (T) starts to deviate from the typical Curie–Weiss behavior (~ 460 K for BZT-40 film), is often considered the onset of the dynamic behavior of local polarization and termed as Burns temperature [17]. Table I summarizes the values of γ, Δ, *and* T_m (at 500 kHz), for BZT thin films with various Zr contents.

It is apparent from Table I that with the increase in Zr contents there is a systematic decrease in Δ, and T_m, whereas the degree of relaxation is much more evident in $Ba(Zr_{0.40}Ti_{0.60})O_3$ thin film. Beyond T_d, BZT thin films are paraelectric, in other words it is the highest temperature below which one begins to detect a local polarization

TABLE I Summary of the fitting parameters of the dielectric anomaly of BZT thin films.

Composition	γ (500 kHz)	Δ (*K*) (500 kHz)	T_m (*K*) (100 kHz)
$Ba(Zr_{0.30}Ti_{0.70})O_3$	1.7	83	213
$Ba(Zr_{0.40}Ti_{0.60})O_3$	1.8	76	170
$Ba(Zr_{0.60}Ti_{0.40})O_3$	1.6	65	154
$Ba(Zr_{0.70}Ti_{0.30})O_3$	1.3	35	147

in nano regions [17]. T_m is only located in the temperature range where a continuous change in the polarization distribution from polar nano-regions to a polar clusters takes place. In this temperature range the local polarization direction is determined by a fixed arrangement of atoms and no cooperative coupling exists between individual polar nano-regions so that an externally applied electric field could reverse the local polarization. In other words, with the application of external electric field rapid orientation of dipole moments in these nano-regions takes place however, once the field is removed the orientation is lost. Usually relaxor behavior is associated with the disorder in the occupation of the equivalent positions by different ions. For example in case of classic relaxor $Pb(Mg_{1/2}Nb_{2/3})O_3$ it is the disorder in the occupation of Mg^{+2} and Nb^{+5} ions in the *B* site of ABO_3 perovskite lattice. On the other hand in case of $Pb_{1-x}La_x(Zr_yTi_{1-y})O_3$ relaxors, it is the disorder in *A* site (between Pb^{+2} and La^{+3} ions) that yield the observed relaxor behavior. In this respect, it is interesting to note that in $Ba(Zr_xTi_{1-x})O_3$ system, in spite of Zr and Ti are isovalent ions, beyond an optimal content of Zr it exhibits typical relaxor behavior. In any classic relaxor system, in the vicinity of T_m, number of nano-size polar-regions exist each of which has a net polarization (P_s). Each of these polar regions has characteristic relaxation time (τ) decided by the local field configuration in them and their characteristic size. As the size of the PR decreases the energy barrier that separates the switching its polarization states also decreases and when the barrier height becomes comparable to thermal energy (kT), the direction of P_s fluctuates with temperature [1, 4]. Viehland *et al.* [2] proposed that the short-range interactions between the PR control the fluctuation of P_s leading to it's freezing at a characteristics temperature. Like the spin glass systems the dielectric relaxation in relaxors can be described by the well-known Vogel-Fulcher relation [18]:

$$f = f_o \exp[-E_a/k_B(T_m - T_f) \qquad (2)$$

where f is the measurement frequency, f_o is the pre-exponential factor, E_a is the activation energy for polarization fluctuations of an isolated cluster, k_B is the Boltzmann constant, T_m is the temperature corresponding to the dielectric maxima, and T_f is the static freezing temperature. From Equation 2 as T_m approaches the characteristic freezing temperature $T_f, f \rightarrow 0$, in other words the kinetics of polarization fluctuation becomes extremely sluggish. From the frequency dependent (10 kHz to 1 MHz) dielectric anomaly we estimated the T_m and corresponding frequency (*f*) for all relaxor BZT thin films with different Zr contents. The data were fitted according to Vogel Fulcher relation (Equation 2) using a non-linear curve-fitting program and the parameters f_o, E_a, and T_f were extracted as a function of Zr contents. Fig. 3 shows the experimental data and the fitted line for BZT-40 thin film. Similar fitting was performed for other BZT compositions and the results are tabulated in Table II. For all the BZT thin film compositions in this study, a remarkable fit of the experimental data with the Vogel-Fulcher relation, suggests that the observed relaxor behavior is analogous to a spin glass system with polarization fluctuation above the static freezing temperature T_f. As shown in Table II, within the fitting error, it seems that the freezing temperature (T_f) reduces with the increase in Zr contents whereas the activation energy reduces in the range of 20–30 meV in the films with higher Zr contents (60.0–70.0 at%).

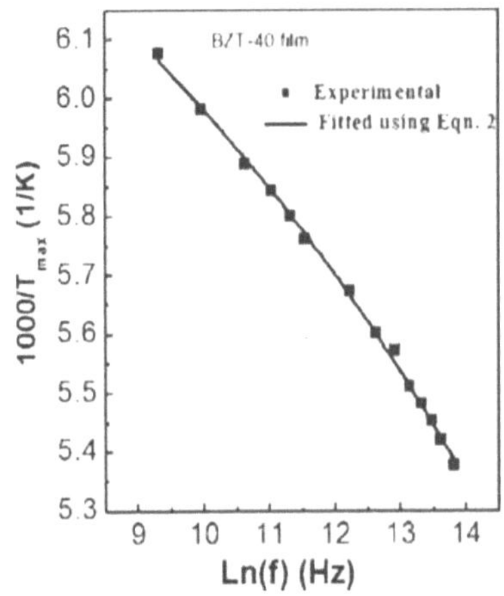

Figure 3 Variation of the inverse of T_m of $Ba(Zr_{0.40}Ti_{0.60})O_3$ thin film with frequency. The solid line is fitted according to the Vogel-Fulcher equation (see text).

TABLE II Summary of the Vogel-Fulcher fitting parameters of BZT thin films

Composition	Freezing temperature (T_f) (*K*)	Activation energy (E_a) (eV)	Pre-exponential factor (f_o) (Hz)
$Ba(Zr_{0.30}Ti_{0.70})O_3$	151(±13)	0.05 (±0.002)	0.036×10^{10}
$Ba(Zr_{0.40}Ti_{0.60})O_3$	118(±7)	0.06 (±0.005)	2.69×10^{10}
$Ba(Zr_{0.60}Ti_{0.40})O_3$	121(±4)	0.02 (±0.004)	0.01×10^{10}
$Ba(Zr_{0.70}Ti_{0.30})O_3$	94(±5)	0.03 (±0.006)	0.015×10^{10}

3.2. Micro-Raman scattering

From the dielectric studies of BZT single crystals it has been reported that on heating the rhombohedral to orthorhombic (T_2) and orthorhombic to tetragonal (T_3) transition temperatures of $BaTiO_3$ is increased whereas the Curie temperature (T_c) (tetragonal to cubic phase transition) is decreased with the increase in Zr content. All these three transition temperatures merge at about 15 at% Zr contents and the transition temperature decreases with further increase in Zr contents. Based on these studies a tentative phase diagram of $Ba(Zr_xTi_{1-x})O_3$ (up to $x \leq 0.2$) has been constructed which indicates that at room temperature $Ba(Zr_{0.20}Ti_{0.80})O_3$ has rhombohedral crystal structure [8]. The structural modification of BTO thin films doped with various Zr contents was investigated by micro-Raman scattering measurements and as reported in our earlier works [19], up to 10.0 at% Zr doped BTO films, the following facts can be ascertained from the temperature dependent Raman spectra, (i) the cubic-tetragonal transition temperature (T_c) progressively reduces with the increase in Zr contents, (ii) the tetragonal to orthorhombic transition temperature remains difficult to identify from the measured spectra, and (iii) the tetragonal (orthorhombic) to rhombohedral transition temperatures increase with increased Zr contents. Unlike the corresponding bulk compositions, the characteristic phase transition temperatures could not be identified from temperature dependent capacitance measurements of the respective thin film samples due to the very broad nature of dielectric anomaly. Beyond 10.0 at% Zr doping, the low temperature (77 K) phase is rhombohedral, with the increase in measurement temperature; no further phase transition could be detected. As for example in case of 20.0 at% Zr doped BTO thin films we have observed a frequency independent broad dielectric anomaly with the temperature corresponding the dielectric constant maxima is around 260 K (data not shown). The structural evolution of this film is elucidated in the temperature dependent Raman spectra (Fig. 4a). The anti-resonant mode about 125 cm^{-1} has been associated to the vibration of Zr ions in oxygen cage [13]. The low temperature (77 K) phase is identified rhombohedral, and also it is interesting to note that even at high temperature almost all the modes related to the rhombohedral structure are present although the intensities are reduced and the FWHM are increased with temperature. In case of $BaTiO_3$ thin film we observed the existence of second order Raman peaks beyond the transition temperature, which indicates a disordered cubic structure. This could be due to the fact that even above transition temperature Ti^{+4} ions do not assume the octahedral position of the perfect ABO_3 unit cell. However, the fact that all the modes in case of BZT-20 films remain relative sharp and intense (as compared to $BaTiO_3$), it clearly indicates that far above the temperature corresponding to the dielectric maxima (T_m) the thin film retains its low temperature rhombohedral structure. In case of $Ba(Zr_{0.40}Ti_{0.60})O_3$ thin films the observed modes are due to rhombohedral crystal structure at all temperature ranges (Fig. 4b). The relatively sharper modes irrespective of the temperature of measurements are indicative of polar nano-size regions in these thin films [20]. A plateau is observed in the high frequency region which consists an extra mode at ~780 cm^{-1} (marked by small arrow in Fig. 4b) in addition to the $A_1(LO_3)$ line at 720 cm^{-1}. The appearance of

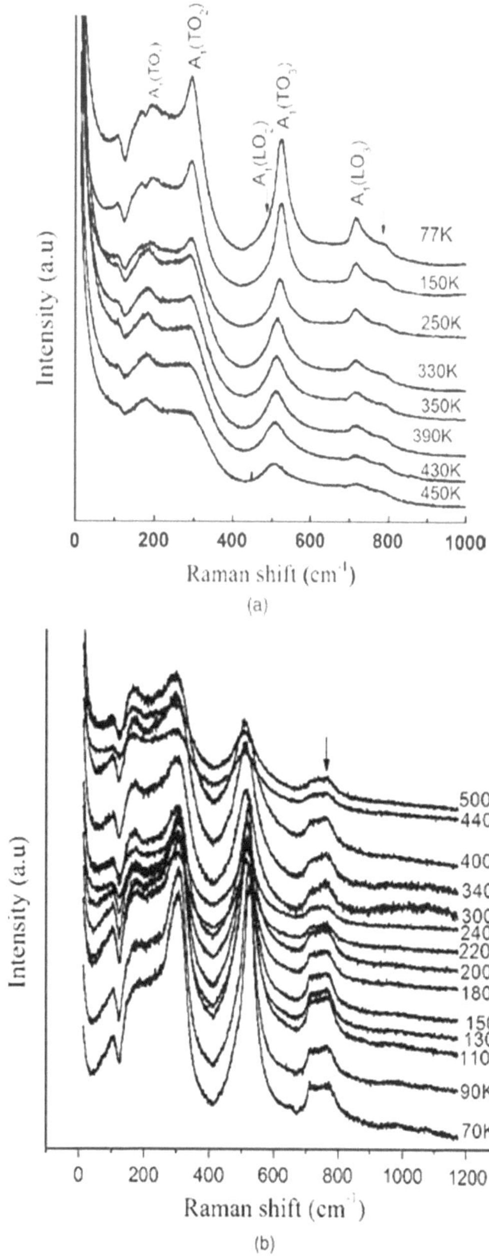

Figure 4 Temperature variation of the micro-Raman spectra of (a) $Ba(Zr_{0.20}Ti_{0.80})O_3$ and (b) $Ba(Zr_{0.40}Ti_{0.60})O_3$ thin films.

such plateau has been correlated with the relaxor nature of BZT ceramics [13]. As presented in the previous section, from the dielectric measurements, we have observed that BZT-20 film exhibits a typical diffuse phase transition behavior (plateau was indistinct) whereas the relaxor nature was distinctly confirmed in 25.0, 30.0 and 40.0 at% Zr doped thin films where the plateau became progressively well developed in the respective Raman spectra.

4. Discussions

The ABO_3 perovskite relaxor systems can be broadly classified into three categories: (i) solid-solution systems involving a polar and non-polar material (e.g. $BaTiO_3$: $BaZrO_3$), (ii) polar material incorporated with aliovalent dopant either at *A* (e.g. $Pb_{1-x}La_x(Zr_yTi_{1-y})_{1-x/4}O_3$) or *B* site ($Ba(Nb_xTi_{1-x})O_z$), and (iii) complex perovskite oxide compounds with ions with mixed valence at *B* sites (e.g. $Pb(Mg_{1/3}Nb_{2/3})O_3$, $Pb(Sc_{1/2}Ta_{1/2})O_3$). The common aspect of all the relaxor materials in these three categories is the intrinsic disorder of constituent cations either in *A* or *B* site and a broad temperature dependent dielectric anomaly loosely termed as diffused phase transition (DPT). Extensive studies have been made to understand the underlying mechanism of relaxor behavior in various complex perovskite systems (category III) and several models have been proposed to explain the observed relaxor behavior which is summarized as follows: As proposed by Smolenski in case of the ABO_3 perovskite, the diffuse nature of dielectric maxima is related to the cation disorder in the '*B*' site of the lattice [21]. This model fails due to the observed fact that many ferroelectrics that exhibit DPT characteristics are not all necessarily exhibit relaxor characteristics. The relaxor nature is proposed to be due to short-range cation order in polar nano-regions in a disorder matrix [22]. In these nano-scale clusters the dipole moments thermally fluctuates between equivalent directions [1]. The broad distribution of relaxation times for cluster orientation originates from the distribution of the potential barriers separating the different orientational states. Each nano-polar region was considered having no interaction among each other and in this respect the material system was considered analogous to super-paramagnetic system and thus the relaxor material system termed as superparaelectric [4]. However, the frequency dispersion of the dielectric maxima (T_m) do not follow a simple Debye type relaxation and in that respect it was postulated that the said polar nano-regions are interactive among themselves unlike the case of a superparamegnetic material [23]. In other words, the relaxors behave like interacting ordered nano-polar regions in a matrix with cation disorder and in this respect the relaxor systems have been proposed to behave like a spin glass state with similar interacting superparamagnetic clusters [24]. In PMN system, a short range cooperative interaction between these superparaelectric clusters was considered by Viehland *et al.* to explain the freezing of the superparaelectric moments into dipolar glassy state with long range polarization ordering at lower temperature [2]. Qian and Bursill [25] on the other hand assumed that the nanometer scale chemical defects are the source of random fields, which control the dynamics of the polar clusters in relaxor systems. The broad dispersive behavior of the temperature dependent dielectric anomaly has been explained by considering changes in the cluster size and correlation length as function of temperature. Despite of extensive studies in last two decades the dielectric relaxation phenomena in bulk relaxor ceramics remain unclear. As compared to category II and III based relaxor systems, the underlying mechanism resulting the relaxation behavior of category I based relaxor systems have been less studied especially in thin film form. The relaxor behavior was in fact first reported by Smolenskii et al in one of such systems namely Sn doped $BaTiO_3$, followed by Hf, Ce, Y, Zr doped $BaTiO_3$ ceramics (beyond a critical dopant contents) have also been reported to exhibit typical relaxor behavior [26].

In the present work the observed relaxor behavior in BZT thin films (with Zr content $\geq$25.0 at%) can be hypothesized as follows: In BZT solid solution system, the *B* site is randomly occupied by either Ti^{+4} or Zr^{+4} ions. $BaZrO_3$ is cubic and non-ferroelectric whereas $BaTiO_3$ is ferroelectric with tetragonal crystal structure at room temperature. The long-range polarization ordering due to Ti-O dipole-dipole interaction (yielding so called domain structure) in $BaTiO_3$ is hindered due to Zr^{+4} substitution. The primary support of this hypothesis is reflected in the measured room temperature polarization hysteresis as a function of Zr content (Fig. 5). The figure shows that as the Zr contents are systematically increased the hysteresis loops become slimmer which indicates that the long range polarization ordering is broken with Zr doping. From XRD analyses we have found that within the detection limit of the X-Ray diffractometer used, up to 70.0 at%, Zr^{+4} ions are completely soluble in $BaTiO_3$ thin films without the formation of any secondary phase/(s). The solubility of Zr in $BaTiO_3$ lattice has been confirmed by systematic increase of the lattice parameter of BZT films with the increase in Zr contents [9]. Within the limit of this solid solubility, no cationic order between Zr^{+4} and Ti^{+4} cations are expected in $Ba(Zr_xTi_{1-x})O_3$ thin films. The long range polarization order of BTO is marginally disturbed when Zr content is less (say 5.0 at%), however, in case of 20.0 at% Zr doped film, due to the abundance of Zr modified oxygen octahedra the long range periodicity of Ti modified oxygen octahedra is disturbed and the polar regions are developed in a weakly polar matrix. The observed slimmer polarization hysteresis of BZT-20 thin films probably indicates that the dimension of the polar regions are smaller than the size of the micron size domains in BTO thin film, however, they are large enough to have necessary dipolar cooperation among neighboring unit cells to exhibit

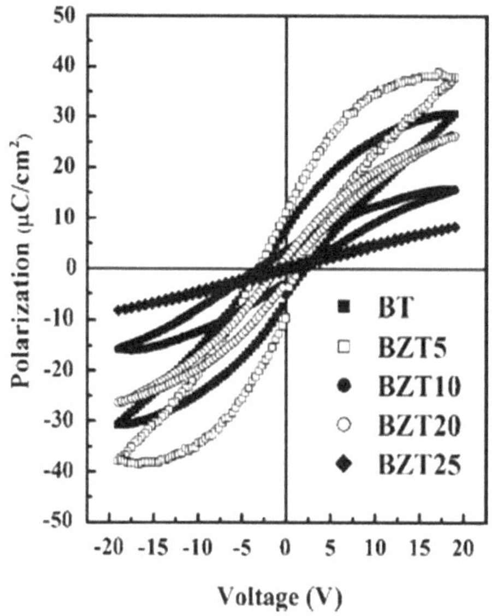

Figure 5 Polarization hysteresis of $Ba(Zr_xTi_{1-x})O_3$ ($0 \leq x \leq 0.25$) thin films measured at room temperature.

polarization hysteresis. With the increase in Zr content the size of the polar regions are decreased and the Zr rich regions are increased and as a result the intermediate layer in the vicinity of the polar regions becomes more and more non-polar. It is noticed that the mode at $\sim$125 cm^{-1} present in all Zr rich (>5.0 at%) BZT thin film of the measured temperature. Smaller the sizes of the PR regions larger are the temperature fluctuations in them [1]. The thermal energy (kT) results the polarization fluctuations in the polar regions. In other words the polarization in these PR are dynamic in nature and the thin film starts behaving like a relaxor. Since, Zr and Ti ions are in same valence state and also the magnitude of their ionic radii are not very different; substitution of Ti by Zr is not expected to yield very strong random field at relatively lower Zr contents. However, a mutual interaction between the polar and non-polar regions is expected and as the fraction of Zr rich regions are increased. The nature of this interaction may be electrostatic or elastic in nature. Unless they are far apart the electrostatic interaction is plausible through dipolar interaction among the PR. Moreover, it has been proposed that PR regions create a complicated pattern of normal and shear strains in the PE surroundings. The mutual interaction among the polar regions depends on their arrangements, spacing, as well as direction of spontaneous polarization in them [27]. The mutual interaction among the polar regions, as well as interaction (electro-static as well as elastic) among the polar regions and non-polar regions could be the source of random field, which becomes significant with the increase of Zr contents. As shown in Table I, the decrease of T_m with the increase of Zr contents could be due to the increase of the magnitude this random field [28]. Additionally, in the case of BZT thin films, microstructure could also induce random field distribution. As for example the grain boundary could contribute random electric field in addition to that produced by the mutual interaction among the polar and non-polar regions. We have observed that the grain size of the BZT film decreases with the increase of Zr contents [15]. Smaller grain size and higher concentration of grain boundary would in turn increase the random field strength and thereby T_m is reduced with the increase in Zr contents. The broadening of the dielectric anomaly could also be related to the relative fraction of the polar regions. As the fraction of the polar regions reduces with the increase of the Zr contents, the broadening (Δ) also decreases (see Table I). The degree of relaxation (γ) is also related to the mean size of the polar regions. As mentioned earlier, with the increase in Zr contents the size of the polar regions are influenced, and γ is maximum at a critical Zr contents of about 40.0 at% and beyond that it decreases as the mean size of the polar regions changes.

The frequency dispersion of the dielectric anomaly of BZT thin films relaxors follow the well-known Vogel-Fulcher relation (Equation 2). In Equation 2, E_a represents the activation energy of the polarization fluctuation. On cooling, the polarization fluctuation minimizes due to the development of short range ordering between the neighboring polar regions. The dispersion of the size of these polar regions decides the spectrum of the relaxation time. With the analogy of a magnetic spin glass system, E_a is believed to be a product of the volume of the polar region (V_{PR}) and the anisotropic energy of polarization direction (K_{anis}). K_{anis} represents the barrier for rotation of the polarization vector between the allowable directions of a specific crystal structure. Assuming K_{anis} does not change significantly for BZT thin film relaxors with varying Zr contents (30.0 to 70.0 at%) the reduction of the activation energy E_a from 50–60 meV (for 30.0 to 40.0 at% Zr doped films) to 20–30 meV (for 60.0–70.0 at% Zr doped films) indicates that the volume content of the polar regions (V_{PR}) reduce with the increase in Zr contents (Table II). As the temperature is reduced a strong correlation among the polar regions may develop which would tend to drive the system towards a long-range ferroelectric state. The parameter kT_f can be considered as a measure of the interaction energy required for the correlation among the polar regions. As shown in Table II, the systematic reduction of T_f (thereby kT_f) probably indicates that the tendency for long range polar ordering reduces in BZT relaxor films with the increase in Zr contents. The order of the pre-exponential factor is about 2–4 orders of magnitude smaller than that reported for PMN bulk relaxor [2]. The difference could be due to difference in sizes and nature of the polar regions in these two systems.

To get further understanding about the nature of the polar regions we have performed detailed analyses of the recorded Raman spectra of BZT thin films. As pre-

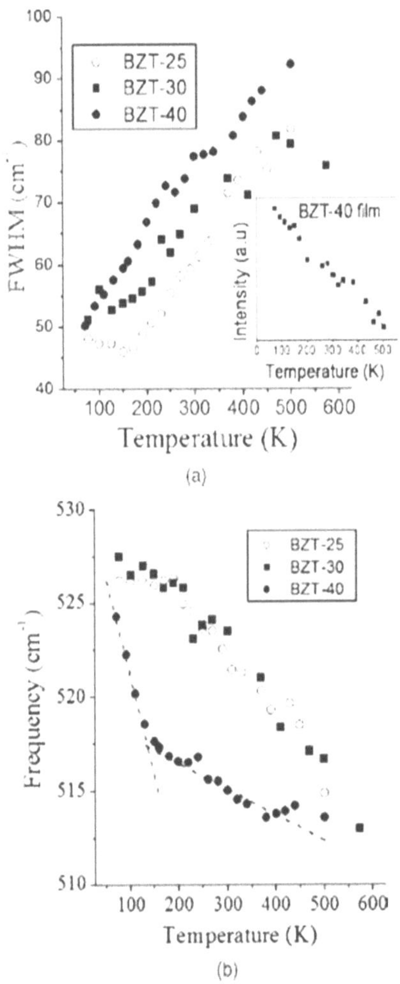

Figure 6 The temperature variation of (a) the FWHM and (b) the frequencies of TO mode of BZT thin films with various Zr contents. The inset of Fig. 6a shows the variation of the integrated intensity of TO mode as a function of temperature for BZT 40 thin film.

sented earlier the observed Raman modes in these thin films result from the coupling of the hard phonon polar modes with the fluctuating polarization in the polar regions. For relaxor ceramics such as 9/65/35 PLZT as well as $Ba(Zr_{0.40}Ti_{0.60})O_3$, it was shown that the integrated intensity of TO mode at ~520 cm^{-1} exhibits a continuous increase upon cooling followed by a plateau region where the intensity does not increase upon further cooling. The temperature corresponding to the initiation of this plateau region correlates well with the corresponding freezing temperatures of these relaxors as determined from the Vogel Fulcher analyses. No appreciable change in FWHM as well as frequency of this mode was observed in relaxor BZT ceramics. The increase of the TO mode intensity has been correlated to the size of the polar regions, both in PLZT and BZT relaxor ceramics. We have performed similar analyses in relaxor BZT thin films and the TO mode has been fitted with a damped harmonic oscillator type phonon function corrected for the Bose-Einstein factor. Fig. 6 shows the temperature variation of (a) the FWHM and (b) the frequencies of TO mode of BZT thin films with various Zr contents. The inset of Fig. 6a shows the variation of the integrated intensity of TO mode as a function of temperature for BZT-40 thin film. Several interesting features are apparent from these plots. First, unlike the bulk BZT-40 relaxor, the integrated intensity is continuously increased with the reduction of temperature and no plateau is observed below the characteristic freezing temperature (T_f). Secondly, with the increase in Zr contents ($\geq$25.0 at%), irrespective of the temperature, the magnitudes of FWHM are increased whereas the mode frequencies are systematically decreased. Thirdly, with the decrease in measurement temperature, there is a continuous decrease of the FWHM and increase of the TO mode frequency for all BZT thin film relaxor compositions. The behavior of the fitted TO Raman mode correlates well with the hypothesis presented earlier in this section for BZT relaxor thin films. Thus, the continuous increase of the integrated intensity of the TO mode with the reduction in temperature is indeed related to the growth of polar regions due to increased mutual interaction among them with cooling. For BZT 40 thin films, probably the clustering of the polar regions continue below the freezing temperature yielding a long rage polarization order. The change in the magnitudes of the half width and the mode frequency reflects the size of the polar regions: lower FWHM indicates bigger polar regions whereas bigger the polar regions larger are their mode frequencies. Viewing in this light one can identify that irrespective of the temperature, with the increase in Zr contents the size of the polar regions are continuously reduced and also for all BZT relaxor thin films, the size of the polar regions increases with the reduction of temperature. Although not very prominent in 25.0 and 30.0 at% Zr doped compositions; in case of BZT-40 thin film, a distinct change in the slope of frequency (as marked by dashed lines in Fig. 6b) is apparent. The temperature corresponding to this slope change matches well with the freezing temperature for this composition (see Table II) and the rapid increase of the frequency below the characteristic freezing temperature probably indicates the occurrence of a long range polar ordering below the freezing temperature. Fig. 7 shows the polarization hysteresis of BZT-40 thin films measured at different temperature. The appearance of polarization hysteresis below the characteristic T_f of BZT thin film indeed indicate the induction of long range polar order. It remains unclear whether in relaxor BZT ceramics undergoes a transition from a relaxor phase to a ferroelectric state at T_f [12, 13]. Our data indicates such a transition occurs in BZT-40 thin films.

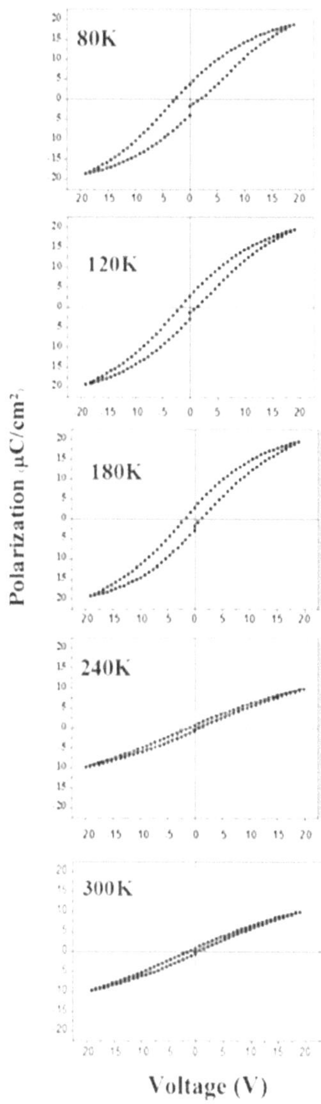

Figure 7 Polarization hysteresis of $Ba(Zr_{0.40}Ti_{0.60})O_3$ thin films measured in the temperature range of 80–300 K.

5. Conclusions

The present work was devoted to understand the relaxor behavior of sol-gel derived $Ba(Zr_xTi_{1-x})O_3$ (BZT) thin films ($0.30 \leq x \leq 0.70$). Beyond 25.0 at% Zr substitution, BZT thin films exhibit typical relaxor characteristics with a diffuse temperature dependent dielectric anomaly coupled with the shift of both the dielectric constant and loss tangent towards higher temperature with the increase in measurement frequencies. It was found that the temperature corresponding to the peak of dielectric anomaly (T_m) as well as the degree of broadening (Δ) systematically reduced with the increase in Zr content, whereas the degree of relaxation (γ) had a maximum (~1.8) at 40.0 at% Zr doped BZT film. The long range polarization ordering is disturbed with Zr substitution in Ti site of the barium titanate (BTO) lattice to yield polar region (Ti rich) whose sizes are reduced with the increase in Zr contents. The dielectric relaxation in these thin films is attributed due to the presence of these polar regions in a non-polar host. The frequency dependence of the dielectric anomaly follows Vogel-Fulcher (VF) relation indicating the interaction among these nano-polar regions followed by a cooperative freezing similar to that of spin-glass systems and classic complex oxide relaxor ferroelectrics. The activation energy (E_a) and the freezing temperature (T_f) as a function of Zr contents were estimated from the VF fitting of the dielectric maxima vs frequency data. From the fact that both E_a and T_f reduce with the increase in Zr contents, it has been argued that the volume of the polar regions and the interaction energy for cooperative freezing reduce with the increase in Zr contents. The existence of these polar regions is ascertained on the basis of the analyses of micro-Raman spectra. The temperature dependent Raman spectra indicated a rhombohedral structure of these polar regions (irrespective of the Zr contents) at liquid nitrogen temperature and no apparent structural transition was observed upon heating far above the temperature corresponding to the dielectric maxima (T_m). From the variation of the frequency and the full width at half maxima of the TO mode of Raman spectra for relaxor BZT films it was supported that (i) with the increase of Zr contents the size of the polar regions are reduced and (ii) when Zr content is fixed, the size of the polar regions increase with cooling yielding a long range polar ordering. Thus, in case of BZT-40 thin film, a stable polarization hysteresis was obtained below its freezing temperature. Detailed micro-Raman analyses in conjunction with temperature dependent transmission electron microscopy measurements are in progress to further investigate the relaxor behavior in these films.

Acknowledgements

This work was supported in parts by NSF-DMR 305588 and NSF-ID 100097018 grants.

References

1. L. E. CROSS, *Ferroelectrics* **76** (1987) 241.
2. D. VIEHLAND, S. J. JANG, L. E. CROSS and M. WUTTIG, *J. Appl. Phys.* **68** (1990) 2916.
3. V. WESTPHAL, W. KLEEMANN and M. D. GLINCHUK, *Phys. Rev. Lett.* **68** (1992) 847.
4. L. E. CROSS, *Ferroelectrics.* **151** (1994) 305.
5. G. A. SAMARA, *J. Phys.: Condens. Matter.* **15** (2003) R367.
6. Z. YU, C. ANG, R. GUO and A. S. BHALLA, *J. Appl. Phys.* **92** (2002) 2655.
7. Idem. *ibid.*, **92** (2002) 1489.
8. Z. YU, R. GUO and A. S. BHALLA, *ibid.* **88** (2000) 410.

9. A. DIXIT, S. B. MAJUMDER, R. S. KATIYAR and A. S. BHALLA, *Appl. Phys. Lett.* **82** (2003) 2679.
10. J. ZHAI, X. YAO, L. ZHANG and B. SHEN, *ibid.* **84** (2004) 3136.
11. P. SCIAU and A. M. CASTAGNOS, *Ferroelectrics,* **270** (2002) 259.
12. P. SCIAU, G. CALVANIN and J. RAVEZ, *Solid State Comm.* **113** (2000) 77.
13. R. FARHI, M. E. MARSSI, A. SIMON and J. RAVEZ, *J. Eur. Phys. B* **9** (1999) 599.
14. Y. HOTTA, G. W. J. HASSINK, T. KAWAI and H. TABATA, *Jpn. J. Appl. Phys.* **42** (2003) 5908.
15. A. DIXIT, S. B. MAJUMDER, A. SAVVINOV, R. GUO and A. S. BHALLA, *Mat. Lett.* **56** (2002) 933.
16. H. T. MARTIRENA and J. C. BURFOOT, *Ferroelectric.* **7** (1974) 151.
17. G. BURNS and F. H. DACOL, *Phys. Rev. B.* **28** (1983) 2527.
18. H. VOGEL, *Z. Phys.* **22** (1921) 645; G. FULCHER, *J. Am. Ceram. Soc.* **8** (1925) 339.
19 A. DIXIT, S. B. MAJUMDER, P. S. DOBAL, R. S. KATIYAR and A. S. BHALLA, *Thin Solid Films* **447–448** (2004) 284.
20. I. G. SINY, R. S. KATIYAR and A. S. BHALLA, *Ferroelectric Review.* **2** (2000) 51.
21. G. SMOLENSKI and A. AGRANOVSKAYA, *Sov. Phys. Solid State.* **1** (1960) 1429.
22. C. RANDALL, A. BHALLA, T. SHROUT and L. E. CROSS, *J. Mater. Res.* **5** (1990) 829.
23. L. NEEL, *Compt Rend. Acad. Sci.* **228** (1949) 664.
24. A. MORGONWNIK and J. MYDOSH, *Solid State Commun.* **47** (1983) 321.
25. H. QIAN and L. A. BURSILL, *Int. J. Mod. Phys. B* **10** (1996) 2007.
26. C. ANG, Z. JING and Z. YU, *J. Phys. Condens. Matter.* **14** (2002) 890 and references therein.
27. V. A. ISUPOV, *Physics of the Solid State* **45** (2003) 1107.
28. M. D. GLINCHUK and R. FARHI, *J. Phys.: Condens. Mater* **8** (1996) 6985.

JOURNAL OF MATERIALS SCIENCE 41 (2006) 97–106

Magnetoelectric coupling, efficiency, and voltage gain effect in piezoelectric-piezomagnetic laminate composites

SHUXIANG DONG, JIE-FANG LI, D. VIEHLAND
Department of Materials Science & Engineering, Virginia Tech, Blacksburg, VA 24061

The magnetoelectric (ME) effect of piezoelectric-magnetostrictive laminate composites, which is a product tensor, has been studied. Based on piezoelectric and piezomagnetic constituent equations, the longitudinal-mode vibration and equivalent circuits have been derived. The effective magnetoelectric coupling coefficient, voltage-gain, and output efficiency have been determined. Our results show: (i) that there is an extreme high voltage gain effect of >260 under resonance drive: the induced ME voltage is much higher than the input voltage to the coils for magnetic excitation; (ii) that there is an optimum ratio of the piezoelectric to piezomagnetic layer thicknesses, which results in maximum effective magnetoelectric coupling; and (iii) that the maximum output efficiency of magnetoelectric laminate at resonance drive is ~98%, if eddy currents are neglected. This high ME voltage gain effect offers potential for power transformer applications.

1. Introduction

The magnetoelectric effect is a polarization response to an applied magnetic field *H*, or conversely a spin response to an applied electric field *E* [1]. Ferroelectromagnetic materials have been studied [2–5] such as $BiFeO_3$ and $Pb(Fe_{1/2}Nb_{1/2})O_3$; however to date, a single phase material with a high inherent coupling between spin and polarization has yet to be found.

Magneto-electric behavior has also been studied as a composite effect in multi-phase systems consisting of both piezoelectric and magnetostrictive materials [6–32]. Piezoelectric/ magnetostrictive composites have been the topic of numerous investigations, both experimentally and analytically. Various composite connectivities of the two phases have been studied including: 3-3 (i.e., ceramic-ceramic particle composite [6, 7]; ceramic, rare-earth iron alloys and polymer composites [8, 9]) and 2-2 (laminate composites [10–32]). These studies have confirmed the existence of magnetoelectric effects in composites; however, the magnitude of the coupling was low for both connectivities.

Analytical and experimental investigations have focused on 2-2 type laminate composites of magnetostrictive/piezoelectric bi-materials operated at both low and resonance frequency ranges [9–17]. Recently, we have designed and prototyped several resonance-type laminate geometries, and a strong resonance ME effect and ME voltage gain effect were observed [11, 15–17]. Resonance-type magnetoelectric devices are needed in applications as high-power solid-state devices. Such applications require high voltage gains and high output efficiencies. In this paper, we will further treat the resonance ME effect using an analytical approach. Our approach is based on an equation of motion to couple the two constitutive equations of the piezoelectric and piezomagnetic bi-materials, and an equivalent circuit method. This method is significantly different than prior analytical method [8,17–19, 22, 23]. Our analysis was developed for a long-plate-type magnetostrictive/piezoelectric laminate, operated in a longitudinal-mode vibration. The purpose was to extend the analytical approach for a resonance type ME transformer. The effective ME coupling coefficient, voltage gain, and output efficiency at resonance have been derived. The results show the presence of extremely high magnetoelectric voltage gain effects, suitable for solid-state transformer applications.

2. Analysis of Piezoelectric-magnetostrictive laminate composites under longitudinal excitation

Fig. 1 shows the schematic of the composite geometry used in this investigation. It is a long plate type

0022-2461 © *2006 Springer Science + Business Media, Inc.*
DOI: 10.1007/s10853-005-5930-8

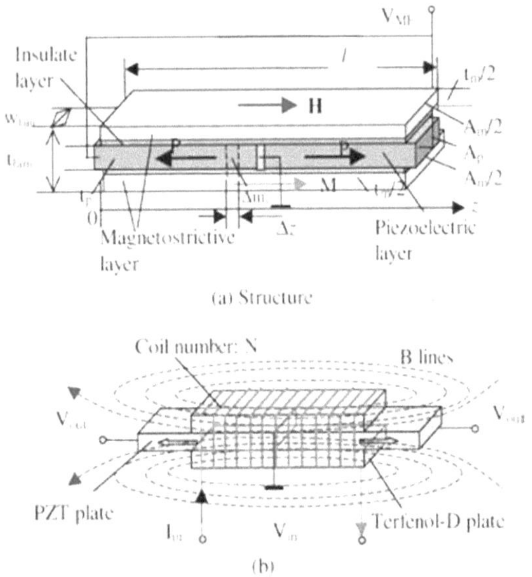

Figure 1 (a) Schematic of the geometry of the magnetostrictive/piezoelectric laminate composite, the polarization direction of the two piezoelectric plates is along the longitudinal direction, the magnetic field is applied along longitudinal direction; and (b) schematic and operational principle of the ME transformer. An N-turn coil carrying current I_{in} around the magnetoelectric laminate is used to produce a magnetizing field.

piezoelectric/magnetostrictive laminate composite, in which the piezoelectric layer (PZT) is sandwiched between two magnetostrictive ones (Terfenol-D, $Tb_{1-x}Dy_xFe_{2-y}$). More complicated geometries of this general type, such as a thin multi-layer type, are possible, but that given in Fig. 1a readily allows for equivalent circuit analysis. Conductive magnetostrictive layers are separated from a piezoelectric one by thin insulating layers. Thus, eddy currents are effectively eliminated. This long-type ME laminate design intensifies the longitudinal direction vibrations along which fields are applied. The two piezoelectric layers are longitudinally polarized reversely (push-pull type [11]), which maximizes the voltage and power outputs.

The working principle is as follows. A harmonic ac magnetic field H is applied along the longitudinal direction of the composite. This causes the two magnetostrictive layers to shrink/expand in response to H. The magnetostrictive strain acts upon the piezoelectric layer that is bonded between the two magnetostrictive layers, causing the piezoelectric layer to strain, producing a voltage output. This transduction of magnetic to electrical energy is what we designate as the magnetoelectric coupling effect.

A solenoid with *N* turns around the laminate that carries a current of I_{mag} (or I_{in}) was used to excite an ac magnetic field H_{ac}, as shown in Fig. 1b. An input ac voltage applied to the coils was V_{mag} (or V_{in}), and its frequency was *f*. This excites a H_{ac} of the same frequency *f*, along the longitudinal direction of the laminate. When the frequency of H_{ac} is equal to the resonance frequency of the laminate, the magnetoelectric coupling effect is so strong that the output ME voltage V_{out} induced in the piezoelectric layer is much higher than V_{mag}. Thus, under resonant drive, there is a high voltage gain, due to the magnetoelectric effect.

To obtain a maximum magnetoelectric voltage gain, the polarization direction of the piezoelectric layer was chosen to be along its length direction under longitudinal vibration, as shown in Fig. 1a. This is because both the piezoelectric constant g_{33} and electromechanical coupling coefficient k_{33} in the longitudinal mode are 2× that of the transverse g_{31} and k_{31} coefficients [33]. Consequently, higher output voltages V_{out} can be obtained from the magnetoelectric laminate. Magnetostrictive Terfenol-D also has the highest coupling in the longitudinal mode [34]. Thus, to obtain the maximum magnetoelectric effect, the magnetization direction was chosen to be along its length direction under longitudinal vibration, as shown in Fig. 1a.

2.1. Constitutive equations of the laminate composite

Application of *H* along the length direction of the laminate excites a longitudinal ($d_{33,m}$) mode in the magnetostrictive layer. Two sets of constitutive linearized equations are required to describe the coupled responses of the piezoelectric and magnetostrictive layers. These are [33]

(Piezoelectric constitutive equation)

$$\varepsilon_{3p} = s_{33}^D \sigma_{3p} + g_{33,p} D_3; \quad E_3 = -g_{33,p}\sigma_{3p} + \beta_{33}^T D_3 \tag{2.1.1}$$

(Piezomagnetism constitutive equations)

$$\sigma_{3m} = \frac{1}{s_{33}^B}\varepsilon_{3m} - \lambda_{33} B_3; \quad H_3 = -\lambda_{33}\varepsilon_{3m} + \nu_{33}^s B_3 \tag{2.1.2}$$

$$\nu_{33}^s = \frac{1}{\mu_{33}^s}; \quad \mu_{33}^s = \mu_{33}^T(1 - k_{33,m}^2);$$

$$s_{33}^B = s_{33}^H(1 - k_{33,m}^2); \quad k_{33,m}^2 = \frac{d_{33,m}^2}{s_{33}^H \mu_{33}^T};$$

$$\lambda_{33} = \frac{d_{33,m}}{s_{33}^H \mu_{33}^s}$$

where ε_{3p} and ε_{3m} are the longitudinal piezoelectric and piezomagnetic strains; D_3 is the dielectric displacement; s_{33}^D is the elastic compliance of the piezoelectric material under constant *D;* $g_{33,p}$ and $d_{33,m}$ are the piezoelectric and piezomagnetic constants; σ_{3p} and σ_{3m} are the longitudinal stresses in the piezoelectric and magnetostrictive layers; β_{33}^T is the dielectric stiffness under constant stress;

ν_{33}^{s} is the magnetic stiffness (reluctivity) under constant strain; μ_{33}^{s} and μ_{33}^{T} are the magnetic permeabilities under constant strain and stress; s_{33}^{B} and s_{33}^{H} are the elastic compliances of the magnetostrictive layer under constant B_3 and H_3; $k_{33,p}$ and $k_{33,m}$ are the coupling coefficients of the piezoelectric and magnetostrictive layers; λ_{33} is the magnetostrictive coefficient; and B_3 is the magnetization.

Upon considering the insulating layer in the laminate, we have to introduce a third constitutive equation:

$$\varepsilon_{3I} = \frac{1}{Y^I}\sigma_{3I} \qquad (2.1.3)$$

where ε_{3p}, σ_{3p} and Y^I are longitudinal strain, stresses and Young's modulus of the insulating layer, respectively.

For Terfenol-D, although strain due to magnetostriction is λH^2, under appropriate magnetic biases pseudo-linear piezomagnetic equations can be used to express their magnetostrictive performances [21, 34]. The optimum magnetic bias will have Terfenol-D working in its optimum magnetostrictive state, but operating in its pseudo-linear piezomagnetic range. In addition, the constitutive equationns. (2.1.1), (2.1.2) and (2.1.3) do not account for loss factors. Significant energy dissipation and nonlinearity in both piezoelectric and magnetostrictive materials are known, in particular under resonant operation. Accordingly, a better analysis would include a mechanical quality factor Q_m to account for dissipation.

2.2. Magneto-elastic-electric coupling

Under an applied H_{ac}, a longitudinal vibration mode is excited by magnetostrictive effect in the Terfenol-D layers of the laminate. Under harmonic motion along $\hat{z}$ (longitudinal direction), it will be supposed that all small mass units Δm_i in the magnetostrictive, piezoelectric and insulation layers of the laminate at z have the same displacement $u(z)$ or strain ε_z, given as

$$u_p(z) = u_m(z) = u_I = u(z)$$

or

$$\varepsilon_{3p}(z) = \varepsilon_{3m}(z) = \varepsilon_{3I} = \varepsilon_3(z) = \frac{\partial u(z)}{\partial z} \qquad (2.2.1)$$

This follows from Fig. 1a by assuming that the layers in the laminate act only in a coupled manner, without any sliding between layers. Following Newton's Second Law, we then have

$$\sum_i \Delta m_i \frac{\partial^2 u}{\partial t^2} = \sum_i \sigma_{3i} A_i \qquad (2.2.2)$$

where i denotes the layer number, and A_i cross-section area of ith layer. We can then use this motion equation to couple the two piezoelectric and piezomagnetism equations (2.1.1) and (2.1.2).

We have previously derived a solution to the equation of motion for the laminate design shown in Fig. 1 given as

$$u(z) = \frac{\dot{u}_1}{j\omega}\cos kz + \frac{\dot{u}_2 - \dot{u}_1 \cos kl}{j\omega \sin kl}\sin kz. \qquad (2.2.3)$$

$$k^2 = \frac{\omega^2}{\bar{\upsilon}^2};\ \bar{\upsilon}^2 = \left(\frac{n_m}{s_{33}^B} + \frac{n_p}{s_{33}^D} + n_I Y^I\right)/\bar{\rho}; \qquad (2.2.4)$$

where $u(z)$ is the mechanical displacement, $\dot{u}_1$ and $\dot{u}_2$ are the mechanical displacement velocities (i.e., mechanical currents) at the two ends ($z = 0$ and $z = l$) of the laminate; $\bar{\upsilon}$ is the average sound velocity in the laminate; $n_m = \frac{A_m}{A_{lam}}$, $n_p = \frac{A_p}{A_{lam}}$, $n_I = \frac{A_I}{A_{lam}}$ are geometrical factors describing the cross-sectional area ratio (or volume fraction) of the magnetostrictive, piezoelectric and insulation layers, respectively; $A_{lam} = A_m + A_p + A_I$ is cross-sectional area of the laminate; and $\bar{\rho} = n_m\rho_m + n_p\rho_p + n_I\rho_I$ is the average mass density of the laminate. In addition, the forces acting on the end faces of the laminate (at $z = 0$ and $z = l$) can be determined as

$$F_1 = F(0) = -\left(\frac{A_p}{s_{33}^D} + \frac{A_m}{s_{33}^B} + Y^I A_I\right)\frac{\dot{u}_2 - \dot{u}_1\cos kl}{j\bar{\upsilon}\sin kl} + \frac{A_p g_{33,p}}{s_{33}^D}D_3 + \frac{\lambda_{33}}{j\omega N}V_{mag} \qquad (2.2.5)$$

$$F_2 = F(l) = -\left(\frac{A_p}{s_{33}^D} + \frac{A_m}{s_{33}^B} + Y^I A_I\right)\frac{\dot{u}_2\cos kl - \dot{u}_1}{j\bar{\upsilon}\sin kl} + \frac{A_p g_{33,p}}{s_{33}^D}D_3 + \frac{\lambda_{33}}{j\omega N}V_{mag}. \qquad (2.2.6)$$

The strain in the piezoelectric layer caused by the magnetostrictive layers produces a corresponding voltage and current in the piezoelectric one. By eliminating σ in Equation 2.1.1, the electric field across the piezoelectric layers can be determined as

$$E_3 = -\frac{g_{33,p}}{s_{33}^D}\varepsilon_{3,p} + \bar{\beta}_{33}D_3 \qquad (2.2.7)$$

where,

$$\bar{\beta}_{33} = \beta_{33}^T\left(1 + \frac{g_{33,p}^2}{s_{33}^D\beta_{33}^T}\right).$$

The output voltage V_{out} produced between the middle and one end of the piezoelectric layer due to the magnetostrictive strain can then be determined as

$$V_{\text{out1}} = \int_{l/2}^{l} E_3 dz = \frac{-g_{33,\text{p}}}{j\omega s_{33}^{D}}(\dot{u}_2 - \dot{u}_0) + \bar{\beta}_{33} D_3 l/2$$

or

$$V_{\text{out2}} = \int_{l/2}^{0} -E_3 dz = \frac{g_{33,\text{p}}}{j\omega s_{33}^{D}}(\dot{u}_1 - \dot{u}_0) - \bar{\beta}_{33} D_3 l/2 \tag{2.2.8}$$

where $\dot{u}_0$ is mechanical displacement velocity at middle point of the laminate. For a $\lambda/2$ resonator, the middle is a nodal position, i.e., $\dot{u}_0 = 0$, and $\dot{u}_1 \equiv -\dot{u}_2$. Correspondingly, the induced dielectric displacement current I_{disp1} from one end to the middle or center in one half piezoelectric layer is

$$\text{I}_{\text{disp1}} = \text{j}\omega \text{C}_0 V_{\text{out1}} + \varphi_p(\dot{u}_2 - \dot{u}_0)$$

or

$$I_{\text{disp2}} = -j\omega \text{C}_0 V_{\text{out2}} + \varphi_p(\dot{u}_1 - \dot{u}_0)$$

Because the two output end electrodes of the piezoelectric layer are connected together, and the middle electrode is common ground, the induced current from the whole piezoelectric layer should be

$$\text{I}_{\text{out}} = \text{I}_{\text{disp1}} + (-\text{I}_{\text{disp2}}) = \text{j}\omega(2\text{C}_0) V_{\text{out}} + \varphi_p(\dot{u}_2 - \dot{u}_1)$$

$$C_0 = \frac{A_p}{\bar{\beta}_{33} l/2}, \quad \varphi_p = \frac{A_p g_{33,\text{p}}}{s_{33}^{D} \bar{\beta}_{33} l/2} \tag{2.2.9}$$

where C_0 is the clamped capacitance, and φ_p is elasto-electric coupling factor for the piezoelectric layer.

The forces exerted on the two end faces of the laminate at $z = 0$ and $z = l$ can now be related to the exciting voltage V_{mag}, the mechanical displacement velocities $\dot{u}_1$ and $\dot{u}_2$, and the output voltage V_{out} from the piezoelectric layer, given as

$$F_1 = Z_1 \dot{u}_1 + \left(Z_2 + \frac{\varphi_p^2}{j\omega(-2C_0)}\right)(\dot{u}_1 - \dot{u}_2) + \varphi_\text{p} V_{\text{out}} + \varphi_\text{m} V_{\text{mag}}$$

$$F_2 = -Z_1 \dot{u}_2 + \left(Z_2 + \frac{\varphi_p^2}{j\omega(-2C_0)}\right)(\dot{u}_1 - \dot{u}_2) + \varphi_p V_{\text{out}} + \varphi_\text{m} V_{\text{mag}}$$

$$Z_1 = j\overline{\rho v} A_{\text{lam}} \cdot tg\left(\frac{kl}{2}\right),$$

$$Z_2 = \frac{\overline{\rho v} A_{\text{lam}}}{j \sin(kl)}, \quad \varphi_m = \frac{\lambda_{33}}{j\omega N}; \tag{2.2.10}$$

where Z_1 and Z_2 are the mechanical impedances of the laminate; and φ_m and φ_p are the magneto-elastic and elasto-electric coupling factors. In addition, the input current I_{mag} to coils excites a "mechanical current"$(\dot{u}_1 - \dot{u}_2)$ in the laminate. From Equation 2.1.2b and using Faraday's law, the coupling relation between I_{mag}, V_{mag} and $(\dot{u}_1 - \dot{u}_2)$ can be derived as

$$\begin{aligned} I_{\text{mag}} &= \frac{\lambda_{33}}{j\omega N}(\dot{u}_1 - \dot{u}_2) + \frac{V_{\text{mag}}}{j\omega L^S} \\ L^s &= Am\mu_{33}^{s} N^2/l; \end{aligned} \tag{2.2.11}$$

where L^s is the clamped inductance, and N is the coils number.

Equations 2.2.9–2.2.11 completely describe the magneto(-elastic-)electric coupling between the magnetostrictive and piezoelectric layers under an exciting current I_{mag} or voltage V_{mag}, via the mechanical displacement velocities $\dot{u}_1$ and $\dot{u}_2$.

2.3. Magnetoelectric equivalent circuit under resonance drive

Based upon the magnetoelectric coupling equations 2.2.9–2.2.11, a magnetoelectric equivalent circuit can be derived for the longitudinal vibration mode of the geometry shown in Fig. 1. This equivalent circuit is shown in Fig. 2a.

Under free boundary conditions, the force acting on the end faces are $F_1 = 0$ and $F_2 = 0$. Thus, we can short the two channels to ground. Under this condition, the circuit of Fig. 2a is simplified to that shown in Fig. 2b. Note that Z in the Fig. 2b is: $Z = \overline{\rho v} A_{\text{lam}}/j2\tan(kl/2)$. In this simplified circuit, the input current I_{mag} (or voltage V_{mag}) excites a "mechanical current"$I_c = \dot{u}_1 - \dot{u}_2$, via the magneto-elastic coupling factor ϕ_m. Subsequently, I_c induces an output voltage V_{out}, via the elasto-electric coupling.

Assuming the laminate composite to be a $\lambda/2$-resonator, operating in a length extensional mode, the series angular resonance frequency is $\omega_s = \frac{\pi \bar{v}}{l}$. Under resonant drive the mechanical impedance Z in the equivalent circuit of Fig. 2b can be approximated by a Taylor series expansion of the frequency $f(\omega)$ about ω_s, given as

$$\begin{aligned} f(\omega) &= \frac{Z_0}{j2\tan\frac{\omega l}{2\bar{v}}} = f(\omega_s) + f'(\omega_s)(\omega - \omega_s) \\ &+ \frac{1}{2} f''(\omega_s)(\omega - \omega_s)^2 + \ldots \approx -\frac{\pi Z_0}{4 j\omega_s}(\omega - \omega_s) \\ &= -2jL_\text{m}(\omega - \omega_s) \end{aligned} \tag{2.3.1}$$

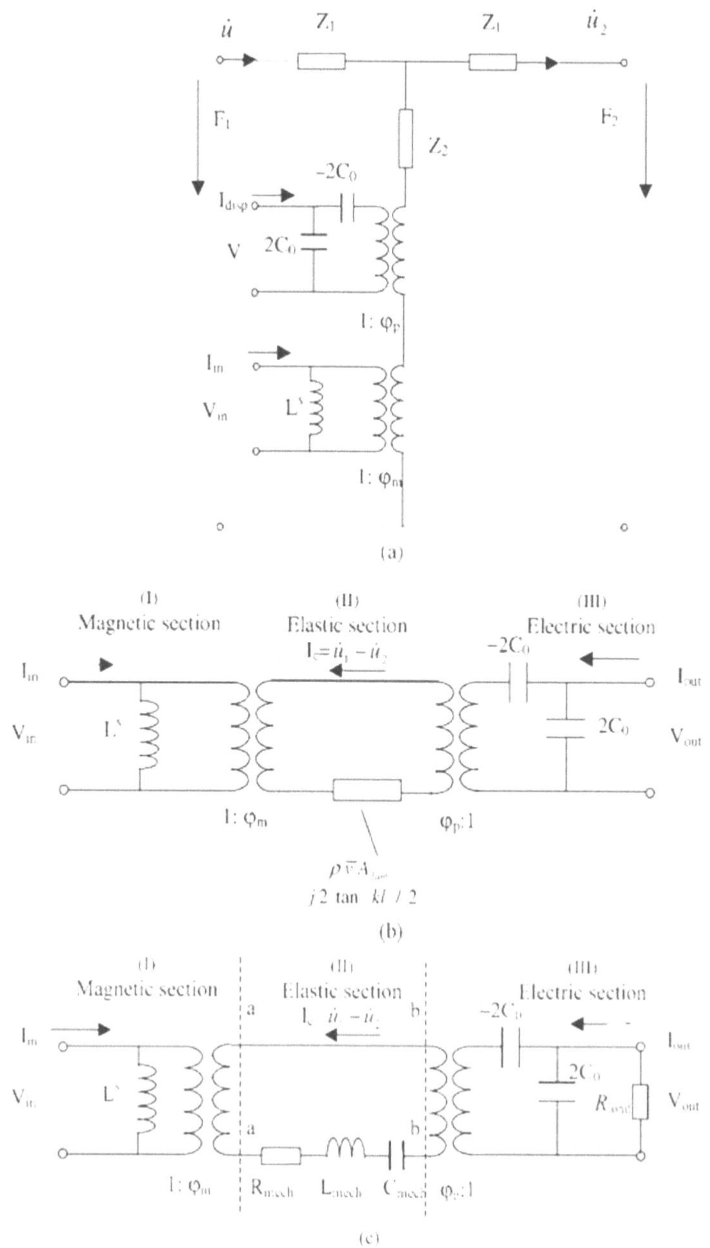

Figure 2 Magnetoelectric equivalent circuits. (a) Magnetoelectric equivalent circuit for a longitudinal-extension vibration mode under applied ac voltage V_{in} and (b) under free-free boundary condition, and (c) under resonance drive with load R_{Load}.

Then, compared with the impedance expansion of a series L_{m}, C_{m} circuit, we finally obtain

$$\omega_s^2 = \frac{1}{L_{\text{mech}}C_{\text{mech}}}; \quad L_{\text{mech}} = \frac{\pi Z_0}{8\omega_s};$$
$$C_{\text{mech}} = \frac{1}{\omega_s^2 L_{\text{mech}}}; \quad Z_0 = \overline{\rho v} A_{\text{lam}} \qquad (2.3.2)$$

where L_{mech} and C_{mech} are the motional mechanical inductance and capacitance.

At resonance, this equation yields an approximation for the mechanical impedance Z. The mechanical quality factor Q_{m} of the laminate under resonance drive is finite, due to dissipation [35]. This limitation of the vibration amplitude must also be included, in order to accurately predict the resonant response. Finite values of Q_{mech} result in an effective motional mechanical resistance $R_{\rm mech}$ of

$$R_{\rm mech} = \frac{\omega_s L_{\rm mech}}{Q_{\rm mech}} = \frac{\pi Z_0}{8 Q_{\rm mech}}. \quad (2.3.3)$$

Accordingly, the equivalent circuit of the laminate under resonance drive is given in Fig. 2c, where $R_{\rm Load}$ is an external load.

3. Calculating results and discussion

3.1. Magnetoelectric coupling

The magnetoelectric coupling effect is a product property between the magneto-elastic and elasto-electric coupling coefficients. We can define the effective magnetoelectric coupling $k_{\rm mag\text{-}elec}$ (eff) in terms of energy (U) as

$$k^2_{\rm mag\text{-}elastic}({\rm eff}) \equiv \frac{U^{\rm out}_{\rm electrical}}{U^{\rm stored}_{\rm elastic}} \times \frac{U^{\rm stored}_{\rm elastic}}{U^{\rm input}_{\rm magnetic}}$$

$$\approx k^2_{\rm mag\text{-}elastic}({\rm eff}) \times k^2_{\rm elasto\text{-}elec}({\rm eff}) \quad (3.1.1)$$

where $k_{\rm mag\text{-}elastic}$ (eff) and $k_{\rm elasto\text{-}elec}$ (eff) are the magneto-elastic and elasto-electric coupling factors.

3.1.1. Efffective magneto-elastic coupling factor

By setting the elasto-electric coupling factor $\varphi_{\rm p}$ to zero (i.e., no piezoelectric phase), the laminate shown in Fig. 1a becomes a magneto-elastic transducer or resonator. From Fig. 2c, the magnetic-elastic coupling factor is then given as [34]

$$k^2_{\rm mag\text{-}elastic}({\rm eff}) = \frac{\omega_p^2 - \omega_s^2}{\omega_p^2}; \quad (3.1.2)$$

where $\omega_{\rm p}$ and $\omega_{\rm s}$ are parallel resonance and series resonance frequencies of the magneto-elastic resonator with infinite $Q_{\rm mech}$ and zero load. From the circuit in Fig. 2c (b-b in short circuit), $\omega_{\rm p}$ and $\omega_{\rm s}$ can be related to the input admittance $Y_{\rm e}$, as

$$k^2_{\rm mag\text{-}elec}({\rm eff}) = \frac{64 s^B_{33} k^2_{33,\rm m} g^2_{33\rm p} n_{\rm m} n_{\rm p}}{\pi^2 \bar{\beta}_{33}\left(n_{\rm m} s^D_{33} + n_{\rm p} s^B_{33} + n_I Y^I s^B_{33} s^D_{33}\right)[\pi^2/2\left(n_{\rm m} s^D_{33} + n_p s^B_{33} + n_I Y^I s^B_{33} s^D_{33}\right) + 8 n_{\rm m} k^2_{33\rm m} s^D_{33}]}$$

$$Y_e = \frac{1}{j\omega L^s} + \frac{\varphi^2_{\rm m}}{R_{\rm mech} + j\omega L_{\rm mech} + \frac{1}{j\omega C_{\rm mech}}}. \quad (3.1.3)$$

At resonance, the input admittance $Y_{\rm e}$ is purely resistive, i.e, $Y_{e(\rm imaginary)} = 0$ for $R_{\rm mech} = 0$. This gives $\omega_s^2 = \frac{1}{L_{\rm mech} C_{\rm mech}}$ and $\omega_p^2 = \frac{1}{L_{\rm mech}}\left(\frac{1}{C_{\rm mech}} + \frac{k^2_{33,m}}{C^B}\right)$. Following equation (3.1.2), we obtain the magneto-elastic coupling factor as

$$k^2_{\rm mag\text{-}elastic}({\rm eff}) = \frac{C_{\rm mech}}{C^B + C_{\rm mech} k^2_{33,\rm m}} k^2_{33,\rm m} \quad (3.1.4)$$

where $C^B = (1 - k^2_{33,\rm m}) C^{\rm H}$ is the elastic compliance of the magnetostrictive layers at constant B, and $C^{\rm H} = \frac{l s^{\rm H}_{33}}{A_m}$ is the compliance at constant H.

3.1.2. Effective elasto-electric coupling factor

To obtain $k^2_{\rm elasto-electric}$ (eff), the magnetic section of the equivalent circuit is electrically shorted (i.e., $\phi_{\rm m} = 0$, and a-a is shorted in Fig. 2c). Thus, the magnetoelectric equivalent circuit becomes the standard equivalent circuit of a piezoelectric resonator, near its resonance frequency $\omega_{\rm r}$. The electromechanical coupling is [33]

$$k_{\rm elasto-electric}({\rm eff}) = \frac{\omega_p^2 - \omega_s^2}{\omega_p^2}; \quad (3.1.5)$$

where $\omega_{\rm p}$ and $\omega_{\rm s}$ are parallel resonance and series resonance frequencies of the elastic-electric vibrator with infinite $Q_{\rm mech}$ and zero external load. From the circuit in Fig. 2c, $\omega_{\rm p}$ and $\omega_{\rm s}$ can be related to the output admittance $Y_{\rm o}$, as

$$Y_o = j\omega(2C_0) + \frac{\varphi_p^2}{R_{\rm mech} + j\omega L_{\rm mech} + \frac{1}{j\omega C'_{\rm mech}}}; \quad (3.1.6)$$

where $\omega_{\rm p}$ and $\omega_{\rm s}$ can be obtained as $\omega_r^2 = \frac{1}{L_{\rm mech} C_{\rm mech}}$ and $\omega_s^2 = \frac{1}{L_{\rm mech} C'_{\rm mech}}$; and$C'_{\rm mech} = \frac{C_{\rm mech} \cdot (2C_0)}{(2C_0) - C_{\rm mech} \varphi_p^2}$. Thus, the effective elasto-electric coupling factor is

$$k^2_{\rm elasto-electric}({\rm eff}) = \frac{C_{\rm mech} \varphi_p^2}{2C_0}. \quad (3.1.7)$$

3.1.3. Effective magneto-electric coupling factor

The effective magnetoelectric coupling factor $k_{\rm mag-elec}$(eff) can now be obtained by inserting equations (3.1.4) and (3.1.7) into (3.1.1), and simplifying to

$$s^B_{33} = s^H_{33}\left(1 - k^2_{33,m}\right), s^D_{33} = s^E_{33}\left(1 - k^2_{33,p}\right) \quad (3.1.8)$$

If $n_{\rm m} = 0$ or $n_{\rm p} = 0$, then $k^2_{\rm mag\text{-}elec}({\rm eff}) = 0$. Thus, there is an optimum geometric parameter $n_{\rm opt}$ at which$k^2_{\rm mag\text{-}elec}$(eff) is maximum. This occurs at $\frac{\partial k^2_{\rm mag\text{-}elec}({\rm eff})}{\partial n_m} = 0$ (or $\frac{\partial k^2_{\rm mag\text{-}elec}({\rm eff})}{\partial n_p} = 0$). Assuming the

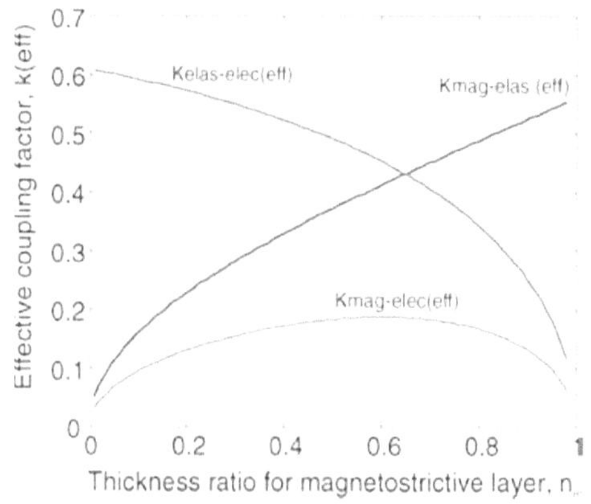

Figure 3 Effective magneto-elastic ($K_{mag-elas,\ eff}$), elasto-electric ($K_{elas-elec,\ eff}$), and magnetoelectric coupling factors ($K_{mag-elec,\ eff}$).

insulating layer to be thin, the optimum geometric parameter for the magnetostrictive layer is then

$$n_{m,opt} = \frac{1}{1+\gamma(1+8k_{33m}^2/\pi^2)^{1/2}} \tag{3.1.9}$$

where $\gamma = \frac{s_{33}^D}{s_{33}^B}$ is a ratio of the compliance constants of the piezoelectric and magnetostrictive layers.

Fig. 3 shows a calculation of $k_{\text{mag-elec}}$ (eff) as a function of the geometric parameter n_m for a laminate of piezoelectric PZT-8 (APC-841) and magnetostrictive Terfenol-D. When $n_m = 0$, this laminate composite contains no magnetostrictive material, thus$k_{\text{mag-elec}}$ (eff) is zero. Correspondingly, when $n_m = 1$, the laminate contains only magnetostrictive material, and again$k_{\text{mag-elec}}$ (eff) is zero. Between these two geometric limits, a maximum value of$k_{\text{mag-elec}}$ (eff) ~ 0.2 is found, as can be seen in Fig. 3 near $n_m \sim 0.61$.

3.2. Magnetoelectric voltage gain under resonance drive

A magnetoelectric voltage gain was found by analysis of the equivalent circuit in Fig. 2c. Assuming that the circuit is unloaded and by applying Ohm's law to the mechanical loop (elastic section), the voltage gain (V_{out}/V_{in}, and $V_{in} = V_{mag}$) can be estimated as

$$\frac{\varphi_p V_{out}}{\varphi_m V_{in}} = \left| \frac{\varphi_p^2/j\omega(2C_0)}{R_m + j\omega L_m + 1/j\omega C_m + 1/j\omega(-2C_0) + 1/j\omega(2C_0)} \right|$$

or

$$\frac{V_{out}}{V_{in}} = \left| \frac{\varphi_p \varphi_m}{j\omega(2C_0)} \frac{1}{R_m + j\omega L_m + 1/j\omega C_m} \right| \tag{3.1.11}$$

Under resonant drive, where $\omega_s = \left(1/L_{mech}C_{mech}\right)^{1/2}$, the maximum voltage gain V_{gain1} is

$$V_{gain1,max} = \frac{4Q_{mech}\varphi_m\varphi_p}{\pi\omega_s C_0 Z_0}. \tag{3.1.12}$$

From this relationship, it can be seen that the maximum voltage gain is strongly dependent on mechanical quality factor Q_m. Calculations for a Terfenol-D/PZT laminate composite based on Equation 3.1.11 show that the maximum voltage gains at resonance are 1.54×10^5 for $Q_m = 1000$, 1.55×10^4 for $Q_m = 100$, and 7.7×10^3 for $Q_m = 50$, respectively. These calculated values are far higher than experimental data.

However, we can further simplify the equivalent circuit from Fig. 2c into Fig. 4a, and it can be seen that the ratio of output voltage to input voltage (voltage gain V_{gain2}) is proportional to the impedance ratio (according to Ohm's Law). Thus, the modified voltage gain is

$$V_{gain2} = \left|\frac{V_{out}}{V_{in}}\right| = \alpha \left| \frac{\frac{1}{j\omega C}}{R_1 + j\omega L_1 + \frac{1}{j\omega C_1} + \frac{1}{j\omega C}} \right| \tag{3.1.13}$$

where $C = C_0\varphi_m^2/2\varphi_p^2$; $R_1 = R_{mech}/\varphi_m^2$; $L_1 = L_{mech}/\varphi_m^2$; $C_1 = \varphi_m^2 C'_{mech}$. α is a dimensionless ratio factor, relating dc magnetic bias. If we choice $\alpha = 1$, the maximum voltage gain V_{gain2} under resonance drive is

$$V_{gain2,max} = \frac{4Q_{mech}\varphi_p^2}{\pi\omega_s C_0 Z_0} \tag{3.1.14}$$

From this relationship, it can be seen that the maximum voltage gain is mainly related to the piezoelectric section of the equivalent circuit in Fig. 4a. The voltage gain is directly proportional to Q_{mech} and $\phi_p{}^2$ (or $g_{33,p}^2$) in piezoelectric layer. This is because the output voltage is generated by this section. The function of the magnetic section of the circuit is to transducer the magnetic energy into a mechanical vibration. The piezoelectric one subsequently transduces this vibration to an electric output.

Fig. 4b shows the calculated voltage gain V_{gain} as a function of frequency for $Q_m = 100$, 500 and 1400. These calculations were performed using Equations 3.1.11 and

3.1.13, assuming a Terfenol-D/PZT laminate length of 70 mm, width of 10 mm, and thickness of 6 mm. The voltage gain for $Q_m = 100$ was only ~20. However, for $Q_{mech} = 500$, the gain was ~100. A typical value of Q_{mech} for PZT-8 is 1400; using this value, a maximum voltage gain of 280 can be estimated. Calculation values using Equation 3.1.13 are much close to our measured voltage gain ones.

This voltage gain is significantly larger than that of other voltage gain devices, such as electromagnetic and piezoelectric transformers [36, 37]. The ME voltage gain effect is quite purposeful for power electronics, such as transformer application.

3.3. Efficiency

To estimate the efficiency of magnetoelectric transduction, the equivalent circuit in Fig. 4a was converted by an impedance method to that shown in Fig. 5a. In so doing, it was assumed that the losses in the laminate are only mechanical, i.e., electrical losses (eddy current loss) were neglected. Assuming a load of Z_{Load}, the magnetoelectric efficiency of the laminate at resonance is

$$\eta = \frac{P_{out}}{P_{in}} = \frac{I^2 \cdot \mathrm{Re}(Z_{Load})}{I^2 \cdot \mathrm{Re}(Z_{in})} = \frac{R_{Load}}{R_{Load} + (\varphi_m/\varphi_p)^2[1 + (2\omega_s C_0 R_{Load})^2]R_1}. \tag{3.1.15}$$

By setting $\delta\eta/\delta R_{Load} = 0$, the optimum load can be estimated as $R_{Load,opt} = \frac{1}{2\omega_s C_0}$. When $R_{Load} = R_{Load,\,opt}$, the circuit has maximum output power. Under resonant operation, the maximum efficiency is given as $\eta_{max} = \frac{\varphi_p^2}{\varphi_p^2 + \frac{\pi Z_0 C_0 \omega_s}{Q_{mech}}}$. Clearly, a higher Q_{mech} results in a higher transduction efficiency.

Fig. 5b shows the calculated value of η as a function of R_{Load} for different values of Q_m. η can be seen to vary significantly with R_{Load}. A maximum efficiency η_{max} was found for $R_{Load} \approx 2 \times 10^6$ ohms. The value of η_{max} at this R_{Load} is dependent on Q_m. For $Q_m = 100$, η_{max} was less than 90%; however, for $Q_{mech} = 1000$, η_{max} was ≈98%. However, for a bulk Terfenol-D material operated of a

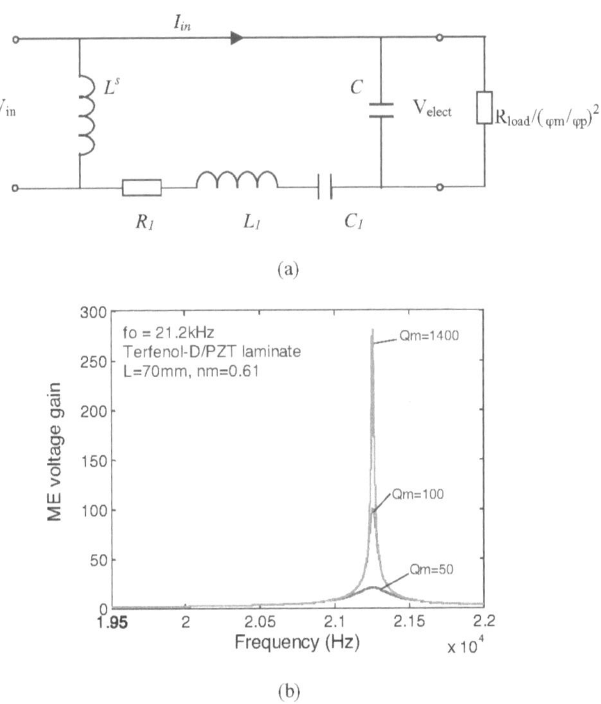

Figure 4 Simplified equivalent circuit and voltage gain calculated using Equation 3.1.13. (a) Equivalent circuit, and (b) calculated voltage gain.

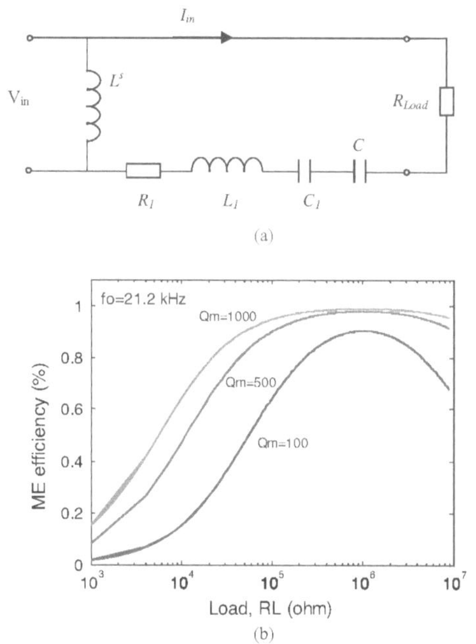

Figure 5 Simplified equivalent circuit illustrating magnetoelectric efficiency, and magnetoelectric efficiency calculated using equation 3.1.15. (a) Equivalent circuit, and (b) calculated efficiency. (Eddy current effects and electric loss in stored energy elements, L^s, C_1 and C', are neglected). In Fig.5a, $R'_{\text{Load}} = \frac{R_{\text{Load}}}{[1+(2\omega C_0 R_{\text{Load}})^2](\varphi_m/\varphi_p)^2}$; $C' = \frac{[1+(2\omega C_0 R_{\text{Load}})^2]}{\omega^2(2C_0)^2 R_{\text{Load}}^2}(\frac{\varphi_m}{\varphi_p})^2(2C_0)$.

high frequency of >20 kHz, the eddy current losses will be serious, resulting in that the effective magnetoelectric efficiency far lower than the theoretical value. To overcome the eddy current losses, a thin multi-layer design is necessary for high-power magnetoelectric transformer applications.

4. Experiments

We designed and fabricated one long-type of Terfenol-D/PZT magneto-electric laminates with dimensions of 70 mm in length, 10 mm in width, and 6 mm in thickness. The Terfenol-D layers were grain-oriented in the length direction, and the PZT layer (PZT-8) was polarized in length direction too. The PZT plate was laminated between two Terfenol-D plates and insulated with thin glass layers using an epoxy resin, and cured at 80°C for 3–4 h under load. The induced voltages across the two electrodes of PZT layer under a drive by coils around the laminate were then measured using an oscilloscope. Fig. 6 shows the measured voltage gain, the ratio of induced voltage V_{out} to inputted voltage V_{in} to coils, of our ME transformer as a function of the drive frequency f. A maximum voltage gain of ~260 was found at a resonance frequency of 21.3 kHz. In addition, at the resonance state, the maximum voltage gain of the ME transformer was strongly dependent on an applied dc magnetic bias H_{dc}. For $H_{\text{dc}} \approx 300$ Oe, our prototype exhibited a maximum voltage gain of ≈300, which is quite close to the predicted value using (3.1.14).

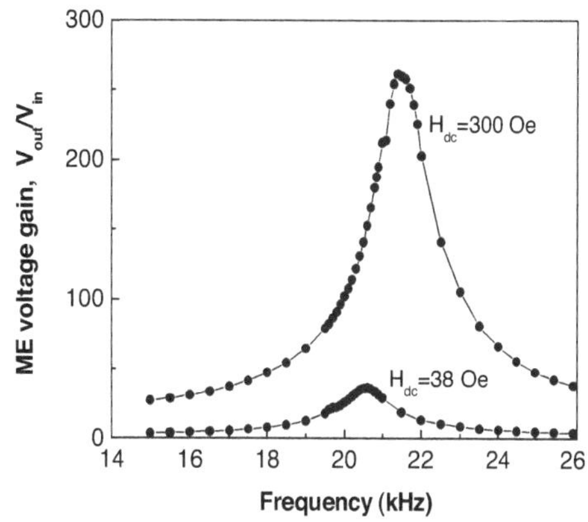

Figure 6 Measured magneto-electric voltage gain effect under resonance drive.

Compared with conventional electromagnetic transformers, our ME transformer (i) does not require secondary coils with a high-turns ratio in order to obtain a step-up voltage output; and (ii) it has significantly higher voltage gains and a notably wider frequency bandwidth due to low Q_{m} of Terfenol-D materials. For example, the voltage gain and frequency bandwidth for a typical piezoelectric transformer are only 10–100 and 0.1 kHz [36, 37], which are far lower than the 300 and 1.5 kHz observed for our ME transformer, respectively.

5. Summary

Effective magneotelectric couplings, output efficiencies and voltage gains have been analyzed based on a ME equivalent circuit method for laminate composites of piezoelectric PZT and magnetostrictive Terfenol-D. The analysis predicts: (i) there is an optimum geometric parameter n_{opt} which maximizes $k_{\text{mag-elec}}$ (eff); (ii) a high voltage gain occurs under resonance drive, exceeding values of 280 for $Q_{\text{mech}} = 1400$; (iii) a maximum efficiency occurs when the external load equals the impedance of the static capacitance, $\eta_{\max} \sim 98\%$ for $Q_{\text{mech}} = 1000$ for a multi-thin-layer design; and (iv) higher values of Q_{mech} in piezoelectric material result in higher voltage gains and maximum efficiencies. Our experimental results confirm that the ME laminate has a strong voltage gain effect. This is an important finding, as there are many potential applications for high-power solid-state ME transformers.

Acknowledgments

This research was supported by the Office of Naval Research under grants N000140210340, N000140210126, and MURI N000140110761.

References

1. L. D. LANDAU and E. LIFTSHITZ, "Electrodynamics of Continuous Media" (Pergamon Press, Oxford, 1960) P.119.
2. DI MATTEO and S. JANSEN, AGM, *Phys. Rev. B* **66** (2002) 100402.
3. I. E. DZYALOSHINSKII, *Sovit Phys.*-JETP **37** (1959) 81 [*Sov. Phys JETP* **10** (1960) 628].
4. L. WIEGELMANN, A. A. STEPANOV, I. M. VITEBSKY, A. G. M. JANSEN and P. WYDER, *Phys. Rev. B* **49** (1994) 10039.
5. J. WANG, J. B. NEATON, H. ZHENG, V. NAGARAJAN, S. B. OGALE, B. LIU, D. VIEHLAND, V. VAITHYANATHAN, D. G. SCHLOM, U. V. WAGHMARE, N. A. SPALDIN, K. M. RABE, M. WUTTIG and R. RAMESH, *Sci.* **299** (2003) 1719.
6. J. RYU, A. VAZQUEZ-ARAZO, K. UCHINO and H. KIM, *J. Electro.* **7** (2001) 17.
7. T. WU and J. HUANG, *International J. of Solids and Strus.* **37** (2000) 2981.
8. C. NAN, M. LI and J. HUANG, *Phys. Review B* **63** (2001) 144415. C. NAN, *Phys. Rev. B.* **50** (1994) 6082.
9. C. NAN, N. CAI, Z. SHI, J. ZHAI, G. LIU and Y. LIN, *ibid.* **71** (2005) 014102.
10. M. BICHURIN, D. A. FILIPPOV and V. M. PETROV, *ibid.* **68** (2003) 132408.
11. S. X. DONG, J. CHENG, J.F. LI and D. VIEHLAND, *Appl. Phys. Lett.* **83** (2003) 4812.
12. H. YU, M. ZENG Y. WANG, *et al., ibid.* **86** (2005) 032508.
13. M. ZENG, J.G. WAN, Y. WANG, *et al. J. Appl. Phys.* **95** (2004) 8069.
14. D. A. FILIPPOV, M. BICHURIN, V. PETROV, *et al., Phys. Solid State* **46** (2004) 1674.
15. S. X. DONG, J. F. LI and D. VIEHLAND, *Appl. Phys. Lett.* **85** (2004) 5305.
16. *Idem., ibid.,* **85** (2004) 3534.
17. *Idem., ibid.* **84** (2004) 4188.
18. M. AVELLANEDA and G. HARSHE, *J. Inte. Mate. Systems and Struc.* **5** 501 (1994).
19. J. RYU, A. VAZQUEZ-ARAZO, K. UCHINO and H. KIM, *Jpn. J. Appl. Phys.* **40** (2001) 4948.
20. G. SRINIVASAN, E. RASMUSSEN, B. LEVIN and R. HAYES, *Phys. Rev. B* **65** (2002) 134402.
21. G. SRINIVASAN, E. RASMUSSEN, J. GALLEGOS, R. SRINIVASAN, Y. BOKHAN and V. LALETIN, *ibid. B* **64** (2001) 214408.
22. J. RYU, S. PRIYA, A. VAZQUEZ-ARAZO, K. UCHINO and H. KIM, *J. Am. Ceram. Soc.* **84** (2001) 2905.
23. G. SRINIVASAN, V. LALETIN, R. HAYES, N. PUDDUBNAYA, E. RASMUSSEN and D. FEKEL, *Solid State Commu.* **124** (2002) 373.
24. M. BICHURIN, V. PETROV and G. SRINIVASAN, *J. Appl. Phys.* **92** (2002) 7681.
25. K. MORI and M. WUTTIG, *Appl. Phys. Lett.* **81** (2002) 100.
26. S. X. DONG, J.F. LI and D. VIEHLAND, *IEEE Transactions on Ultrasonics, Ferroelectrics, and Freque. Control* **50**(1) (2003) 253.
27. *Idem., J. Appl. Phys.* **96** (2004) 3382.
28. S. X. DONG, J. ZHAI, Z. XING, J.F. LI and D. VIEHLAND, *Appl. Phys. Lett.* **86** (2005) 102901.
29. S. X. DONG, J.F. LI and D. VIEHLAND, *ibid.* **85** (2004) 2307.
30. *Idem., IEEE Transactions on Ultrasonics, Ferroelectrics, and Frequency Control* **50** (2003) 1236.
31. *Idem., ibid.* **51** (2004) 794.
32. Idem., *J. Appl. Phys.* **97** (2005) 103902.
33. W. MASON, "Physical Acoustics, Principle and Methods" (New York Academic Press 1964) p. 263.
34. G. ENGDAHL, "Magnetostrictive Materials Handbook" (Academic Press, ISBN: 0-12-238640-X, 2000).
35. W. G. CADY, "Piezoelectricity, An Introduction to theory and applications of Electromechanical Phenomena in Crystal" (Dover Publications, New York, 1964).
36. A. VAZQUEZ-CARAZO, in Fifth International Conference on Intelligent Materials (Smart System and Nanotechnology, University Park, PA, USA 2003).
37. C. ROSEN, "Elect. Components Symposium," 7th (Washington, D.C., 1956) p.205.

JOURNAL OF MATERIALS SCIENCE 41 (2006) 107–116

Piezoresponse force microscopy and recent advances in nanoscale studies of ferroelectrics

A. GRUVERMAN
North Carolina State University, Raleigh, NC 27695-7920
E-mail: Alexei_Gruverman@ncsu.edu

S. V. KALININ
Oak Ridge National Laboratory, Oak Ridge, Tennessee 37831

In this paper, we review recent advances in piezoresponse force microscopy (PFM) with respect to nanoscale ferroelectric research, summarize the basic principles of PFM, illustrate what information can be obtained from PFM experiments and delineate the limitations of PFM signal interpretation relevant to quantitative imaging of a broad range of piezoelectrically active materials. Particular attention is given to orientational PFM imaging and data interpretation as well as to electromechanics and kinetics of nanoscale ferroelectric switching in PFM.

1. Introduction

Rapid development of electronic devices based on ferroelectric thin films has necessitated studies of ferroelectric properties at the nanoscale. Fortunately, this need came at the time when new techniques for nanoscale characterization of materials became available. Scanning probe microscopy (SPM) has emerged as a powerful tool for high-resolution characterization of virtually all types of materials, including metals, semiconductors, dielectrics, polymers and biomolecules. A number of papers and books on scanning probe methods have been published, providing an introduction to the basic SPM principles and advanced applications [1–5]. SPM techniques have revolutionized the field of ferroelectricity, for the first time providing an opportunity for nondestructive visualization of domain structures in ferroelectric thin films at the nanoscale. SPM made possible nanoscale mapping of the surface potential, evaluation of local electromechanical properties and nonlinear dielectric constant measurements, i.e. it provides crucial information on the ferroelectric materials properties with unprecedented spatial resolution. SPM has also opened new venues in nanoscale domain patterning for such applications as high-density data storage [6, 7] and ferroelectric lithography [8, 9]. Steadily increasing number of research papers indicates a growing importance of SPM in the field of ferroelectricity (Fig. 1). Research groups in US, Europe and Asia are actively using SPM for high-resolution characterization of ferroelectric materials both in bulk and thin layer forms [10].

Among the SPM techniques for the nanoscale characterization of ferroelectrics, by far the most popular one is piezoresponse force microscopy (PFM) [6–10]. In this paper, we review recent advances in PFM with respect to nanoscale ferroelectric research, summarize the basic principles of PFM, illustrate what information can be obtained from PFM experiments and delineate the limitations of PFM signal interpretation relevant to quantitative imaging of a broad range of piezoelectrically active materials.

2. Principle of piezoresponse force microscopy

Piezoresponse force microscopy is based on the detection of local piezoelectric deformation of a ferroelectric sample induced by an external electric field (hence the name "piezoresponse"). Depending on the relative orientations of the applied field and the polarization vector, sample deformation can be in the form of elongation, contraction or shear. For the converse piezoelectric effect, the field-induced strain S_j can be expressed as [11]:

$$S_j = d_{ij} E_i \tag{1}$$

where d_{ij} are components of the piezoelectric tensor (in reduced Voigt notation) and E_i is the applied field.

0022-2461
DOI: 10.1007/s10853-005-5946-0

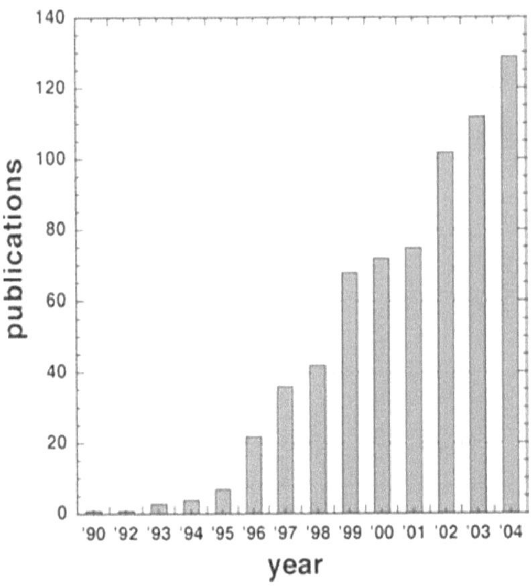

Figure 1 Number of publications on SPM studies of ferroelectrics per year.

Using the thermodynamic approach it can be shown that in a single-domain ferroelectric the piezoelectric coefficient relates to the spontaneous polarization P_s via the following expression [12, 13]:

$$d_{ij} = \varepsilon_{im} Q_{jmk} P_{sk}, \qquad (2)$$

where ε_{im} is the dielectric constant and Q_{jmk} is the electrostriction coefficient.

The linear coupling between the piezoelectric and ferroelectric parameters infers that the domain polarity can be determined from the sign of the field-induced strain. Application of the uniform electric field along the polar direction results in the elongation of the domain with polarization parallel to the applied field and in the contraction of the domain with opposite polarization. The field-induced strain in this case can be written as:

$$S = \frac{\Delta Z}{Z} = \pm d_{33} E \qquad (3)$$

where ΔZ is the sample deformation and Z is the sample thickness. Equation 3 can be rewritten as:

$$\Delta Z = \pm d_{33} V \qquad (4)$$

where V is an applied voltage. The $\pm$ sign reflects the piezoelectric coefficients of opposite sign for antiparallel domains. Thus, opposite domains can be visualized by monitoring their voltage-induced surface displacement.

Due to its extremely high vertical sensitivity, nanoscale topography variations can be routinely measured in SPM. However, domain imaging based on detection of static piezoelectric deformation is difficult to implement unless a sample has a very smooth surface. The reason is simple: the static cantilever deflection due to the piezoelectric deformation will be superimposed on the deflection signal due to the surface roughness, which renders static piezoresponse domain imaging in samples with rough surfaces highly problematical.

Based on Equation 4 one might conclude that the electrically induced topographic contrast between opposite domains can be infinitely enhanced by increasing the imaging voltage. However, there is a strict limitation imposed on this parameter: to perform *nondestructive* visualization of domain structure, the imaging voltage should be kept below the coercive voltage of the ferroelectric sample. In addition, a high imaging voltage will lead to an increased contribution of the electrostatic signal to the tip-sample interaction, which in some cases can obscure the domain image. Given that a typical value of the coercive field in a 200-nm-thick Pb(Zr, Ti)O3 ferroelectric film is approximately 50 kV/cm, the imaging voltage should not exceed 1 V, otherwise the imaging process will change the domain structure by inducing the polarization reversal. In a PZT film with the d_{33} constant of about 200 pm/N the surface displacement induced by an external voltage of 1 V will be only 0.2 nm. Obviously, such a displacement could not be reliably detected in ferroelectric films, where topographic features can be on the order of several nanometers. The static approach can be applied in some limited cases, for example, to ferroelectric samples with carefully polished surfaces, relatively high values of piezoelectric constants and coercive fields.

A problem of low sensitivity of a static piezoresponse mode has been circumvented by employing a dynamic piezoresponse imaging method based on the voltage-modulation approach, which allowed sensitivity to be increased by three orders of magnitude [14–17]. In this approach, an ac modulation (imaging) voltage $V = V_0 \cos \omega t$ is applied to the ferroelectric sample and surface displacement is measured using a standard lock-in technique by detecting the vertical vibration of the cantilever (Fig. 2a), which follows sample surface oscillation. A domain map can be obtained by scanning the surface while detecting the first harmonic component of the normal surface vibration (vertical piezoresponse, or VPFM):

$$\Delta Z = \Delta Z_0 \cos(\omega t + \varphi) \qquad (5)$$

where $\Delta Z_0 = d_v V_0$ is a vibration amplitude, d_v is effective piezoelectric constant and φ is a phase difference between the imaging voltage and piezoresponse, which provides information on the polarization direction. With the modulation voltage applied to the probing tip, positive domains (polarization vector oriented downward) will vibrate in phase with the applied voltage so that $\varphi(+) = 0°$, while vibration of negative domains (polarization vector oriented upward) will occur in counter phase: $\varphi(-) =$

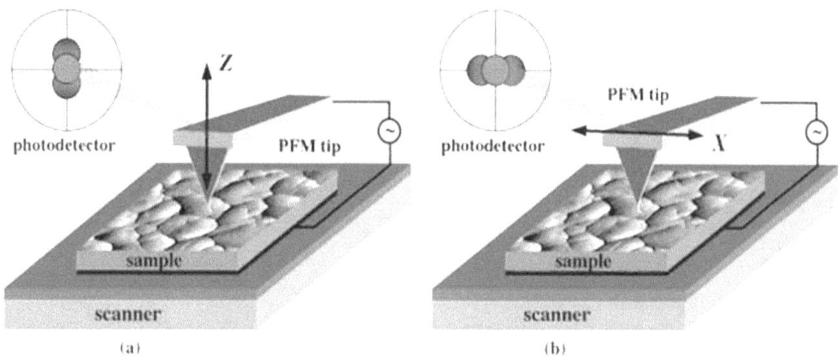

Figure 2 Schematics of the vertical (a) and lateral (b) PFM signal detection.

180°. Note that while Equation 4 is rigorous for a uniform field case, the field below the SPM tip is highly non-uniform. The rigorous solution of this problem has been given by Kalinin and Karapetian [18, 19] and it was shown that for transversally isotropic materials (e.g. *c* domain in tetragonal perovskite ferroelectrics, poled polymers, etc.) $d_v \approx d_{33}$, recovering the early assumptions in PFM data interpretation [16].

It should be noted that despite apparent simplicity of the PFM method, quantification of the PFM data, particularly in the case of thin films, is nontrivial due to the complexity of the tip-sample interaction which involves not only electromechanical but also electrostatic components. Experimental conditions, such as driving voltage, frequency, loading force, cantilever force constant, tip apex radius, ambient environment, as well as physical properties of the samples (thickness, dielectric constants, orientation, defect structure, crystallinity, electrode material) should be taken into account to avoid misinterpretation of the PFM results [20]. However, qualitative PFM domain imaging is extremely robust and currently PFM is one of the main tools for high-resolution characterization of ferroelectric crystals and thin films [21].

One of the significant features of the dynamic PFM method is that it also allows delineation of domains with polarization parallel to the sample surface (*a*-domains) [22–24]. In lateral PFM (LPFM) [22], *a*-domains are visualized by detecting the torsional vibration of the cantilever (Fig. 2b). Application of the modulation voltage across the sample generates sample vibration in the direction parallel to its surface due to the piezoelectric shear deformation. This surface vibration, translated via the friction forces to the torsional movement of the cantilever, can be detected in the same way as the normal cantilever oscillation in vertical PFM. For the uniform field for *a* domain, the amplitude of the in-plane oscillation is given by:

$$\Delta X_0 = d_{15} V_0 \qquad (6)$$

while polarization direction can be determined from the phase signal since oscillation phases of opposite *a*-domains differ by 180°. It should be noted, however, that quantification of the shear piezoelectric coefficients in LPFM is a challenging problem that is complicated by the tip-surface tribology, inhomogeneous field distribution and mechanical clamping effects.

Note that Equation 6 can be used only when the in-plane polarization vector is perpendicular to the physical axis of the cantilever. However, in general case the in-plane electromechanical response vector can be oriented arbitrarily with respect to the cantilever. To perform complete three-dimensional (3D) reconstruction of polarization, an advanced approach – vector PFM – based on combination of VPFM and LPFM has been developed.

3. Vector PFM

Electromechanical response of the surface to the applied tip bias in general case is a *vector* having three independent components (PR_x, PR_y, PR_z). To obtain complete information on materials properties, all three components are required. The original VPFM approach allowed only the measurement of PR_z component. Lateral PFM allows PR_y component to be detected simultaneously (the cantilever is oriented along the *x*-axis). Combined lateral and vertical PFM imaging had been used by Eng *et al.* to reconstruct surface crystallography in a barium titanate crystal [23]. Rodriguez *et al.* have applied this approach to 3D polarization reconstruction in micrometer-size PZT capacitors [25]. In both cases, sequential acquisition of two LPFM images at two orthogonal orientations of the sample with respect to the cantilever, further referred to as *x*-LPFM and *y*-LPFM imaging, has been accomplished in the samples with well-defined crystallographic orientation. In general case of an arbitrarily oriented sample, only semi-quantitative information on materials properties can be obtained unless relative sensitivities of LPFM and VPFM are carefully calibrated.

An example of 2D vector PFM image is illustrated in Fig. 3, which shows VPFM and LPFM images of a strontium bismuth tantalate (SBT) thin film (Fig. 3a and b,

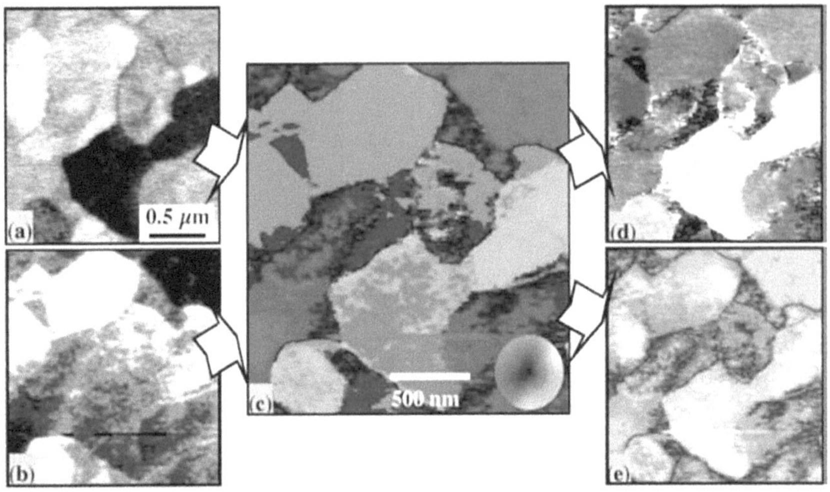

Figure 3 (a) Vertical and (b) lateral PFM images of the SBT thin film. (c) Vector representation of 2D PFM data; (d) Angle and (e) amplitude images derived from image the 2D map in (c).

respectively). To represent vector PFM data, the VPFM and LPFM images are normalized so that the intensity changes between −1 and 1, i.e. vpr, $lpr \in (-1, 1)$. Using commercial software [26] 2D vector data (vpr, lpr) is converted to the amplitude/angle pair, $A_{2D} = \mathrm{Abs}(vpr + I\,lpr)$, $\theta_{2D} = \mathrm{Arg}(vpr + I\,lpr)$. In Fig. 3c, color corresponds to the orientation, while intensity corresponds to the magnitude. Notably, this vector PFM image shows that color is virtually uniform inside most of the grains, suggesting that they exist in a single domain state, while polarization orientation changes between the grains. This information can be represented in the scalar form by plotting separately phase θ_{2D}, and magnitude, A_{2D}, as illustrated in Fig. 3d and e, respectively.

To achieve the full potential of vector PFM, quantitative interpretation of the electromechanical response data is required. The units of PFM response are (nm/V), similar to the dimensionality of piezoelectric constant tensor, d_{ij}. In the uniform z-oriented electric field, response in z-direction is given by the piezoelectric constant d_{33}. In LPFM, under similar conditions the signal is given by the shear components of piezoelectric constant tensor, $xPR_l = d_{35}$ and $yPR_l = d_{34}$, since components d_{31} and d_{32} result in axially symmetric deformation of material that does not contribute to displacement at the center [27]. For the non-uniform field similar to that below the SPM tip, this approximation is no longer rigorous and the effective piezocoefficient will be a complex function of mechanical and electromechnical constants of material. Although the exact analytical solutions are not available for these cases, following the analogy with VPFM, the assumption $xPR_l = d_{35}$ and $yPR_l = d_{34}$ are expected to provide a good first approximation for the description of the LPFM data, even though analytical solutions or numerical simulations are required to prove this conjecture.

Note that this interpretation applies when the piezoelectric tensor is given in the laboratory coordinate system. However, it is conventional to represent the piezoelectric constant tensor in the coordinate system related to the orientation of crystallographic axis of the material, d^0_{ij}. In this case, the intrinsic material symmetry limits the number of non-zero components and allows the material-specific parameters to be tabulated. These coordinate systems are related by three Euler rotation angles, ϕ, θ, and ψ. The relationship between the d_{ij} tensor in the laboratory coordinate system and the d^0_{ij} tensor in the crystal coordinate system is [28]:

$$d_{ij} = A_{ik} d^o_{kl} N_{lj} \qquad (7)$$

where the elements of the rotation matrices N_{ij} and A_{ij} as a function of Euler angles are given in Ref. [28]. For example, for *a* domain in tetragonal ferroelectric material with polarization in the (010) direction oriented perpendicular to the cantilever axis, the relationship between coordinate systems is given by $\phi = 0$, $\theta = \pi/2$, $\psi = 0$ and from Equation 7 $vPR = 0$, $xPR_l = d^0_{15}$ and $yPR_l = 0$, consistent with the early assumption of $xPR_l = d^0_{15}$ for in-plane domains in tetragonal ferroelectrics [24].

To summarize, Equation 7 fully describes the relationship between material properties and vector PFM data and allows semi-quantitative assessment of d^0_{ij} provided that the crystallographic orientation of the sample is known. Alternatively, if the elements of the d^0_{ij} tensor are known then reconstruction of local crystallographic orientation from vector PFM data can be performed.

4. Electromechanics of ferroelectric switching in PFM

Scanning force microscopy provides a unique opportunity for controlling the ferroelectric properties at the nanoscale and direct studies of the domain structure evolution under an external electric field, which cannot be matched by previously available techniques. A conductive probing tip can be used not only for domain visualization but also for modification of the initial domain structure. Application of a small dc voltage between the tip and bottom electrode generates an electric field of several hundred kilovolts per centimeter, which is higher than the coercive voltage of most of ferroelectrics, thus inducing local polarization reversal. This approach was suggested for such applications as ultrahigh density data storage [6, 7] and ferroelectric lithography [8, 9]. These applications require thorough understanding of both thermodynamics and kinetics of the switching process as described in the following two sections.

The driving force for the 180° polarization switching process in ferroelectrics is change in the bulk free energy density [29, 30]:

$$\Delta g_{\text{bulk}} = -\Delta P_i E_i - \Delta d_{i\mu} E_i X_\mu \tag{8}$$

where P_i, E_i, X_μ, and $d_{i\mu}$, are components of the polarization, electric field, stress and piezoelectric constants tensor, correspondingly, $i = 1, 2, 3$, and $\mu = 1, .., 6$. The first and second terms in Equation 8 describe ferroelectric and ferroelectroelastic switching, respectively. For materials, such as $LiNbO_3$ and lead zirconate-titanate (PZT), the signs of the corresponding free energy terms are opposite and the polarities of the domains formed by ferroelectric and ferroelectroelastic switching are opposite, thus providing an approach to distinguish these switching mechanisms.

The free energy of the nucleating domain is

$$\Delta G = \Delta G_{\text{bulk}} + \Delta G_{\text{wall}} + \Delta G_{\text{dep}} \tag{9}$$

where the first term is the change in bulk free energy, $\Delta G_{\text{bulk}} = \int \Delta g_{\text{bulk}} dV$, the second term is the domain wall energy, and the third term is the depolarization field energy. In the Landauer model of switching, the domain shape is approximated as a half ellipsoid with the small and large axis equal to r_d and l_d, correspondingly (Fig. 4a). The domain wall contribution to the free energy in this geometry is $\Delta G_{\text{wall}} = b r_d l_d$, where $b = \sigma_{\text{wall}} \pi^2/2$ and σ_{wall} is the direction-independent domain wall energy. The depolarization energy contribution is $\Delta G_{\text{dep}} = c r_d^4 / l_d$, where

$$c = \frac{4\pi P_s^2}{3\varepsilon_{11}} \left[\ln\left(\frac{2l_d}{r_d} \sqrt{\frac{\varepsilon_{11}}{\varepsilon_{33}}} \right) - 1 \right] \tag{10}$$

has only a weak dependence on the domain geometry [31].

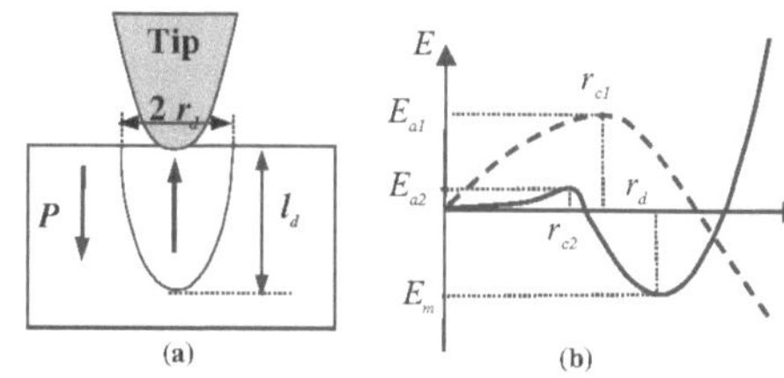

Figure 4 (a) Domain geometry during tip-induced switching. (b) Free energy as a function of the lateral domain size. Dashed line - in a uniform electric field; Solid line – in a tip-induced electric field.

In the uniform field case, the free energy surfaces as a function of l_d, r_d, has a saddle point character and the domain grows indefinitely once the critical size corresponding to activation barrier for nucleation E_a is reached (Fig. 4b). The critical domain size and activation energy for nucleation can be obtained from minimization of Equation 10 as $r_c = 0.83\, b/a$, $l_c = 1.86\, bc^{1/2} a^{-3/2}$ and $E_a = 0.518\, b^3 c^{1/2} a^{-5/2}$, where $a = 4\pi P_s E/3$. For typical ferroelectric materials, such as $BaTiO_3$ ($\sigma = 7$ mJ/m^2, $P_s = 0.26$ C/m^2, $\varepsilon_{11} = 2000$, $\varepsilon_{33} = 120$ [32]), in the uniform field $E = 10^5$ V/m the corresponding values are $E_a = 2.4 \times 10^5$ eV and $l_c = 16.4$ μm, $r_c = 0.264$ μm. Thus, for relatively weak fields corresponding to experimental coercive fields, homogeneous domain nucleation is impossible, which explains why in typical ferroelectric materials domain nucleation occurs at the surface or at the interface defects.

The opposite is true for the tip-induced switching, when the small radius of curvature of the tip results in large (10^6–10^9 V/m) electric fields localized at the tip apex. The corresponding domain free energy can be determined from electroelastic field distribution generated by the PFM tip as

$$\Delta G_{\text{bulk}} = \int_V \Delta g_{\text{bulk}}(\vec{r}) dV = 2\pi \int_0^{l_d} dz \int_0^{r(z)} \Delta g_{\text{bulk}}(r, z) r dr \tag{11}$$

where $r(z) = r_d \sqrt{1 - z^2/l_d^2}$. An initial insight into the PFM switching phenomena can be obtained using point charge models that are applicable if domain sizes l_d, $r_d \gg R$, a, where R is the tip radius and a is the contact radius and provided that the singularity at the origin is weak enough to ensure convergence of the integral in Equation 11. For ferroelectric switching induced by a point charge q_s located on the surface, the integral in Equation 11 can be taken analytically: $\Delta G_{\text{bulk}} = d r_d l_d / (l_d + \gamma r_d)$, where $d = 2P_s q_s / (\varepsilon_0 + \sqrt{\varepsilon_{11}\varepsilon_{33}})$ and $\gamma = \sqrt{\varepsilon_{33}/\varepsilon_{11}}$. In this point charge approximation, domain size and energy are $r_e = 0.342 d^{2/3} (bc)^{-1/3}$, $l_e = 0.2d/b$ and $E_m = -0.205 d^{5/3} (bc)^{-1/3}$ [33, 34].

The activation energy for domain nucleation in this approximation is zero, due to the infinite field at the origin. This analysis predicts that domain shape in the switching process follows the invariant relation $r_e^3/l_e^2 = b/c$.

The applicability of the point charge approximation to the thermodynamics of domain switching on the large length scales is limited by the contribution of the electrostatic fields produced by the conical part of the tip, which decay much slower than that produced by the point charge. At the smaller length scales comparable to the tip radius of curvature, the thermodynamics of switching process requires exact electroelastic field structure to be taken into account. In this case, it was shown that domain nucleation requires certain threshold bias of the order of 0.1–1 V, corresponding to non-zero activation energy for nucleation (of the order of $\sim kT$) [29, 35].

In the last several years, a number of reports have become available on the high-order ferroic switching in PFM [30, 35]. Due to the rapid decay of corresponding electroelastic fields, the use of Equation 11 in the point charge/force approximation results in singularity in the contact area, necessitating exact electroelastic field structure to be taken into account. Using rigorous electroelastic solutions, it was shown [35] that for higher order ferroic switching (e.g. ferroelectroelastic), the domain size is limited by the tip-sample contact area, thus allowing precise control of domain size. Finally, the non-linear effects could have severe impact in PFM and should be taken into account at high frequencies [36].

5. PFM spectroscopy

One of the most important applications of PFM is local piezoelectric spectroscopy, i.e., measurements of local hysteresis loops at the $\sim$10 nm level [24, 37]. PFM hysteresis loops readily provide information on local electromechanical activity and coercive voltage variations between dissimilar grains. Currently, there are two main approaches to the hysteresis loop measurements in PFM that differ by the mode of voltage application: either step voltage or pulse voltage. Depending on the dielectric properties of the ferroelectric sample and domain stability one or another approach provides more reliable results [38]. However, generally quantitative interpretation of PFM spectroscopy presents a complex problem. As has been discussed previously, a major challenge of quantitative PFM characterization of ferroelectric thin films stems from the inhomogeneous distribution of the tip-generated field and random grain orientation [39, 40]. The vector PFM approach in conjunction with the local switching experiments can be used to analyze the effect of grain crystallographic orientation on the local hysteresis loop parameters [41]. Below, the inhomogeneous distribution of the SPM tip-generated field has been taken into account to quantify the piezoresponse signal of individual grains using the step mode of hysteresis loop measurement.

Fig. 5 shows surface topography along with the VPFM and LPFM images of the strontium bismuth tantalate (SBT) film [42]. Local piezoelectric hysteresis loop measurements [21] show that grains that exhibit strong VPFM contrast also have distinctive hysteresis loops (Fig. 6a), indicative of a large out-of-plane polarization component. On the other hand, the VPFM hysteresis loop for grain 3, which exhibits gray contrast in VPFM, is linear, indicative of non-ferroelectric nature of the grain or purely in-plane polarization corresponding to (001) grain orientation [43]. The LPFM loop of the same grain, shown in Fig. 6b, exhibits clear hysteresis behavior, which is consistent with the in-plane orientation of the polarization vector.

To analyze the hysteresis loop shape in PFM, the vertical surface displacement under the applied tip bias can be calculated as [41, 44]:

$$-\tilde{A}_{\text{piezo}} = \alpha \int_0^\infty d_{33} E_z dz = \alpha d_{33} \left\{ \int_0^l E_z dz - \int_l^\infty E_z dz \right\} \tag{12}$$

where l is the growing domain length. Integration yields $A_{\text{piezo}} = \alpha d_{33}\{V(0) - 2V(l)\}$, where $V(0)$ is the potential on the surface and $V(l)$ is the potential at the domain boundary. It is shown in Ref. [18] that in the strong indentation regime, for distance l from the center of the contact area larger than contact radius a, the potential distribution inside the material can be approximated using

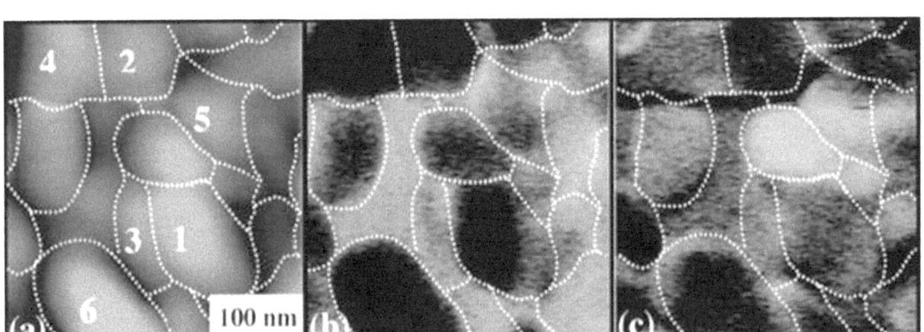

Figure 5 Surface topography (a), vertical (b), and lateral (c) PFM signal of SBT thin film. Z-scale in (a) is 20 nm. (Reprinted with permission from [41]).

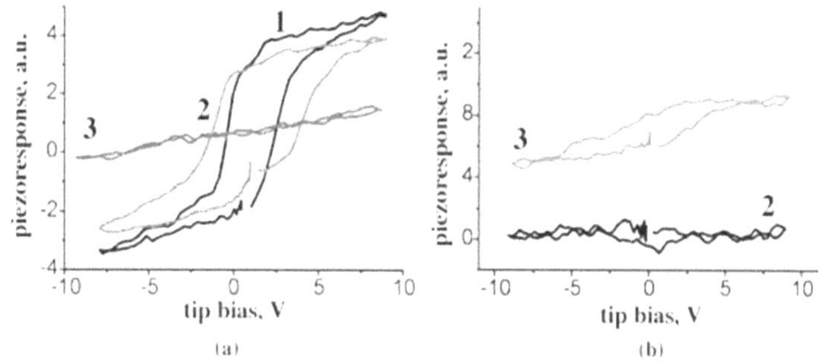

Figure 6 (a) VPFM and (b) LPFM piezoelectric hysteresis loops for grains in Fig. 5. For clarity, vertical loops are shown only for grains 1, 2, and 3. (Reprinted with permission from [41]).

the point-charge model. In this case, assuming that in the point-charge approximation the domain size is related to the biasing voltage as $l(V_{dc}) \sim V_{dc}$ [33, 34], it can be shown that the shape of the PFM hysteresis loop should follow the functional form

$$PR = \alpha d_{33}\{1 - \eta / V_{dc}\} \qquad (13)$$

where PR is the piezoresponse amplitude, $PR = A_{piezo}/V_{ac}$. To compare the experimental PFM loop shape with Equation 13, adjustment for the capacitive cantilever-surface interaction is introduced by subtracting the linear loop for the grain 3 from the hysteresis loops for the ferroelectric grains. A corrected hysteresis loop and a corresponding fit by Equation 13 are shown in Fig. 7a, illustrating excellent agreement of the experimental data with Equation 13. Generally, however, care should be taken while subtracting the linear contribution of the cantilever-surface interaction as it might differ for differently oriented grains. This is particularly important for materials with high anisotropy of the dielectric constants, such as barium titanate.

The same formalism has been extended to determine the effect of grain orientation on local coercive voltage. The longitudinal piezoelectric coefficient in SBT in the [011] plane at the angle θ from the (010) axis can be estimated as $d_{zz}(\theta) \approx d_{33} \cos^3 \theta$. Similarly, the lateral piezoresponse coefficient is $d_{zx}(\theta) = d_{31} \cos\theta$. Both vertical and lateral responses decrease with the deviation from the polar (010) direction. The relationship between the PFM coercive bias and crystallographic orientation of the grain can be estimated from the work of switching, which is proportional to $\mathbf{E} \cdot \mathbf{P} = E\,P \cos(\theta)$, where $\mathbf{E}$ is electric field and $\mathbf{P}$ polarization vector. Therefore, coercive bias is expected to increase with deviation angle from polar axis as $V_{coer}(0)/\cos(\theta)$, where $V_{coer}(0)$ is coercive bias for the (010) grain. Comparison of the angular dependence of piezoresponse signal and coercive bias suggests that for the off-axis orientation of the grains the response decreases and coercive bias increases, in agreement with experimental results illustrated in Fig. 7b. In the limiting case of the (001)-oriented grain in SBT piezoresponse and a coercive field become zero and infinity, respectively. It is worthwhile mentioning that establishing correlation between local and macroscopic coercive voltages requires statistical approach that involves local spectroscopy of a number of grains as well as careful studies of switching behavior of the macroscopic capacitor that incorporates these grains - mission still to be accomplished.

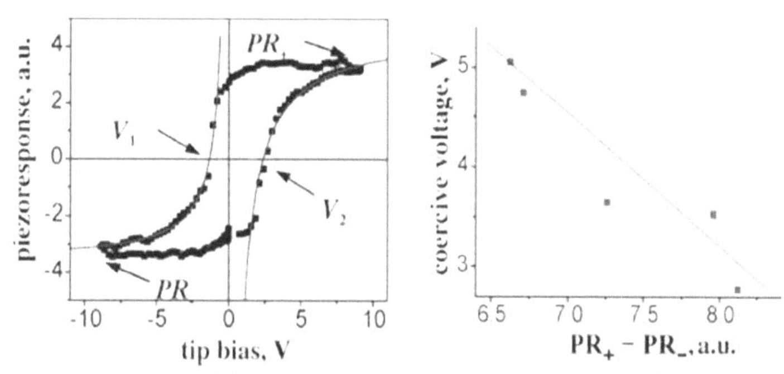

Figure 7 (a) Corrected hysteresis loop and corresponding fit by Equation 13 for positive and negative tip biases. (b) Correlation between maximal switchable polarization and coercive voltage $V_1 + V_2$. (Reprinted with permission from [41]).

6. Domain growth kinetics in PFM

Analysis in Section 4 describes the equilibrium domain shape governed by the extent of electroelastic field created by the tip. However, well outside of the tip-sample contact area, the domain shape and size are controlled by the kinetic effects as discussed below.

When an electric field is applied opposite to the polarization direction of a single-domain ferroelectric capacitor, the switching mechanism involves several steps: nucleation of multiple domains, their forward growth, subsequent sideways expansion and coalescence [45]. In PFM, switching involves nucleation of a single domain as was discussed in the previous sections. High spatial resolution of the PFM approach allows direct investigation of the nanoscale domain growth. However, poor time resolution, which is determined by the time required for image acquisition (~several minutes) makes *in situ* measurements of domain dynamics during fast switching processes difficult. While PFM can be readily used to investigate slow polarization relaxation processes with characteristic times of the order of minutes and above, it is a challenge to deduce the mechanism of domain transformation when polarization reversal occurs in a matter of microseconds and faster.

This problem is usually circumvented by studying the domain structure dynamics in a quasi-static regime using *step-by-step switching*. This method has been previously used at the macroscopic level in classical switching experiments on correlating the domain structure evolution and transient current in ferroelectric crystals [45] and later was applied to thin films [6, 46]. In this approach, partial reversal of polarization is generated by applying a voltage pulse shorter than the total switching time with subsequent piezoresponse imaging of the resulting domain pattern. By applying a sequence of voltage pulses of successively increasing duration and acquiring piezoresponse images after each pulse a consistent picture of time dependent behavior of domain structure can be obtained providing information on the domain wall velocity, its spatial anisotropy and its field dependence. To avoid data misinterpretation due to spontaneous backswitching between the pulses, stability of the produced intermediate patterns should be checked by acquiring domain images at different time intervals after single pulse application. To describe the sidewise expansion of the domain it is necessary to take into account the field dependence of the domain wall velocity and the spatial distribution of the electric field generated by the probing tip.

Fig. 8 shows the PFM amplitude and phase images of an array of 9 domains fabricated in a lithium niobate crystal by applying negative 10 ms voltage pulses of various magnitudes in the range from 20 to 70 V [47]. The PFM contrast is the same across the 180° domain boundaries, which appear as dark lines in the amplitude image, suggesting that the fabricated domains extend from the top to the bottom interface. Fig. 9 shows the time dependence of the domain radius for three different pulse amplitudes, which follows logarithmic law [48]. In addition, it has been found that the domain size follows linear voltage dependence [49]. This behavior suggests that the domains in Fig. 8 represent different stages of the switching kinetic process and do not correspond to the equilibrium state domains [33]. Notably, results shown in Fig. 8 suggest that the kinetics of the sidewise domain growth can be described by a universal scaling curve $g(t) = r(V, t)/V$ illustrated in inset in Fig. 9. It is suggested that this universal scaling behavior is directly related to the field dependence of domain wall velocity and field distribution inside the material. Here, we analyze the kinetics of the sidewise domain growth using the classical activation model of the wall motion in the tip-generated field assuming a weak indentation regime [50].

To calculate the tip-generated field distribution, the tip was modeled as a charged sphere with radius R at the distance δ from the sample surface. The normal component of this electric field at a section of the 180° domain wall with the sample surface at a distance r from the tip

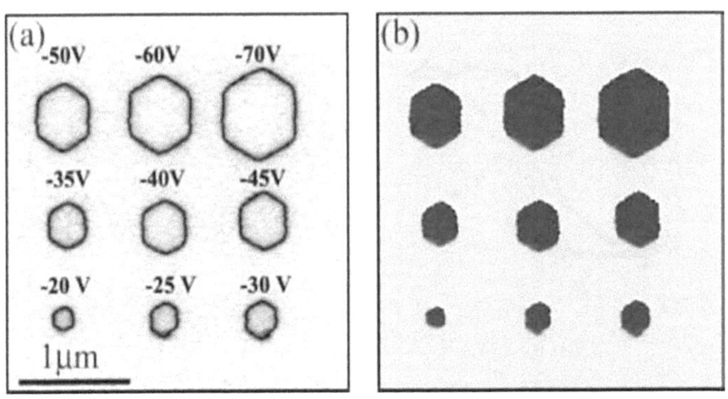

Figure 8 VPFM (a) amplitude and (b) phase images of ferroelectric domains fabricated in lithium niobate by 10 ms voltage pulses of various amplitudes. (Reprinted with permission from [47]).

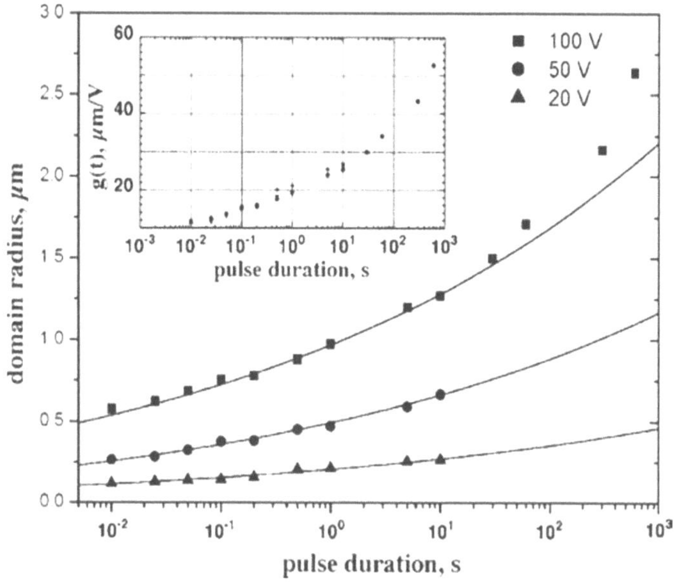

Figure 9 Domain radius versus the pulse duration for various pulse magnitudes lithium niobate. Inset shows a scaling curve $g(t) = r(V, t)/V$ calculated using all data points from the main plot in Fig. 9. (Reprinted with permission from [47]).

is calculated using an expression for the electric potential from Mele [51]:

$$E(r) = \frac{C_t V_t}{2\pi\varepsilon_0\sqrt{\varepsilon_c\varepsilon_a}+1}\sqrt{\frac{\varepsilon_a}{\varepsilon_c}}\frac{R+\delta}{((R+\delta)^2+r^2)^{3/2}} \quad (14)$$

where C_t and V_t are the tip capacitance and bias, respectively, and ε_a and ε_c are the dielectric constants along the non-polar and polar axes of the sample, respectively. In the present study, we used the following values: $R = 50$ nm, $\delta = 1$ nm, $\varepsilon_a = 85$, $\varepsilon_c = 30$. The tip capacitance was calculated to be 1.6×10^{-17} F [47]. The set of data in Fig. 9 has been fitted using the following expression for the time dependence of the domain radius assuming an exponential field dependence of the wall velocity [52]:

$$t = \int \frac{dr}{v(r)} = \int \frac{dr}{e^{-\alpha/E(r+r_0)}} \quad (15)$$

where $v\ (r)$ is a local wall velocity and α is the activation field. The meaning of the fitting parameter r_0 can be understood as follows. Underneath the probing tip, the generated field is much larger than the local coercive field and the domain growth develops as a nonactivated process. The spatial inhomogeneity of the tip-generated field results in a transition from the nonactivated to the activated process. Therefore, r_0 can be considered as the domain radius at which the activation type of the wall motion begins. The r_0 value was found to be 17 nm for the applied voltage of 20 V and 110 nm for the 100 V voltage. The activation energy was found to decrease with an increase in the applied voltage from $\sim 2\times10^3$ kV/cm for 20 V to $\sim$50 kV/cm for 100 V. It should be noted that a strong decrease of the external field with the distance from the tip as well as non-local tip effect might result in a different mechanism of domain wall motion that may explain a less adequate fitting of $r(t)$ for large ($r > 1.5\ \mu$m) domains in Fig. 9.

7. Conclusion

Rapid development of ferroelectric-based devices generated a strong need for extensive investigation of the nanoscale properties of ferroelectric materials. Application of piezoresponse force microscopy to ferroelectrics opened new possibilities not only for their high-resolution imaging of domain structures, but also for quantitative characterization and control of ferroelectric properties at the nanoscale. Clearly, future will evidence broad application of this technique for ferroelectrics, as well as for more broad class of piezoelectric and electrostrictive materials. It is also expected that PFM will find wide application in studying the electromechanical coupling in biological materials that will open new horizons for understanding the mechanisms that govern growth and regeneration of living tissues.

Acknowledgments

AG acknowledges financial support of the National Science Foundation (Grant No. DMR02-35632). Research performed in part as a Eugene P. Wigner Fellow and staff member at the Oak Ridge National Laboratory, managed by UT-Battelle, LLC, for the U.S. Department of Energy under Contract DE-AC05-00OR22725 (SVK). Support from ORNL SEED funding is acknowledged (SVK).

References

1. D. SARID, Scanning Force Microscopy with Applications to Electric, Magnetic and Atomic Forces, Oxford Series in Optical and Imaging Sciences (University Press, Oxford, 1991).
2. R. WIESENDANGER, Scanning Probe Microscopy and Spectroscopy: Methods and Applications (University Press, Cambridge, 1994).
3. D. A. BONNELL, Scanning Probe Microscopy and Spectroscopy: Theory, Techniques, and Applications (John Wiley & Sons, October 2000).
4. V. TSUKRUK, Advances in Scanning Probe Microscopy (Macromolecular Symposia 167), (John Wiley & Sons, July 2001).
5. A. DE STEFANIS and A. A. G. TOMLINSON, Scanning Probe Microscopies: From Surfaces Structure to Nano-Scale Engineering, Trans Tech Publications, Ltd.; April 2001.
6. T. TYBELL, P. PARUCH, T. GIAMARCHI and J.-M. TRISCONE, *Phys. Rev. Lett.* **89** (2002) 097601.
7. K. TERABE, M. NAKAMURA, S. TAKEKAWA, K. KITAMURA, S. HIGUCHI, Y. GOTOH and Y. CHO, *Appl. Phys. Lett.* **82** (2003) 433.
8. S. V. KALININ, D. A. BONNELL, T. ALVAREZ, X. LEI, Z. HU, J. H. FERRIS, Q. ZHANG and S. DUNN, *Nano Letters* **2** (2002) 589.
9. S. V. KALININ, D. A. BONNELL, T. ALVAREZ, X. LEI, Z. HU, R. SHAO and J. H. FERRIS, *Adv. Mat.* **16** (2004) 795.
10. See, for example, references in A. Gruverman, in "Encyclopedia of Nanoscience and Nanotechnology", edited by H.S.Nalwa, (American Scientific Publishers, Los Angeles, Vol. 3, 2004) pp. 359–375.
11. W. G. CADY, Piezoelectricity: An Introduction to the Theory and Applications of Electromechanical Phenomena in Crystals (Dover Publications, New York, 1964).
12. A. F. DEVONSHIRE, *Philos. Mag.* **40** (1949) 1040.
13. A. F. DEVONSHIRE, *Adv. Phys.* **3** (1954) 85.
14. P. GUTHNER and K. DRANSFELD, *Appl. Phys. Lett.* **61** (1992) 1137.
15. K. FRANKE, J. BESOLD, W. HAESSLE and C. SEEGEBARTH, *Surf. Sci. Lett.* **302** (1994) L283.
16. A. GRUVERMAN, H. TOKUMOTO, S. A. PRAKASH, S. AGGARWAL, B. YANG, M. WUTTIG, R. RAMESH, O. AUCIELLO and V. VENKATESAN, *Appl. Phys. Lett.* **71** (1997) 3492.
17. T. HIDAKA, T. MARUYAMA, M. SAITOH, N. MIKOSHIBA, M. SHIMIZU, T. SHIOSAKI, L. A. WILLS, R. HISKES, S. A. DICAROLIS and J. AMANO, *Appl. Phys. Lett.* **68** (1996) 2358.
18. S. V. KALININ, E. KARAPETIAN and M. KACHANOV, *Phys. Rev. B* **70** (2004) 184101.
19. E. KARAPETIAN, M. KACHANOV and S. V. KALININ, *Phil. Mag.*, in print
20. S. V. KALININ and D. A. BONNELL, *Phys. Rev.* **B 65** (2002) 125408.
21. Nanoscale Characterization of Ferroelectric Materials, edited by M. Alexe and A. Gruverman (Springer-Verlag, Berlin, 2004).
22. L. M. ENG, H.-J. GÜNTHERODT, G. ROSENMAN, A. SKLIAR, M. ORON, M. KATZ and D. EGER, *J. Appl. Phys.* **83** (1998) 5973.
23. L. M. ENG, H.-J. GUNTHERODT, G. A. SCHNEIDER, U. KOPKE and J. M. SALDANA, *Appl. Phys. Lett.* **74** (1999) 233.
24. A. ROELOFS, U. BÖTTGER, R. WASER, F. SCHLAPHOF, S. TROGISCH and L. M. ENG, *Appl. Phys. Lett.* **77** (2000) 3444.
25. B. J. RODRIGUEZ, A. GRUVERMAN, A. I. KINGON, R. J. NEMANICH and J. S. CROSS, *J. Appl. Phys.* **95** (2004) 1958.
26. Mathematika 5.0, Wolfram Research.
27. S. V. KALININ, B. J. RODRIGUEZ, S. JESSE, J. SHIN, A.P. BADDORF, P. GUPTA, H. JAIN, D.B. WILLIAMS and A. GRUVERMAN, unpublished.
28. R. E. NEWNHAM, Properties of Materials: Anisotropy, Symmetry, Structure (Oxford University Press, 2005).
29. M. ABPLANALP, PhD thesis, Swiss Federal Institute of Technology, Zurich (2001).
30. M. ABPLANALP, J. FOUSEK and P. GUNTER, *Phys. Rev. Lett.* **86** (2001) 5799.
31. Equation 10 is valid only for $l_d > r_d$. To avoid this limitation, used here was the expression for the demagnetization factor for prolate ellipsoid from J. A. OSBORN, *Phys. Rev.* **67** (1945) 351.
32. F. JONA and G. SHIRANE, Ferroelectric Crystals, (Dover Publications, New York, 1993).
33. M. MOLOTSKII, *J. Appl. Phys.* **93** (2003) 6234.
34. M. MOLOTSKII, A. AGRONIN, P. URENSKI, M. SHVEBELMAN, G. ROSENMAN and Y. ROSENWAKS, *Phys. Rev. Lett.* **90** (2003) 107601.
35. S. V. KALININ, A. GRUVERMAN, B. J. RODRIGUEZ, J. SHIN, A. P. BADDORF, E. KARAPETIAN and M. KACHANOV, *J. Appl. Phys.* **97**, 074305 (2005).
36. B. D. HUEY, in "Nanoscale Phenomena in Ferroelectric Thin Films," edited by S. Hong (Kluwer Academic Publishers, 2004).
37. M. ALEXE, A. GRUVERMAN, C. HARNAGEA, N. D. ZAKHAROV, A. PIGNOLET, D. HESSE and J. F. SCOTT, *Appl. Phys. Lett.* **75** (1999) 1158.
38. S. HONG, J. WOO, H. SHIN, J. U. JEON, Y. E. PAK, E. L. COLLA, N. SETTER, E. KIM and K. NO, *J. Appl. Phys.* **89** (2001) 1377.
39. C. HARNAGEA, A. PIGNOLET, M. ALEXE and D. HESSE, *Integrated Ferroelectrics* **38** (2001) 23.
40. C. HARNAGEA, PhD thesis, Martin-Luther-Universität Halle Wittenberg, Halle, 2001.
41. S. V. KALININ, A. GRUVERMAN and D. A. BONNELL, *Appl. Phys. Lett.* **85** (2004) 795.
42. Shown here are PFM images representing the $A \cos \theta$ signal, where A is piezoresponse amplitude and θ is phase.
43. A. GRUVERMAN, A. PIGNOLET, K. M. SATYALAKSHMI, M. ALEXE, N. D. ZAKHAROV and D. HESSE, *Appl. Phys. Lett.* **76** (2000) 106.
44. C. S. GANPULE, V. NAGARJAN, H. LI, A. S. OGALE, D. E. STEINHAUER, S. AGGARWAL, E. WILLIAMS, R. RAMESH and P. DE WOLF, *Appl. Phys. Lett.* **77** (2000) 292.
45. E. FATUZZO and W. J. MERZ, Ferroelectricity (North-Holland, Amsterdam, 1967).
46. O. LOHSE, S. TIEDKE, M. GROSSMANN and R. WASER, *Integrated Ferroelectrics* **22** (1998) 123.
47. B. J. RODRIGUEZ, R. J. NEMANICH, A. KINGON, A. GRUVERMAN, S. V. KALININ, K. TERABE, X. Y. LIU and K. KITAMURA, *Appl. Phys. Lett.* **86** (2005) 012906.
48. A. AGRONIN, Y. ROSENWAKS and G. ROSENMAN, *Appl. Phys. Lett.* **85** (2004) 452.
49. For pulse duration of 10 ms this linear behavior holds at least up to 250 V.
50. S. V. KALININ, A. GRUVERMAN, J. SHIN, A. P. BADDORF, E. KARAPETIAN and M. KACHANOV, cond-mat/0406383.
51. E. J. MELE, *Am. J. Phys.* **69** (2001) 557.
52. R. C. MILLER and G. WEINREICH, *Phys. Rev.* **117** (1960) 1460.

JOURNAL OF MATERIALS SCIENCE 41 (2006) 117–127

Dielectric response of polymer relaxors

BOŻENA HILCZER, HILARY SMOGÓR, JANINA GOSLAR
Institute of Molecular Physics, Polish Academy of Sciences, M. Smoluchowskiego 17, 60-179, Poznań, Poland

Dielectric response of vinylidene fluoride type ferroelectric polymers is dominated by that of segmental motions in the amorphous phase in temperature range 200–300 K and contributions related to the local mode and ferroelectric–paraelectric transition in the crystalline phase of the polymer at higher temperatures. Diffuse and frequency-dependent dielectric anomaly observed in fast electron irradiated polyvinylidene fluoride-trifluoroethylene P(VDF/TrFE) has been related to relaxor-like behavior induced in the semicrystalline ferroelectric copolymers. As random field and the response of polar nanosize clusters determine the relaxor behavior the effects of disorder and fast electron irradiation (below and above T_C) on the three contributions to the dielectric response of PVDF, P(VDF/TrFE)(75/25) and P(VDF/TrFE)(50/50) are shown. The processes involved in radiation-induced functionalization of PVDF-type polymers are discussed on the basis of results of ESR, IR and Raman spectroscopy studies.

1. Introduction

Ferroelectrics, exhibiting also pyroelectric and piezoelectric properties are attractive for applications since both the sensing and actuating functions can be realized in the same element. After the discovery of ferroelectricity in polyvinylidene fluoride (PVDF) type polymers in 1980 [1–3] the materials have received particular attention for many reasons. One of the most important features is that the polymers can be easily produced in the form of large thin flexible sheets and in a variety of shapes. Moreover, low dielectric permittivity ε and elastic stiffness of PVDF-type polymer films at room temperature, resulting in high voltage sensitivity, as well as their low acoustic impedance, fitting well to that of water and tissue, make them attractive for application in electromechanical transducers using the piezoelectric constants d_{31} and d_{33} [4, 5]. As the permittivity and dielectric losses are involved in figures of merit of pyroelectric sensors, PVDF-type polymers with high pyroelectric activity and low ε and tan δ values are used in pyroelectric applications [4–6].

The vacuum dipole moment of vinylidene fluoride unit $\mu(\text{VDF}) = 7.07 \cdot 10^{-30}$ Cm originates from the distribution of positively charged hydrogen atoms and negatively charged fluorine atoms and lies in the plane of C–C bond. In the $[-CH_2-CF_2-]_n$ polymer the dipoles are attached to the main chain and can adopt various orientations resulting in several molecular conformations of the chain. The chains can be packed in various manner forming polymorphous modifications of PVDF [7, 8]. Both the conformations of the polymer chains and their packing in the unit cell are determined by the processing conditions. Nonpolar polymorph II with antiparallel packing of trans-gauche-trans-gauche′ (TGTG′) chain conformation is obtained on cooling molten PVDF to room temperature. Chains with planar-zigzag all-trans conformations packed in a parallel manner form ferroelectric PVDF I modification with dipole moments perpendicular to the chain axis. The ferroelectric PVDF polymorph can be obtained from melt-crystallized polymer by mechanical orientation or poling in high electric field. Polar modification PVDF III, consisting of a parallel packing of TTTGTTTG′ conformation is obtained by annealing the polymorph II at ~430 K. It should be noted that PVDF is a semicrystalline polymer with the degree of crystallinity of ~50%.

The long-range interactions of dipoles in PVDF I crystalline modification are rather strong since the ferroelectric ordering persists, depending on processing conditions, to the temperatures just below the melting point or the polymer melts before it undergoes a transition to the paraelectric phase [9–12]. The long-range forces can be weakened by incorporation into the PVDF chain units with dipole moment different from that of vinylidene fluoride. Trifluoroethylene (TrFE) was found to

0022-2461 © *2006 Springer Science + Business Media, Inc.*
DOI: 10.1007/s10853-005-5949-x

be the most interesting unit to functionalise the properties of PVDF [8, 9]. The dipole moment of trifluoroethylene unit μ(TrFE) = 5.74 · 10^{-30} Cm is lower by ~20% than that of vinylidene fluoride and goes out of the plane of the –C–C– bond defined as the (x, y) plane: μ_x(TrFE) = 0.37 · 10^{-30} Cm, μ_y(TrFE) = 3.77 · 10^{-30} Cm, μ_z(TrFE) = 4.30 · 10^{-30} Cm, as shown in Fig. 1 [13]. Vinylidene fluoride and trifluoroethylene copolymerise in a random copolymer at any VDF/TrFE ratio and the Curie point of the copolymers is shifted linearly towards lower temperatures with increasing contents of TrFE [14].

The chemical heterogeneity in the polymer chain is increased in terpolymers of VDF with TrFE and chlorotrifluoroethylene (CTrFE) or chlorofluoroethylene (CFE) units. The dipole moment of chlorotrifluoroethylene μ(CTrFE) = 2.14 · 10^{-30} Cm with projections μ_x(CTrFE) = 1.58 · 10^{-30} Cm, μ_y(CTrFE) = 1.40 · 10^{-30} Cm and μ_z(CTrFE) = 0.03 · 10^{-30} Cm in rectangular coordinates with C–C bond in the (x, y) plane [13] differs considerably from the dipole moments of VDF and TrFE (Fig. 1). Statistical disorder in terpolymers based

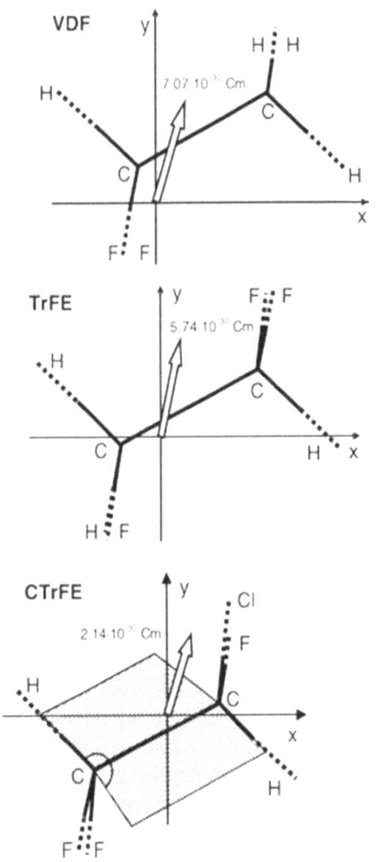

Figure 1 Schematic of spatial dipole moment arrangement of vinylidene fluoride (VDF), trifluoroethylene (TrFE) and chlorotrifluoroethylene (CTrFE) units [13].

on the zigzag PVDF chain was found to exhibit diffuse and dispersive dielectric anomaly described by Vogel–Fulcher law, characteristic of ferroelectric relaxors [15–17]. Similar relaxor-like behaviour was observed earlier in P(VDF/TrFE) copolymers irradiated with fast electrons [18–21] and the authors claimed on a giant electrostrictive response and high elastic energy density in the materials. Moreover, Zhang and his co-workers pointed at an ability to tailor the properties of polymer relaxors to the requirements of transducer and actuator technology due to their strain anisotropy [20, 21]. The volume strain and either the longitudinal or transverse strain, parallel or perpendicular to the direction of electric field applied, may be used in various applications. Also the reliability of the devices can be improved due to the high strain anisotropy that can be achieved in polymer relaxors and can reduce the unwanted mechanical coupling between different directions.

As the properties of electron irradiated P(VDF/TrFE) were interesting for applications in modern transducers and actuators, radiation-induced functionalization of the copolymers has been studied repeatedly [22–31] it should be observed however, that the physical origin of relaxor behaviour in PVDF-type polymers has not been demonstrated clearly. Recently, Chen stressed that the dielectric response of electron irradiated P(VDF/TrFE)(50/50) copolymer and terpolymer of P(VDF/TrFE/CTFE)(65/35/9) is dominated by the response of non-crystalline regions and termed the behaviour as "dielectric relaxor" [32]. We proposed to consider PVDF to be similar to the PSN and PST-type relaxors with a coexistence of both the short-range polar order in the amorphous phase and long-range polar order in the crystalline phase of the polymer [30]. Here we would like to present in detail changes in the dielectric response of PVDF as well as P(VDF/TrFE)(75/25) and (50/50) copolymers caused by fast electron irradiation and discuss processes involved in radiation-induced functionalization.

2. Experimental procedures

As we would like to discuss the effect of fast electron irradiation on PVDF-type polymers with different ferroelectric ordering (different temperatures of ferroelectric-paraelectric transitions) we studied dielectric response of PVDF, P(VDF/TrFE)(75/25) and P(VDF/TrFE)(50/50) films. PVDF and P(VDF/TrFE)(50/50) films were 10–80 μm thick and radially oriented by hot-pressing at p = 600 MPa and T = 450 K. The P(VDF/TrFE)(75/25) copolymer samples with thickness of 25 μm were biaxially oriented and corona-charged.

The samples were irradiated with electrons of energy E^* = 0.5–1.5 MeV in a Van de Graaff accelerator with doses D^* = 0.5–4.3 MGy both below and above the Curie point. The current density of the electrons amounted to 0.1 $\mu A/cm^2$. Calculated energy losses of 0.5 MeV electrons yield the extrapolated range of ~0.8 mm, thus the incident electrons of $E^* = 0.5$–1.5 MeV passed through the samples studied [31, 33]. The samples were transferred from the high vacuum of the accelerator into the air and the radiation damage to the polymers was studied.

Electron spin resonance (ESR) technique with CW X-Band Radiopan SE/X-2547 spectrometer was used to study free radicals and their decay at room temperature. The ESR spectra were recorded from a few days up to 2 months after termination of irradiation. Number of spins in the samples was determined by ESR with respect to the Ultramarine Blue Standard sample having $6 \cdot 10^{15}$ spins of the $S_3^{\bullet -}$ radicals.

After the decay of free radicals we studied the concentration of isolated and conjugated C=C bonds in the samples by IR spectroscopy on Perkin-Elmer FT-IR 1725X spectrometer. IR and Raman spectroscopy was used also to characterize the conformational order/disorder in the samples. NIR Raman spectra were obtained with Bruker IFS66 FRA 106 spectrometer in back scattering geometry at room temperature. The samples were excited by Nd:YAG laser radiation of 1064 nm wavelength and 200 mW power.

Dielectric response of the films was measured in the frequency range 100 Hz–1 MHz by using computer controlled HP-4284A LCR Meter later than 2 months after termination of irradiation. The samples with gold sputtered electrodes were placed in an Oxford Instruments Cryostat CF 1240 where the temperature was changed from 100 to 450 K at a rate of 1 K/min.

3. Results

3.1. Radiation damage

Modification of the dielectric properties of P(VDF/TrFE) by fast electron irradiation was observed for the first time by Odajima *et al.* who reported a broadening of the dielectric anomaly, a downward shift of the Curie point and a decrease of the thermal hysteresis at the ferroelectric-paraelectric transition in irradiated copolymers [34]. Daudin, Legrand *et al.* [35–40] studied systematically the effect of electron irradiation on the properties of (70/30) and (60/40)P(VDF/TrFE) copolymers. The authors confirmed radiation-induced changes observed by Odajima *et al.* and found a small upward shift of the glass transition caused by electron irradiation. Moreover, their UV and XPS spectroscopy studies revealed double C=C bonds in irradiated samples and a decrease in the degree of crystallinity was observed by XRD [38]. They discussed also various contributions responsible for the downward shift of the Curie temperature and considered the defect-induced decrease of the dipolar energy to be most important [39]. The decrease in the polarization as well as the decrease in the crystallinity of irradiated P(VDF/TrFE) copolymers were also reported by Bharti *et al.* [26].

It has been pointed out by many authors that the radiation-chemistry of fluoropolymers differs somewhat from that of their hydrocarbon analogues [41]. This is attributed to the fact that though the fluorine atom provides the strongest bond with carbon atom (dissociation energy of C–F bond amounts to 530.5 kJ/mol whereas that of the C–H bond is equal to 429.7 kJ/mol) irradiation breaks both C–H and C–F bonds leading to formation of in-chain radicals and elimination of hydrogen fluoride. Of course the C–C bonds are also susceptible to radiation-induced scissions producing end-chain radicals. Let us discuss the effect of electron irradiation on PVDF-type polymers in the high vacuum of Van de Graaff accelerator. The cleavage of C–H and C–F bonds by fast electrons can produce two kinds of in-chain and two kinds of end-chain radicals in PVDF:

$$\begin{aligned}&(\mathrm{I})\text{–}CF\text{–}C^{\bullet}H\text{–}CF_2\text{–}+(\mathrm{II})\text{–}CH_2\text{–}C^{\bullet}F\text{–}CH_2\text{–}\\&+(\mathrm{III})\text{–}CH_2\text{–}C^{\bullet}F_2+(\mathrm{IV})\text{–}CF_2\text{–}C^{\bullet}H_2.\end{aligned}$$

In random copolymers P(VDF/TrFE)(75/25) and P(VDF/TrFE)(50/50) one can expect additional three kinds of in-chain and two kinds of end-chain radicals:

$$\begin{aligned}&(\mathrm{I})\text{–}CF_2\text{–}C^{\bullet}H\text{–}CF_2\text{–}+(\mathrm{II})\text{–}CH_2\text{–}C^{\bullet}F\text{–}CH_2\text{–}\\&+(\mathrm{III})\text{–}CH_2\text{–}C^{\bullet}F_2+(\mathrm{IV})\text{–}CF_2\text{–}C^{\bullet}H_2\\&+(\mathrm{V})\text{–}CH_2\text{–}C^{\cdot}F\text{–}CFH\text{–}\\&+(\mathrm{VI})\text{–}CH_2\text{–}C^{\bullet}F\text{–}CHF+(\mathrm{VII})CF_2\text{–}C^{\bullet}F\text{–}CF_2\text{–}\\&+(\mathrm{VIII})\text{–}CF_2\text{–}C^{\bullet}FH+(\mathrm{IX})\text{–}CHF\text{–}C^{\bullet}F_2.\end{aligned}$$

When the irradiated samples are transferred from the high vacuum into the air some alkyl radicals become oxygenated forming alkoxy or alkyl peroxy radicals [42]. Electron spin resonance studies yield information on the kinds of free radicals introduced by fast electrons and on the free radical decay. It should be noted however, that the detailed analysis of the ESR spectra of electron irradiated PVDF-type polymers is difficult since the kinetics of the radical formation and decay is unknown [41]. The chain-end radicals can undergo radical recombination reactions with other radicals which may lead to branching structures whereas desorption of HF is responsible for the formation of double C=C bonds [41].

Fig. 2 shows room temperature ESR spectra of PVDF and P(VDF/TrFE)(75/25) and (50/50) copolymers irradiated at $T^* = 300$ K with the dose $D^* = 1.25$ MGy of 1 MeV incident electrons recorded at different times after termination of irradiation. The multicomponent spectrum of PVDF with the separation of outermost peaks of

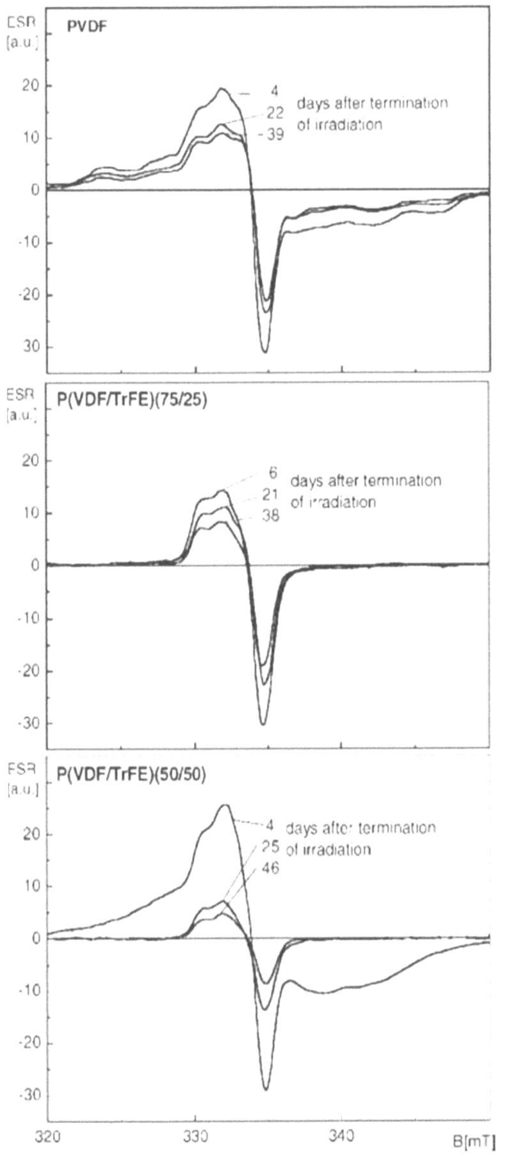

Figure 2 Room temperature ESR spectra of PVDF, P(VDF/TrFE)(75/25) and P(VDF/TrFE)(50/50) irradiated at 300 K with a dose of 1.25 MGy of 1 MeV electrons recorded at different time after termination of irradiation.

about 23 mT consists of overlapping lines of the hyperfine structure, which have been assigned to the chain-end $-CF_2-C^{\bullet}H_2$ and in-chain $-CF_2-C^{\bullet}H-CF_2-$ alkyl radicals A large asymmetric peak in the centre of the spectrum is due to peroxy radicals either in-chain or end-chain [44]. The same kind of radicals was identified in the ESR spectra of irradiated P(VDF/TrFE)(50/50) copolymer, whereas in the case of P(VDF/TrFE)(75/25) copolymer the peroxy radicals are formed mainly as a result of rapid transformation of alkyl radicals to the peroxy ones in the presence of the air. The principal g-tensor values of this oxygenated radicals were estimated as $g_{\|} = 2.0327$, $g_{\perp} = 2.0081$ and $\langle g \rangle = 2.0163$, and are characteristic of peroxy radicals produced from the alkyl radicals $-CF_2-C^{\bullet}F-CF_2-$ [42]. Though the integral intensity of the ESR spectra of irradiated PVDF decreases with time the signals originating of both primary and oxygenated radicals are observed in the spectra after 2 months of storage at room temperature. The concentration of free radicals in the copolymers decays within ~2 months to a constant value that depends on the dose and temperature of irradiation and on the TrFE content [28, 31]. An example of room temperature decay of the free radical density in P(VDF/TrFE)(50/50) samples irradiated at $T^* = 300$ K with the same dose of electrons of incident energy $E^* = 1$ MeV and $E^* = 1.5$ MeV is shown in Fig. 3. It can be seen that the final concentration of free radicals is independent of the energy E^* of the incident electrons. The kinetics of free radical creation and decay is however, dependent on the temperature T^* of irradiation and in general is unknown for PVDF-type polymers [41]. We found that samples irradiated at high temperatures show low-intensity radical signal (Fig. 4) with $g = 2.0175$ and with peak–peak line width of 1.4 mT. The signal is two orders of magnitude lower than that of samples irradiated at 300 K and is attributed to polyenyl radicals $CH_2-(CF{=}CH)_n-C^{\bullet}F-CH_2-CFH$ [44]. The unpaired electron in the radicals is delocalised over conjugated C=C bonds and we assessed the maximum number of C=C linkage from our IR studies to amount to $n = 5$.

The decay of free radicals and desorption of HF results in formation of isolated and conjugated C=C bonds, which can be revealed by IR spectroscopy. Examples of IR spectra of P(VDF/TrFE)(50/50) copolymer irradiated with various doses of 1 MeV electrons are shown in Fig. 5. The band at 1754 cm^{-1} is attributed to vibrations of isolated C=C bonds and the bands of conjugated C=C bonds appear at smaller wavenumbers [45].

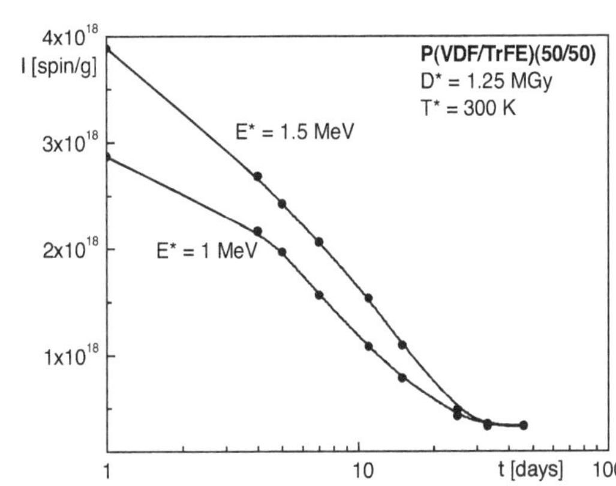

Figure 3 Room temperature decay of the density of free radicals in P(VDF/TrFE)(50/50) copolymer irradiated at $T^* = 300$ K with the dose $D^* = 1.25$ MGy of 1 and 1.5 MeV electrons.

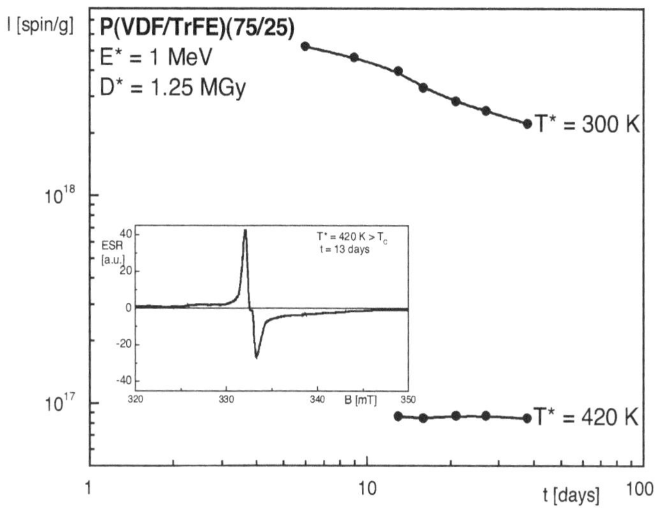

Figure 4 Room temperature decay of the density of free radicals in P(VDF/TrFE)(75/25) copolymer irradiated with the dose $D^* = 1.25$ MGy of 1 MeV electrons at 300 and 420 K; (Inset) ESR spectrum of the copolymer irradiated at $T^* = 420$ K recorded 13 days after termination of irradiation.

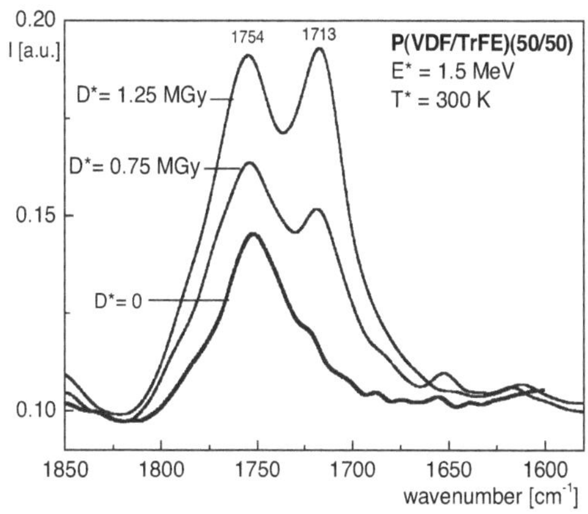

Figure 5 IR spectra of P(VDF/TrFE)(50/50) non-irradiated and irradiated at $T^* = 300$ K with various doses of 1.5 MeV electrons.

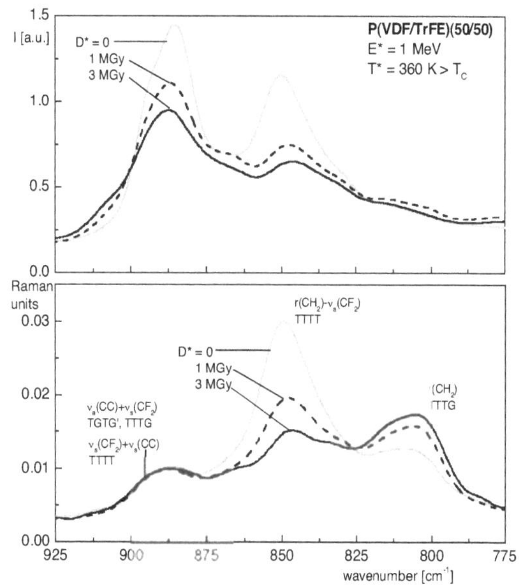

Figure 6 Room temperature IR and Raman spectra of P(VDF/TrFE)(50/50) non-irradiated and irradiated at $T^* = 360$ K with various doses of 1 MeV electrons.

Tashiro has shown that vibrational spectra are useful in characterization of conformational disorder in P(VDF/TrFE) copolymers [8]. We have used vibrational spectroscopy to show that fast electron irradiation of the copolymers results in a reduction of the coherence of long-range ferroelectric all-trans conformation and an increase of the concentration of TTTGTTTG′ and TGTG′ conformations [28, 29, 31]. The effect is illustrated in Fig. 6 where radiation-induced decrease of the intensity of all-trans band at $\sim$850 cm^{-1} is visible in IR and Raman spectra, whereas the band intensity at $\sim$810 in Raman spectrum, characteristic of TTTGTTTG′ conformation, increases.

3.2. Dielectric response

The scission of the PVDF-type chains by double C=C bonds, defects and conformational disorder introduced by fast electron irradiation affect the physical properties of the polymers. The radiation-induced modification in dielectric dispersion and absorption is clearly visible in the case of PVDF, where the Curie temperature is far above the glass transition (Fig. 7). The dielectric response in low-temperature range is characteristic of segmental

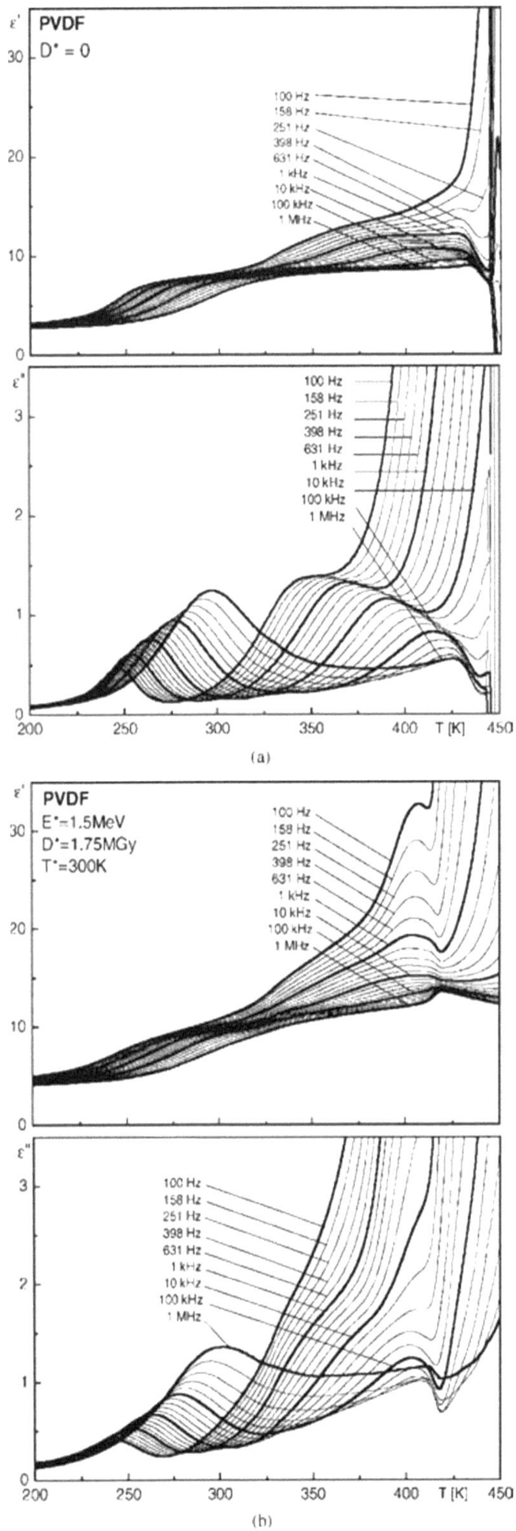

Figure 7 Dielectric response of PVDF films, non-irradiated (a) and after irradiation with the dose $D^* = 1.75$ MGy of 1.5 MeV electrons at $T^* = 300$ K (b).

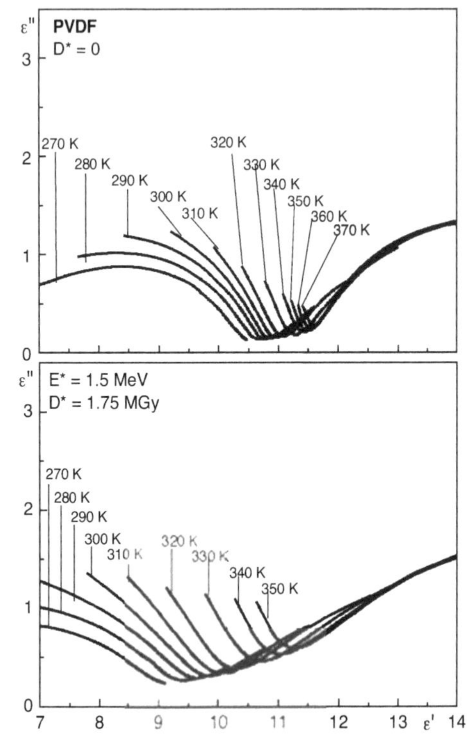

Figure 8 Cole-Cole plots of PVDF films, non-irradiated and after irradiation with the dose $D^* = 1.75$ MGy of 1.5 MeV electrons at $T^* = 300$ K.

motions of the polymer, i.e., freezing of dipolar motions in the amorphous phase of the semicrystalline polymer. The process is strongly temperature dependent and the frequency and temperature dependence of the imaginary part of the permittivity $\varepsilon''(f, T)$ yield the temperature variation of the relaxation times τ (T). The temperature dependence of the relaxation time can be described by Vogel–Fulcher relation:

$$\tau = \tau_{\mathrm{VF}} \exp[B/(T - T_{\mathrm{VF}})],$$

where T_{VF} is a characteristic temperature at which all relaxation times diverge and the relaxation time distribution becomes infinitely broad. At higher temperature dielectric response is dominated by a local mode in the crystalline phase of the polymer. The response due to wide-angle oscillations of dipoles attached to the main chain, followed by their rotation with chain cooperation is described by Arrhenius law [11]:

$$\tau = \tau_{\mathrm{A}} \exp[\Delta H/kT].$$

Torsional and rotational motions of dipoles attached to the main chain leading to changes from all-trans to TGTG′ are responsible for the dielectric anomaly, characteristic of the ferroelectric-paraelectric transition observed just before dynamical melting of the polymer, where a minimum in the permittivity appears. As is seen in Fig. 7b, fast electron irradiation results in a downward shift of the Curie point and in a considerable broadening and decreasing of the anomaly at T_C. Changes appear also in the response characteristic of the local mode and segmental mode of the polymer: the response of the local mode becomes smeared out and the dispersion increases. It is clearly visible in Fig. 8, where radiation-induced changes in the Cole–Cole plot are shown.

The comparison of $\varepsilon'(f, T)$ dependences of electron-irradiated PVDF and non-irradiated P(VDF/TrFE)(75/25) (Figs 7 and 9) indicates that the effect of introduction of units with dipole moment lower than that of the VDF into the PVDF chain is similar to that of electron irradiation. Radiation-induced modification of the dielectric response of the P(VDF/TrFE)(75/25) copolymer, shown also in Fig. 9, consists in further decrease and broadening of the permittivity maximum and lowering of the Curie temperature. Dielectric relaxation of P(VDF/TrFE)(75/25)

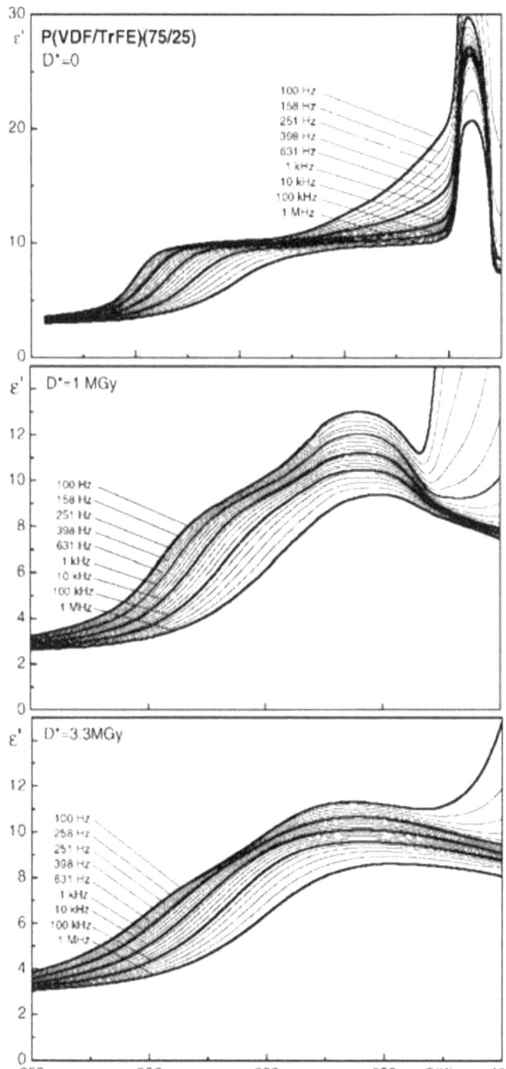

Figure 9 $\varepsilon'(T, f)$ dependences of non-irradiated P(VDF/TrFE)(75/25) film and the copolymer films irradiated at 420 K with various doses D^* of 1 MeV electrons.

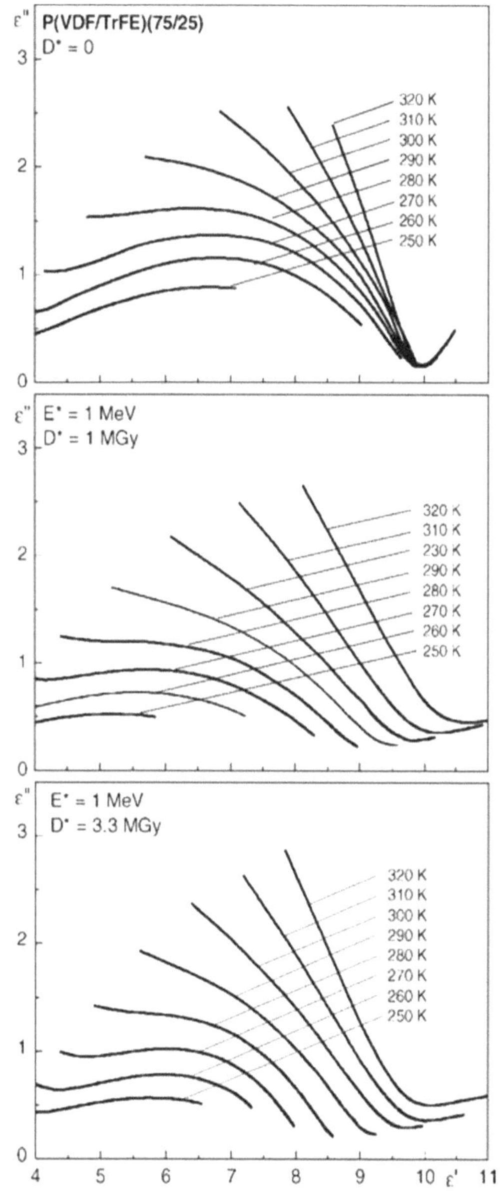

Figure 10 Cole-Cole plots for non-irradiated P(VDF/TrFE)(75/25) film and the copolymer films irradiated at 420 K with various doses D^* of 1 MeV electrons.

copolymer samples are well described by the Cole-Cole relation in the low temperature range, as shown in Fig. 10.

Dielectric anomaly related to the ferroelectric-paraelectric transition appears in radially-oriented P(VDF/TrFE)(50/50) film at ~342 K, i.e., close to the glass transition anomaly, thus electron irradiation, even with low doses, results in a diffuse and dispersive dielectric response, characteristic of ferroelectric relaxors (Fig. 11). Radiation-induced changes in temperature dependence of the real and imaginary part of the permittivity of P(VDF/TrFE) copolymers are clearly visible in Fig. 12. Dielectric anomalies related to the glass transition and the ferroelectric–paraelectric transition merge in a broad maximum for P(VDF/TrFE)(50/50) irradiated with 1 MeV electrons using the dose as small as 0.75 MGy, whereas a single broad permittivity maximum appears in P(VDF/TrFE)(75/25) after irradiation with the dose of 3.3 MGy. Our results indicate that much lower doses of irradiation are needed to achieve relaxor-like dielectric response with high permittivity values for copolymers with higher intrinsic statistical disorder (for the (50/50) copolymer the dose $D^* \approx 0.75$ MGy, whereas for (75/25) the dose $D^* \approx 3$ MGy).

Changes in temperature variation of the relaxation times τ obtained from $\varepsilon''(f, T)$ dependences measured for P(VDF/TrFE)(75/25) irradiated with 1 MeV electrons are shown in Fig. 13. The solid lines represent the fits to the Vogel–Fulcher equation and the fitting parameters, B and $T_{VF,}$ are plotted in Fig. 14 versus the dose of irradiation. The figure contains also the parameters obtained for PVDF and P(VDF/TrFE)(75/25) copolymer irradiated with 1.5 MeV electrons. The activation energy B of micro-

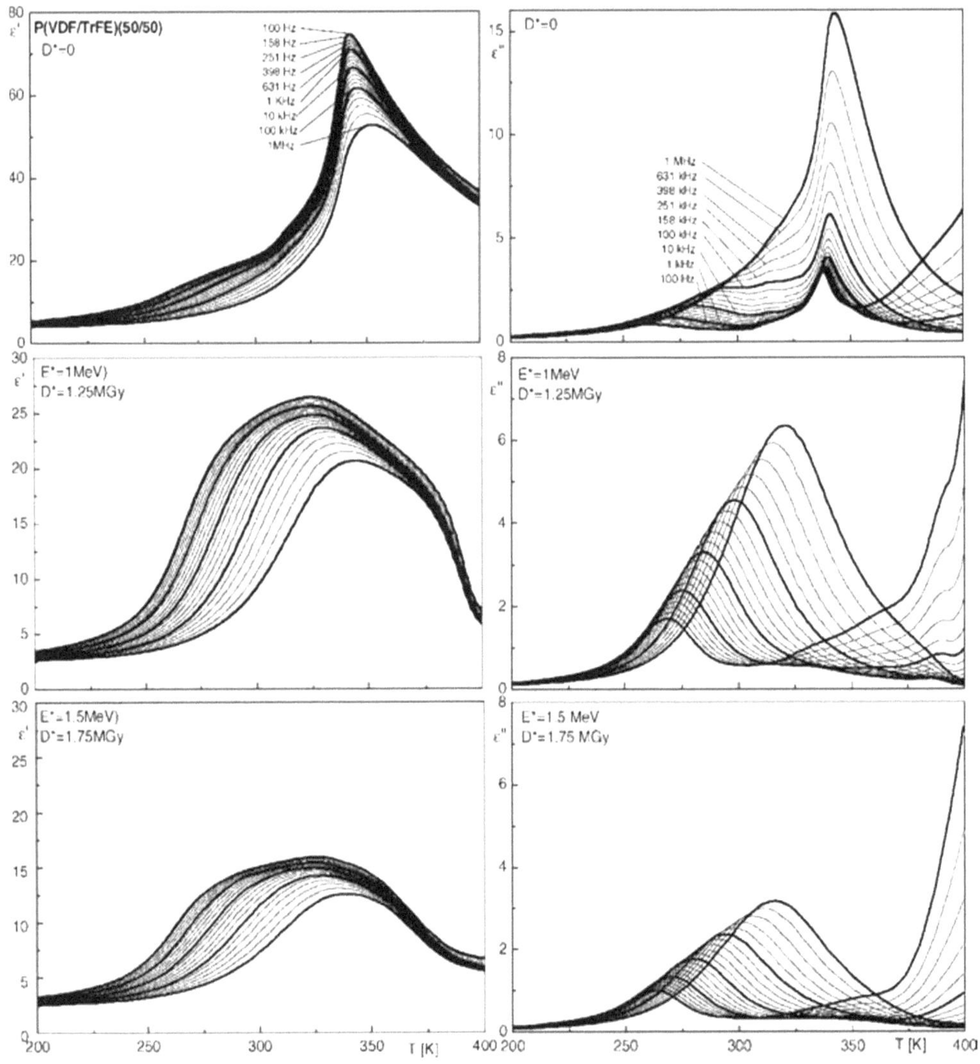

Figure 11 Dielectric response of P(VDF/TrFE)(50/50) copolymer, non-irradiated and after fast electron irradiation at $T^* = 300$ K.

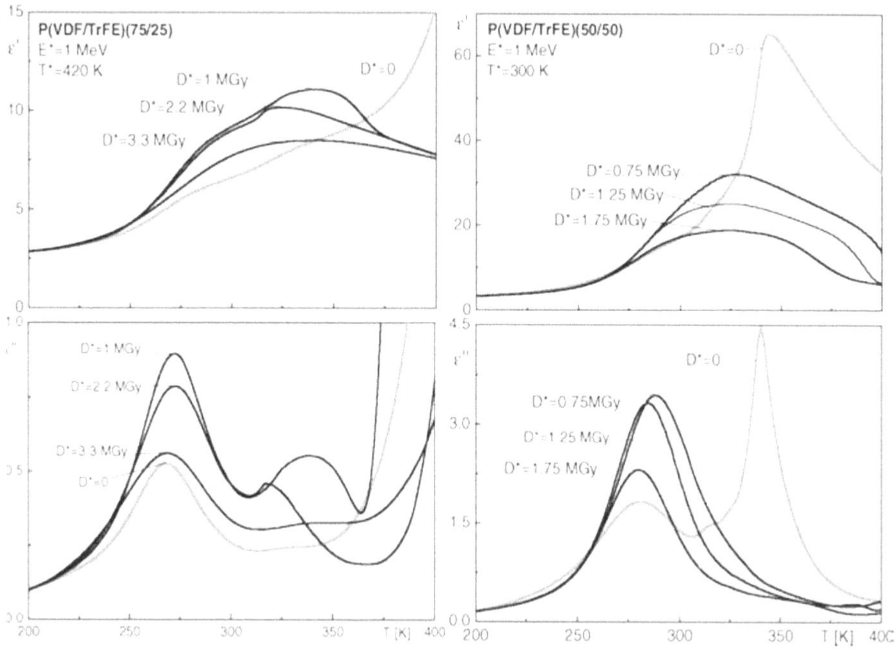

Figure 12 Temperature dependence of the real and imaginary part of dielectric permittivity measured at the frequency of 10 kHz for P(VDF/TrFE) copolymers irradiated with various doses D^* of 1 MeV electrons at various temperatures T^*.

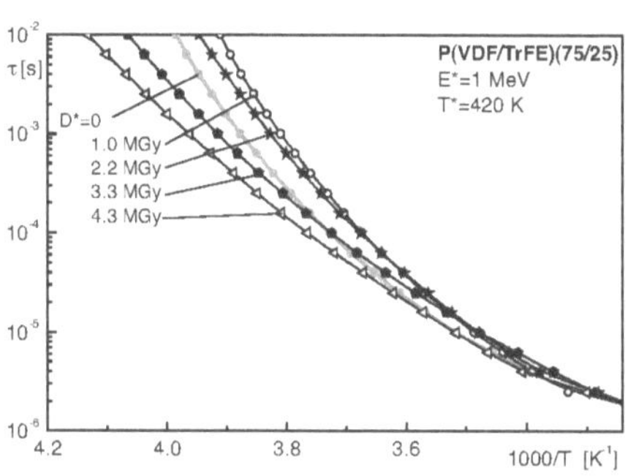

Figure 13 Dielectric relaxation times τ vs. the inverse temperature for P(VDF/TrFE)(75/25) irradiated at $T^* = 420$ K with various doses D^* of 1 MeV electrons.

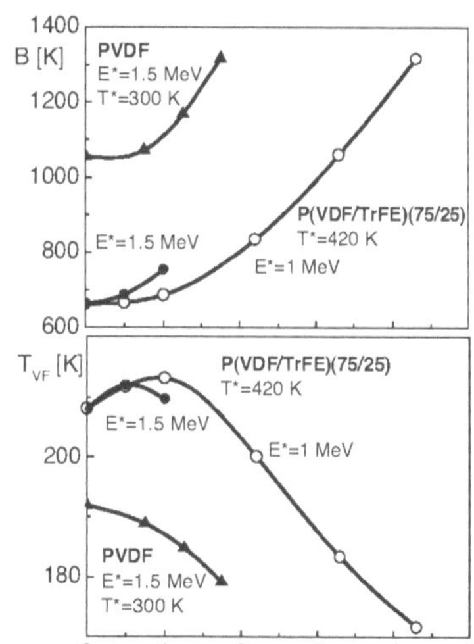

Figure 14 Dose dependences of the parameters B and T_{VF} obtained on fitting the temperature variation of dielectric relaxation times to Vogel–Fulcher equation for PVDF and P(VDF/TrFE)(75/25) irradiated with electrons of incident energy E^* at the temperature T^*.

Brownian dipolar motions in the amorphous phase of the copolymer increases and the static freezing temperature T_{VF} shifts towards lower temperatures for irradiation with doses higher than 1.5 MGy. However, it should be observed that the values of the parameters are affected by an overlapping of the Curie point dielectric anomaly with that of the glass transition and thus the radiation-induced changes in B and T_{VF} can be considered qualitative only. Moreover, the interpretation of the experimental data of samples irradiated above the Curie point is complicated due to the fact that heating the samples at temperatures in the ferroelectric–paraelectric transition region leads to conformational changes [8] and in consequence, changes in the dielectric response [24].

4. Conclusions

Dielectric response of semicrystalline PVDF-type ferroelectric polymers contains contribution of the response related to dipolar freezing in the amorphous phase and contributions of the local mode and changes from ferroelectric all-trans conformation to non-polar TGTG′ conformation in the crystalline phase. The aim of the paper was to show the way in which the dielectric response can be transformed in a single, diffuse and dispersive dielectric anomaly, characteristic of ferroelectric relaxors. The Curie point anomaly appears in PVDF at temperatures ~150 K higher than the dynamic glass transition but the Curie point can be shifted downward by loading the planar zigzag chains with units having the dipole moment different than that of VDF [8, 14] or by introducing random field in form of radiation-induced defects [34–40]. It has been shown that relaxor-like dielectric anomaly can be obtained in P(VDF/TrFE) copolymers by fast electron irradiation [18–27] or by loading either with CTrFE or with CFE [15–17].

We studied modification of the dielectric response of PVDF and P(VDF/TrFE) copolymers by fast electron irradiation as well as radiation damage to the materials by ESR, IR and Raman spectroscopy. The results show that three processes are involved in electron-induced functionalization of the ferroelectric polymers. Fast electron irradiation in a Van de Graaff accelerator breaks the C–H, C–F and C–C bonds and creates various types of in-chain and end-chain alkyl radicals in the polymer films. Some of the radicals become oxygenated when the samples are transferred from the high vacuum of the accelerator into the air. The kinetics of formation and decay of alkyl and alkyl peroxy radicals is determined mainly by the VDF content in the sample and by the temperature and dose of irradiation. As the hydrogen fluoride is desorbed the decay of the radicals is accompanied by creation of isolated and conjugated C=C bond in the PVDF-type chain and polyenyl radicals with unpaired electron delocalized over several C=C linkage appear. Such defects as double C=C bonds, polyenyl radicals and non-polar TGTG′ conformations are sources of random field, which in the case of P(VDF/TrFE) copolymers adds to the random field due to intrinsic statistical disorder. Moreover, it should be observed that C=C bonds present in the polyvinylidene chains break the coherence of ferroelectric all-trans conformation into regions of shorter trans-sequences, which can be thought as polar nanoclusters.

The radiation-induced functionalization of the ferroelectric polymers of PVDF-type may be considered as consisting of two effects. The first one is similar to that observed in crystalline ferroelectrics [46] where the downward shift of the Curie temperature is related to a decrease in the density of ferroelectrically active dipoles and random field (defects) result in rounding and decrease of the Curie point anomaly. Thus the relaxor-like dielectric response with a single broad and dispersive anomaly appears in semicrystalline ferroelectric polymers as a result of defect-induced merging of the Curie point anomaly with the response of the amorphous phase. The second effect is related to a contribution of the response of polymer chain fragments with short trans coherence (polar nanoclusters) to the dielectric response of irradiated samples and causes an increase of the dielectric dispersion.

Acknowledgment

The work was supported by the Grant 2PO3B 121 24 from the Committee of Scientific Researches in Poland and the Centre of Excellence for Magnetic and Molecular Materials for Future Electronics within the European Commission Contract No. G5MA-CT-2002-04049.

References

1. T. FURUKAWA, M. DATE, E. FUKADA, Y. TAJITSU and A. CHIBA, *Jpn. J. Appl. Phys.* **19** (1980) L109.
2. T. FURUKAWA, M. DATE and E. FUKADA, *J. Appl. Phys.* **51** (1980) 1135.
3. T. YAGI, M. TATEMOTO and J. SAKO, *Polymer J.* **12** (1980) 209.
4. T. T. WANG, J. M. HERBERT and A. M. GLASS (Eds.), "The application of ferroelectric polymers". (Blackie, Glasgow, Scotland 1988).
5. K. UCHINO, "Ferroelectric Devices". (Marcel Dekker, Inc., New York 2000) p. 156.
6. S. B. LANG, *Key Eng. Mater.* **92–93** (1994) 83.
7. A. J. LOVINGER, in "Developments in crystalline polymers-1", edited by. J. C. Bassett (Appl. Sci. Pub. Ltd., London, 1982), p.195.
8. K. TASHIRO, in "Ferroelectric polymers — chemistry, physics and applications", edited by. H. S. Nalwa, (Marcel Dekker, New York, 1995) p. 62.
9. N. KOIZUMI, *Key Eng. Mater.* **92–93** (1994) 161.
10. R. G. KEPLER, in "Ferroelectric polymers — chemistry, physics and applications", edited by. H. S. Nalwa, (Marcel Dekker, New York, 1995) p. 1183.
11. B. HILCZER, J. KUłEK, E. MARKIEWICZ and M. KOSEC, *J. Non-Crystalline Solids* **305** (2002) 167.
12. T. FURUKAWA, *Adv. Colloid Interf. Sci.* **71–72** (1997) 183.
13. J. MAłECKI, Private information (MP3 calculations of the dipole moments).
14. H. OHIGASHI and T. HATTORI, *Ferroelectrics* **171** (1995) 11.
15. T. C. CHUNG and A. PETCHSUK, *Ferroelectr. Lett.* **28** (2001) 136.
16. H. XU, Z.-Y. CHENG, D. OLSON, T. MAI, Q. M. ZHANG and G. KAWAMOS, *Appl. Phys. Lett.* **78** (2001) 2360.
17. V. BOBNAR, B. VODOPIVEC, A. LEVSTIK, M. KOSEC, B. HILCZER and Q. M. ZHANG, *Macromolecules* **36** (2003) 4436.
18. Q. M. ZHANG, V. BHARTI and X. ZHAO, *Science* **280** (1998) 2101.
19. X. ZHAO, V. BHARTI, Q. M. ZHANG, T. ROMOTOWSKI, F. TITO and R. TING, *Appl. Phys. Lett.* **73** (1998) 2054.
20. Z.-Y. CHENG, T.-B. XU, V. BHARTI, S. WANG and Q. M. ZHANG, *ibid.* **74** (1999) 1901.
21. Z.-Y. CHENG, V. BHARTI, T.-B. XU, S. WANG and Q. M. ZHANG, *J. Appl. Phys.* **86** (1999) 2208.
22. T. KARAKI, I. CHOU and L. E. CROSS, *Jpn. J. Appl. Phys.* **39** (2000) 5668.

23. Y. TANG, X.-Z. ZHAO, H. L. W. CHAN and C. I. CHOY, *Appl. Phys. Lett.* **77** (2000) 1713.
24. H. SMOGÓR, B. HILCZER and S. WARCHOŁ, *Ferroelectrics* **258** (2001) 291.
25. B. HILCZER, H. SMOGÓR, T. PAWŁOWSKI, S. WARCHOŁ and M. NOWICKI, *ibid.* **261** (2001) 139.
26. V. BHARTI, G. SHANTHI, H. XU, Q. M. ZHANG and K. LIANG, *Mater. Lett.* **47** (2001) 107.
27. D. ZHANG, X. YAO, X. CHEN, B. SHEN and L. ZHANG, *Ferroelectrics* **264** (2001) 21.
28. B. HILCZER, H. SMOGÓR, J. GOSLAR and S. WARCHOŁ, *Rad. Effects & Def. Solids* **158** (2003) 349.
29. H. SMOGÓR, B. HILCZER, C. Z. PAWLACZYK, J. GOSLAR and S. WARCHOŁ, *Ferroelectrics* **294** (2003) 191.
30. V. A. STEPHANOVICH, M. D. GLINCHUK, E. V. KIRICHENKO and B. HILCZER, *J. Appl. Phys.* **94** (2003) 5937.
31. B. HILCZER, H. SMOGÓR, J. GOSLAR and T. PAWŁOWSKI, *Ferroelectrics* **298** (2004) 113.
32. A. CHEN, in Abstracts of the Conference "Fundamental Physics of Ferroelectrics 2004" (Colonial Williamsburg, February 2004), p. 44.
33. R. R. ROY and R. D. REED, "Interactions of Phonons and Leptons with Matter" (Academic Press, New York 1968) p. 113.
34. A. ODAJIMA, Y. TAKASE, T. ISHIBASHI and K. YUASA, *Jpn. J. Appl. Phys.*, **Suppl. 24** (1985) 881.
35. B. DAUDIN, M. DUBUS and J. F. LEGRAND, *J. Appl. Phys.* **62** (1987) 994.
36. B. DAUDIN, M. DUBUS, F. MACCHI and J. F. LEGRAND, *Nuclear Instrum. & Meth. in Phys. Res. B* **32** (1988) 177.
37. F. MACCHI, B. DAUDIN and J. F. LEGRAND, *Ferroelectrics* **109** (1990) 303.
38. F. MACCHI, B. DAUDIN, A. ERMOLIEFF, S. MARTHON and J. F. LEGRAND, *Radiation Effects & Defects in Solids* **118** (1991) 117.
39. B. DAUDIN, J. F. LEGRAND and F. MACCHI, *J. Appl. Phys.* **70** (1991) 4037.
40. J. F. LEGRAND, B. DAUDIN and E. BELLET-ALMARIC, *Nucl. Instrum. & Meth. Phys. Res. B* **105** (1998) 177.
41. J. S. FORSYTHE and D. J. T. HILL, *Prog. polym. sci.* **25** (2000) 101.
42. B. RANBY and J. F. RABEK, "ESR Spectroscopy in Polymer Research", (Springer Verlag, Berlin 1977), p. 254.
43. Z. B. ALFOSSI, (Ed.), "Peroxyl Radicals", (John Willey & Sons Ltd., New York, 1997), p. 377.
44. N. BETZ, E. PETERSOHN and A. LE MOËL, *Radiat. Phys. Chem.* **47** (1996) 411.
45. B. SCHRADER, (Ed.), "Infrared and Raman Spectroscopy; Methods and Applications". (VCH Publishers Inc., New York, 1995) p. 190.
46. B. HILCZER, *Key Eng. Mater.* **101–102** (1995) 95.

JOURNAL OF MATERIALS SCIENCE **41** (2006) 129–136

The relaxor enigma — charge disorder and random fields in ferroelectrics

WOLFGANG KLEEMANN
Angewandte Physik, Universität Duisburg-Essen, D-47048 Duisburg, Germany

Substitutional charge disorder giving rise to quenched electric random-fields (*RF* s) is probably at the origin of the peculiar behavior of *relaxor* ferroelectrics, which are primarily characterized by their strong frequency dispersion of the dielectric response and by an apparent lack of macroscopic symmetry breaking at the phase transition. Spatial fluctuations of the *RF* s correlate the dipolar fluctuations and give rise to polar nanoregions in the paraelectric regime as has been evidenced by piezoresponse force microscopy (PFM) at the nanoscale. The dimension of the order parameter decides upon whether the ferroelectric phase transition is destroyed (*e.g.* in cubic $PbMg_{1/3}Nb_{2/3}O_3$, PMN) or modified towards *RF* Ising model behavior (*e.g.* in tetragonal $Sr_{1-x}Ba_xNb_2O_6$, SBN, $x \approx 0.4$). Frustrated interaction between the polar nanoregions in cubic *relaxors* gives rise to cluster glass states as evidenced by strong pressure dependence, typical dipolar slowing-down and theoretically treated within a spherical random bond-*RF* model. On the other hand, freezing into a domain state takes place in uniaxial *relaxors*. While at T_c non-classical critical behavior with critical exponents $\gamma \approx 1.8$, $\beta \approx 0.1$ and $\alpha \approx 0$ is encountered in accordance with the *RF* Ising model, below $T_c \approx 350$ K *RF* pinning of the walls of frozen-in nanodomains gives rise to non-Debye dielectric response. It is relaxation- and creep-like at radio and very low frequencies, respectively.

1. Introduction

Relaxor ferroelectrics include a large group of solid solutions, mostly oxides, with a perovskite or tungsten bronze structure. In contrast to ordinary ferroelectrics (FE) whose physical properties are quite adequately described by the Landau-Ginzburg-Devonshire theory [1]. *relaxors* possess the following main features: (i) a significant frequency-dependence of the electric permittivity, (ii) absence of both spontaneous polarization and structural macroscopic symmetry breaking, (iii) FE-like response arising after field cooling to low temperature [2].

Very high response coefficients and an enhanced width of the high response regime around the "ordering" temperature T_m, ("Curie range") make *relaxors* popular systems for applications as piezoelectric/electrostrictive *actuators* and *sensors* (*e.g.* scanning probe microscopy, ink jet printer, adaptive optics, micromotors, vibration sensors/attenuators, Hubble telescope correction, ...) and as *electro-* or *elasto-optic* and *photorefractive elements* (segmented displays, modulators, image storage, holographic data storage, ...).

When reflecting on the occurrence of *relaxor* behavior in perovskites, there appear to be two essential ingredients:

i. existence of *lattice disorder*,
ii. existence of *polar nanoregions* at temperatures much higher than T_m.

The first ingredient can be taken for granted, since *relaxor* behavior in these materials does not occur in the absence of disorder. The second ingredient is manifested in many experimental observations common to all perovskite *relaxors*, as will be discussed later.

The following physical picture has emerged for *relaxors* and seems to be widely accepted. Chemical substitution and lattice defects introduce extra charges or dipolar entities in mixed ABO_3 perovskites. At very high temperatures, thermal fluctuations are so large that there are no well-defined dipole moments. However, on cooling, the presence of these dipolar entities manifests itself as small polar nanoregions below the so-called Burns

0022-2461 © *2006 Springer Science + Business Media, Inc.*
DOI: 10.1007/s10853-005-5954-0

temperature, T_d [3]. These regions grow as the correlation length, r_c, increases with decreasing T, and finally, two different situations may arise. If the regions become large enough (macrodomains) so as to percolate (or permeate) the whole sample, then the sample will undergo a static, cooperative FE phase transition at T_c. On the other hand, if the nanoregions grow with decreasing T, but do not become large enough or percolate the sample, then they will ultimately exhibit a dynamic slowing down of their fluctuations at $T \leq T_m$ leading to an isotropic *relaxor* state with random orientation of the polar domains.

A matter of dispute is still the physical significance and the very origin of the Burns temperature. Very probably it is not a usual phase transitions temperature. It might rather be considered as a so-called Griffiths temperature, which signifies the onset of weak singularities in a diluted ferroic system below the transition temperature of the undiluted system [4, 5]. However, the sharp onset of weak singularities is not at all confirmed in *relaxor* systems. We rather believe that the temperature regime in which the domains grow in size is continuous and merely determined by the correlating forces due to the underlying quenched random field (*RF*) distribution [6] as will be discussed in the next chapter.

2. Polar nanoregions

While many researchers believe that the above mentioned ingredients for *relaxor* behavior to appear are more or less independent, we have argued [6] that the primary cause of *relaxor* behavior is the charge disorder, which is at the origin of the occurrence of polar nanoregions and their fluctuations within the highly polarizable lattice [7]. In order to describe disordered systems and to explore their basic thermodynamic behavior simple spin models are frequently used. The model Hamiltonian

$$H = -\sum_{\langle ij \rangle} J_{ij} S_i S_j - \sum_i h_i S_i \qquad (1)$$

accounts for random interactions (or random bonds, *RB*s), J_{ij}, between nearest neighbor spins S_i and S_j, and for quenched random fields (*RF*s), h_i acting on the spins S_i. While the *RB*s are at the origin of spin glass behavior [8], *RF*s may give rise to disordered domain states provided that the order parameter has continuous symmetry [9]. This is easily shown with the help of energy arguments considering both the bulk energy decrease by fluctuations of the *RF*s and the energy increase due to the formation of domain walls. A remarkable exception, which does not necessarily lead to a disordered ground state, is the random-field Ising model (*RFIM*) system in $d = 3$ dimensions. Owing to its discontinuous spin symmetry, atomically thin domain walls are expected, which are energetically unfavorable. For this reason the *3d RFIM* is expected to exhibit long-range order below the critical temperature T_c. However, as a tribute to the *RF*s new criticality due to a $T = 0$ fixed point [10] and strongly decelerated critical dynamics are encountered [11].

Unfortunately, the original idea [9] to realize a ferromagnetic *RFIM* by doping with random magnetic ions fails. Their spin dynamics always couples to that of the host system such that their dipolar fields cannot be regarded as quenched ones. Regrettably, there is no chance to dope a ferromagnet with *magnetic monopoles*, which might readily provide quenched local magnetic fields. Bearing this in mind, the situation should be much more favorable for the electric counterparts to ferromagnets, where electric charges may take the role of field-generating monopoles. Indeed, in FE systems electric charge disorder should easily give rise to quenched *RF*s. This was proposed previously in order to understand the peculiar *relaxor* behavior of the archetypical solid solution $PbMg_{1/3}Nb_{2/3}O_3$ (PMN) [12]. Unfortunately, apart from the expected extreme slowing-down of the *RFIM* [11], which is closely related to the relaxor-typical huge polydispersivity [2, 12], no *RFIM* criticality was observed. This is a consequence of the high pseudo-spin dimension of the polarization order parameter, $\boldsymbol{P}$. It has eight easy $\langle 111 \rangle$ directions in the cubic unit cell and thus quasi-continuous symmetry [6]. Clearly the search for an appropriate uniaxial FE (one-component order parameter $\pm P_z$, *i.e.* $n = 2$) with charge disorder seems advisable in order to materialize a proper ferroic *3d RFIM* system. Only recently [13] the uniaxial relaxor crystal $Sr_{0.61-x}Ce_xBa_{0.39}Nb_2O_6$ (SBN*61*:Ce, $0 \leq x < 0.02$) has been found to fulfil the conditions of a ferroic *RFIM* (see Chap. 4).

Evidence for the existence of polar nanoregions well above T_m has come from high resolution TEM which also showed the growth of these regions with decreasing T [14]. The evidence is also prominently reflected in certain properties of these systems. To provide the context, recall that for relaxors in the absence of electrical bias there are random + and − fluctuations of the dipolar polarization so that $\sum P_d = 0$, *i.e.*, there is no measurable remanent polarization. However, $\sum P_d^2 \neq 0$, and we then expect the existence of these polar regions to be manifested in properties which depend on P^2, *e.g.*, electrostriction which is reflected in the thermal expansion and the quadratic electro-optic effect, which is reflected in the refractive index, or birefringence. Indeed, both of these properties have provided quantitative measures of this polarization for the relaxors. The manifestation of the presence of polar nanodomains in strong relaxors in terms of the electro-optic effect was first demonstrated by Burns and Dacol [3] in measurements of the T dependence of the refractive index, n. For a normal ABO_3 FE crystal, starting in the high-temperature PE phase, n decreases linearly with decreasing T down to T_d at which point n deviates from linearity. The deviation is proportional to the square of the polarization and increases as the polarization evolves with decreasing T. If the FE transition is of first order, then there is a discontinuity in n at T_c followed by the expected deviation. This qualitative picture is representative of the behavior of many perovskite FEs. However, in the case of relaxors, Burns and Dacol observed deviations from linear $n(T)$ well above T_m. These deviations can be quantitatively described by the relationship [3, 14]

$$\Delta n = \frac{(\Delta n_{11} + \Delta n_{12})}{3} = \left(\frac{n_0^3}{2}\right)\left[\frac{(g_{33} + 2g_{13})}{3}\right] P_d^2 \quad (2)$$

where the Δn's are the changes in the parallel and perpendicular components of n, n_o is the index in the absence of polarization (P_d), and the g_{ij} are the quadratic electro-optic coefficients. The temperature at the onset of the deviation from linear $n(T)$, $T_d = 620$ K in the case of PMN, is the Burns temperature. Above this temperature, thermal fluctuations are so large that there are no well-defined dipolar regions or clusters. These regions nucleate at T_d by taking advantage of the statistical fluctuations of the *RF*s and grow on lowering T. Diffuse scattering and HRTEM results indicate that in PMN these regions grow from 2 to 3 nm in size above 400 K to $\approx$10 nm at $\approx$160 K ($T_m \approx 230$ K) [2]. Thus, they are much smaller than typical FE domains, which are orders of magnitude larger. The small size of these nanoregions explains why they cannot easily be detected by diffraction measurements and the bulk structures of PMN and most strong mixed ABO_3 relaxors remain cubic to both x-ray and neutron probes and to long wavelength photons down to lowest temperatures.

Evidence for the existence of polar nanoregions well above T_m in relaxors has also been deduced from the T dependence of the susceptibility. As noted earlier, it is well established that $\chi'(T)$ in the high temperature, cubic PE phase of ABO_3 FEs follows the Curie–Weiss law $\chi' = C/(T-\theta)$, where θ is the Curie–Weiss temperature, over a wide temperature range. However, $\chi'(T)$ of relaxors shows large deviations from this law for $T > T_m$. Deviation from Curie–Weiss response sets in upon nucleation of the polar nanoregions at T_d, and this deviation grows with decreasing T as the size of the regions and their correlations increase. It can be described by a modified Curie-Weiss law [15]

$$\chi = \frac{C[1 - q(T)]}{[T - \theta(1 - q(T))]}, \quad (3)$$

which relates $\chi(T)$ below T_f to the local "spin glass" order parameter q, which is a function of temperature [16, 17]. While $q \to 0$ above at $T > T_d$, it increases with decreasing temperature below T_d because of increased dipolar correlations [16]. In such a case the local order parameter due to correlations between neighboring polar domains of polarization P_i and P_j is $q = \langle P_i P_j \rangle^{1/2}$. The "universal relaxor polarization" regime of relaxors, where an unusual T dependence of the susceptibility, $\chi' = C/(T-T_0)^2$, is observed [17], is essentially the regime where the Edwards-Anderson order parameter q reveals a marked temperature dependence.

3. Cubic relaxors

The archetypical perovskite-like lead-containing *relaxor* $Pb(Mg_{1/3}Nb_{2/3})O_3$ (PMN, ABO_3 space group $Pm\bar{3}m$) system has been known for fifty years [12]. As-grown PMN single crystals exhibit excellent crytalline properties with small mosaic spread ($\omega \leq 0.01°$). However, on the nanometer scale there is a significant degree of chemical and structural disorder [18]. Fig. 1 shows a random distribution of B site ions on an enlarged unit cell. Moreover, atoms on the B sites are occasionally short-range ordered within quenched chemical nanodomains with $Fm\bar{3}m$ symmetry [19]. The polar order parameter of PMN is directed along one of the eight rhombohedral $\langle 111 \rangle$ directions. Hence, an eight state Potts model might be applicable to this case. Fig. 2 shows the evolution of domains as simulated in a two-dimensional 4-state Potts model with respective planar *RF*s as a function of temperature [20]. At high temperatures, $k_BT/J = 10$, it is seen that the domains image the *RF* distribution, while on decreasing the temperature down to $k_BT/J = 0.2$ the domains become coarse grained and merely image the local fluctuations of the *RF* distribution. An appropriate means to evidence the nanoregions even when being dynamic, *i.e.* above any

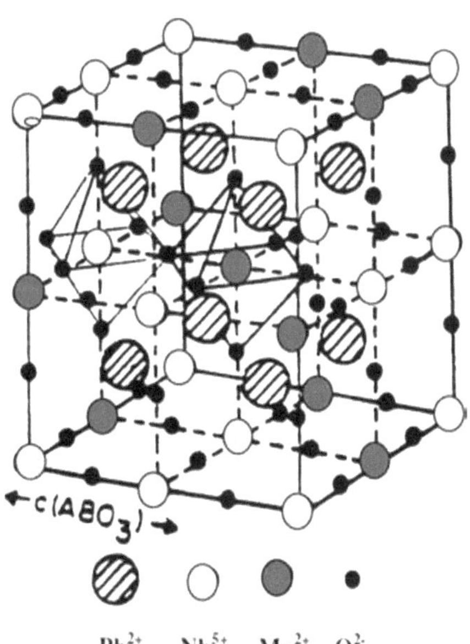

Figure 1 Random ionic distribution in PMN (Pb^{2+} = hatched circles, O^{2-} = small solid circles, Nb^{5+} = large solid circles, Mg^{2+} = open circles).

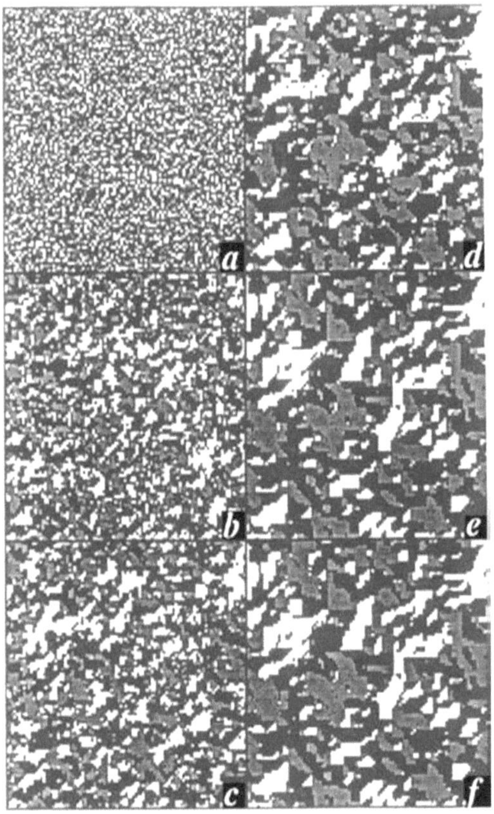

Figure 2 Domain distribution with polarizations $\pm P_x$ and $\pm P_y$ as indicated by different gray tones in a 4-state random field Potts model at temperatures $k_B T/J = 10$ (a), 1 (b), 0.8 (c), 0.6 (d), 0.3 (e) and 0.2 (f) (from [20]).

transition or freezing temperature, is the optical second-harmonic generation, SHG, as evidenced for both PMN [21] and SBN [22]. In both cases the SHG intensity starts to grow well above the transition temperatures.

Since the order parameter of PMN is close to be continuous, an equilibrium phase transition into a long-range ordered FE phase is excluded [9]. However, Blinc *et al.* [23] developed another route towards an ordered low-T phase. Based on the existence of polar nanoregions and their above described correlations, $q = \langle P_i P_j \rangle^{1/2}$ [16], they proposed a *spherical random bond RF* (*SRBRF*) theory. Here polar clusters of any size fulfilling the spherical constraint are considered as randomly interacting "*superspins*" which undergo a transition into a cluster glass state. Theory has been solved for infinitely ranged interactions in mean-field approximation. Experimental tests by means of ^{93}Nb *NMR* reveal that the *RF*s are Gaussian distributed and that the Edwards-Anderson glass order parameter is finite below $T \approx 300$ K. Hence, no equilibrium static glassy freezing can be expected. Nevertheless, the preponderance of the random bonds with respect to the *RF*s clearly favors a glassy scenario which can be tested by measuring the (truncated) divergence of the nonlinear susceptibility, χ_3. More rigorously [24], the anisotropy parameter $a_3 = \chi_3/\chi_1^4$ has to maximize when approaching T_g on cooling, where χ_1 denotes the linear susceptibility. This has, indeed, been observed on both PMN and the related *relaxor* lead lanthanum zirconate-titanate (PLZT) [24].

A remark concerning the magnitude of the *RF*s seems in order. We are convinced that the primary function of the *local RF*s due to built-in charge disorder is to form energetically favored nanoregions [6]. These interact in a glass-like manner and form a spherical "*superspin*" glass [16, 23]. The glass transition, secondly, becomes smeared owing to *effective RF*s, h_i entering the *superspin* Hamiltonian, Equation 1. Since cluster spins containing N atomic spins experience only the fluctuations of the *local RF*s, the magnitudes of the *effective RF*s are reduced by factors $1/N^{1/2}$. This is why only weak smearing effects are observed. It should finally be remarked that the *SRBRF* theory has by far not yet generally been accepted. Still the origin of the nanoregions and their transition into a glassy state are disputed and not yet understood from a rigorous theoretical point of view. Clearly, the *RF* model simplifies the situation, since it neglects the possible relevance of bond disorder and the randomness of the quadrupolar degrees of freedom, which may give rise to structural glass behavior.

4. Uniaxial relaxors

In contrast to the cubic family related to PMN the polarization of the strontium-barium niobate family, $Sr_xBa_{1-x}Nb_2O_6$, (SBN), is a single component vector directed along the tetragonal c direction, which drives the symmetry point group from paraelectric parent 4/mmm to polar 4 mm at the phase transition into the low-T long-range-ordered polar phase as determined by X-ray diffraction [25]. Since SBN is tetragonal on the average, it belongs to the Ising model universality class rather than to the Heisenberg one as proposed for PMN-like system [6]. Assuming the presence of quenched random fields (*RF*s), available theory predicts the existence of a phase transition into long-range order within the *RF* Ising model (*RFIM*) universality class [9] preceded by giant critical slowing-down above T_c [11]. Only recently the SBN system has been found to fulfil the above necessary condition [13] and the ferroic RFIM seems to be materialized at last [26].

When explaining the unusual *relaxor* behavior, again the appearance of fluctuating polar precursor clusters at temperatures $T > T_c$ has to be considered as the primary signature of the polar *RFIM* [2, 3, 6]. Acting as precursors of the spontaneous polarization, which occurs below T_c, they have been evidenced in various zero-field cooling (ZFC) experiments comprising linear birefringence [27], linear susceptibility [28], dynamic light scattering

[29] and Brillouin scattering [30]. After freezing into a metastable domain state at $T < T_c$ the clusters were also directly observed with the help of high resolution piezoresponse force microscopy, PFM [31] (Fig. 3).

One of the major achievements provided by the discovery of the FE *RFIM* is the possibility to study the complete set of critical exponents on a ferroic system for the first time after their prediction [9, 10]. While most of the exponents compare well with predictions from theory and simulations [32], a remarkable deviation is found for the order parameter exponent, where $\beta = 0.14$ as determined by ^{93}Nb nuclear resonance [33] (Fig. 4) clearly deviates from the prediction $\beta \approx 0$ [32]. However, our value comes close to that observed recently on the standard *RFIM* system, the dilute uniaxial antiferromagnet $Fe_{1-x}Zn_xF_2$, $x = 0.15$, in an external magnetic field [34]. Further, the most disputed value, namely the specific heat exponent $\alpha \approx 0$ [35] clearly complies with the logarithmic divergence as found on $Fe_{1-x}Zn_xF_2$ [10], which still lacks theoretical confirmation.

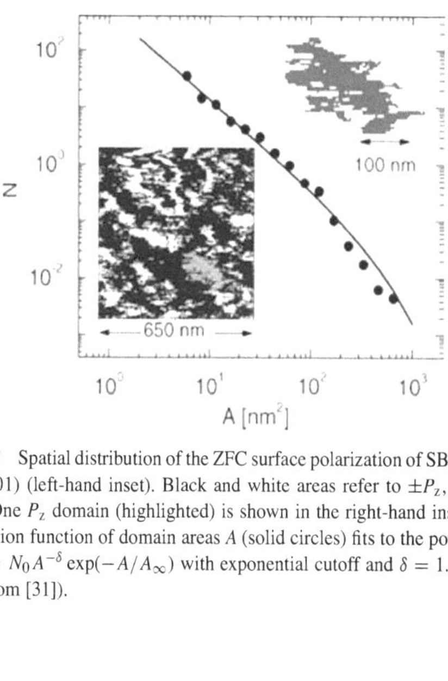

Figure 3 Spatial distribution of the ZFC surface polarization of SBN*61*:Ce ($x = 0.01$) (left-hand inset). Black and white areas refer to $\pm P_z$, respectively. One P_z domain (highlighted) is shown in the right-hand inset. The distribution function of domain areas A (solid circles) fits to the power law $N(A) = N_0 A^{-\delta} \exp(-A/A_\infty)$ with exponential cutoff and $\delta = 1.5$ (solid line) (from [31]).

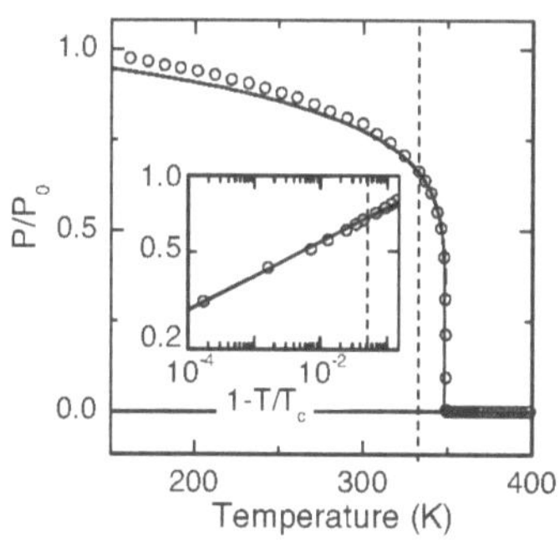

Figure 4 Order parameter of SBN as measured by NMR techniques displaying criticality with an exponent $\beta = 0.16$ (from [33]).

5. Domain dynamics in uniaxial relaxors

Domains in FE crystals are well-known to have a considerable influence on the value of their complex dielectric susceptibility, $\chi^* = \chi' - i\chi''$, and related quantities [36]. Owing to its mesoscopic character the domain wall susceptibility strongly reflects the structural properties of the crystal lattice. This is most spectacular in crystals with inherent disorder, where the domain walls are subject of stochastic pinning forces and $\chi*$ is highly polydispersive due to a wide distribution of Debye-type response spectra [37, 38],

$$\chi^*(\omega) \propto \frac{\ln(1/\omega\tau_0)^{2/\Theta}}{(1 + i\omega\tau)}, \tag{4}$$

where τ_0 and τ (with $\tau > \tau_0$), ω and $\Theta \approx 0.8$ (in $d = 3$) are relaxation times, the angular frequency and a roughness exponent, respectively.

More generally, the dynamic behavior of domain walls in random media under the influence of a periodic external field gives rise to hysteresis cycles of different shape depending on various external parameters. According to recent theory [39] on disordered ferroic (ferromagnetic or FE) materials, the polarization, $\boldsymbol{P}$, is expected to display a number of different features as a function of T, frequency, $f = \omega/2\pi$, and probing *ac* field amplitude, E_0. They are described by a series of dynamical transitions between different "phases", whose order parameter $Q = (\omega/2\pi)\oint P dt$ reflects the shape of the P *vs.* E loop being either zero or non-zero. When increasing the *ac* amplitude, E_0, the polarization displays four regimes. First, at very low fields, $E_0 < E_\omega$, only "*relaxation*" with $Q = 0$, but no macroscopic motion of the walls occurs at finite frequencies, $f > 0$. Second, within the range $E_\omega < E_0 < E_{t1}$, a thermally activated drift motion ("*creep*") is expected, while above the depinning threshold E_{t1} the "*sliding*" regime is encountered within $E_{t1} < E_0 < E_{t2}$. In both regimes $Q \neq 0$ is encountered. Finally, for $E_0 > E_{t2}$ a complete reversal of the polarization ("*switching*") occurs in the whole sample in each half of the period, $\tau = 1/f$, hence, $Q = 0$. It should be noticed that all transition fields, E_ω, E_{t1} and E_{t2}, are expected to depend strongly on both T and f [39].

We have shown [40] that two different non-Debye responses corresponding to the field regions $E_0 < E_\omega$ ("*relaxation*") and $E_\omega < E_0 < E_{t1}$ ("*creep*") occur in the low-f dispersion of the uniaxial *relaxor* crystal $Sr_{0.61-x}Ce_xBa_{0.39}Nb_2O_6$ (*SBN*:Ce, $x = 0.0066$) in the vicinity of its FE transition temperature, $T_c = 320$ K. It shows both characteristics in adjacent frequency regimes. While the well-known relaxational ln(1/f) characteristic of relaxing domain wall segments in a weak random field [38] applies to "high" frequencies, $f > 100$ Hz, an alternative $1/f^{\beta}$ dependence is observed in the "low"-f regime, $f < 1$ Hz. In order to understand the latter behavior, we introduce polydispersivity via a broad distribution of wall mobilities, μ_w, which describe the viscous motion of the walls in the creep regime, where they overcome a large number of potential walls due to a high density of pinning defects. As a characteristic of irreversibility the walls stop when switching off the field. Within this concept the rapid individual Debye-type relaxation processes are averaged out on the long-time scale of a creep experiment. $(1/f)^{\beta}$ behavior at low frequencies has recently also been reported on the *relaxor*-type crystal $PbFe_{1/2}Nb_{1/2}O_3$ [41].

Dielectric response data were taken on a Czochralski-grown very pure crystal of *SBN*:Ce (size 0.5 × 5 × 5 mm^3) with probing electric-field amplitudes of 200 V/m applied along the polar *c* axis. A wide frequency range, $10^{-5} < f < 10^{6}$ Hz, was supplied by a Solartron 1260 impedance analyzer with a 1296 dielectric interface. Different temperatures were chosen both below and above T_c and stabilized to within ±0.01 K. Fig. 5 shows representative data of χ' (curve 1) and χ'' vs. f (curve 2) taken at T = 294 K. They illustrate the main features of the dielectric dispersion of zero-field-cooled (*ZFC*) *SBN*:Ce: (i) the dielectric response strongly increases below $f_{min} \approx 25$ Hz (marked by the dotted line); (ii) neither saturation of χ' nor a peak of χ'' are observed in the infra-low-frequency limit, where (iii) the magnitude of χ'' exceeds that of χ' by one order of magnitude; (iv) a Cole-Cole-type plot of χ'' vs. χ' is characterized by a positive curvature at frequencies $f<f_{min}$ (Fig. 5; inset), which is opposite to the conventional Debye-type one; (v) at higher frequencies, $f > f_{min}$, χ'' increases again in a power-law-like fashion (straight line in a log-log presentation), while χ' changes its curvature and gently bends down.

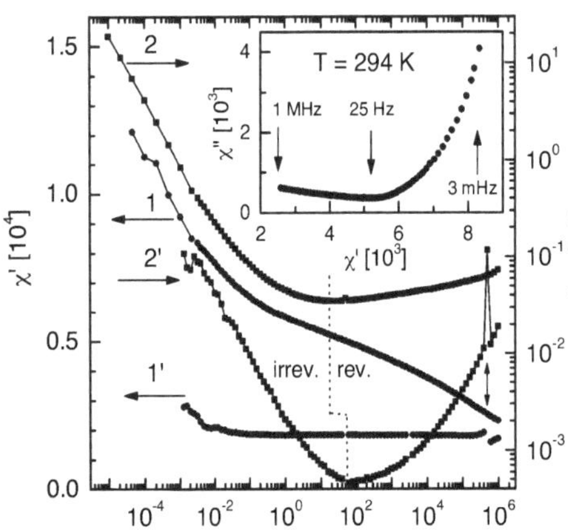

Figure 5 Dielectric spectra of χ' and χ'' vs. f of unpoled (curves 1 and 2) and poled (curves 1′ and 2′) *SBN*:Ce taken at T = 294 K. Solid lines are guides to the eye and the vertical dotted line separates different response regimes. A piezoelectric anomaly at f = 0.5 MHz is marked by a double arrow. The inset shows χ'' vs. χ' (from [40]).

The dominating domain-wall nature of the response is evidenced by its drastic reduction when poling the sample with E = 350 kV/m from above T_c into a near-single domain state as shown by the curves 1′ and 2′ in Fig. 5. Despite its decrease by two orders of magnitude χ'' reveals, again, a symmetric increase on both sides of $f_{min} \approx 65$ Hz (dotted line), which becomes power-law-like in the asymptotic low- and high-f regimes, respectively. This applies also to χ' (curve 1′) after subtracting a background corresponding to the minimum at f = 65 Hz. Interestingly, a sharp piezoelectric resonance of both χ' and χ'' is observed at $f_{min} \approx 0.5$ MHz after poling. This is typical of the near-single domain state, which activates a piezoelectric resonance.

The high-f response of both the *ZFC* and the *FC* states confirms many of the characteristics predicted by Equation 4. Inspection shows that χ' decreases linearly on a linear-log scale prior to the steeper decrease at $f > 10^4$ Hz, while χ'' obeys linearity on a log-log scale. Clearly, the ω prefactor strongly suppresses χ'' close to f_{min} when compared with χ'. Upon increasing f the same factor determines the positive curvature of χ'' despite the competing In(1/f) contribution (curve 2′ in Fig. 5). Simultaneously, χ' is bent down in a dispersion step-like fashion.

While the high-frequency dispersion regime is attributed to polarization processes due to the reversible motion of domain-wall segments experiencing restoring forces, *viz*., relaxation, the low-frequency response is due to the irreversible viscous motion of domain-walls. They experience memory-erasing friction by averaging over numerous pinning centers in a creep process. The latter type of motion becomes possible for at least two reasons: screening of depolarization fields by free charges in the bulk or at the surface and/or pinning of the domain-walls at quenched random fields, which is believed to be due to quenched charge disorder in the special case of *SBN*:Ce [13, 40].

Dielectric domain response under the action of an external electric field is readily modeled by considering the average polarization, $P(t) = (2P_s/D)\, x(t)$, of a regular stripe domain pattern of up and down polarized regions carrying spontaneous polarization, $\pm P_s$, and having an average width D. It arises from a sideways motion of walls perpendicular to the field direction by a distance x. Starting with $P(0) = 0$ at $x(0) = 0$, the favorable domains enhance their total width by an amount $2x$ until reaching (in principle) the limit $P = P_s$ for $x \to D/2$. By assuming viscous motion of the walls one obtains the rate equation

$$\dot{P}(t) = \left(\frac{2P_s}{D}\right)\mu_w E(t), \tag{5}$$

where the wall velocity $\dot{x}(t) = \mu_w E(t)$ involves the wall mobility μ_w and the driving field $E(t)$. Assuming constant mobility at sufficiently weak fields and disregarding the depinning threshold one finds

$$P(t) = \left[\frac{2\mu_w P_s}{i\omega\varepsilon_0 D} + \chi_\infty\right]\varepsilon_0 E_0 \exp(i\omega t). \quad (6)$$

under a harmonic field, $E(t) = E_0\exp(i\omega t)$. In Equation 6 the second term refers to "instantaneous" response processes due to reversible domain-wall rearrangements occurring on shorter-time scales (see above).

The above relations are expected to hold in the limit of small displacements x, before the walls are stopped either by depolarizing fields (in conventional FEs) or by new domain conformations under the constraint of strong random fields (in disordered FEs). Weak periodic fields thus probe a linear *ac* susceptibility

$$\chi_w^*(\omega) = \chi_\infty\left(1 + \frac{1}{i\omega\tau_w}\right), \quad (7)$$

with $\chi_\infty/\tau_w = (2\mu_w P_s/\varepsilon_0 D)$. The "relaxation" time τ_w denotes the time in which the interface contribution to the polarization equals that achieved instantaneously, $\Delta P = \varepsilon_0\chi_\infty E$.

Since the electric fields used in our experiments (E_0 = 200 V/m) are well below the coercive field, $E_c \approx 150$ kV/m, we have to account for the nonlinearity of v vs E in the creep regime, where thermal excitation enables viscous motion below the depinning threshold $E_{crit} \approx E_c$. Approximating this regime roughly by a power law $v \propto E^\delta$, $\delta > 2$, Equation 7 may be modified phenomenologically by introducing a Cole-Davidson-type exponent $\beta < 1$,

$$\chi_w^*(\omega) = \chi_\infty\left[1 + \frac{1}{(i\omega\tau_{eff})^\beta}\right], \quad (8)$$

similarly as used in the case of polydispersive Debye-type relaxation [42]. Here τ_{eff} denotes an effective relaxation time.

It has to be remarked that our approach neglects the hysteretic properties of the *ac* susceptibility, which are not contained in our adiabatic approach, Equation 6. This deficiency has been overcome in a recent approach based on a statistical model [43], where Equation 8 was deduced from the periodic motion of the domain walls in a randomly pinning medium on the basis of the quenched Edwards-Wilkinson equation. A similar result was recently obtained within a Rayleigh loop approach [44].

Decomposition of Equation 8 yields

$$\chi'(\omega) = \chi_\infty[1 + \cos(\beta\pi/2)/(\omega\tau_{eff})^\beta] \quad \text{and}$$
$$\chi''(\omega) = \chi_\infty \sin(\beta\pi/2)/(\omega\tau_{eff})^\beta \quad (9)$$

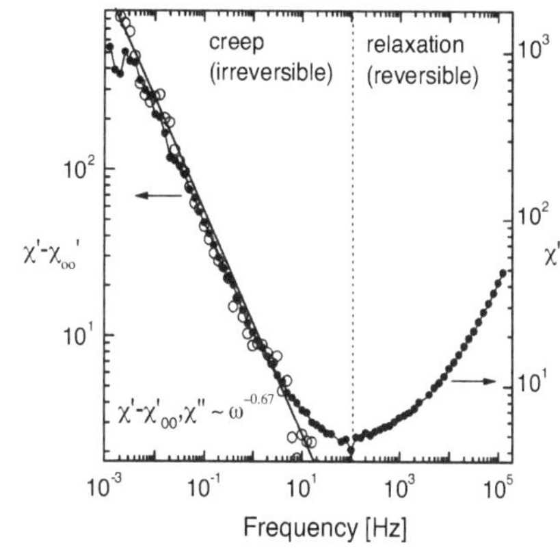

Figure 6 Dielectric spectra of $\chi'-\chi'_\infty$ (open circles), where $\chi'_\infty = 1820$, and χ'' (solid circles) vs. f of poled *SBN*:Ce taken at $T = 294$ K. The solid line is a best fit to Equation 6 and Equation 7 with $\beta = 0.67$ (from [45]).

such that

$$\chi''/(\chi' - \chi_\infty) = \tan(\beta\pi/2). \quad (10)$$

The power law-type spectral dependencies of χ' and χ'' are well supported by our experiments. While the unpoled sample exhibits an exponent $\beta \approx 0.2$ (not shown), *i.e.* large polydispersivity, the poled sample yields $\beta \approx 0.67$ for both components of $\chi*$ (Fig. 6) [45]. Obviously the polydispersivity is largely suppressed at low domain wall densities. This seems to show that polydispersivity is less affected by the nonlinearity in the creep regime, $v \propto E^\delta$ with $0 < \delta < 1$ in first approximation, than by the mutual wall interactions in the nanodomain regime [31]. Very satisfactorily, also the Cole-Cole plot, Equation 10, which is another independent test of the ansatz, Equation 8, reveals a very similar exponent, $\beta \approx 0.69$.

It should be noticed that the monodispersive relation, Equation 7, satisfies the Kramers-Kronig relationships, since $\chi'' \propto 1/\omega$ is a purely conductive contribution due to ohmic-like domain wall sliding and $\chi' = \chi'_\infty$ is constant. This is, however, no longer satisfied for $\beta < 1$, Equation 8. Hence, the spectral features displayed in Fig. 6 must necessarily change at very low frequencies. Here we conjecture - in accordance with the theory of dynamic phase transitions in random media [39] – that monodispersivity, *i.e.* the sliding regime, should be attained asymptotically when approaching the static limit. This has recently been confirmed on the quantum-ferroelectric *relaxor* $SrTi^{18}O_3$ in its domain state below $T_c = 25$ K [46].

6. Conclusion

The enigma of *relaxor* ferroelectrics seems to come close to be deciphered—50 years after the discovery of this remarkable material class [12]. Based on a vast amount of experimental and theoretical evidence it could be shown that the inherent charge disorder and its quenched random electric field distribution must be at the very origin of *relaxor* behavior. The primary action of RFs is their correlating force onto the order parameter, which stabilizes polar nanoregions against thermal fluctuations [20]. This has been evidenced very clearly by high resolution PFM both on the uniaxial *relaxor* SBN [31] and, very recently, also on the cubic *relaxor*-like compound, $PMN_{0.8}$-$PT_{0.2}$ [47].

However, since the ferroelectric phase transition in cubic *relaxors* like PMN is necessarily destroyed by arbitrarily weak RFs [9], random interactions between the different constituents of the solid solutions become relevant in these compounds. That is why cluster glassy scenarios are probably most appropriate for their description near to and below the freezing temperature [16, 23]. More research is needed to clarify the applicable model(s). The situation is much clearer in uniaxial relaxors like SBN, which is widely accepted to represent the first ferroic materialization of the 3*d* random-field Ising model [26]. Despite the clarity of this model and its consequences, future research is yet needed, *e.g.*, for understanding details of the critical behavior when comparing experimental and theoretical results.

Acknowledgments
Work supported by the Deutsche Forschungsgemeinschaft via Forschungsschwerpunkt "Strukturgradienten in Kristallen".

References
1. M. E. LINES and A. M. GLASS, in "Principles and Applications of Ferroelectrics and Related Materials" (Clarendon, Oxford, 1977).
2. L. E. CROSS, *Ferroelectrics* **76** (1987) 241.
3. G. BURNS and F. H. DACOL, *Solid State Commun.* **48** (1983) 853; *Phase Trans.* **5** (1985) 261.
4. R. B. GRIFFITHS, *Phys. Rev. Lett.* **23** (1968) 69.
5. CH. BINEK and W. KLEEMANN, *Phys. Rev. B* **51** (1995) 12888.
6. V. WESTPHAL, W. KLEEMANN and M. D. GLINCHUK, *Phys. Rev. Lett.* **68** (1992) 847.
7. S. B. VAKHRUSHEV, B. E. KVYATKOVSKY, A. A. NABEREZHOV, N. M. OKUNEVA and B. B. TOPERVERG, *Ferroelectrics* **90** (1989) 173; G. SCHMIDT, H. ARNDT, G. BORCHARDT, J. V. CIEMINSKI, T. PETZSCHE, K. BORMANN, A. STERNBERG, A. ZIRNITE and A. V. ISUPOV, *Phys. Stat. Solidi A* **63** (1981) 501.
8. K. BINDER and A. P. YOUNG, *Rev. Mod. Phys.* **58** (1986) 801.
9. I. IMRY and S. K. MA, *Phys. Rev. Lett.* **35** (1975) 1399.
10. D. P. BELANGER and A. P. YOUNG, *J. Magn. Magn. Mater.* **100** (1991) 272.
11. D. S. FISHER, *Phys. Rev. Lett.* **56** (1986) 416; A. T. OGIELSKI and D. A. HUSE, *Phys. Rev. Lett.* **56** (1986) 1298.
12. G. A. SMOLENSKI and V. A. ISUPOV, *Dokl. Acad. Nauk SSSR* **97** (1954) 653.
13. W. KLEEMANN, J. DEC, P. LEHNEN, T. H. WOIKE and R. PANKRATH, in: "Fundamental Physics of Ferroelectrics 2000", edited by R. E. Cohen, *AIP Conf. Proc.* **535** (2000) 26.
14. G. A. SAMARA, *Solid State Physics* **56** (2001), edited by H. Ehrenreich and F. Spaepen (Academic Press, New York, 2001) p. 240 and references therein; *J. Phys.: Cond. Matter* **15** (2003) R367.
15. D. SHERRINGTON and S. KIRKPATRICK, *Phys. Rev. Lett.* **35** (1975) 1972.
16. D. VIEHLAND, M. WUTTIG and L. E. CROSS, *Ferroelectrics* **120** (1991) 71; *Phys. Rev. B* **46** (1993) 8003.
17. A. BOKOV and Z. G. YE, *ibid.* **66** (2002) 064103.
18. P. BONNEAU *et al., Solid State Chem.* **91** (1991) 350; L. E. CROSS, *Ferroelectrics* **151** (1994) 305; A. D. HILTON *et al., J. Mater. Sci.* **25** (1990) 3461; T. EGAMI *et al., Ferroelectrics* **199** (1997) 103; B. DKHIL *et al., Phys. Rev. B* **65** (2002) 4104.
19. E. HUSSON, M. CHUBB and A. MORELL, *Mat. Res. Bull.* **23** (1988) 357.
20. H. QIAN and L. A. BURSILL, *Int. J. Mod. Phys. B* **10** (1996) 2027.
21. Y. UESU, H. TAZAWA and K. FUJISHIRO, *J. Kor. Phys. Soc.* **29** (1998) S703.
22. P. LEHNEN, J. DEC, W. KLEEMANN, TH. WOIKE and R. PANKRATH, *Ferroelectrics* **268** (2002) 113.
23. R. BLINC, J. DOLINSEK, A. GREGOROVIC, B. ZALAR, C. FILIPIC, Z. KUTNJAK, A. LEVSTIK and R. PIRC, *Phys. Rev. Lett.* **83** (1999) 424.
24. Z. KUTNJAK, C. FILIPIC, R. PIRC, A. LEVSTIK, R. FARHI and M. EL MARSSI, *Phys. Rev. B* **59** (1999) 294.
25. J. R. OLIVER, R. R. NEURGAONKAR and L. E. CROSS, *J. Appl. Phys.* **64** (1988) 37.
26. W. KLEEMANN, J. DEC, P. LEHNEN, R. BLINC, B. ZALAR and R. PANKRATH, *Europhys. Lett.* **57** (2002) 14.
27. P. LEHNEN, W. KLEEMANN, TH. WOIKE and R. PANKRATH, *Eur. Phys. J.B* **14** (2000) 633.
28. J. DEC, W. KLEEMANN, V. BOBNAR, Z. KUTNJAK, A. LEVSTIK, R. PIRC and R. PANKRATH, *Europhys. Lett.* **55** (2001) 781.
29. W. KLEEMANN, P. LICINIO, TH. WOIKE and R. PANKRATH, *Phys. Rev. Lett.* **86** (2001) 6014.
30. F. M. JIANG and S. KOJIMA, *Phys. Rev. B* **62** (2000) 8572.
31. P. LEHNEN, W. KLEEMANN, TH. WOIKE and R. PANKRATH, *ibid.* **64** (2001) 224109.
32. A. A. MIDDLETON and D. S. FISHER, *ibid.* **65** (2002) 134411.
33. R. BLINC, A. GREGOROVIC, B. ZALAR, R. PIRC, J. SELIGER, W. KLEEMANN, S. G. LUSHNIKOV and R. PANKRATH, *ibid.* **64** (2001) 134109.
34. F. YE, L. ZHOU, S. LAROCHELLE, L. LU, D. P. BELANGER, M. GREVEN and D. LEDERMAN, *Phys. Rev. Lett.* **89** (2002) 157202.
35. Z. KUTNJAK, W. KLEEMANN and R. PANKRATH, *Phys. Rev. B.* (submitted)
36. J. FOUSEK and V. JANOVEC, *Phys. Stat. Sol. (a)* **13** (1966) 105.
37. L. B. IOFFE and V. M. VINOKUR, *J. Phys. C* **20** (1987) 6149.
38. T. NATTERMANN, Y. SHAPIR and I. VILFAN, *Phys. Rev. B* **42** (1990) 8577.
39. T. NATTERMANN, V. POKROVSKY and V. M. VINOKUR, *Phys. Rev. Lett.* **87** (2001) 197005.
40. W. KLEEMANN, J. DEC, S. MIGA, TH. WOIKE and R. PANKRATH, *Phys. Rev. B* **65** (2002) 220101R.
41. Y. PARK, *Solid State Commun.* **113** (2000) 379.
42. A. K. JONSCHER, in "Dielectric Relaxation in Solids" (Chelsea Dielectric Press, London, 1983).
43. A. A. FEDORENKO, V. MUELLER and S. STEPANOW, *Phys. Rev. B* **70** (2004) 224104.
44. D. DAMJANOVIC, S. S. N. BHARADWAJA and N. SETTER, *Adv. Mater.* (in print).
45. W. KLEEMANN, J. DEC and R. PANKRATH, *Ferroelectrics* **286** (2003) 21.
46. J. DEC, W. KLEEMANN and M. ITOH, *Ferroelectrics* **298** (2004) 163.
47. V. V. SHVARTSMAN and A. L. KHOLKIN, *Phys. Rev. B* **69** (2004) 014102.

JOURNAL OF MATERIALS SCIENCE **41** (2006) 137–145

Properties of ferroelectric ultrathin films from first principles

IGOR A. KORNEV*
Physics Department, University of Arkansas, Fayetteville, Arkansas 72701, USA; Novgorod State University, Russia
E-mail: ikornev@uark.edu

HUAXIANG FU, LAURENT BELLAICHE
Physics Department, University of Arkansas, Fayetteville, Arkansas 72701, USA

Advances in first-principles computational approaches have, over the past ten years, made possible the investigation of basic physical properties of simple ferroelectric systems. Recently, first-principles techniques also proved to be powerful methods for predicting finite-temperature properties of solid solutions in great details. Consequently, *bulk* perovskites are rather well understood nowadays. On the other hand, one task still remains to be accomplished by *ab-initio* methods, that is, an accurate description and a deep understanding of ferroelectric *nanostructures*. Despite the fact that nanometer scale ferroelectric materials have gained widespread interest both technologically and scientifically (partly because of novel effects arising in connection with the reduction of their spatial extension), first-principles-based calculations on ferroelectric nanostructures are rather scarce. For instance, the precise effects of the substrate, growth orientation, surface termination, boundary conditions and thickness on the finite-temperature ferroelectric properties of *ultrathin films* are not well established, since their full understandings require (i) microscopic insights on nanoscale behavior that are quite difficult to access and analyze via experimental probes, and (ii) the development of new computational schemes. One may also wonder how some striking features exhibited by some bulk materials evolve in the corresponding thin films. A typical example of such feature is the morphotropic phase boundary of various solid solutions, where unusual low-symmetry phases associated with a composition-induced rotation of the spontaneous polarization and an enhancement of dielectric and piezoelectric responses were recently discovered. In this paper, recent findings resulting from the development and use of numerical *first-principles-based* tools on ferroelectric ultrathin films are discussed.

1. Introduction

Ferroelectric materials are of unique importance for a variety of existing and potential device applications. Examples include piezoelectric transducers and actuators, nonvolatile ferroelectric memories, and dielectrics for microelectronics and wireless communication [1–3]. An important class of ferroelectric materials are the perovskites ABO_3. The perovskite crystal structure ABO_3 can be regarded as a three-dimensional network of corner sharing BO_6 octahedra, with the B ions in the center of the octahedra. In a cubic perovskite, the A site is twelvefold surrounded by oxygen ions. As the temperature is reduced, many of these compounds undergo a phase transition and develop a switchable spontaneous electric polarization, thus becoming ferroelectric. Most of the perovskite compounds that are of greater interests are not simple systems, but rather complex solid solutions with the general formula $A(B', B'')O_3$, i.e., with two kinds of B atoms

*Author to whom all correspondence should be addressed.

0022-2461 © *2006 Springer Science + Business Media, Inc.*
DOI: 10.1007/s10853-005-5962-0

since their physical properties may be tailored by varying the concentration of B atoms. The $Pb(Zr,Ti)O_3$ (PZT) alloys, in particular, are sensitive to a change of the Ti concentration and are currently used, e.g., in the U.S. Navy transducers and actuators because of their large piezoelectricity [4]. Other examples include the $Pb(Mg,Nb)O_3$ and $Pb(Zn,Nb)O_3$ compounds which exhibit such extraordinarily high values of the piezoelectric constants when alloyed with $PbTiO_3$ [5] that they could result in a new generation of piezoelectric "wunderbar" devices [6]. A particularly relevant compositional region in solid solutions is the morphotropic phase boundary (MPB). This compositional region separates tetragonal and rhombohedral phases of various solid solutions and has been recently shown to possess unusual low-symmetry phases—that are associated with a composition-induced rotation of the spontaneous polarization and an enhancement of dielectric and piezoelectric responses [7–13].

Ferroelectric simple systems and solid solutions have been intensively studied both experimentally and theoretically in the past 5 years. Consequently, their properties are fairly well understood nowadays. On the other hand, understanding the physical properties in ferroelectric-based *nanostructures* has become one of the most intriguing and fundamental problems of modern physics. Advances in materials growth and characterization, which provide a driving force for ferroelectric-related theoretical study have, in the past decade, made possible the investigation of basic physical processes in new ferroelectric materials on a nanometer scale [14–20].

Physical properties at nanometer scales in low dimensional ferroelectric structures are attractive fundamentally, as well as technologically. As a matter of fact, there is a rapidly growing interest in ferroelectrics and ferroelectric/semiconductor integrated structures because of their potential applications in advanced microsystems (such as high-density nonvolatile ferroelectric memories), a new generation of gate dielectrics for faster and smaller computers, and tunable microwave devices (such as filters and phase shifters in smart communication and radar systems). Much of the recent activity in ferroelectrics has been motivated by the desire for developing miniaturized actuators for micro-electro-mechanical systems applications, which mostly involves nanostructures in *thin-film* forms. In search for multi-Gbit ferroelectric memories, the thickness of a ferroelectric film is also of a great concern. Auciello *et al.* [21] mention that thicknesses of less than 200 nm are necessary for future nonvolatile ferroelectric random-access memory, while Scott [22] discusses the possibility of thicknesses as low as 30 nm for some particular device applications.

The constantly decreasing size of such devices down to a nanoscale gives rise to many fundamental questions regarding the stability of different physical properties in these low-dimensional structures. One of the most important phenomena observed in ferroelectric films is indeed the size effect. As a matter of fact, Curie temperature, switching rate, coercivity as well as the field response all depend strongly on the *thickness* of ferroelectric films. Moreover, it was long thought that there is a critical size on the order of hundreds of angstroms below which ferroelectricity would disappear entirely [23].

The situation changed after experiments provided unambiguous evidence for ferroelectric ground states in 40 Å-thick perovskite oxide films [24] and in crystalline copolymer films as thin as 10 Å [25]. Similarly, and very recently, Fong *et al.*[26] provided the experimental evidence that ferroelectricity persists down to vanishingly small sizes. It thus appear that the answer to the crucial question of whether or not there is a critical thickness for ferroelectricity is still unsettled [24, 27–30]. Similarly, the precise effects of surface on the thermodynamic properties of ferroelectric nanostructures are open for discussion [29, 31–35]. This leads to a growing and general theoretical interest in the properties of nanoscale ferroelectric systems. Furthermore, considering the fact that phase transitions in a system are greatly affected when the object is confined to sizes that are comparable to the correlation length, one may thus wonder how some striking features—such as the existence of a MPB, exhibited by some bulk perovskite materials—evolve in the corresponding low-dimensional structures.

An intriguing problem in these films also concerns their polarization patterns. For instance, the various following patterns have recently been predicted or observed: out-of-plane *monodomains* [24, 27, 28, 36], 180° out-of-plane *stripe domains* [29, 36, 37], 90° multidomains that are oriented *parallel* to the film [38], and *microscopically-paraelectric* phases [28]. The fact that different patterns have been reported for similar *mechanical* boundary conditions supports a concept discussed in Refs. [28, 39, 40], namely that they arise from different *electrical* boundary conditions. A reactive atmosphere can indeed lead to a partial compensation of surface charges in films with nominal ideal open-circuit (OC) conditions [36], while metallic or semiconductor electrodes "sandwiching" films do not always provide *ideal* short-circuit conditions (SC)—resulting in a non-zero internal field [28]. The *degree* of surface charges' screening in thin films can thus vary from one experimental set-up to another, possibly generating *different* polarization patterns [28, 36].

Finally, multidomains and their formation mechanism in low-dimensional structures are still not well established. *Atomic-scale* details of multidomains are also scarce in ferroelectric thin films. One may also wonder if uncompensated depolarizing fields can yield ferroelectric phases that do *not* exist in the corresponding bulk material. Candidates for these latter anomalies are films made of alloys with a composition lying near their MPB, because of the easiness of rotating their polarization [8–10, 41].

One reason behind that lack of knowledge is that thin films are difficult to synthesize in a good quality form, and their characterization is by no means straightforward. Similarly, realistically simulating thin films is a challenge. Technically speaking, the main theoretical methods used/developed so far in the study of low-dimensional ferroelectrics fall into two categories, namely: (1) approaches based on the Landau-Ginzburg theory [42] versus (2) first-principles-based schemes.

In the Landau-Ginzburg approach, the free energy is expressed as a functional of the order parameter only, and all other degrees of freedom have been integrated out in the derivation of the theory starting from a microscopic Hamiltonian. In this approach, solids are often treated as continuum. Moreover, since the spatial inhomogeneity of the order parameter is important for low-dimensional structures, the analytical solution of corresponding Euler-Lagrange equations is barely possible [43]. Furthermore, some of the parameters needed in the expansion of the free-energy functional can only be estimated from the experimental data available.

Although continuum theories, like the Landau-Ginzburg theory, are expected to be valid only on length scales much larger than a lattice constant [42], it is noteworthy to realize that they are used in the analysis of the experimental results for the thinnest films and heterostructures down to the nanoscale. This is partly because the ferroelectric correlation length is about one lattice constant [31, 44]. The following argument is somewhat more intuitive than definite. If the size of a ferroelectric nanostructure is much larger than the correlation length of the order parameter fluctuations, results can be interpreted as bulk ferroelectricity affected by the presence of surface, which imply that a Landau-Ginzburg-type theory is valid.

Within this framework, many of the characteristics of nanoscale ferroelectrics can be faithfully reproduced (see, e.g., [29, 38, 43, 45], and references therein). However, microscopic foundations of phenomenological approaches are somewhat arbitrary and atomic-scale mechanisms are difficult to assess. A similar method makes use of the Ising model in a transverse field (IMTF) and the relationships between the IMTF and Landau-Ginzburg parameters have been established in Refs. [46, 47].

An alternative approach to continuum models are first-principles-based calculations that require no experimental input. Some phenomena in low-dimensional ferroelectrics can be understood by using *direct* first-principles methods. Although first-principles methods are rather accurate, their large computational cost currently prevents them from being used to study complex phenomena and/or systems. For instance, it is unlikely that direct first principles will capture the complexity of real disordered materials such as $Pb(Zr_{0.5}Ti_{0.5})O_3$ (PZT) solid solutions. Moreover, since these methods usually deal with the ground states of solids at zero temperature, the understanding of the sequence of phase transitions and their temperatures on a material-specific basis is hard to achieve. On the other hand, first-principles-*derived* computational approaches have proved to be powerful methods in recent years for understanding complex materials, as well as for predicting the finite-temperature physical properties of hypothetical (and "wunderbar") compounds that have not yet been synthesized [48]. In particular, effective Hamiltonian methods with parameters obtained by fitting direct first-principles computations [8, 9, 49, 50] provide insight into the role of atomic species in ferroelectrics, leading to the design of new materials [48, 51, 52].

The main purpose of the present review is to describe the understanding of physical properties of ferroelectric ultrathin films that has been recently gained using direct first-principles and first-principles-based approaches. More precisely, we selected some papers and summarize their results (we apologize for papers not included in this article and that we may have missed). A particular emphasis will be put to indicate the mechanical and electrical boundary conditions adopted in these studies since ferroelectrics should be very sensitive to such conditions. Among the other topics to be discussed here are the role of size and surface effects on properties of ultrathin films made of insulating perovskites.

2. Recent works on ultrathin films from first principles (in chronological order)

2.1. "Microscopic model of ferroelectricity in stress-free $PbTiO_3$ ultrathin films" by P. Ghosez and K.M. Rabe, *Appl. Phys. Lett.* **76** (2000); 2767 "Ferroelectricity in $PbTiO_3$ Thin Films: A First Principles Approach" by K.M. Rabe and P. Ghosez, *J. Electroceram.* **4:2/3**, (2000) 379

In these two articles, first-principles-derived effective Hamiltonian calculations have indicated the possibility of retaining the ferroelectric ground state at very small thicknesses [27, 53]. More precisely, these authors applied the first-principles-based effective Hamiltonian approach to study the ferroelectric instability in ultrathin films, and predicted that *stress-free* (0 0 1) $PbTiO_3$ films under *short circuit* electrical boundary conditions (i.e., corresponding to the perfectly screened surface charges or equivalently to perfect electrodes) as thin as three unit cells [27] and one unit cell [53] remain to posses a stable polarization.

The effective Hamiltonian used for thin films was modified with respect to the one of the bulk by terminating the interatomic short-range force constants at the surface, changing the effective dipole-dipole interaction resulting from the perfect electrodes, and adding corrections to preserve global translational symmetry and charge neutrality. The authors also found that the perpendicular direction (i.e., along the [0 0 1]-z-axis) is more favorable than a direction lying in the (0 0 1) plane for the spontaneous polarization. Furthermore, while in the interior the polarization approaches the bulk value even for very thin films, it is significantly enhanced at the surface. The stability of the ferroelectric state relative to the paraelectric state was also predicted to increase with decreasing thickness. The results of Rabe and Ghosez are consistent with a related study on periodic slabs using direct first-principles simulations [31]. As a matter of fact, this latter study predicted ferroelectric ground states for various *stress-free* ultrathin ABO_3 slabs under the condition of vanishing internal electric field (i.e., under short-circuit boundary conditions).

To understand the ferroelectric instability and surface polarization enhancement in perpendicularly polarized films, a *simplified* model fitted to the effective Hamiltonian results has been considered [27]. It was obtained by (i) suppressing the periodic boundary conditions on the short-range terms, (ii) including the short-range surface corrections, (iii) changing in the effective dipole–dipole interaction to take into account the electrical boundary conditions. This "simple" model revealed that the combined effects of contributions (i) and (ii) were found to compete with the suppression of ferroelectricity by the dipolar contribution (iii).

It was also stressed that the first-principles-based results can be directly related to those of previous phenomenological studies. In particular, a microscopic interpretation of the so-called extrapolation length λ was proposed [53]. In the framework of the modified phenomenological Landau-Ginzburg theory [54] that is usually adopted to study the size effects in ferroelectric thin films, a surface energy term $\sim (P_+^2 + P_-^2)/\lambda$ is added to the usual expansion of the free energy, where λ is the "extrapolation length" which can be positive (corresponding to suppression) or negative (corresponding to enhancement of the polarization at the surface) and $P_\pm$ being the spontaneous polarization at the surfaces. Rabe and Ghosez [53] found that for short-circuit electrical and zero-stress mechanical boundary conditions, the truncation of the short range interactions at the surface leads to an enhancement of surface polarization, which results in a negative λ.

Note that, in principle, from layer-by-layer profiles of the polarization given in Refs. [27, 53] the value of λ and the correlation length can be easily estimated assuming that the deviation of the polarization from the bulk value decays with distance exponentially [54]. In the approaches used in Refs. [27, 31, 53], the epitaxial stress due to the lattice mismatch between a given substrate and the film, which can significantly change the ground state of the film was not studied. Moreover, the effects of a real metal-perovskite interface on both the atomic relaxation and polarization were also left out in these pioneering studies.

2.2. "Surface effects and ferroelectric phase transitions in $BaTiO_3$ ultrathin films" by Tinte and M.G. Stachiotti, *Phys. Rev. B* **64** (2001) 235403

Tinte and Stachiotti [29] used a first-principles-derived approach to study (0 0 1) $BaTiO_3$ thin films for different mechanical strains.They indicate that their electrical boundary conditions are not short-circuit (they are in fact close to be ideal open-circuit). Their starting point is an atomistic approach based on a shell model for $BaTiO_3$ bulk with parameters obtained from first-principles calculations.

It was first examined if the model describes properly the static surface properties of $BaTiO_3$. It is well known that a reduction of the system size enhances quite generally the relative importance of the surface boundaries. This is particularly true in the case of ferroelectric films. Existence of the vacuum surrounding films will cause surface-induced atomic relaxations and cell-shape changes near the film surfaces due to truncation of periodicity. Surfaces of ferroelectric materials have been studied using first principles methods [31–35]. Tinte and Stachiotti demonstrated that the model developed for the bulk material indeed is successful for describing surface properties, such as structural relaxations and surface energies for Ti-terminated surface. For the BaO surface, the description was less accurate. In agreement with direct first-principles results [31, 33, 34], they also show that ionic motions on surfaces could indeed dominate the bulk energetics for thin slabs.

Calculations for thin films were carried out in a periodic slab geometry; a vacuum region of 20 Å separating the periodic slabs was introduced to minimize the interaction between periodic images. It was found that the results are insensitive to the vacuum region size. (The reason is that the net *out-of-plane* polarization of the slab is zero). Technically, a (0 0 1) TiO_2 -terminated stress-free slab of 28 Å thick containing 8 TiO_2 layers and 7 BaO layers was chosen to perform some case studies indicated in the following.

Finite-temperature molecular dynamics simulations on stress-free films revealed a significant non-zero polarization parallel to the slab surface, while the polarization component perpendicular to the surface is zero at all temperatures, a direct result of huge internal polarizing fields along the growth axis and associated with open-circuit conditions. It was found that the polarization profile at the surface depends on the surface termination (i.e., TiO_2 vs BaO terminated), and in-plane ferroelectricity is strongly enhanced at the TiO_2-terminated surface and suppressed at the BaO-terminated surface, a fact that is in agreement with results [27, 31, 32, 53] from direct first-principles.

According to the obtained phase diagram of the (0 0 1) $BaTiO_3$ stress-free film, the transition temperatures for the paraelectric–⟨1 0 0⟩ polarized and the⟨100⟩ − ⟨1 1 0⟩ polarized phases coincide with the cubic–tetragonal and tetragonal–orthorhombic transition temperatures of the bulk material.

The atomistic model was also used to investigate the strain effects on the ferroelectric properties of the film assuming that the internal elastic strain fields are homogeneous, so the 2D clamping holds throughout the ultrathin film. It was shown that at low temperature when the film is *compressively strained*, the polarization vector starts rotating towards the z-axis. In this case, a strain induced *multi-domain* ferroelectric state with an out-of-plane orientation of polarization can be stabilized. Another valuable microscopic information provided by this article is the fact that the 180° domain wall is centered on a Ba-O plane, i.e., the atomic displacements have odd symmetry across (and vanish on) the BaO plane, which indicates that the domain boundary is indeed very sharp, its width being of approximately one lattice constant, in agreement with Refs. [44, 55].

The authors have also done simulations on slabs of different widths using the same mechanical boundary condition (compressive strain of -1.5%; for this value of strain the in-plane polarization of the slab was zero) to determine the temperature at which the out-of-plane polarization of each domain vanishes. They predicted from these simulations that the strain effect produced by the presence of a hypothetical compressive substrate stabilizes stripe domains in a "not short-circuited" TiO_2-terminated $BaTiO_3$ film as thin as 20 Å.

2.3. "Critical thickness for ferroelectricity in perovskite ultrathin films" by J. Junquera and P. Ghosez, *Nature* **422** (2003) 506

Using the direct first-principles Siesta code [56], Junquera and Ghosez [28] have been able to simulate the structure of a realistic ferroelectric capacitor made of an (0 0 1) ultrathin film of $BaTiO_3$ between two metallic $SrRuO_3$ electrodes. This film was also mimicked to be under a specific compressive strain ($\sim -2\%$) by assuming that its in-plane lattice constant is the one of $SrTiO_3$. It was shown that as the film thickness decreases, the electrical boundary conditions move further away from the ideal short-circuit conditions. In other words, ultrathin films can have a depolarizing field, even when sandwiched between metallic electrodes, with this field increasing in magnitude when the size decreases. As a result of this depolarizing field, Junquera and Ghosez predicted a critical thickness for ferroelectricity to be about 6 unit cells ($\approx$ 24 Å). One has to remember, however, that the pioneering work of Junquera and Ghosez deals with paraelectric versus normally-polarized single domain ferroelectric thin films. In other words, the computational burden of direct first-principles techniques did not allow the authors to investigate the possibility of *multi-domain* formation as an alternative of their results.

2.4. "Ab initio study of the phase diagram of epitaxial $BaTiO_3$" by Oswaldo Diéguez, Silvia Tinte, A. Antons, Claudia Bungaro, J. B. Neaton, Karin M. Rabe and David Vanderbilt, *Phys. Rev. B* **69** (2004) 212101

One particular feature of films is the presence of large strain in the plane of the film due to the lattice mismatch between a substrate and the thin film. If such mismatch exists, the ferroelectric material inside the film will feel a compressive or tensile strain, which will have a drastic effect on the direction of the polarization. For instance, the polarization may lie in the layers or may have a component perpendicular to such layers.

Diéguez *et al.*'s paper [57] addresses this issue by reporting the temperature-misfit strain phase diagram for epitaxial (0 0 1) $BaTiO_3$, as predicted by direct first-principles and effective-Hamiltonian approaches [57]. The form of the strain tensor$\{\eta_\mu\}$ (in Voigt notation) is relevant to two cases of interest, namely stress-free versus epitaxially strained (0 0 1) films. In the former case, all the components of strain tensor fully relax. On the other hand, the second situation is associated with the freezing of three in-plane components of strain tensor due to the lattice mismatch δ between the film and the substrate, i.e. $\eta_6 = 0$ and $\eta_1 = \eta_2 = \delta$—while the other components relax during the simulations [38, 57, 58].

Direct first-principles calculations were carried out using the VASP software package [59, 60]. Systematic optimizations of the five-atom unit cell in the six possible phases considered in Ref. [58] have been performed. Namely, the following phases have been optimized by relaxing the atomic positions and the out-of-plane cell vector: p (P4/mmm) with zero polarization; c (P4mm) with polarization along the z direction; aa (Amm2) with xy-in-plane polarization, and $P_x = P_y$; a (Pmm2) with polarization along the x direction; ac (Pm) with xz-in-plane polarization, and $P_x \neq P_z$; r (Cm) with all non zero components of the polarization, and $P_x = P_y \neq P_z$, where P_x, P_y and P_z are the Cartesian components of the spontaneous polarization along the [1 0 0], [0 1 0] and [0 0 1] pseudo-cubic directions, respectively. The finite temperature studies of epitaxial $BaTiO_3$ were carried out using the effective Hamiltonian approach of Ref. [49, 50].

The result obtained in Ref. [57] is qualitatively different from that computed previously [58] using a Landau-Ginzburg-Devonshire theory with parameters fitted at temperatures in the vicinity of the bulk phase transitions. The *ab initio* calculations predicted that at $T = 0$ K epitaxial $BaTiO_3$ film undergos a sequence of c (P4mm)

— *r* (Cm) — *aa* (Amm2) phase transitions as strain grows. Finite-temperature simulations using effective-Hamiltonian predicted a temperature-strain phase diagram similar to Pertsev *et al.*'s [58] at high temperature, but without the *ac* phase at low temperature. Unlike Ref. [58], the resulting phase diagram is symmetric with respect to the zero misfit strain and shows that all phase transitions are of second-order. In this work, effects of *interfaces, thickness and surfaces* have not been considered since the simulated material is in fact a periodic bulk under mechanical constraints rather than a film with a finite thickness.

2.5. "Ultrathin films of Ferroelectric solid solutions under a residual depolarizing field" by I. Kornev, H. Fu and L. Bellaiche, *Phys. Rev. Lett.* **93**, (2004) 196104

In Ref. [61], Pb-O terminated (0 0 1) $Pb(Zr_{0.5}Ti_{0.5})O_3$ thin films "sandwiched" between non-polar systems (to simulate, e.g., air, vacuum, electrodes and/or non-ferroelectric substrates) were modeled. Such low-dimensional structures are mimicked by large periodic supercells that are elongated along the *z*-direction, and that contain a few number of PZT-layers. The non-polar region outside the film is much larger than the thickness of the film, to allow well-converged results for the film properties [29].

The authors of this article develop an effective Hamiltonian approach for which the total energy of thin film (under uncompensated depolarizing fields) is written as a sum of one term depending on bulk parameters and a second term that mimics the effects of an *internal* electric field—that arises from the *partial or full screening* of polarization-induced charges at the *surfaces*—on the films properties. This second term is directly proportional to a β parameter that characterizes the *strength* of the *total* electric field inside the film. Specifically, $\beta = 0$ corresponds to ideal OC conditions for which the depolarizing field has its maximum magnitude (when polarizations lie along the *z*-axis), while an increase in β lowers this magnitude. The value of β resulting in a vanishing total internal electric field is dependent on the supercell geometry, and in particular on the number of its non-polar layers [31]. This study thus allowed the investigation of electrical boundary conditions on properties of ultrathin films.

Extensive calculations were also made to clarify the effects of different mechanical boundary conditions on ferroelectric properties of (0 0 1) PZT ultrathin films. A rich variety of ferroelectric phases, including unusual triclinic and monoclinic states has been found, as well as, peculiar laminar nanodomains depending on the interplay between electrical and mechanical boundary conditions [61].

Specifically, the results obtained in Ref. [61] for *stress-free* ultrathin films, that are under *short-circuit* electrical boundary conditions, are remarkably similar to those of Ref. [27]. For instance, the polarization at the surfaces is significantly enhanced with respect to the bulk, and increases as the film thickness decreases. Under *stress-free* conditions, the film has a spontaneous polarization *aligned along the* *z*-axis for (large) values of β that correspond to a screening of at least 98% of the polarization-induced surface charges. On the other hand, when β becomes smaller, the internal field along the growth direction would be too strong to allow an out-of-plane component of the local mode [29]. As a result, the polarization aligns along an *in-plane*$\langle 0\ 1\ 0 \rangle$ direction. The most striking result for stress-free PZT films is the polarization path when going from out-of-plane to in-plane. In this case, the polarization continuously rotates and passes through *low-symmetry* monoclinic and triclinic phases. It was also revealed that only in-plane components of the polarization appear in films under a *tensile* strain in the presence of depolarizing fields. Conversely, a large enough *compressive* strain annihilates the (in-plane) components of of the polarization for *any* β.

Another interesting result of this study [61] is the occurrence of *nanodomain* structures in ultrathin films under *compressive strain. Ab initio* simulation [28] has indicated that realistic electrode materials may not have sufficient carrier density to completely screen the depolarizing field for ultrathin single-domain ferroelectric films. In particular, spontaneous electric polarization can not be sustained in single-domain $BaTiO_3$ films of thickness smaller than 7 unit cells with $SrRuO_3$ electrodes. Another possibility to eliminate the depolarizing field and stabilize the ferroelectric phase is the appearance of a stripe domain structure in ultrathin films [26, 36]. Recently, investigation of the atomistic structure of the 180° — and 90° —domain boundaries in the ferroelectric perovskite compound $PbTiO_3$ have been performed using a first-principles ultrasoft-pseudopotential approach [55], establishing the geometry of the domain walls at the atomic level and calculating the creation energy of the domain walls. The domain wall was found to be very narrow, with a width in the order of the lattice constant. The quantitative support for the calculations has been provided by the experiment, that, in ultrathin films, 180° —stripe domains have a small equilibrium period and produce in the *X*-ray scattering pattern the characteristic features of ferroelectric phase [26, 36]. Similar calculations have been presented for the interactions between vacancies and ferroelectric domain walls, confirming the tendency of these defects to migrate to, and pin, the domain walls [62].

The morphology of nanodomains predicted in Ref. [61] contrasts with the two "simplest" pictures of *out-of-plane* 180° domains found in magnetic films, namely the flux-closure domain structures [42]—for which surfaces have solely *in-plane* magnetizations while the dipoles of the inside layers are either parallel or antiparallel *to the* *z*-axis—

and the open-stripe structures [42]—for which all the layers, including the ones at the surfaces, exhibit dipoles that are either parallel or antiparallel to the *z-axis*. The laminar nanodomains "only" exist for *compressive strain* and a *large-enough depolarizing field*, consistent with the experimental conditions of Ref. [36] and the theoretical findings of Refs. [29, 38]. The predicted period Λ of the laminar domains is $\simeq 8$ in-plane lattice constants. This agrees well with the measurements of Ref. [36] yielding $\Lambda = 37$ Å for 20 Å-thick $PbTiO_3$ films. The stripe domain sizes observed in this study are intriguingly close to the ultimate limit of a single-molecule ferroelectric memory element. Interestingly, under compressive strain conditions and *partially* compensated depolarizing fields, "bubble" nanodomains were found to propagate throughout the entire thickness of the film. This situation is similar to that in which a bubble domain is formed in a ferromagnetic material with the easy-magnetization axis normal to the film surface by applying an external magnetic field. The phenomenon observed could provide a potential route toward ferroelectric bubble memory.

One criticism that can be raised regarding this study is the neglect of surface/interface effects that go beyond the surface charges (e.g., chemical bonding) in the modeling. Also, the misfit dislocations, which are a cause of stress relaxation and considerable strain reduction were not considered. A simple answer for that problem would be to use a temperature-dependent substrate effective lattice parameter allowing for the possible presence of misfit dislocations (or the thermal expansion difference between the substrate and the film) at the interface. Another point of the model presented in Ref. [61] is the fact that periodic boundary conditions was used in the *z*-direction of the supercell, which are not quite efficient due to the long-range nature of the dipole-dipole interaction which lead to artificial electrostatic interactions between the periodically repeated images of the film (to minimize the interaction between periodic images, a vacuum region of approximately 150Å separating the periodic images has been used).

2.6. "Ferroelectricity in $Pb(Zr_{0.5}Ti_{0.5})O_3$ thin films: critical thickness and 180° stripe domains" by Zhongqing Wu, Ningdong Huang, Zhirong Liu, Jian Wu, Wenhui Duan, Bing-Lin Gu, and Xiao-Wen Zhang, *Phys. Rev. B* **70** (2004) 104108

In Ref. [63], Monte Carlo simulations on the basis of a first-principles-derived Hamiltonian were made on the (0 0 1) $Pb(Zr_{0.5}Ti_{0.5})O_3$ thin films, under ideal open-circuit conditions and for different mechanical boundary conditions.

It is shown that the ferroelectricity in thin films depends critically on the strain constraint imposed by the substrate. For stress-free conditions, a nonzero polarization with an in-plane direction always exists in the system, i.e. it does not disappear even in the *monolayer* films. Under a compressive strain of −2% the out-of-plane polarization exhibits a strong dependence on the film thickness. Their calculations also reveal that the critical thickness appears only when the strain imposed by the substrate is strong enough to suppress the in-plane polarization. Above a critical thickness of about 3 unit cells (under the compressive strain of −2%), the out-of-plane polarization forms periodic 180°—stripe domains to screen the depolarizing field, and the domain period increases with the film thickness. Below the critical thickness, the stripe domain structure disappears, which suggests that the ferroelectricity can be suppressed even in the absence of the depolarizing electrostatic field. The competition between short-range and dipole–dipole interaction is revealed to be responsible for these phenomena.

2.7. "Properties of $Pb(Zr,Ti)O_3$ ultrathin films under stress-free and open-circuit electrical boundary conditions" by Emad Almahmoud, Yulia Navtsenya, Igor Kornev, Huaxiang Fu, and L. Bellaiche, *Phys. Rev. B* **70**, (2004) 220102(R)

Almahmoud *et al.*'s [64] investigated finite-temperature properties of (001) PZT stress-free ultrathin films under open-circuit electrical boundary conditions, that are PbO-terminated and have a Ti composition around 50%.

In Ref. [64] three additional terms (each with its own parameter) have been added to the effective Hamiltonian of bulk alloy to mimic explicit interactions between films and the vacuum. Two parameters quantify how the existence of the vacuum affects the *out-of-plane* components of the local modes and inhomogeneous strains near the surface. The third parameter characterizes the change, with respect to the bulk, of the short-range interaction between the *in-plane* components of the local modes near the surface. This approach was successfully tested, at low temperature, against direct first-principles calculations for ultrathin films. These simulations point out that the phase transition sequence can dramatically differ between bulks and thin films.

In particular, the authors found that (1) such films do not have any critical thickness under which ferroelectricity would disappear; (2) their polarization lies along directions perpendicular to the growth direction because of the huge depolarizing fields; (3) surface effects, and especially vacuum-induced changes of short-range interaction, significantly affect their local and macroscopic properties; (4) these ultrathin films exhibit a morphotropic phase boundary where the polarization continuously rotates, in a (0 0 1) plane between a ⟨0 1 0⟩ and⟨1 1 0⟩ direction, as the Ti composition decreases at small temperature; (5) such rotation leads to large piezoelectric and dielectric coefficients; and (6) the nature of phase transitions can change when going from bulks to ultrathin films.

3. Concluding remarks

Recent works have shown that it is now possible to compute the physical properties of ferroelectric nanostructures from *direct* first-principles and first-principles-based techniques. An understanding (at a microscopic level) of ferroelectric-based nanostructures starts to emerge thanks to these approaches. In particular, it is now becoming clear how the surface, interface, thickness, electrical and mechanical boundary conditions affect the properties of ultrathin films.

It is reasonable to believe that while *ab initio* efforts will continue to focus on thin films, they may also be further extended to other ferroelectric nanostructures (e.g., dots, wires, etc...) [30, 65]. The anticipated insights may guide the synthesis of greatly improved materials and the realization of devices with new and/or improved capabilities.

Acknowledgments

This work was supported by ONR grants N 00014-01-1-0365, N 00014-04-1-0413 and N 00014-01-1-0600 and NSF grants DMR-9983678 and DMR-0404335.

References

1. R. RAMESH, S. AGGARWAL and O. AUCIELLO, *Mat. Sc. Eng.* **32** (2001) 191.
2. N. SETTER (Ed.), in "Piezoelectric materials in devices: extended reviews on current and emerging piezoelectric materials, technology, and applications" (EPFL Swiss Federal Institute of Technology, 2002).
3. J. SCOTT, in "Ferroelectric memories" (Springer Verlag, Berlin, 2000).
4. K. UCHINO, in "Piezoelectric Actuators and Ultrasonic Motors" (Kluwer Academic Publishers, Boston, 1996).
5. S.-E. PARK and T. R. SHROUT, *J. Appl. Phys.* **82** (1997) 1804.
6. R. F. SERVICE, *Science* **275** (1997) 1878.
7. B. NOHEDA, *Appl. Phys. Lett.* **74** (1999) 2059.
8. L. BELLAICHE, A. GARCIA and D. VANDERBILT, *Phys. Rev. Lett.* **84** (2000) 5427.
9. L. BELLAICHE, A. GARCIA and D. VANDERBILT, *Ferroelectrics* **266**, 41 (2002).
10. B. NOHEDA, *Current Opinion in Solid State and Materials Science* **6** (2002) 27.
11. H. FU and R. E. COHEN, *Nature* **403** (2000) 281.
12. I. GRINBERG, V. R. COOPER and A. M. RAPPE, *Nature* **419** (2002) 909.
13. Z. WU and H. KRAKAUER, *Phys. Rev. B* **68** (2003) 014112.
14. A. OHTOMO, D. A. MULLER, J. L. GRAZUL and H. Y. HWANG, *Nature* **419** (2002) 378.
15. D. WOLPERT, K. KOROLEV, S. SACHS, J. KNAB, W. COX, J. CERNE, A. G. MARKELZ, T. ZHAO, R. RAMESH and B. H. MOECKLY, *Physica E: Low-dimensional Systems and Nanostructures* **19** (2003) 236.
16. E. D. MISHINA, V. I. STADNICHUK, A. S. SIGOV, Y. I. GOLOVKO, V. M. MUKHOROTOV, S. NAKABAYASHI, H. MASUDA, D. HASHIZUME and A. NAKAO, *Physica E: Low-dimensional Systems and Nanostructures* **25** (2004) 35.
17. D. G. SCHLOM, J. H. HAENI, J. LETTIERI, C. D. THEIS, W. TIAN, J. C. JIANG and X. Q. PAN, *Materials Science and Engineering B* **87** (2001) 282.
18. A. LIN, X. HONG, V. WOOD, A. A. VEREVKIN, C. H. AHN, R. A. MCKEE, F. J. WALKER and E. D. SPECHT, *Applied Physics Letters* **78** (2001) 2034.
19. Y. WANG, C. GANPULE, B. T. LIU, H. LI, K. MORI, B. HILL, M. WUTTIG, R. RAMESH, J. FINDER, Z. YU, *et al. Appl. Phys. Lett.* **80** (2002) 97.
20. K. EISENBEISER, J. M. FINDER, Z. YU, J. RAMDANI, J. A. CURLESS, J. A. HALLMARK, R. DROOPAD, W. J. OOMS, L. SALEM, and S. BRADSHAW, *et al., Appl. Phys. Lett.* **76** (2000) 1324.
21. O. AUCIELLO, J. F. SCOTT and R. RAMESH, *Physics Today* **51** (1998) 22.
22. J. F. SCOTT, *Ann. Rev. Mat. Sci.* **28** (1998) 79.
23. M. E. LINES and A. M. GLASS, "Principles and Applications of Ferroelectrics and Related Materials" (Clarendon Press, 1977).
24. T. TYBELL, C. H. AHN and J.-M. TRISCONE, *Appl. Phys. Lett.* **75** (1999) 856.
25. A. V. BUNE, V. M. FRIDKIN, S. DUCHARME, L. M. BLINOV, S. P. PALTO, A. V. SOROKIN, S. G. YUDIN and A. ZLATKIN, *Nature* **391** (1998) 874.
26. D. D. FONG, G. B. STEPHENSON, S. K. STREIFFER, J. A. EASTMAN, O. AUCIELLO, P. H. FUOSS and C. THOMPSON, *Science* **304** (2004) 1650.
27. P. GHOSEZ and K. M. RABE, *Appl. Phys. Lett.* **76** (2000) 2767.
28. J. JUNQUERA AND P. GHOSEZ, *Nature* **422** (2003) 506.
29. S. TINTE AND M. G. STACHIOTTI, *Phys. Rev. B* **64** (2001) 235403.
30. H. FU and L. BELLAICHE, *Phys. Rev. Lett.* **91** (2003) 257601.
31. B. MEYER and D. VANDERBILT, *Phys. Rev. B* **63** (2001) 205426.
32. B. MEYER, J. PADILLA and D. VANDERBILT, *Faraday Discussions* **114** (1999) 395.
33. R. E. COHEN, *J. Phys. Chem. Sol.* **57** (1996) 1393.
34. R. COHEN, *Ferroelectrics* **194** (1997) 323.
35. L. FU, E. YASCHENKO, L. RESCA and R. RESTA, *Phys. Rev. B* **60** (1999) 2697.
36. S. K. STREIFFER, J. A. EASTMAN, D. D. FONG, C. THOMPSON, A. MUNKHOLM, M. V. R. MURTY, O. AUCIELLO, G. R. BAI and G. B. STEPHENSON, *Phys. Rev. Lett.* **89** (2002) 067601.
37. A. KOPAL, T. BAHNIK and J. FOUSEK, *Ferroelectrics* **202** (1997) 267.
38. Y. L. LI, S. Y. HU, Z. K. LIU and L. Q. CHEN, *Appl. Phys. Lett.* **81** (2002) 427.
39. R. R. MEHTA, B. D. SILVERMAN and J. T. JACOBS, *J. Appl. Phys.* **44** (1973) 3379.
40. J. JUNQUERA, O. DIEGUEZ, K. M. RABE, P. GHOSEZ, C. LICHTENSTEIGER and J.-M. TRISCONE, in "Fundamental Physics of Ferroelectrics" (NISTIR, Gaitherburg. Colonial Williamsburg, VA, 2004), pp. 86–87.
41. I. A. KORNEV AND L. BELLAICHE, *Phys. Rev. Lett.* **91** (2003) 116103.
42. L. D. LANDAU and E. M. LIFSCHITZ, in "Electrodynamics of Continuous Media." (Pergamon Press, 1984).
43. M. D. GLINCHUK, E. A. ELISEEV, V. A. STEPHANOVICH and R. FARHI, *J. Appl. Phys.* **93** (2003) 1150.
44. V. ZHIRNOV, *Sov. Phys. JETP* **35** (1958) 1175.
45. N. A. PERTSEV, V. G. KUKHAR, H. KOHLSTEDT and R. WASER, *Phys. Rev. B* **67** (2003) 054107.

46. M. G. COTTAM, D. R. TILLEY and B. ZEKS, *J. Phys. C: Solid St. Phys.* **17** (1984) 1793.
47. Y. WANG, W. ZHONG and P. ZHANG, *Phys. Rev. B* **53** (1996) 11439.
48. A. M. GEORGE, J. INIGUEZ and L. BELLAICHE, *Nature* **413** (2001) 54.
49. W. ZHONG, D. VANDERBILT and K. RABE, *Phys. Rev. Lett.* **73** (1994) 1861.
50. W. ZHONG, D. VANDERBILT and K. RABE, *Phys. Rev. B* **52** (1995) 6301.
51. I. A. KORNEV and L. BELLAICHE, *Phys. Rev. Lett.* **89** (2002) 115502.
52. A. AL-BARAKATY and L. BELLAICHE, *Appl. Phys. Lett.* **81** (2002) 2442.
53. K. RABE and P. GHOSEZ, *Journal of Electroceramics* **4** (2000) 379.
54. R. KRETSCHMER and K. BINDER, *Phys. Rev. B* **20** (1979) 1065.
55. B. MEYER and D. VANDERBILT, *Phys. Rev. B* **65** (2002) 104111.
56. J. M. SOLER, E. ARTACHO, J. D. GALE, A. GARCÍA, J. JUNQUERA, P. ORDEJÓN and D. SÁNCHEZ-PORTAL, *J. Phys.: Cond. Matter* **14** (2002) 2745.
57. O. DIEGUEZ, S. TINTE, A. ANTONS, C. BUNGARO, J. B. NEATON, K. M. RABE and D. VANDERBILT, *Phys. Rev. B* **69** (2004) 212101.
58. N. PERTSEV, A. ZEMBILGOTOV and A. TAGANTSEV, *Phys. Rev. Lett.* **80** (1998) 1988.
59. G. KRESSE and J. FURTHMULLER, *Phys. Rev. B* **54** (1996) 11169.
60. G. KRESSE and J. HAFNER, *Phys. Rev. B* **47** (1993) 558.
61. I. KORNEV, H. FU and L. BELLAICHE, *Phys. Rev. Lett.* **93** (2004) 196104.
62. L. HE and D. VANDERBILT, *Phys. Rev. B* **68** (2003) 134103.
63. Z. WU, N. HUANG, Z. LIU, J. WU, W. DUAN, B.-L. GU and X.-W. ZHANG, *Phys. Rev. B* **70** (2004) 104108.
64. E. ALMAHMOUD, Y. NAVTSENYA, I. KORNEV, H. FU and L. BELLAICHE, *Phys. Rev. B* **70** (2004) 220102(R).
65. I. NAUMOV, L. BELLAICHE and H. FU, *Nature (London)* **432** (2004) 737.

JOURNAL OF MATERIALS SCIENCE 41 (2006) 147–153

Fredholm integral equation of the Laser Intensity Modulation Method (LIMM): Solution with the polynomial regularization and L-curve methods

SIDNEY B. LANG
Department of Chemical Engineering, Ben-Gurion University of the Negev, 84105 Beer Sheva, Israel
E-mail: lang@bgu.ac.il

The Laser Intensity Modulation Method (LIMM) is widely used for the determination of the spatial distribution of polarization in polar ceramics and polymers, and space charge in non-polar polymers. The analysis of experimental data requires a solution of a Fredholm integral equation of the 1st kind. This is an ill-posed problem that has multiple and very different solutions. One of the more frequently used methods of solution is based upon Tikhonov regularization. A new method, the Polynomial Regularization Method (PRM), was developed for solving the LIMM equation with an 8th degree polynomial using smoothing to achieve a stable and optimal solution. An algorithm based upon the L-curve method (LCM) was used for the prediction of the best regularization parameter. LIMM data were simulated for an arbitrary polarization distribution and were analyzed using PRM and LCM. The calculated distribution function was in good agreement with the simulated polarization distribution. Experimental polarization distributions in a poorly poled sample of polyvinylidene fluoride (PVDF) and in a $LiNbO_3$ bimorph, and space charge in polyethylene were analyzed. The new techniques were applied to the analysis of 3-dimensional polarization distributions.

1. Introduction

1.1. LIMM experiment

LIMM was first suggested by Lang and Das-Gupta in 1981 [1] and described in detail in 1986 [2]. A review of the current implementation was presented in 2004 [3]. The experimental technique is as follows. LIMM samples are prepared as thin plates or sheets with their flat surfaces normal to the polar axis. These surfaces are coated with very thin opaque electrodes. The sample may be either freely suspended or attached to a grounded metal plate by an electrically conductive paste or cement. The surface of the sample is exposed to a laser beam whose intensity is modulated sinusoidally by means of an acoustooptic modulator or a built-in modulator in the laser. The energy of the laser beam is absorbed at the electrode and heat diffuses into the sample as temperature waves. The depth of penetration of the waves is greater for low modulation frequencies and less for high ones. This produces a spatially non-uniform time-varying temperature distribution that interacts with the unknown polarization or space charge distribution to produce a pyroelectric current. The real and the imaginary components of the current generated at each of 50 to 100 different frequencies are amplified and a lock-in amplifier is used to determine the amplitude and phase relative to that of the modulated intensity of the laser beam.

1.2. Analysis of LIMM data

The relative polarization distribution, $\beta(z)$, is found from a solution of the LIMM equation [2, 4, 5]:

$$I(\omega) = \frac{A}{L}\int_{z_1}^{z_2} \beta(z)\frac{\partial T(z,\omega)}{\partial t}dz \tag{1}$$

where

$$\beta(z) = \alpha_P P(z) - (\alpha_x - \alpha_\varepsilon)\varepsilon\varepsilon_0 E(z)$$

0022-2461
DOI: 10.1007/s10853-005-5983-8

TABLE I Nomenclature

Parameter	Symbol	Units
A	Laser beam cross-section	m^2
I	Current	A
E	Electric field	Vm^{-1}
k	Thermal conductivity	$Wm^{-1} K^{-1}$
L	Thickness	m
p	Pyroelectric coefficient	$Cm^{-2} K^{-1}$
P	Polarization	Cm^{-2}
P_S	Spontaneous polarization	Cm^{-2}
r	Regularization parameter	Dimensionless
T	Temperature	K
t	Time	s
z	Spatial coordinate	m
z_1, z_2	Spatial coordinates of upper (irradiated) and lower surfaces of sample	m
α	Thermal diffusivity	$m^2 s^{-1}$
α_P	Temperature dependence of polarization	K^{-1}
α_x	Thermal expansion coefficient	K^{-1}
α_ε	Temperature dependence of permittivity	K^{-1}
$\beta(z)$	Polarization distribution function (Equation 1)	$Cm^{-2} K^{-1}$
γ	Scale factor in Equation 2	Dimensionless
ε	Relative permittivity	Dimensionless
ε_0	Permittivity of vacuum	Fm^{-1}
η	Roughness residual in Equation 6	$C^2 m^{-7} K^{-2}$
λ	Coefficient in Equation 2	Dimensionless
ρ	Data fit residual in Equation 5	A^2
σ	Space charge	Cm^{-3}
ω	Radial frequency	$rad\ s^{-1}$

Nomenclature is given in Table I. The term $\partial T/\partial t$ is calculated from the one-dimensional heat conduction equation using boundary conditions appropriate for either a freely-suspended sample or a sample mounted on a substrate [3]. It was shown by a finite-element solution that the use of a one-dimensional analysis instead of the true three-dimensional one introduces negligible error. Equation 1 is then solved for $\beta(z)$ using experimental measurements of $I(\omega)$. If the experimental sample is a polar ceramic or polymer, it is assumed that the polarization is locally compensated by space charge and only $P(z)$ is determined. If the sample is nonpolar, then $P(z) = 0$ and the electric field $E(z)$ is found. From this, the space charge distribution can be determined. There is no way to separate polarization and space charge if both are present. It should be noted that $I(\omega)$ contains experimental errors. Computer roundoff will effectively add errors, even in the case of computer simulated data. Equation 1 is a Fredholm integral equation of the first kind and is an ill-conditioned problem with a large (possibly infinite) number of very different solutions. An illustration of the multiplicity of solutions was given by Phillips [6]. A term such as *sin*(kz) can be added to $\beta(z)$. Provided that k is sufficiently large, the integration will reduce this added factor to less than the experimental errors.

A large number of methods have been proposed for the solution of Equation 1 [3]. The most successful of the methods are various forms of the regularization method [7–11] and the scale transformation [4, 5]. The regularization method utilizes the requirement that the polarization distribution must be smooth. By specifying the degree of smoothing, extreme minima and/or maxima are excluded but moderate ones are permitted. However, its usage requires complex computer programs. The scale transformation method is very easy to implement. It is less accurate than regularization and it does not allow the determination of the polarization in regions closer to the sample electrode than the thermal diffusion length at the highest measured frequency. Recently, Lang [3] proposed a simplified regularization method called the Polynomial Regularization Method (PRM). It is implemented using the computer software program, *Mathematica*® [12]. Analysis of experimental data requires the selection of an appropriate regularization parameter. A technique called the L-curve Method (LCM) is used for that purpose. PRM and LCM are described in the following section.

2. Polynomial regularization method (PRM) and L-curve Method (LCM)

2.1. Polynomial regularization method and simulated data

PRM is implemented by assuming that $\beta(z)$ can be represented by an eighth-degree polynomial:

$$\beta(z) = \lambda_0 + \lambda_1\gamma(z) + \lambda_2\gamma(z)^2 + \lambda_3\gamma(z)^3 + \cdots \tag{2}$$

where

$$\gamma(z) = \frac{Log(z) - Log(z_1)}{Log(z_2) - Log(z_1)}$$

Because LIMM is most useful in determining distributions near the sample surface [13], the distribution is based on a normalized logarithmic scale. Polynomials of degrees greater than eight have been found to give very inaccurate results. Lower degree polynomials gave the general shape of the distributions but did not correspond to them in detail. Eighth-degree polynomials were the most satisfactory. The coefficients λ_j, must be determined. Equation 2 is substituted into Equation 1 and the integral is evaluated numerically to give the real and imaginary parts of $I(\omega)_{calc}$ as a function of the λs. A linear regression (least-squares) solution can be obtained by minimizing the function:

$$\sum_i [I(\omega_i)_{exp} - I(\omega_i)_{calc}]^2 \tag{3}$$

with respect to the λs which are then inserted into Equation 2. This yields one possible polarization distribution but, most likely, an incorrect one. In the regularization technique, the following function is minimized with respect to the λs:

$$\sum_i [I(\omega_i)_{\text{exp}} - I(\omega_i)_{\text{calc}}]^2 + r^2 \int_{z_1}^{z_2} \left(\frac{d^2\beta(z)}{dz^2}\right)^2 dz \quad (4)$$

The regularization parameter, r, smoothes the computed polarization distribution. If $r = 0$, the conventional linear regression solution is found. This solution will have a number of large maxima and minima because of the ill-posed nature of the problem. If r is too large, all of the detail is removed from the computed distribution.

Simulated experimental data will be used to illustrate the PRM. The sample is assumed to be a 25.4-μm thick film of polyvinylidene fluoride (PVDF) with 100-nm thick aluminum electrodes. The lower surface of the sample is in good thermal contact with a thick metallic substrate. A typical frequency range over which data are measured extends from 10 Hz to 100 kHz. Based on the Frequency Range Function proposed by Lang [3], the polarization can be found in the region between 0.01 μm and 10 μm from the laser-irradiated electrode. A polarization distribution, $P(z)$, is assumed, as shown in Fig. 1. This distribution is a purely arbitrary function and does *not* resemble a physically realistic distribution. Then simulated experimental data, $I(\omega)_{\text{exp}}$, are calculated by substituting the distribution into Equation 1. In order that the simulated data will more closely resemble true data, a Gaussian distribution of error is added to each point (standard deviation of 5% of the range of the real and imaginary parts, resp.). The resulting simulated real and imaginary values of current $I(\omega)_{\text{exp}}$ as functions of frequency are shown in Fig. 2. Because the simulated distribution is known, it is possible to examine the closeness of the solution to the true value by varying r. In Fig. 3, the sum of the squares of the differences between $I(\omega)_{\text{exp}}$ and $I(\omega)_{\text{calc}}$ (error of fit) is graphed as a function of r. The minimum error of fit is found for $r = 0.000033$, referred to as the "optimal" value. Calculated distributions for $r = 0$, a very large value of r (0.001) and the optimal value of r (0.000033) are shown in Fig. 4. All three calculated distributions are correct mathematical solutions of Equation 1 but only the distribution corresponding to the optimal value of r reproduces the true distribution.

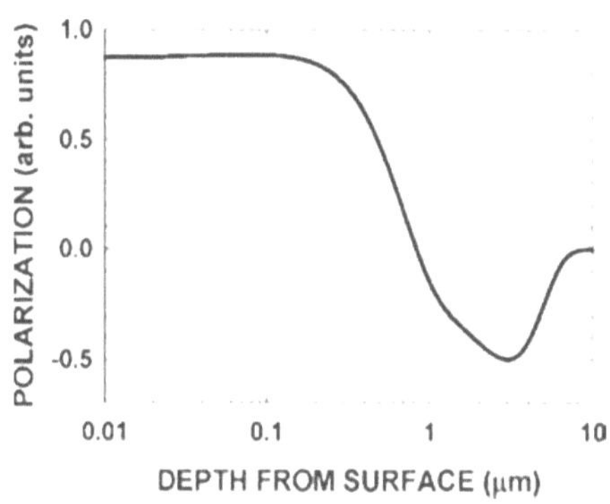

Figure 1 Simulated polarization distribution in 25.4-μm thick PVDF.

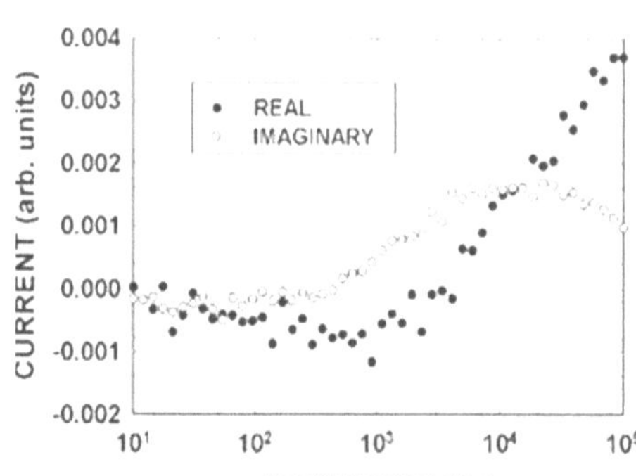

Figure 2 LIMM current versus frequency data calculated from data in Fig. 1 using Equation (1). Random errors added.

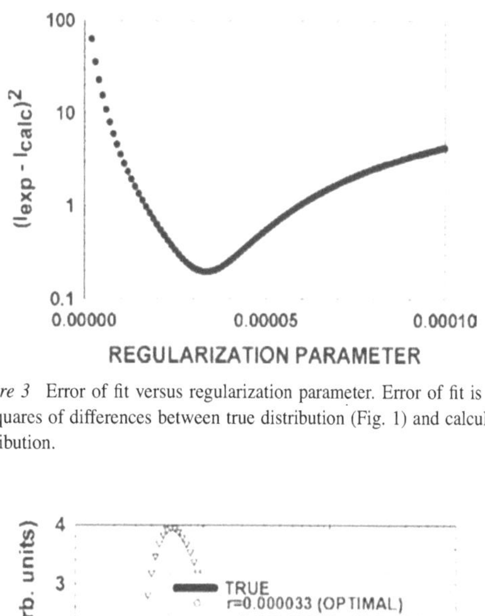

Figure 3 Error of fit versus regularization parameter. Error of fit is sum of squares of differences between true distribution (Fig. 1) and calculated distribution.

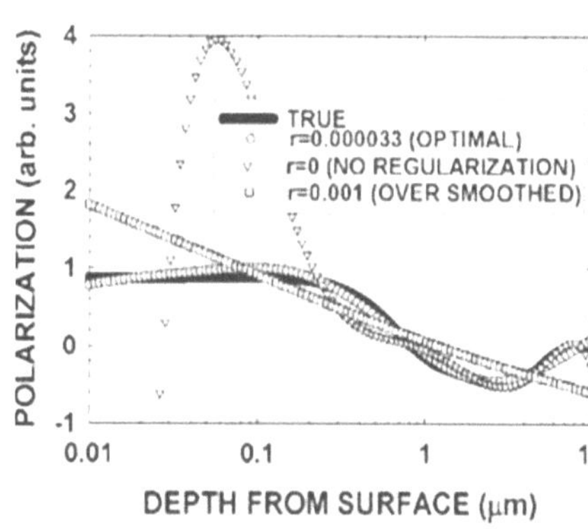

Figure 4 Polarization distributions. True value and those corresponding to no regularization, optimal regularization and oversmoothing.

2.2. L-curve method (LCM) and experimental data

If measured experimental data are analyzed, the optimal value of r cannot be found as illustrated above because the true polarization distribution is unknown. Therefore, a method of selecting a good value of the regularization parameter is required. Several techniques for determination of the regularization parameter are described by Sandner *et al.* [11]. Mellinger [14] has presented an example of the unbiased iterative method. Hansen and O'Leary [15] suggested using the L-curve method for other types of ill-conditioned problems. A range of values of r is selected and two parameters are computed for each value of r. One is the data fit residual:

$$\rho(r) = \sum_{\omega_i} [I(\omega_i)_{\text{exp}} - I(\omega_i)_{\text{calc}}]^2 \tag{5}$$

and the second is the roughness residual:

$$\eta(r) = \int_{z_1}^{z_2} \left(\frac{d^2\beta(z)}{dz^2}\right)^2 dz \tag{6}$$

The data fit residual is a function of the difference between the actual experimental data and values computed for a specific level of r. The roughness residual is a measure of the lack of smoothness in the computed polarization distributions. If $\log(\rho)$ is plotted versus $\log(\eta)$, a curve with an L-shape results. The L-curve for the data in the example above is shown in Fig. 5. The corner corresponds to the point where the calculated distribution $\beta(z)$ changes from domination by large differences between experimental and calculated values to domination by over smoothing. The value of r corresponding to the corner is a good measure of the regularization parameter. It can be found most easily by finding the maximum of the curvature of the function in Fig. 5. The curvature is given by the following formula [15]

$$\frac{\hat{\rho}'\hat{\eta}'' - \hat{\rho}''\hat{\eta}'}{\left[(\hat{\rho}')^2 + (\hat{\eta}')^2\right]^{3/2}} \tag{7}$$

where

$$\hat{\rho} = \log(\rho) \quad \text{and} \quad \hat{\eta} = \log(\eta)$$

The $'$ and $''$ symbols indicate differentiation and double differentiation with respect to r. In order to evaluate Equation 7, the values of $\rho(r)$ and $\eta(r)$ are interpolated with low-order polynomials and then differentiated with respect to r. The maximum value of curvature in this example corresponds to $r = 0.000030$. Fig. 6 shows the true polarization distribution, the one corresponding to the optimal value of r and the one found using the L-curve. The "optimal" and the "L-curve" fits are both in excellent agreement with the true curve.

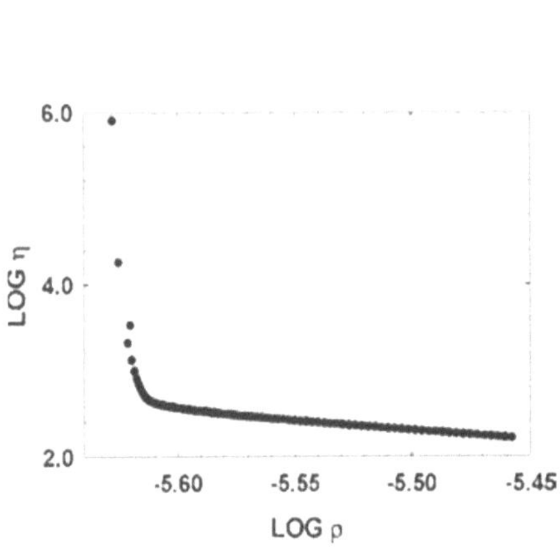

Figure 5 L-curve for simulated data.

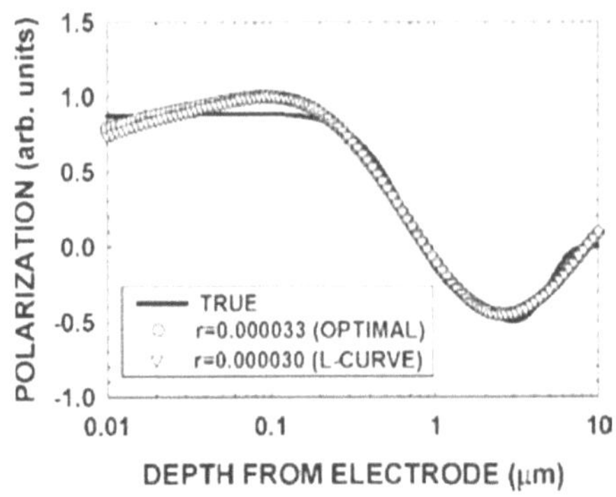

Figure 6 Polarization distributions. True and those corresponding to optimal and L-curve regularization parameters.

3. Experimental studies

Experimental studies on three different materials are presented here.

3.1. Polyvinylidene fluoride

LIMM data were measured on a 25.4-μm thick sample of polyvinylidene fluoride (PVDF). The sample had been weakly poled and then stored for a number of years. The experimental data are shown in Fig. 7. The regularization parameter at the point of maximum curvature in the L-curve plot (Fig. 8) was 0.000235. Although it is difficult to determine the point of maximum curvature by eye, it is easily found by use of Equation 7. The calculated polarization distribution shown in Fig. 9 is very non-uniform as a consequence of the poling and aging conditions.

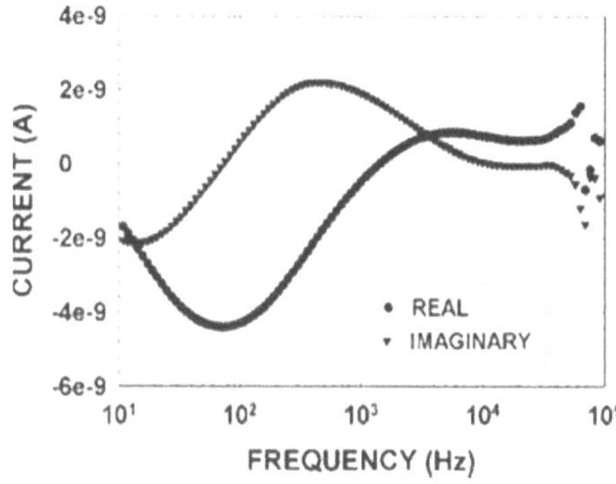

Figure 7 Real and imaginary LIMM data for a poorly-poled sample of PVDF.

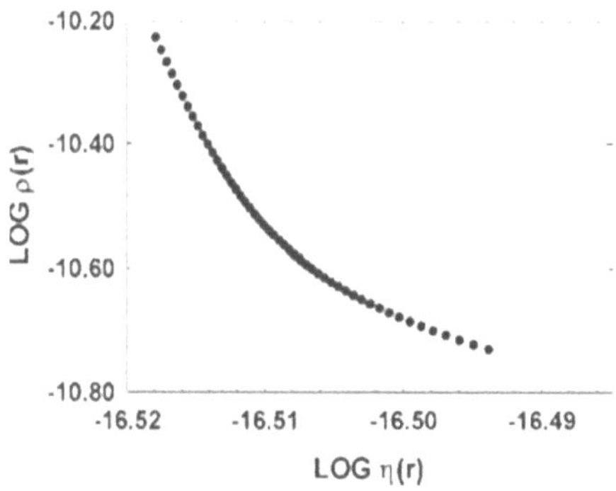

Figure 8 L-curve for PVDF data.

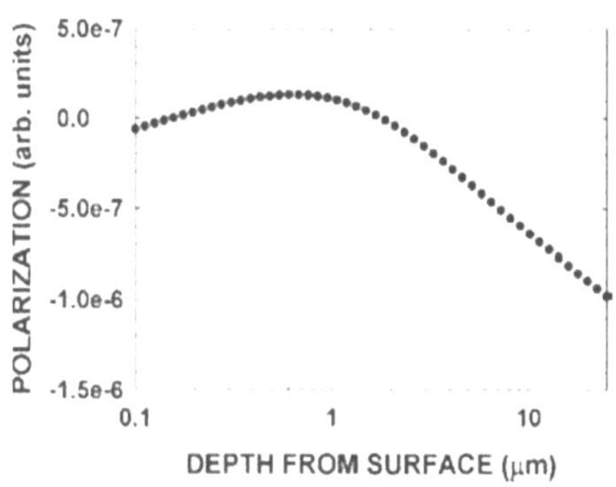

Figure 9 Polarization distribution of PVDF.

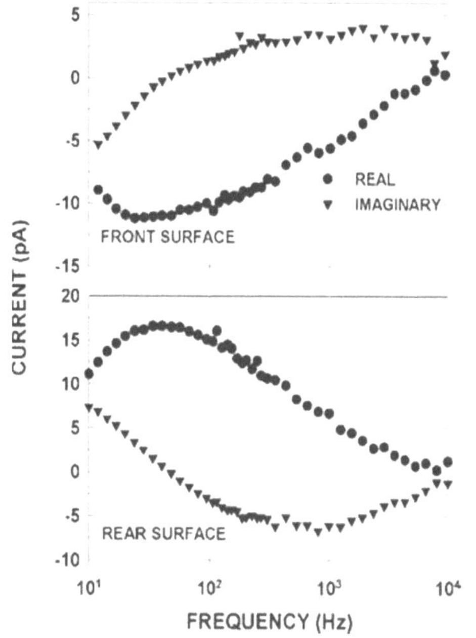

Figure 10 LIMM data for bimorph sample of $LiNbO_3$. Data measured for laser impingement on front and rear surfaces.

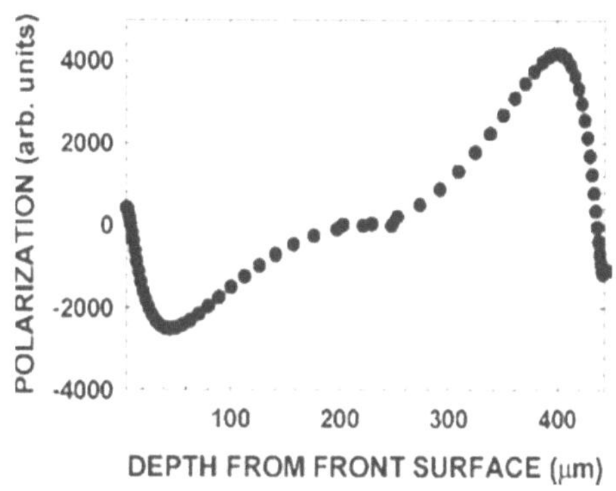

Figure 11 Polarization distribution of bimorph sample of $LiNbO_3$.

3.2. Domain inversion in lithium niobate

Periodic domain inversion has been used to produce wave guides for second harmonic generation in lithium niobate and lithium tantalate [16]. A 0.503-mm thick *z*-cut plate of $LiNbO_3$ was thermally treated in order to partially invert the orientation of the domains [17]. Domain inversion occurred at the +c-surface and extended approximately halfway through the thickness of the sample. LIMM data were measured in order to determine the resulting polarization distribution [18]. The experimental data for the +c and −c surfaces are shown in Fig. 10. The regularization parameters at the points of maximum curvature in the two L-curve plots were 0.0000591 and 0.0000485, respectively. The domain inversion is clearly shown by the calculated polarization distribution (Fig. 11).

3.3. Space charge in cross-linked polyethylene

Samples of 100-μm thick cross-linked polyethylene (XLPE) were poled with a dc electric field of 50 kV mm^{-1} at room temperature for a period of 18 h [19]. XLPE is non-polar although the presence of polar impurities resulted in some polarization which decayed very rapidly. However, the major effect was the deposition of space charge. The LIMM data for measurements on the upper and lower surfaces of the sample are shown in Figs 12 and 13. The thermal expansion coefficient α_x used in the calculations was 200×10^{-6} K^{-1} and α_ε was assumed to equal zero. The corresponding regularization parameter was 0.0000235 for each of the surfaces. The electric field distribution in the sample is shown in Fig. 14. The space charge distribution was calculated from the electric field distribution by use of Gauss's law:

$$\sigma(z) = \varepsilon_0\varepsilon\frac{dE}{dz} \quad (8)$$

where $\varepsilon = 2.3$. The space charge distribution is shown in Fig. 15. Negative space charge was found in regions very close to the electrodes. Because of its limited resolution near sample surfaces, the scale transformation technique for analyzing LIMM data [4] did not reveal this negative charge.

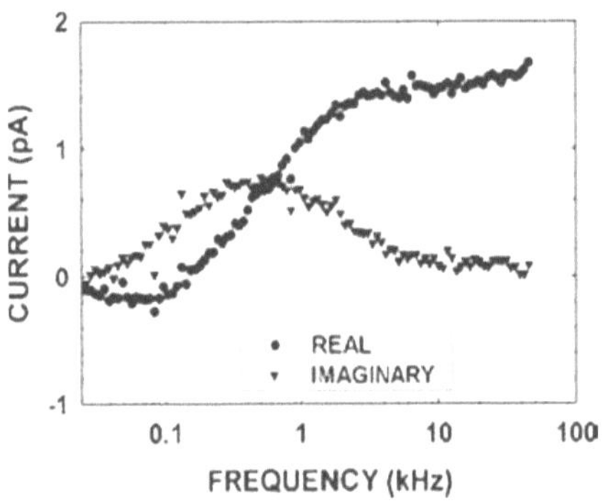

Figure 12 LIMM data for laser impingement on upper surface of cross-linked polyethylene.

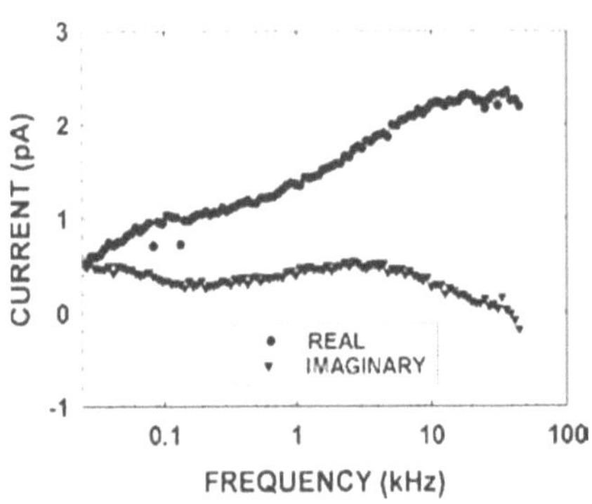

Figure 13 LIMM data for laser impingement on lower surface of cross-linked polyethylene.

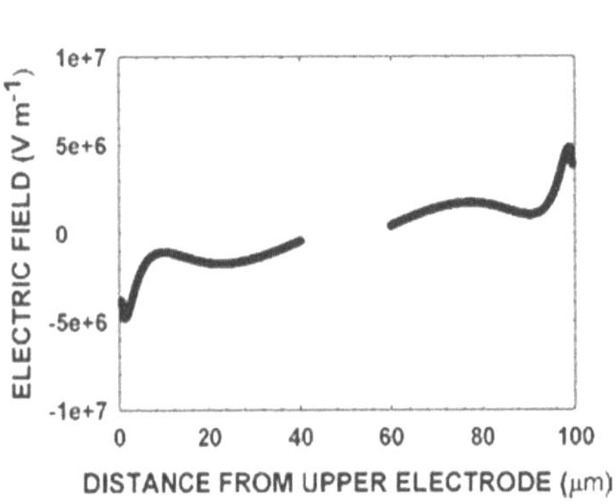

Figure 14 Electric field distribution in cross-linked polyethylene.

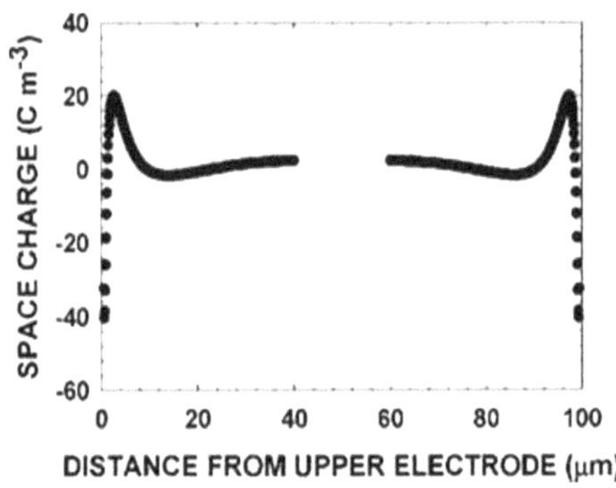

Figure 15 Space charge distribution in cross-linked polyethylene.

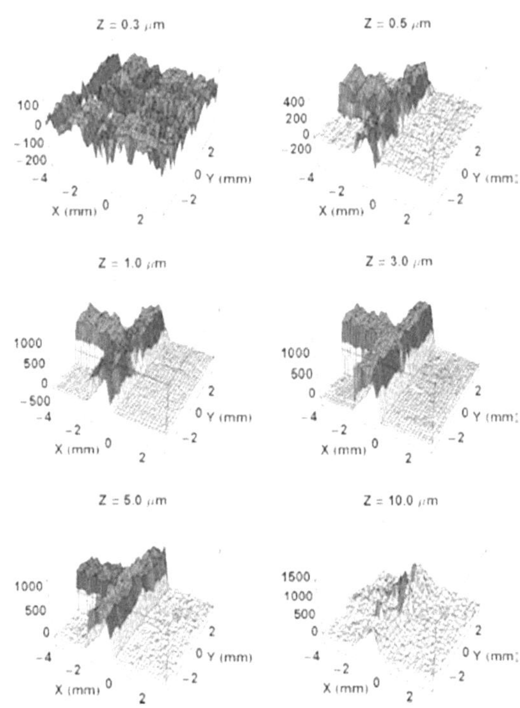

Figure 16 Polarization distributions at various depths in PVDF poled with a T-shaped electrode.

4. Three-dimensional mapping of polarization profiles

The techniques of PRM and LCM were applied to the analysis of three-dimensional polarization data [20]. An 11-μm thick sample of electroded PVDF was affixed to a metallic substrate. It was poled with a dc electric field of 100 MVm^{-1} using a T-shaped electrode. A thermal pulse method of data acquisition rather than LIMM was used. The second-harmonic beam of a *Q*-switched Nd:YAG laser was focused to a spot size with a radius of 200 μm (at the 1/e point) and was scanned over the sample in a 36 by 36 point-raster pattern. The pulse duration was about 5 ns and 30 to 50 pulses were averaged at each point of the scan. The scan points were 200 μm apart covering a 7×7-mm area. A current amplifier was used and the current versus time data were recorded with a digital storage oscilloscope. The recorded data were Fourier-transformed to yield current versus frequency data. This resulted in conventional LIMM data with a frequency range from 50 Hz to 1 MHz. Thr PRM-LCM analysis was carried out on the 1296 data points. Fig. 16 shows the polarization distributions at various depths from the upper surface of the PVDF. The results were noisy at the shallowest depth of 0.3 μm but very clearly showed the T-shaped distribution at greater depths. At a depth of 10 μm, the polarization began to fade into the background. Future studies will use a more strongly focused laser beam.

5. Conclusions

A new method for the solution of the Fredholm integral equation of LIMM is proposed. It utilizes a regularization approach to find an 8th degree polynomial approximation to the polarization distribution. The value of the regularization parameter is found using the L-curve method. Both simulated and experimental LIMM data were analyzed. Very good agreement was found between the calculated and the true distribution of the simulated data. The experimental measurements revealed useful information concerning polarization or space charge. The technique has now been extended to the analysis of three-dimensional data. The calculations were implemented using *Mathematica* [12]. Copies of the computer files developed in this study are available from the author.

References

1. S. B. LANG and D. K. DAS-GUPTA, *Ferroelectrics* **39** (1981) 1249.
2. *Idem.*, *J. Appl. Phys.* **59** (1986) 2151.
3. S. B. LANG, *IEEE Trans. Dielectr. Electr. Insul.* **11** (2004) 3.
4. B. PLOSS, R. EMMERICH and S. BAUER, *J. Appl. Phys.* **72** (1992) 5363.
5. S. BAUER and S. BAUER-GOGONEA, *IEEE Trans. Dielec. Elec. Insul.* **10** (2003) 883.
6. B. L. PHILLIPS, *J. Assoc. Comput. Mach.* **9** (1962) 84.
7. S. B. LANG, *Ferroelectrics* **118** (1991) 343.
8. P. BLOβ, M. STEFFEN, H. SCHAFER, Y. GUO MAO and G. M. SESSLER, *IEEE Trans. Dielec. Elec. Insul.* **3** (1996) 182.
9. P. BLOβ, M. STEFFEN, H. SCHAFER, G. EBERLE and W. EISENMENGER, *ibid.* **3** (1996) 417.
10. S. B. LANG, *ibid.* **5** (1998) 70.
11. T. SANDNER, G. SUCHANECK, R. KOEHLER, A. SUCHANECK and G. GERLACH, *Integr. Ferroelectr.* **46** (2002) 243.
12. WOLFRAM RESEARCH, Inc., Mathematica, Version 5.0, Champaign, IL, USA (2003).
13. C. ALQUIE, C. LABURTHE TOLRA, J. LEWINER and S. B. LANG, *IEEE Trans. Electr. Insul.* **27** (1992) 751.
14. A. MELLINGER, *Meas. Sci. Technol.* **15** (2004) 1347.
15. P. C. HANSEN and D. P. O'LEARY, *SIAM J. Sci. Comput.* **14** (1993) 1487.
16. H. AHLFELDT, J. WEBJORN and G. ARVIDSSON, *IEEE Photon. Tech. Lett.* **3** (1991) 638.
17. V. D. KUGEL and G. ROSENMAN, *App. Phys. Lett.* **23** (1993) 2902.
18. S. B. LANG, V. D. KUGEL and G. ROSENMAN, *Ferroelectrics* **157** (1994) 69.
19. T. PAWLOWSKI, S. B. LANG and R. FLEMING, in Proceedings of the 2004 Conference on Electrical Insulation and Dielectric Phenonema, Boulder, CO, USA (IEEE Service Center, Piscataway, NJ, USA, 2004) p. 93.
20. A. MELLINGER, R. SINGH, M. WEGENER, W. WIRGES, R. GERHARD-MULTHAUPT and S. B. LANG, *Appl. Phys. Lett.* **86** (2005) 82903.

JOURNAL OF MATERIALS SCIENCE 41 (2006) 155–161

Multilayer piezoelectric ceramic transformer with low temperature sintering

LONGTU LI*, NINGXIN ZHANG, CHENYANG BAI, XIANGCHENG CHU, ZHILUN GUI
State Key Laboratory of New Ceramics and Fine Processing, Department of Materials Science and Engineering, Tsinghua University, Beijing 100084, China
E-mail: llt-dms@mail.tsinghua.edu.cn

The low-fired high performance piezoelectric ceramics used for multilayer piezoelectric transformer were investigated. Based on the transient liquid phase sintering mechanism, by doping suitable eutectic additives and optimizing processing, the sintering temperature of the quaternary system piezoelectric ceramics with high piezoelectric properties could be lower to about 960–1000°C. The low-temperature sintering multilayer piezoelectric transformer (MPT) has been developed. Some characteristics of MPT were systemically studied. The measurements include the frequency response of input impedance, frequency response of phase difference between input voltage and current, frequency shifting with load, input impedance changing with load, phase difference between input voltage and current shifting with load, and phase difference between input voltage and vibration velocity. The vibration modes and resonance characters of MPT were measured by a Laser Doppler Scanning Vibrometer. Several kinds of MPT with high voltage step-up ratio, high power density, high transfer efficiency and low cost have been industrially produced and commercialized. It reveals a broad application prospect for back-light power of liquid crystal display and piezo-ionizer etc.

1. Introduction

Low-firing multilayer piezoelectric transformer (MPT) was studied in Beijing Tsinghua University on the 1980's. This kind of MPT has high voltage step-up ratio, low driving voltage, small size, and low cost. Its voltage step-up ratio is ten times higher than that of conventional single-layer transformer. As known, the sintering temperature of normal piezoelectric ceramics is high up to about 1200–1300°C. In order to decrease the consumption of precious metal in internal electrode, the low sintering temperature PZT-based piezoelectric ceramics with high performance have been investigated systematically in Tsinghua University. Several kinds of low-firing high performance piezoelectric ceramics including binary, ternary and quaternary system compositions have been developed respectively. They can be sintered at 820–1000°C and obtained with excellent properties. A new kind of quaternary system piezoelectric ceramics PMN-PZN-PZT can be satisfactorily used for low firing MPT. Recent years, with the rapid development of the electronic information technology which provides a huge market requirement for electronic devices with the increasing of surface mount technology (SMT), which provides more and more conventional components have to be replaced by multilayer chip-type surface mount devices (SMD). Based on this background the chip-type MPT has attracted more attention and interested by many researchers and users. A new type piezoelectric inverter used for the back-light power of liquid crystal display in portable computer and piezo-ionizer has been developed. The MPT and its inverters have many advantages, such as thin shape, small size, high transfer efficiency, anti-interfere for electro-magnetic. The MPT and inverters have been produced in industrial scale and commercialized by Xi'an Konghong Company in China.

*Author to whom all correspondence should be addressed.
0022-2461
DOI: 10.1007/s10853-005-6008-3

2. Low-firing piezoelectric ceramics with high performance

The chemical grade and analytical grade raw materials Pb_3O_4, TiO_2, ZrO_2, Nb_2O_5, ZnO, Li_2CO_3, CdO, MnO_2, $SrCO_3$ were used and powders were prepared by conventional oxide synthesis processing, the samples that were formed into discs having a diameter of 15 mm and thickness of 1mm under the pressure of about 100 Mpa were coated with silver paste and fired. After firing samples were polarized by a DC electric field at 4 kV/mm in 120°C silicon oil. The dielectric and piezoelectric properties were measured by HP4194.

The composition of quaternary system PMN-PZN-PZT ceramics is given as follow:

$$x\mathrm{Pb(Mn_{1/3}Nb_{2/3})O_3} - y\mathrm{Pb(Zn_{1/3}Nb_{2/3})O_3} - z\mathrm{Pb(Z_{ru}T_{iv})O_3} + \mathrm{add.w\%}.$$

where $x+y+z = 1$; $0 \le x \le 0.20$, $0 \le y \le 0.35$, $0.50 \le z \le 0.90$, $u = 0.52$, $v = 0.48$, $0 \le w \le 10$. The technical route of low temperature sintering is the transition liquid phase sintering, low eutectic melting additives as initial phase can promote the sintering in the early and middle stages of sintering and as a final phase can form solid solution to enter the lattice and modify the properties of ceramics. The suitable amount doping can be reacted with PbO to form an eutectic phase and advantage to lower sintering temperature. Both benefit of low firing and high performance ceramics can be obtained. The properties of PMN-PZN-PZT are as follow. $K_p = 0.60$, $d_{33} = 300$ pC/N, $\varepsilon_{33}^T/\varepsilon_0 = 1050$, $Q_m = 1500$, $tg\delta \le 50\times10^{-4}$, sintered at about 960°C–1000°C. This kind of piezoelectric ceramics is satisfied used for MPT.

3. Some characteristics of multilayer piezoelectric transformer

The MPT with Rosen type and central driving type has been studied respectively. The properties of MPT with central driving are listed in the Table I.

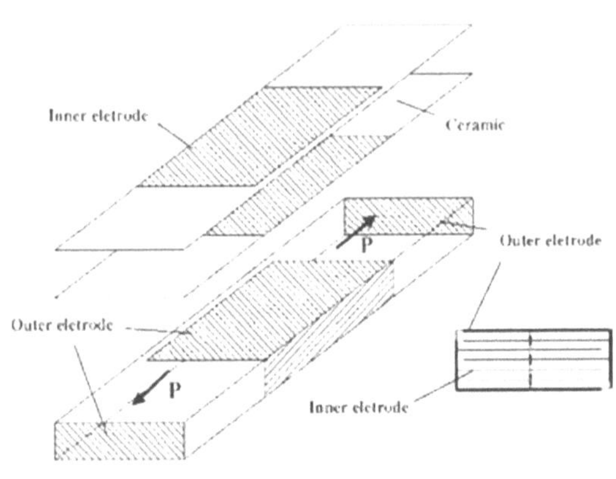

Figure 1 Schematic configuration of MPT.

How to suppress the temperature rise of MPT during operating is one of important technique requirements for practice application, because the transfer efficiency decreases when increasing temperature. Many factors can influence the temperature variation of MPT. The composition of ceramics with low loss, high mechanical quality factor, and low resonance impedance are desired for MPT application. The fabrication processing of MPT is also very important during the sintering of MPT. The densification mismatch behavior between metallic internal electrodes and ceramic layers can induce the residual internal stress and micro-cracks. The induced defects at the interface between electrode and ceramics severely deteriorate the mechanical and electrical properties. Especially, during vibration span, the defects at the interface can generate a dynamic internal stress field and result in the temperature rise. In the present study, the internal electrode (Ag/Pd = 85/15) can be cofired with PMN-PZN-PZT ceramics sintered at about 960–1000°C.

The maximum surface temperature rise is less than 20°C at rated output power. It is indicated that the low sintering temperature MPT with high performance can be obtained. It has been used in back-light power for portable computer. These kinds of MPT and their back-

TABLE I The properties of MPT with central driving

	Specification[a]			
	1503	2505	3006	4509
Resonant frequency (λ/2, kHz)	110	70	54	36
Voltage step-up ratio (load = 100 kΩ)	45	60	70	90
Output power (W)	1.0	3.0	4.0	8.0
Transfer efficiency (%)	≥95	≥95	≥95	≥95
Dimensions (l × w × h) (mm^3)	1.6	2.6	2.6	5.0
Layer thickness (mm) × layer numbers	0.15 × 13	0.20 × 17	0.20 × 17	0.25 × 15
Sintering Temperature (°C)	≤1000	≤1000	≤1000	≤1000

[a]Specification 1503, 2505, 3006, 4509 represent the size of length and width of MPT. Their length and width are 15 mm and 3 mm, 25 mm and 5 mm, 30 mm and 6 mm, 45 mm and 9 mm respectively.

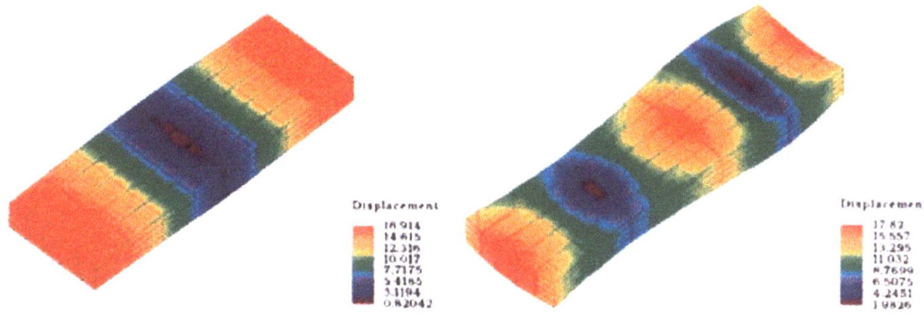

(a) half-wave vibration mode (b) full-wave vibration mode

Figure 2 Finite element analysis for the vibration mode of MPT.

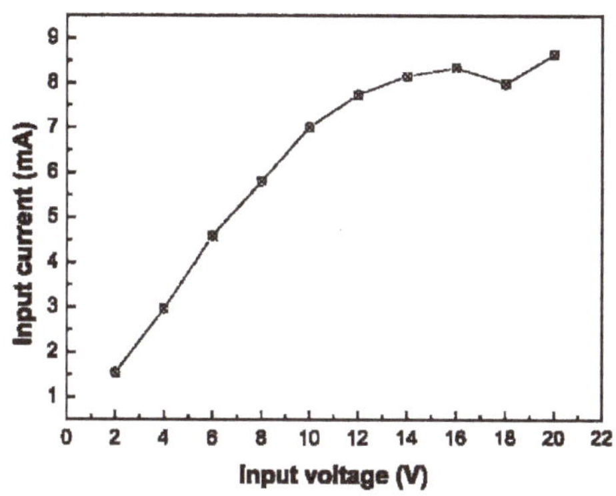

Figure 3 Relationship between input voltage and current for measuring MPT.

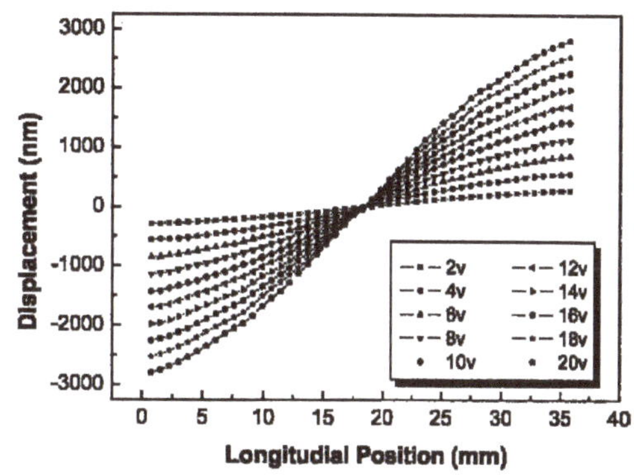

Figure 4 Displacement distribution along the length direction for MPT.

light power and piezo-ionizer have been produced in industrial scale and commercialized by Xi'an Konghong Company in China.

4. Some of resonance characteristics for MPT

The configuration of MPT is shown in Fig. 1. The monolithic multilayer structure contains 17 layers of 0.20 mm in thickness. There is an input electrode strip at the center of the transformer and symmetrical output terminal electrodes at both ends. The mechanical vibration along longitudinal direction can be generated by the electromechanical coupling factors k_{31} and k_{33} under alternating input voltage. For half wave vibration, a nodal line without any vibration appears at its center and such symmetric vibration results in two identical step-up voltages at two output terminals.

Finite Element Analysis (FEA) is a very useful method to simulate the vibration mode of the MPT when subjected to a driving voltage. The symmetric harmonic vibration of the MPT was confirmed by using the FEA method, as shown in Fig. 2. It can be seen that at different harmonic frequency, the MPT vibrates with different vibration mode, the half-wave mode and the full-wave mode in this case. The maximum displacement occurs at the end of MPT. The nodal position located at the central of MPT for half-wave vibration mode.

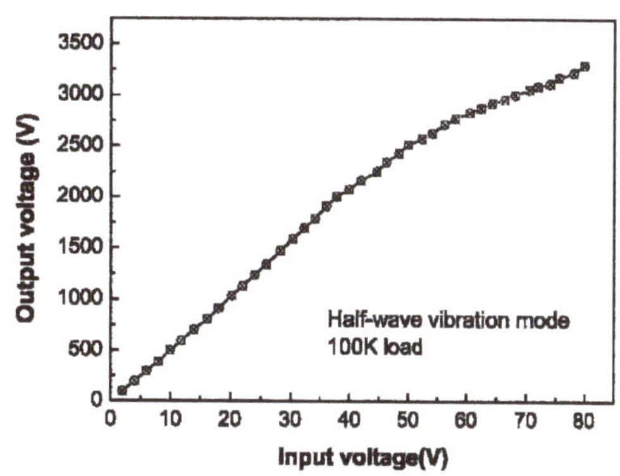

Figure 5 Relationship between input voltage and output voltage for measuring MPT with 100 kΩ load.

Furthermore, the actual vibration modal shapes, including micro vibration displacement, were experimentally

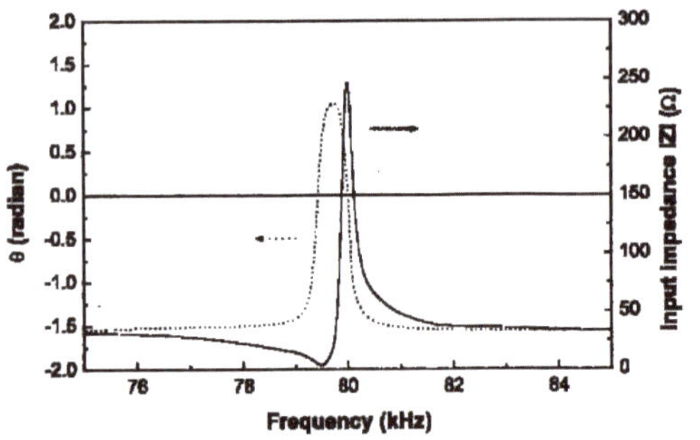

Figure 6 Input impedance and phase difference between input voltage and input current with relation to frequency.

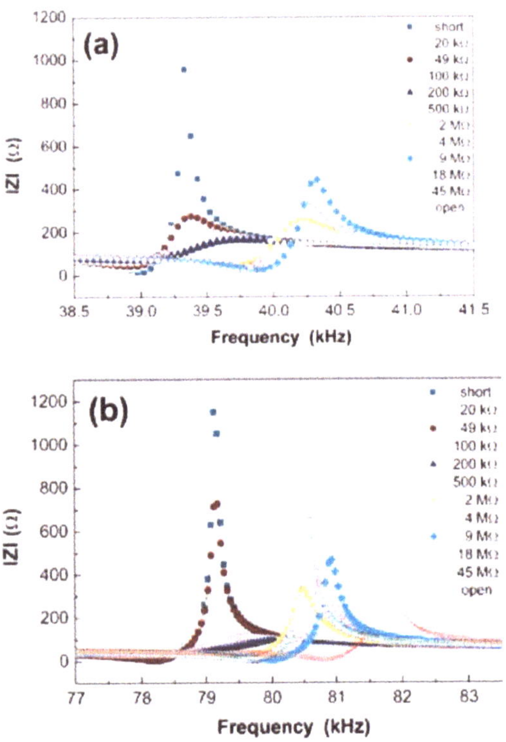

Figure 7 Shifts of input impedance as a function of load at (a) half-wave and (b) full-wave resonant vibration mode, respectively.

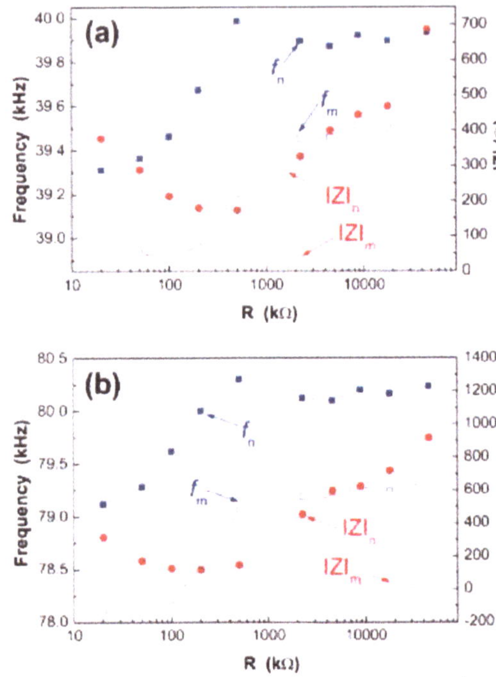

Figure 8 Shifts of frequency and input impedance as a function of load at (a) half-wave and (b) full-wave vibration mode.

measured by using a Laser Doppler Scanning measurement system. Fig. 3 shows the input current value as a function of applied voltage.

The relationship between displacement distribution and input voltage is shown in Fig. 4. It can be seen that the largest displacement occurs at the end of MPT. The displacement increases with increasing of input voltage. Fig. 5 shows the transformation characteristics of the MPT. The output voltage exhibits excellent linear dependence on the input voltage up to 40 V, over which the transformation ratio falls down with the input voltage.

Fig. 6 shows the input impedance and phase difference between input voltage and input current with relation to the frequency. It can be seen that at resonant and antiresonant state, the phase difference is zero and the input impedance reveals pure resistance behavior. Between the two frequencies, the input impedance shows inductance. At off-resonant state, the input impedance shows obvious capacitance characteristics.

The influences of load at the output end of MPT on the input impedance were studied. Fig. 7a and b shows the shifts of input impedance as a function of load at half-wave and full-wave vibration modes respectively. The experimental results show that input impedance depends strongly on loads. When the load increases from short to open circuit, the peak location of input impedance moves toward higher frequency and its magnitude undergoes sharp variation within the two values according to the states of short and open circuit.

Fig. 8a and b shows the variation tendency of input impedance and frequency as a function of load at half-wave and full-wave vibration mode, respectively. The parameters with the subscription "m" refer to values at the local minimal input impedance, while the "n" refers to val-

TABLE II The specification of piezoelectric ceramic transformers

Item	MPT2505	MPT3006	MPT3006	MPT3507	MPT4007	MPT4008	MPT4509
Output power (W)	2.5	3.5	4.0	4.5	5.0	6.5	8.5
Rated output power (W)	2.0	3.0	3.5	4.0	4.5	6.0	8.0
Resonant frequency (kHz)	70 ± 2.0	55 ± 2.0	55 ± 2.0	47 ± 2.0	41 ± 2.0	41 ± 2.0	36.5 ± 2.0
Voltage set-up ratio (at the resonant frequency)	60 ± 10%	65 ± 10%	70 ± 10%	80 ± 10%	67 ± 10%	70 ± 10%	72 ± 10%
Efficiency (at the resonant frequency)	≥ 90%	≥ 90%	≥ 90%	≥ 90%	≥ 90%	≥ 90%	≥ 90%
Input capacitance (nf @ 1 kHz, 1 Vrms)	65 ± 10%	115 ± 10%	135 ± 10%	205 ± 10%	185 ± 10%	190 ± 10%	310 ± 10%
Input voltage (max,Vpp)	28	30	30	30	30	35	35
Input current (max, mArms)	500	600	650	800	700	800	1300
Output voltage (max, Vop)	1500	1700	1800	1900	2000	2300	2600
Output current (max, mArms)	5.5	6.0	7.0	7.3	7.5	8.0	9.2
Temperature rise (max, °C)	25	25	25	25	25	25	25
Operating temperature range (°C)	−10 ∼ 80	−10 ∼ 80	−10 ∼ 80	−10 ∼ 80	−10 ∼ 80	−10 ∼ 80	−10 ∼ 80

ues at local maximal input impedance. The state indicated by "m" refers to the resonant state, while that indicated by "n" refers to the anti-resonant state. An interesting phenomenon is observed about input impedance, $|Z|_m$ varies slightly with loads, while $|Z|_n$ varies with loads strongly, which gives some useful information on the matching impedance of power supply to MPT.

Fig. 9a and b shows the variation of phase difference between input voltage and current as a function of load at half-wave and full-wave vibration modes.

It can be seen that the peaks of phase difference between input voltage and current show symmetric features along the vertical axial-frequency axial. Both the locations of the symmetrical axial and magnitude of the curves change with the load as well as input impedance. The symmetrical axial shifts from low frequency to high frequency as the output circuit changes from short to open, the magnitude of phase difference also shows obviously changes with loads. This phenomenon seems useful to the analysis of operating state of MPT and thus improvement of operating efficiency of MPT.

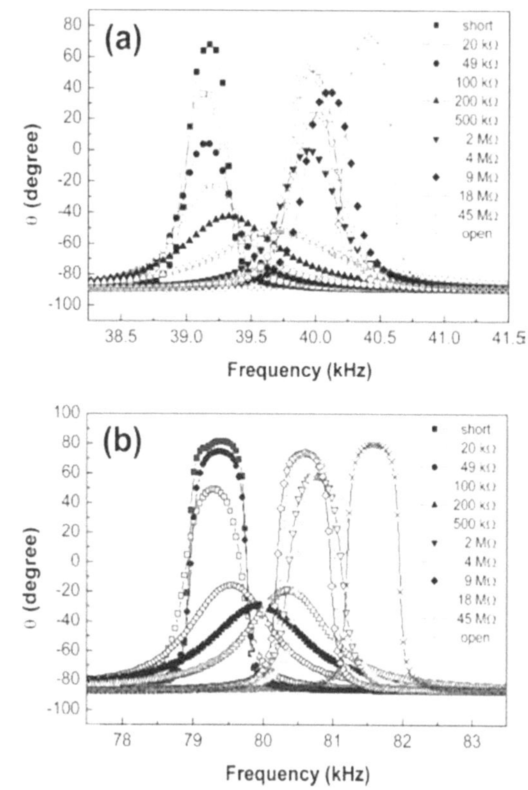

Figure 9 Shifts of phase difference as a function of load at (a) half-wave and (b) full-wave vibration mode, respectively.

The vibration velocity of end part of MPT was measured directly by using an optic fiber interferometer with the type of OFV3000/502. Fig. 10 shows the phase difference between vibration velocity and input voltage decreases with increasing of input voltage for half-wave and full-wave resonance. It can be seen that the phase difference between velocity and input voltage decreases gradually with increasing of input voltage, the phase difference is close to 180° and 0° for open and short circuit respectively.

The low-fired MPT have been industrially produced by Xi'an Konghong Company in China. Table II summaries several kinds of the specification of piezoelectric ceramic transformers.

The multilayer piezoelectric transformers have been used in back light power for LCD. Table III lists the specification of piezo-inverter produced by Konghong Company. The KHI series of piezo-inverter can meet the technical requirements of the wide screen LCD, which include low input voltage and high starting voltage, low power dissipation and high brightness, due to the advantages of high voltage step up ratio, small size, high transferring efficiency, and high safety stability.

TABLE III The specification of piezoelectric inverter

Item	KHI0520	KHI0530	KHI0540	KHI1230	KHI1240	KHI1280	KHI1216	KHI1224	KHI1264
Input voltage (V)	4.5–5.5	4.5–5.5	4.8–5.5	8–18	9–18	10–14	10.8–13	10.8–13	11.4–12
Input current (mA)	340	840	1000	280	360	750	1550	2300	~7500
ON/OFF voltage (V)	1.5–5	1.5–5	1.2–5	1.5–5	1.5–5	1.5–5	1.5–5	1.5–5	1.5–5
Dimming voltage (V)	0–5	0–5	0–5	0–5	0–5	0–3	0–3	0–4	0–3
Output voltage (V)	290	660	730	530	650	650	750	667	1170
Lamp current (mA)	5.0	5.5	6.0	5.0	6.0	6.0	6.0	6.0	4.5
Number of CCFL	1 × 2W	1 × 3W	1 × 4W	1 × 3W	1 × 4W	2 × 4W	4 × 4W	6 × 4W	16 × 4W
Size (mm×mm×mm)	73 × 15.5 × 5.0	73 × 15.5×5.0	73 × 15.5×5.0	110 × 11.7×5.5	95 ×15 × 5.5	137 ×22 × 5.5	160 ×33 × 5.5	180 ×60 × 7.0	320 ×120 × 5.5

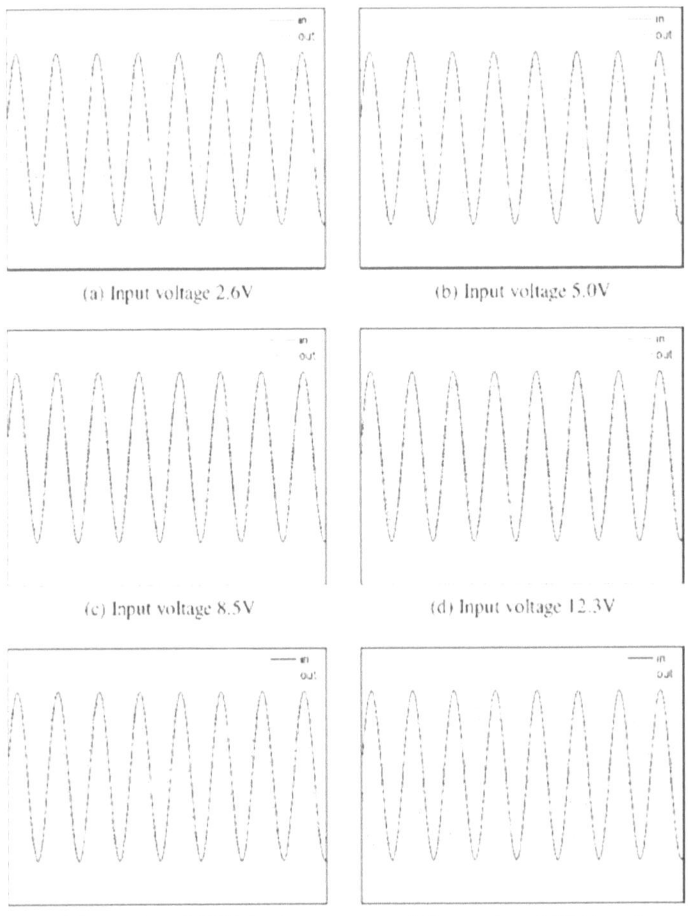

Figure 10 Phase difference between input voltage and end vibration velocity of MPT at full-wave resonance.

5. Summary

PMN-PZN-PZT piezoelectric ceramics doped with suitable eutectic additives, by transient liquid phase sintering, the sintering temperature can be lowered to 960°C–1000°C. This kind of hard type quarternary system piezoelectric ceramics has been used for multilayer piezoelectric ceramic transformer (MPT). The shifts of input impedance and phase difference between input voltage and current as a function of load were investigated. It is found that impedance depends strongly on load. The variation of input impedance and phase difference between input voltage and input current with frequency were also studied. The vibration mode of MPT was tested and analyzed by Laser Doppler Scanning Vibrometer. The phase difference between input voltage and vibration velocity of MPT was studied. All of these investigations on resonant characteristics of MPT are very important and useful for their applications, especially to the design of driving circuit and to the optimum of operating conditions. A series of MPTs has been used for back-light power of LCD and piezo-ionizer etc. It reveals abroad application prospect.

Acknowledgments

Authors would like to thank Mr. Weiti Deng and Wen Gong, and Ms. Enzhu Li for the helpful work on this paper.

References

1. L. T. LI and Z. L. GUI, *Ferroelectrics* **262** (2001) 977.
2. L. T. Li and W. T. Deng, in 37th Electronic Components Conference Proceedings, Boston, May 1987, p. 623.
3. L. T. LI, W. T. DENG, J. H. CHI, Z. L. GUI and X. W. ZHANG, *Ferroelectrics* **110** (1990) 193.
4. L. T. LI, N. X. ZHANG and Z. L. GUI, *Ceramic Transactions* **106** (2000) 373.
5. K. UCHINO, B. KOC, P. LAORATANAKUL and A. VAZQUEZ CARAZO, in "Piezoelectric Materials in Devices", edited by N. Setter (2002) p. 155.
6. J. H. HU, H. L. LI, H. L. W. CHAN and C. L. CHOG, *Sens. Actuator A-Phys.* **88** (2001) 79.
7. J. L. DU, J. H. HU and K. J. TSENG, *Ceram. Int.* **30** (2004) 1797.
8. R. Z. ZUO, L. T. LI, Z. L. GUI, X. B. HU and C. X. JI, *J. Am. Ceram. Soc.* **85** (2002) 787.
9. S. F. WANG, *J. Am. Ceram. Soc.* **76** (1993) 474.

JOURNAL OF MATERIALS SCIENCE 41 (2006) 163–175

A Monte Carlo simulation on domain pattern and ferroelectric behaviors of relaxor ferroelectrics

J.-M. LIU*
Laboratory of Solid State Microstructures, Nanjing University, Nanjing 210093, China
E-mail: liujm@nju.edu.cn

S. T. LAU, H. L. W. CHAN, C. L. CHOY
Department of Applied Physics, The Hong Kong Polytechnic University, Hung Hom, Kowloon, Hong Kong, China

The domain configuration and ferroelectric property of mode relaxor ferroelectrics (RFEs) are investigated by performing a two-dimensional Monte Carlo simulation based on the Ginzburg-Landau theory on ferroelectric phase transitions and the defect model as an approach to the electric dipole configuration in relaxor ferroelectrics. The evolution of domain pattern and domain wall configuration with lattice defect concentration and temperature is simulated, predicting a typical two-phase coexisted microstructure consisting of ferroelectric regions embedded in the matrix of a paraelectric phase. The diffusive ferroelectric transitions in terms of the spontaneous polarization hysteresis and dielectric susceptibility as a function of temperature and defect concentration are successfully revealed by the simulation, demonstrating the applicability of the defect model and the simulation algorithm. A qualitative consistency between the simulated results and the properties of proton-irradiated ferroelectric copolymer is presented.

1. Introduction

The role of defects in ferroelectrics (FEs) has been one of the fundamental issues in physics of ferroelectrics. While most studies related to defects in ferroelectrics deal with domain boundary, twin structure, anti-phase boundary and dislocations etc, a comprehensive knowledge of these defects and their impact on the materials property would be important for materials processing and property optimization [1, 2]. Here, we focus on a special type of crystal defects which are believed to be responsible for the relaxor-like behaviors in some doped ferroelectrics [3–12]. These defects can be either impurity atoms distributed randomly in the lattice or off-center dopant ions which generate the so-called internal random fields or random bonds, or even a frustration of long-range ordering state due to some reasons [9]. Although it remains challenging to identify directly the core configuration and physical behaviors of these defects, extensive studies on relaxor ferroelectrics (RFEs) have provided a sound basis for establishing the essential roles of these defects [7–16]. Therefore, it becomes natural to correlate the existence of lattice defects with the abnormal ferroelectric and dielectric behaviors in RFEs.

It is well known that RFEs exhibit some features not often observed in normal FEs, such as diffusive phase transitions, strong frequency dependence of dielectric susceptibility which corresponds to a broad spectrum of electric-dipole relaxation times, narrow and frequency-dependent hyeteresis in the paraelectric state and weak FE-hysteresis below the FE transition point (Curie point) T_c, in addition to the high dielectric constant and excellent electromechanical performance [3, 4]. It is now believed that these abnormal behaviors are related to the coexistence of two phases in the nano-scale, i.e. well aligned dipolar micro-regions embedded in a matrix of paraelectric (PE) phase upon a decrease of temperature T towards T_c [14–16]. Such a picture of two-phase coexistence was employed in several models to explain the abnormal property of RFEs. The well-documented models include the compositional

*Author to whom all correspondence should be addressed.

0022-2461
DOI: 10.1007/s10853-005-6016-3

inhomogeneity scheme [17], the superparaelectric model [3], and the dipole-glass model [15], which describe the dipole-configuration of RFEs in a framework of standard dipole interactions coupled with an internal random field.

The role of defects mentioned above is considered in the compositional inhomogeneity model in which the type of defects can be impurity atom or dopant ion [12]. For the former, the impurity atoms in the lattice are viewed as disordered (random) static defects coupled locally with the transformational mode which is responsible for a stable dipole. Therefore, the defects may change the magnitude of local dipoles from site to site, by suppressing the dipole moment. For the latter, a model was proposed by Vugmeister *et al.* [9], where a highly polarizable PE host lattice with a displacive dielectric response to electric field E is considered. If this lattice is doped by off-center dopants, a local dipole will appear and the local dopant may occupy one of the crystallographically equivalent off-center sites around the unit cell center, and the resultant dipole moment may align along one of the equivalent vectors. Here it should be pointed out that the off-center displacement would lead to either a suppression or an enhancement of local dipole moment. Thus a consideration of two types of defects in lattice may be necessary: those defects that would suppress local dipoles magnitude (type-I) and those that may suppress or enhance local dipoles in a random manner (type-II). While there is not much direct evidence on the defects associated with the random-field scheme, some interesting experiments on ferroelectric copolymers had been performed recently, where the material was irradiated by ions such that the system was converted from a normal FE phase into two coexisting phases: a normal FE and a RFE phase [18–20]. The irradiated regions may lose their ferroelectricity and can be viewed as containing induced defects.

Recently, a thermodynamic description of the FE lattice with impurity-induced defects was developed by Su *et al.* [12] and Semenovskaya *et al.* [21], who included the effect of local dipole fluctuations induced by impurity ions in the Landau energy. Furthermore, the two-phase coexisted pattern may also depend on mechanisms other than defects, such as the long-range dipole-dipole interaction, gradient energy due to alignment misorientation and long-range elastic interaction [21]. It is necessary to consider these mechanisms in the theoretical approach.

In this paper, we summarize our recent Monte Carlo (MC) simulation on the lattice configuration of electric dipoles and the dielectric property in a FE lattice with local lattice defects [22–24]. The defects coupled into the lattice are identified as two types as mentioned above: one type is those intended to suppress the local dipoles (type-I) and the other supposed to suppress or enhance the dipoles in a random manner (type-II). The effect of both types of defects on the domain configurations and dielectric (ferroelectric) behaviors of RFEs will be simulated. We shall then compare our simulations on the effect of type-I defects with the experimental data on proton-irradiated poly (vinylidene fluoride-trifluoroethylene) 70/30 mol% copolymer (P(VDF-TrFE)) [25]. The reason is clear, since the proton-irradiation results in suppression of electric dipole.

The remaining part of this paper is organized as follows. In Section 2 the thermodynamic model on a defective ferroelectric lattice system with a cubic (square)-tetragonal (rectangle) ferroelectric transition upon temperature decreasing will be presented, followed by the MC-algorithm of simulation. The main results of the simulation and discussion on the significant effect of the two types of defects on the domain configuration and ferroelectric (dielectric) property of the lattice will be given in Section 3 to Section 5, followed by a short conclusion in Section 6.

2. Model and procedure of simulation

The MC simulation is performed on a two-dimensional (2D) $L\times L$ lattice with the periodic boundary conditions applied, where the PE phase takes the square configuration and FE phase the rectangular one. Basically, a three-dimensional lattice with a cubic-tetragonal transition should be employed for a reliable simulation, however, the computational capacity required for such a simulation is extremely big. Moreover, because we do not focus much on the critical phenomena associated with the ferroelectric phase transitions, the effect of finite lattice size may not be significant here to invalidate the main conclusion from the 2D simulation. We once employed a $16\times16\times16$ cubic lattice for a pre-simulation and did not find significant difference of the simulated results (e.g. dielectric susceptibility and ferroelectric property to be defined below) from those we obtained for a 2D lattice of $L\sim64$. In Fig. 1 we present the simulated dielectric susceptibility as a function of temperature T for a 3D lattice with $L = 16$, and two 2D lattices with $L = 64$ and 128, respectively. No significant difference between the several lattices is found in terms of the dielectric behavior. Therefore, all of our simulations reported below will be performed on a 2D lattice. Although the lattice for copolymer P(VDF-TrFE) cannot be viewed as a square in a strict sense, the thermodynamic approach may not be sensitive to the crystal symmetry and details of the lattice structure. Therefore, we ignore the crystallographic difference when we compare the simulated results with the experiments.

2.1. Ginzburg-Landau model of normal ferroelectric lattice

The model starts from the Ginzburg-Landau approach to normal ferroelectric lattice with a square-rectangle transition. For each lattice site, a dipole vector $\boldsymbol{P}$ is imposed with its moment and orientation defined by the energy

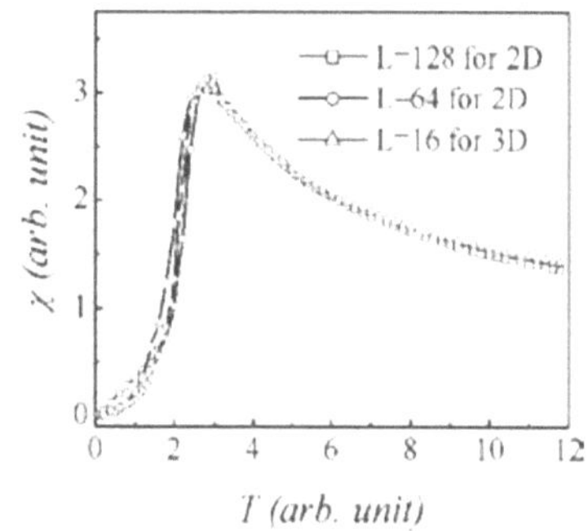

Figure 1 Simulated dielectric susceptibility as a function of temperature T for three lattices (3D lattice with $L = 16$, 2D lattices with $L = 64$ and 128, respectively). $Q = 4$. For details see text.

minimization. We define $\boldsymbol{P} = (P_x(\mathbf{r}), P_y(\mathbf{r}))$ where P_x and P_y are the two components along x-axis and y-axis, respectively. For the energy, we consider the contributions from the Landau double-well potential, the long-range dipole-dipole interaction and gradient energy associated with domain walls. For normal FE crystals the ferroelastic property can not be ignored [26]. However, this effect may not be important for a two-phase coexisted lattice like RFEs, because the actual dipole-ordered regions are nanometers in size and well-separated by the surrounding paraelectric phase whose elastic energy can be ignored.

The Landau double-well potential f_L can be written as [26]:

$$f_L(P_i) = A_1\left(P_x^2 + P_y^2\right) + A_{11}\left(P_x^4 + P_y^4\right) + A_{12}P_x^2P_y^2 + A_{111}\left(P_x^6 + P_y^6\right) \quad (1)$$

where subscript i refers to lattice site i, A_1, A_{11}, A_{12} and A_{111} are the energy coefficients, respectively. For normal FEs, $A_1 > 0$ favors a stable or meta-stable PE phase while a first-order ferroelectric transition will occur if $A_1 < 0$. In fact, one has $A_1 = 1/2\varepsilon_0\chi$ for the PE phase where ε_0 is the vacuum permitivity and χ the dielectric susceptibility of the local lattice. In the present model, the dipole vector is assumed to take one of Q orientations, where Q is treated as a variable. For a square-rectangle transition considered here, $Q = 4$ if one only looks at the Landau energy. However, other orientation states may be chosen too if other dipole interactions are considered. Each dipole is determined by the energy minimization. We shall simulate the effect of Q on the domain configuration, but in most cases we choose Q=4, i.e. four equivalent orientations: [1, 0], [−1, 0], [0, 1] and [0, −1] are allowed, where orientation [1, 0] refers to the horizontal-right x-axis, and [0, 1] refers to the vertical-up y-axis.

If there exists a spatial distribution of the dipoles (refer to either moment or orientation), an additional energy is generated, i.e. the so-called gradients of the polarization field induced by the fluctuation. For a 2D lattice, it can be written as [21, 26]:

$$f_G(P_{i,j}) = \frac{1}{2}G_{11}\left(P_{x,x}^2 + P_{y,y}^2\right) + G_{12}P_{x,x}P_{y,y} + \frac{1}{2}G_{44}(P_{x,y} + P_{y,x})^2 + \frac{1}{2}G'_{44}(P_{x,y} - P_{y,x})^2 \quad (2)$$

where $P_{i,j} = \partial P_i/\partial x_j$. Since parameters G_{11}, G_{12}, G_{44} and G'_{44} are all positive, in most cases this energy term is positive, which means that any dipole fluctuation either in moment or in orientation is not favored. The role of this gradient term is obviously opposite to that of the dipole-dipole interaction to be described below, i.e. ferroelectric ordering is preferred by this term. In addition, the dipole-dipole interaction is long-ranged and it should be considered for an inhomogeneous system. In the SI unit, the energy for site i can be written as [26]:

$$f_{\text{dip}}(P_i) = \frac{1}{8\pi\varepsilon_0\chi}\sum_{\langle j\rangle}\left[\frac{P(r_i)\cdot P(r_j)}{|r_i - r_j|^3} - \frac{3[P(r_i)\cdot(r_i - r_j)][P(r_j)\cdot(r_i - r_j)]}{|r_i - r_j|^5}\right] \quad (3)$$

where $\langle j\rangle$ represents a summation over all sites within a circle region centered at site i with radius R, parameters r_i, r_j, $P(r_i)$ and $P(r_j)$ here should be vectors, r_i and r_j are the coordinates of sites i and j, respectively. In a strict sense, R should be infinite but an effective cut-off at $R = 8$ is taken in our simulation (the as-induced error from this cut-off is less than 2% in terms of the relative accuracy). The total dipole-dipole interaction energy for the whole lattice can be written in the integration form:

$$F_{\text{dip}} = \int d^3r_i f_{\text{dip}}(P_i) = \frac{1}{8\pi\varepsilon_0\chi}\int\int^3 r_i d^3r_j\left[\frac{P(r_i)\cdot P(r_j)}{|r_i - r_j|^3} - \frac{3[P(r_i)\cdot(r_i - r_j)][P(r_j)\cdot(r_i - r_j)]}{|r_i - r_j|^5}\right] \quad (4)$$

It is clearly seen that a minimizing of F_{dip} favors the alignment of dipoles in the head-to-tail form. The energy for an anti-parallel dipole alignment between two neighboring rows is slightly lower than that for a parallel alignment between the two rows. A compatible competition between f_{dip} and f_G is responsible for dominance of either 90°-domain walls or 180°-walls. When f_{dip} is slightly larger than f_G, 180°-walls dominate. Finally, the electrostatic energy induced by an external electric field is:

$$f_E(P_i) = -P_i \cdot E \tag{5}$$

where E is external electric field, and P_i and E are vectors. In our simulation, vector E takes the [1, 0] direction. The Hamiltonian for the system is:

$$H = \sum_{<i>} f_L + f_G + f_{\text{dip}} + f_E \tag{6}$$

where $<i>$ refers to the summation over the whole lattice.

2.2. Effect of defects

Introduction of randomly distributed defects imposes a spatial distribution for the coefficients A_1, A_{11}, A_{12} and A_{111} in the Landau energy Equation 1. Consequently, the other three terms become defect-dependent too. We also assume that only A_1 is affected by the defects and the other three coefficients remain unchanged [21]. That is:

$$\begin{aligned} A_1(r_i) &= A_{10} + b_m \cdot c \\ A_{10} &= \alpha(T - T_0), \quad \alpha > 0 \end{aligned} \tag{7}$$

where $\alpha > 0$ is a materials constant, T is the temperature, A_{10} is the coefficient A_1 in Equation 1, T_0 is the critical temperature for a normal FE crystal with first-order phase transition features, c takes 0 or 1 to represent a perfect site or a defective site, b_m is the coefficient characterizing the influence of defects on T_0 and can be written as:

$$b_m = -\alpha \frac{dT^0(C_0)}{dC_0} \tag{8}$$

where C_0 is the average concentration of defects, which means that L^2C_0 lattice sites are occupied by defects. In the present model, the unit of temperature is scaled by the energy coefficients that appear in Equation 1. The unit of external electric field E_0 is also scaled by the energy coefficients, considering that Equation 5 has the unit of energy. Therefore, the parameters T, P and E do not have units. A quantitative comparison between the present calculation and experimental data is not allowed in the present work.

For a detailed discussion on the parameter b_m, readers may refer to Ref. [21]. It is assumed that both the sign and magnitude of b_m may be defect-dependent and vary from site to site. Take a defective site i as an example to illustrate how to characterize the role of the defect. If b_m is positive, one expects $A_1(r_i)$ at local site i may be positive, which implies a local stable PE state rather than stable FE state, i.e. the defect will suppress the appearance of a local dipole. A FE state is preferred at site i if $b_m<0$, i.e. the local dipole is enhanced by the defect. The higher the value of $|b_m|$, the more significant the effect of the defect at site i. In the present simulation, the magnitude of b_m is randomly taken within $[0.5b_M, b_M]$ where b_M is the given maximum value of b_m. Here, the two types of defects to be considered in the simulation are characterized by the parameter b_m. For type-I defects, $b_m > 0$ for all defects, i.e. all defects will have a role of suppressing the local dipole. For type-II defects, we define another parameter C_p. $C_pC_0L^2$ sites with defects of $b_m<0$ (enhancing the local dipoles) are imposed randomly to the lattice and $(1-C_p)C_0L^2$ sites with defects of $b_m > 0$ (suppressing the local dipoles) are imposed randomly to the lattice. Of course, one may consider the third case where all defects can enhance the local dipole at the defective sites. However, it is expected that this type of defects plays a role opposite to that of the type-I defects. Thus, this case will not be considered here.

2.3. Monte Carlo simulation

Given a set of system parameters, we first simulate the equilibrium domain configuration of the lattice by employing the Metropolis MC algorithm. Subsequently, we present a discussion on the simulation algorithm of dielectric susceptibility under a weak external electric field. For a lattice, each site is assigned a dipole with moment P being chosen in the range of 0–1.0, and orientation being one of the Q states, respectively. Also, a defect is attached to a site and the probability is determined by C_0. A random number R_1 is generated and a defect is attached to this site if $R_1<C_0$, and is not attached otherwise. Here it should be mentioned that the actual defect concentration in the lattice is slightly smaller than C_0 since the Monte Carlo sequence is statistical. However, since $L = 64$ is large enough the as-induced error in the defect concentration from C_0 is negligible. The simulation begins at an extremely high $T = 14.0$ at which no freezing effect occurs within the period of simulation (we choose $T_0 = 4.0$). For a site i chosen at random, f_L, f_G, f_{dip} and f_E are calculated, respectively, and then the value of H is obtained. Now, this site is assigned another dipole with its moment and orientation taken randomly. Subsequently, H is calculated again to compare with the value of H in the original state, and the difference between them is assumed to be ΔH. A probability p is calculated by the Metropolis algorithm:

$$p = \exp\left(-\frac{\Delta H}{kT}\right) \tag{9}$$

where k is the Boltzmann constant. A second random number R_2 is generated and compared with p. If $R_2<p$, the assumed event above is approved and this dipole at site i is imposed, otherwise, the original state remains unchanged. Then one cycle of simulation is completed and a new cycle is initiated until a given number of cycles has been completed. The time of simulation is scaled by the Monte

Carlo step (*mcs*) and one *mcs* represents $L\times L$ cycles. In our simulation, at each temperature, the initial 600mcs runs are done and then the configuration averaging is performed over the subsequent 2500mcs. Note here that for relaxor systems the short time Monte Carlo simulation can give sufficiently accurate agreement with experiments [8]. The data presented below represent an averaging over four runs with different seeds for random number generator of the initial lattice and defect distribution.

Besides the lattice average polarization and the energy terms as shown in Equations 1–4, one can also evaluate the dielectric susceptibility. The susceptibility is evaluated by applying a weak time-varying external field E of amplitude E_0 and frequency ω: $E = E_0 sin(2\pi\omega\cdot t)$. Since E is time-dependent, the kinetic MC algorithm is performed by the sequence of dipole-exchange between two nearest-neighboring sites with the dipole magnitude determined by the local environment instead of dipole flip sequence employed in the Metropolis algorithm. Such an exchange does not imply any exchange of the defect state. It means the defect distribution in the lattice remains unchanged during the simulation. Under an external ac-electric field E, the lattice dielectric susceptibility χ can be written as [8]:

$$\chi' = \frac{C}{NT}\left\langle \sum_i^N \frac{1}{1+(\omega\cdot\tau/\omega_0)^2} \right\rangle$$

$$\chi'' = \frac{C}{NT}\left\langle \sum_i^N \frac{\omega\cdot\tau/\omega_0}{1+(\omega\cdot\tau/\omega_0)^2} \right\rangle \quad (10)$$

where <> represents the configuration averaging, χ' and χ'' are the real and imaginary parts of χ (we focus solely on the real part), ω_0 is the polariton frequency which is a material constant, τ is the averaged time for dipole-exchange between any nearest-neighboring dipole-pair, which is scaled by ω_0, $N = L^2$ and C is a temperature-independent constant. In our simulation here, $\omega_0 = 1$ is assumed for simplification. For any dipole-exchange event, the system energy difference ΔH after and before the assumed exchange is calculated and the probability p approving such an exchange is determined by:

$$p = \exp\left(-\frac{\Delta H}{kT}\right) \quad (11)$$

The simulation is performed by the following procedure. The equilibrium lattice configuration is taken as the input lattice. For a site i, one of its four nearest-neighbors, site j, chosen in random, is paired with site i to perform the dipole-exchange. However, the magnitude of each dipole is re-valued after the assumed exchange. The value of ΔH of this assumed exchange is calculated and the probability for such an exchange is evaluated using Equation 11.

TABLE I System parameters used in the simulation

Parameter	Value	Parameter	Value	Parameter	Value
T_0	4.0	α	1.0	A_{11}	–0.5
A_{12}	9.0	A_{111}	0.8	G_{11}	1.0
G_{14}	0.2	G_{44}	1.0	L	64
b_M	6.0	C_0	0~1.0	R	8

A random number is generated and compared with this probability to decide whether such an assumed exchange is approved or not. This process is repeated until a given number of simulation steps (*mcs*) is reached. For a site i, if m successful dipole-exchange events are counted in M mcs of simulation, as long as M is big enough, time $\tau = M/m$ because τ is the averaged time for dipole-exchange between any nearest-neighboring dipole-pair. Then the dielectric susceptibility is evaluated using Equation 10.

2.4. Choice of system parameters

In the simulation, b_m and C_0 are treated as variables. The other parameters are chosen and the dimensionless normalization of them is done following the work of Hu *et al.* on $BaTiO_3$ system [26]. Such a choice is somewhat arbitrary since we are not focusing on any realistic system in a quantitative sense. These parameters are given in Table I. In the following simulation, the normal ferroelectric lattice is named as the normal lattice, while the lattice with type-I defects or type-II defects is called the defective lattice.

3. Lattice dipole configuration of normal lattice

3.1. Lattice configuration at $Q = 4$

We first look at the simulated dipole configuration of the normal lattice ($C_0 = 0.0$). Fig. 2 shows the simulated patterns at several temperatures for $Q = 4$, where the length and direction of arrows represent the moment and orientation of dipoles. For a clarification of the dipole alignment, only part of each lattice is shown. At high $T = 12.0$ and 6.0, the moment of all dipoles is very small and their alignment is disordered, a typical PE configuration. As T becomes close to T_0 ($T = 4.0$), the dipole moment is still small and no long-range dipole order is found either, although a small dipole-ordered region is observed on the top-right corner.

Once T is below T_0 ($T = 3.0$), the ferroelectric phase transition occurs and the disordered dipole alignment evolves into a long-range ordered structure. A clear ferroelectric multi-domain configuration is formed with the well-predicted head-to-tail dipole alignment and preferred 90° domain walls. Those dipoles on the domain walls are still small in moment and their alignment remains partially disordered. At a low T ($T = 1.0$), the degree of disordering on the walls is significantly suppressed and an almost perfect multi-domain lattice is observed.

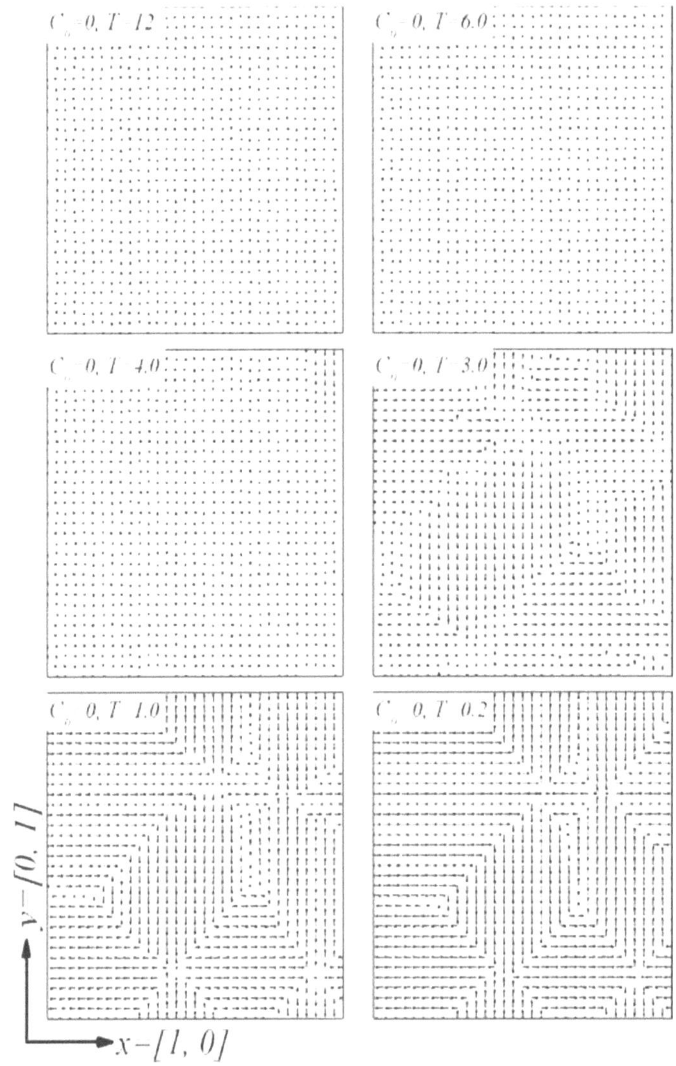

Figure 2 Simulated dipole configuration of normal ferroelectric lattice at different temperatures T ($Q = 4$). $C_0 = 0$.

As for the domain configuration, one may find that most of the domains are 90°-type rather than 180°-type. However, in the dipole configurations shown above, the 180°-domain walls are still observed. One reason for less 180°-domain walls may be due to the energy landscape in which the 90°-walls are preferred than the 180°-walls in the present model. In fact, if smaller G_{11}, G_{12}, G_{44} and G'_{44} are chosen so that a smaller f_G is obtained, i.e. lower domain wall gradient energy is imposed, many more 180°-walls than 90°-walls are observable.

3.2. Lattice configuration at $Q > 4$

When the allowed number of dipole orientations $Q > 4$, the lattice configuration is also simulated, and the simulated dipole alignment patterns at $T = 0.5$ are presented in Fig. 3 ($Q = 4, 8, 16$). For all cases, the lattice is completely disordered at $T > T_0$, i.e. the lattice is in the paraelectric state and all the allowed states of dipole orientation are chosen (patterns not shown here). While almost the same domain pattern is shown for lattices of different Q, the dipole alignment on the 90°-walls is Q-dependent. At $Q = 4$, the 90°-wall is almost perfect with the head-to-tail dipoles aligning perpendicular to each other. For $Q = 8$, most dipoles on the walls have [11] orientations instead of [1,0] or [0,1] orientations. We name these alignment states the distorted states while the four ±[1,0] and ±[0,1] states are viewed as normal states. At $Q = 16$, almost all dipoles on the walls deviate from the [1,0] and [0,1] orientations and prefer the distorted states which have a higher f_L. Therefore, in the Ginzburg-Landau model the four ±[1,0] and ±[0,1]

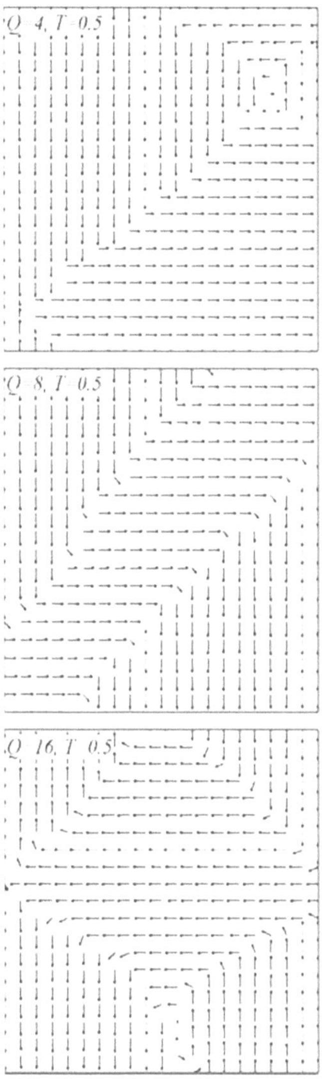

Figure 3 Simulated dipole configuration of normal ferroelectric lattices with different values of Q at $T = 0.5$. $C_0 = 0$.

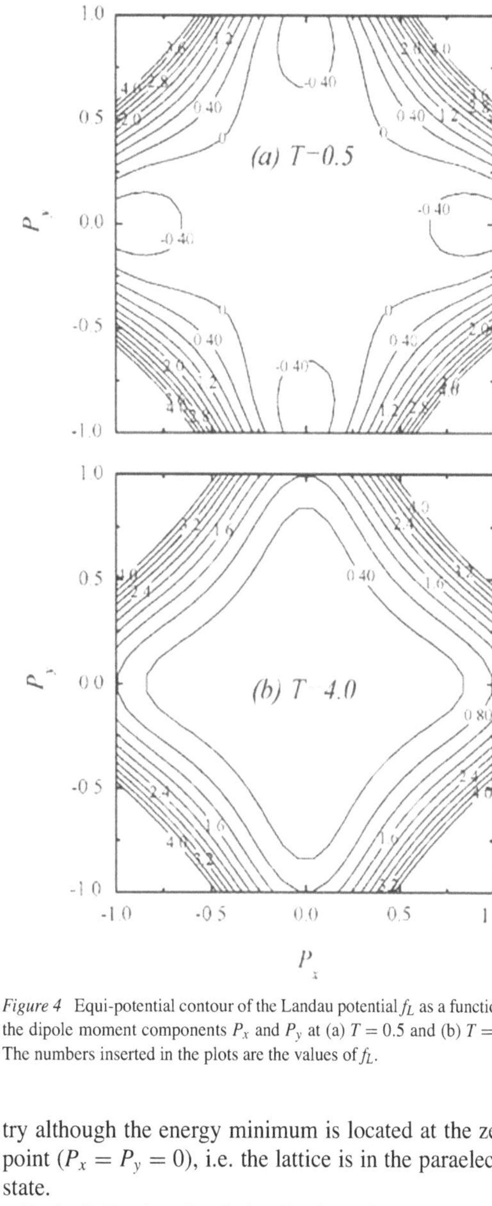

Figure 4 Equi-potential contour of the Landau potential f_L as a function of the dipole moment components P_x and P_y at (a) $T = 0.5$ and (b) $T = 4.0$. The numbers inserted in the plots are the values of f_L.

orientations ($Q = 4$) may not be the unique choices for the dipoles.

As shown above, all dipoles except those on the walls still take one of the four $\pm[1,0]$ and $\pm[0,1]$ orientations no matter how large the value of Q is in the simulation. This behavior can be explained by the symmetry of the Landau potential Equation 1. With the parameters given in Table I, we calculate the Landau potential F_{ld} as a function of (P_x, P_y) at $T = 0.5$ and $T = 5.0$, as shown in Fig. 4a and b, respectively. The square symmetry of $F_{ld}(P_x, P_y)$ at low T, characterized by the four symmetric energy minimals, is clearly shown, which is independent of Q. The result implies that the preferred dipole orientation states are $\pm[0,1]$ and $\pm[1,0]$ irrespective of the value of Q. At high T, $F_{ld}(P_x, P_y)$ still shows the square symmetry although the energy minimum is located at the zero-point ($P_x = P_y = 0$), i.e. the lattice is in the paraelectric state.

In the following simulation for the defective lattice, we fix $Q = 4$ and the effect of varying Q is no longer taken into account.

4. Dipole configuration in defective lattices

4.1. Dipole configuration in a lattice with type-I defects

We look at the dipole alignment in the lattice at different T for $C_0 = 0.5$ and $C_p = 0.0$ (with type-I defects), as shown in Fig. 5. The type-I defects intrinsically suppress the local dipoles. At high T, the lattice is obviously in the PE state and no long-range ordered dipole alignment is seen. With decreasing T, the lattice becomes inhomogeneous

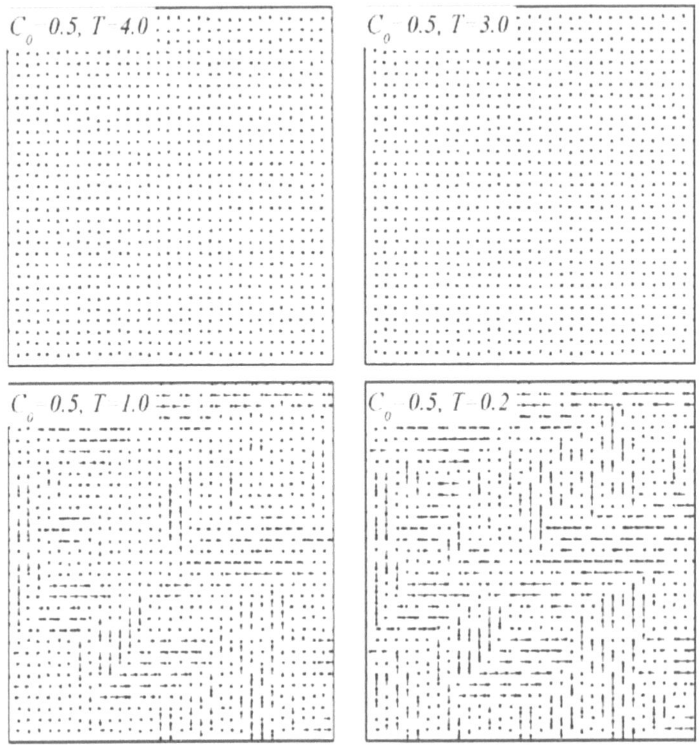

Figure 5 Simulated dipole configuration of a ferroelectric lattice with type-I defects at different temperatures T ($C_0 = 0.5$, $C_p = 0.0$, $Q = 4$). The values of T are inserted in figures.

and some FE ordered regions are observed. With further decreasing of T, these FE ordered regions grow in size and new FE regions appear. At low T, the lattice has a two-phase coexisted microstructure consisting of FE phase embedded in the matrix of PE phase, a typical picture for RFEs.

Given a temperature $T = 1.0$, the lattice dipole configuration with increasing C_0 is shown in Fig. 6. For a small C_0 ($C_0 = 0.2$), the lattice inhomogeneity is already remarkable and the decrease in moment (depolarization effect) of the dipoles along the domain walls becomes significant although the defects are randomly distributed. These regions can be viewed as the non-ferroelectric phase or simply the PE phase. With increasing C_0, the FE regions continue to shrink both in size and in volume fraction. It is seen that only very small dipole clusters exist at $C_0 = 0.5$. As $C_0 = 0.8$, the FE phase in the lattice nearly disappears and the whole lattice becomes paraelectric.

4.2. Dipole configuration in a lattice type-II defects

When type-II defects are introduced into the normal lattice, one sees a dipole configuration quite different from that of the type-I defects. The simulated dipole alignment patterns for $C_0 = 0.5$ ($C_p = 0.5$) at different temperatures are shown in Fig. 7. We indeed find that there are some small-sized areas in which the dipole alignment is ordered at $T = 6.0$, i.e. some local clusters of ordered dipoles form at $T > T_0$. At $T = 4.0$, this clustering tendency becomes more significant. The number and size of the clusters increase with decreasing T. On the other hand, at T just below T_0, we do not see a perfect long-range ordered dipole configuration, while the lattice still consists of areas of ordered dipoles embedded in a matrix of paraelectric phase. With further decreasing of T, a gradual growth and coalescence of these ordered clusters is observed, and the simulated lattice at very low temperature has a configuration approaching that of a normal ferroelectric.

Obviously, the main configuration features described above are similar to those of RFEs in which nano-polar clusters are embedded in a PE matrix over a wide range of temperature both above and below T_0. The feature of diffusive phase transition is clearly reproduced, indicating that the present model works well in describing the microstructure of RFEs. It should be mentioned that in Fig. 7 the long-range ordered configuration is already well developed as T falls to $T = 1.0$, at which a well-defined domain pattern can be identified although there still are some small dipole-disordered zones inside the domains. This configuration can be viewed as the frozen one, which

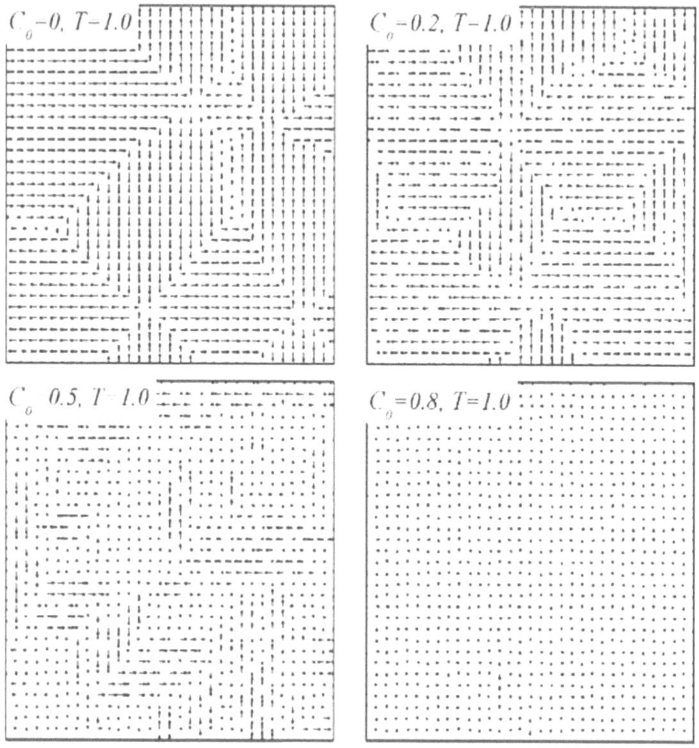

Figure 6 Simulated dipole configuration of a ferroelectric lattice with type-I defects at different defect concentrations C_0 ($T = 1.0$, $C_p = 0.0$, $Q = 4$). The values of C_0 are inserted in the figures.

reflects somehow the freezing behavior of relaxor ferroelectrics with decreasing temperature [21]. However, this freezing behavior is C_0-dependent and the long-range ordered domains disappear when C_0 is high. We simulate the evolution of dipole configuration with different C_0, as shown in Fig. 8 ($T = 3.0$). Multi-domain configuration is clearly seen at $C_0 = 0.0$ and 0.2, but is no longer visible at $C_0 = 0.5$ and 0.8. Consequently, when C_0 becomes larger, the ferroelectric transition is not completed unless T is lower. At any given temperature, the larger the value of C_0, the smaller the ordered clusters.

5. Ferroelectric and dielectric behaviors of defective lattice

5.1. Ferroelectric behaviors

We also study the ferroelectric behaviors of the lattices with the two types of defects, as a function of T and E. The external field was fixed at $\omega = 0.002$ mcs^{-1} and $E_0 = 4.0$ for the simulations in this section. For the lattice with type-I defects, the simulated P-E hysteresis loops at $T = 3.0$ for different C_0 are presented in Fig. 9a. It is seen that the loop shrinks significantly along the P- and E-axis as C_0 increases. At $C_0 = 0.8$ and 1.0, the hysteresis loop becomes very thin, a typical feature of RFEs at a temperature slightly below T_0. Because the loop area A represents the energy dissipated during one cycle of domain reversal driven by the ac-electric field, it can be used to scale the long-range correlation of dipoles in the lattice. In Fig. 9b, the loop area A as a function of T at various C_0 is shown. Given the value of C_0, A increases rapidly with decreasing T, while for a given T, A decreases with increasing C_0. These results are consistent with the behaviors of RFEs.

Nevertheless, for lattices with type-II defects, the simulated ferroelectric behaviors are very different, as shown in Fig. 10a where the P-E hysteresis loops for different C_0 at $T = 3.0$ are given, and in Fig. 10b where the loop area A for different C_0 as a function of T is plotted. As C_0 ($C_p = 0.5$) varies over a broad range, the simulated hysteresis loop does not change much but only shrinks slightly along the P- and E-axis. Even at $C_0 = 1.0$, the lattice polarization P still has a large value. In Fig. 10b, the loop area for $C_0 > 0$ is slightly larger than that for the normal FE lattice at $T > T_0$, although it is slightly smaller than that for the normal FE lattice at $T < T_0$.

The significant difference in ferroelectric behaviors between the lattices with the two different types of defects originates from the fact that, for the lattice with type-II defects, half of the defects enhance the local dipole moment, while for the lattice with type-I defects all defects suppress the local dipole moments. Therefore, for the

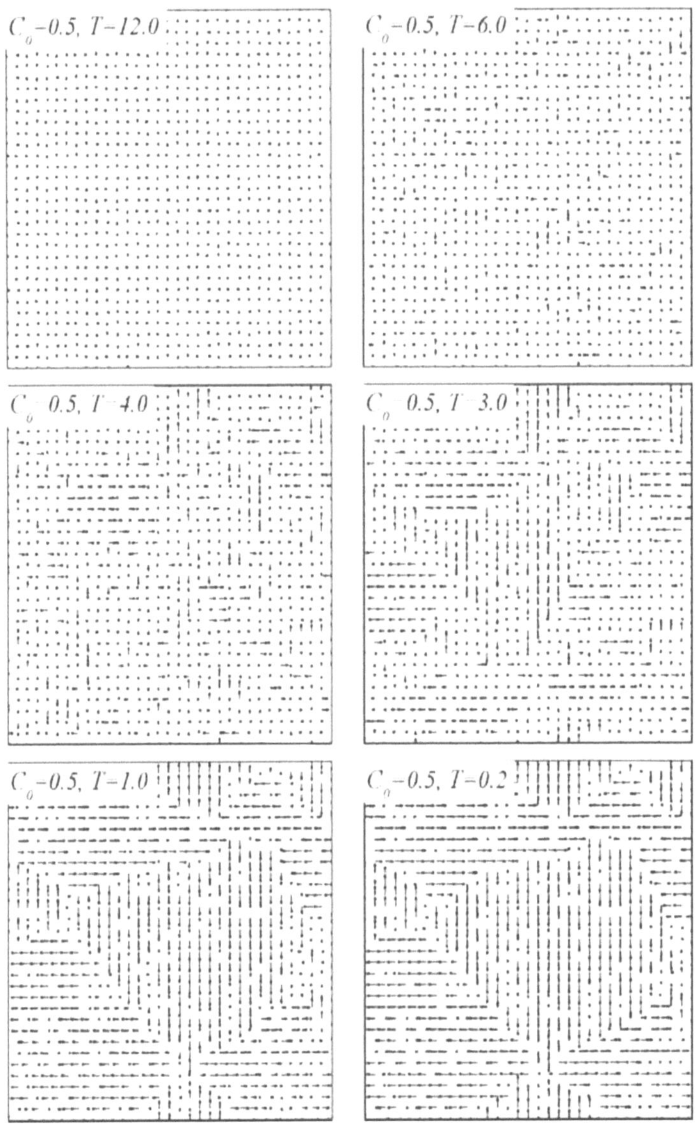

Figure 7 Simulated dipole configuration of a ferroelectric lattice with type-II defects at different temperatures T ($C_0 = 0.5$, $C_p = 0.5$, $Q = 4$). The values of T are inserted in figures.

lattice with type-II defects, some defective sites exhibit even larger dipole moments than those of the normal FE lattice sites at $T > T_0$, although the other sites exhibit smaller dipole moments, as shown in Fig. 8. These results indicate that most RFEs studied experimentally are doped with type-I defects which intrinsically suppress the local dipole moments. This argument will gain further support below in the discussion of the dielectric behavior.

5.2. Dielectric behaviors

The dielectric susceptibility χ as a function of T for different C_0 under a given E ($\omega = 0.002mcs^{-1}$, $E_0 = 4.0$) for lattices with the two types of defects is evaluated, and the results are plotted in Fig. 11a (lattice with type-I defects) and 11b (lattice with type-II defects, $C_p = 0.5$), respectively. It is seen that the introduction of defects affects significantly the dielectric property of the system. At $C_0 = 0$ (normal FE), a Curie-Weiss type single-peaked χ-T relation is generated with $T_c \sim 4.0$. For the lattice with type-I defects, as C_0 increases, the single-peaked χ-T curve shifts toward the low-T side and also downwards slightly. In addition, the normal FE lattice shows a sharp decrease in χ at a temperature just below T_c, but this feature is weakened for the lattice with defects, i.e. the FE transition becomes diffused with increasing C_0. For the

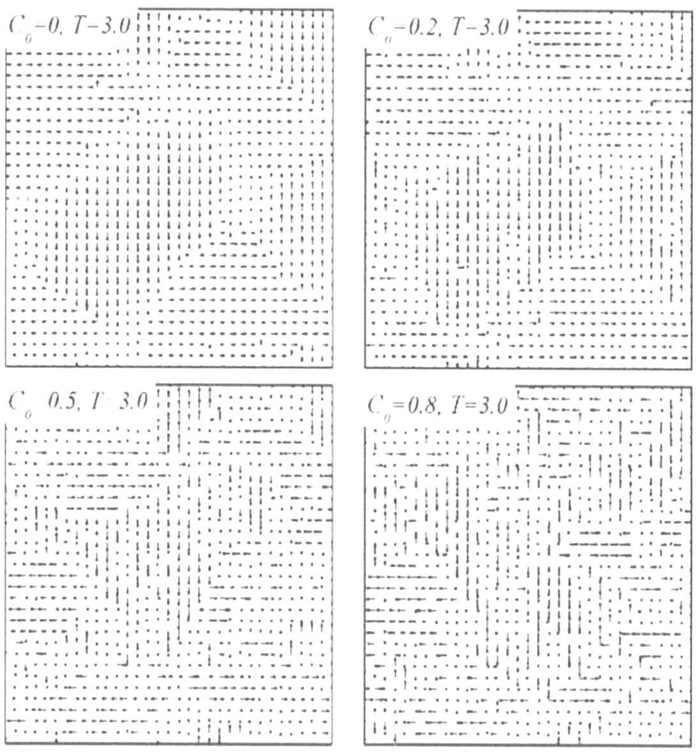

Figure 8 Simulated dipole configuration of a ferroelectric lattice with type-II defects at different defect concentrations C_0 ($T = 1.0$, $C_p = 0.5$, $Q = 4$). The values of C_0 are inserted in the figures.

lattice with type-II defects, with increasing C_0, one sees a remarkable broadening of the dielectric peak around the transition point and the peak position shifts slightly toward the high-T side. Over the high temperature range ($T>T_0$), χ' decreases slightly with increasing C_0, while over the low T range ($T<T_0$) it increases slightly with C_0. When C_0 is extremely large ($C_0 = 1.0$), which means all lattice sites are occupied by the defects, χ' does not exhibit a peak around the transition point and it decreases monotonously with decreasing T. This behavior has never been observed experimentally and its significance is in doubt. Therefore, it is believed that most RFEs under conventional investigations are doped with type-I defects. Finally, the dielectric behaviors for the two-types of defective lattices can be explained qualitatively by considering the energy terms as a function of defect concentration and details of the explanation can be found elsewhere [22, 23].

5.3. Comparison with experiments

We have employed the defect model to simulate the dipole configuration for lattices with two types of randomly distributed defects. A qualitative comparison of these simulated behaviors with experimental data would help provide a justification for the defect model. It has been repeatedly verified that tremendous variations in microstructure and physical property of a variety of ferroelectric polymers may be generated by irradiation with high energy electrons or protons [18–20, 25]. Several experiments revealed that the high-energy particles injected into the copolymers convert the single FE-phase into a two-phase coexisted microstructure with a FE phase and a PE phase. In fact, one may argue that the high energy particle irradiation introduces randomly distributed point-like defects into the sample and disrupts the stability of the FE phase. The microstructural details are determined by the energy level and the dose of the particles. At a given energy level, the defect concentration C_0 is directly related to the dose of irradiation.

The above argument provides justification for the application of the present model to explain the results of the irradiation experiments. While several careful experiments on the effect of irradiation had been performed, in this section, we compare our model simulation with our earlier experiments on the dielectric susceptibility of proton-irradiated poly (vinylidene fluoride-trifluoroethylene) 70/30 mol% copolymer (P(VDF-TrFE)). For details of the experiments, please refer to our earlier report [25].

Fig. 11c shows the measured dielectric constant at 1MHz as a function of T in a cooling run for several samples with the same initial state but irradiated at

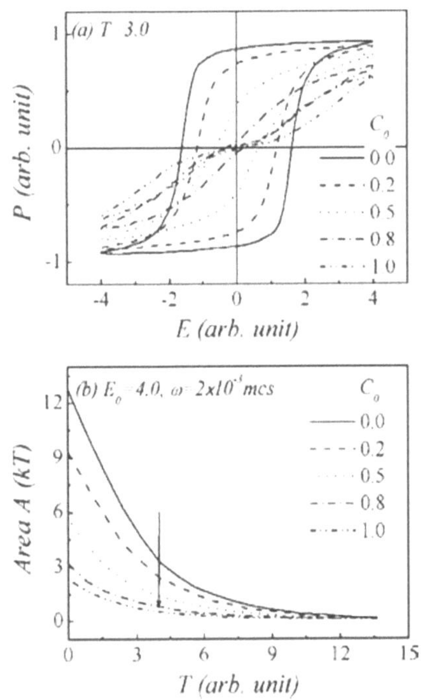

Figure 9 (a) Simulated ferroelectric hysteresis loops for a lattice with type-I defects at different defect concentrations C_0, (b) loop area A at different defect concentrations C_0 as a function of temperature T. $E_0 = 4.0$ and $\omega = 0.002$ mcs^{-1}.

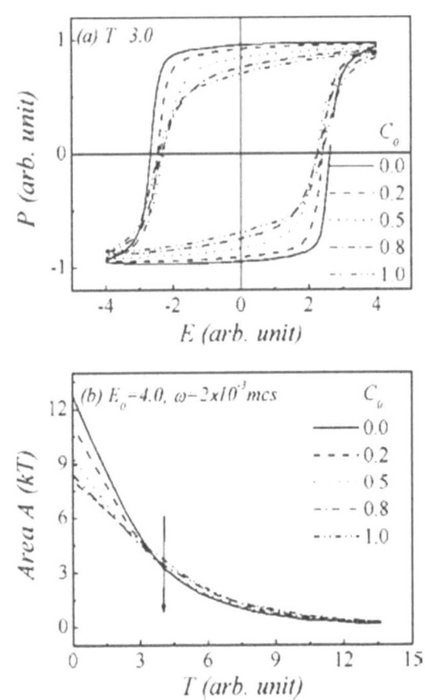

Figure 10 (a) Simulated ferroelectric hysteresis loops for a lattice with type-II defects at different defect concentrations C_0, (b) loop area A at different defect concentration C_0 as a function of temperature T. $C_p = 0.5$, $E_0 = 4.0$ and $\omega = 0.002$ mcs^{-1}.

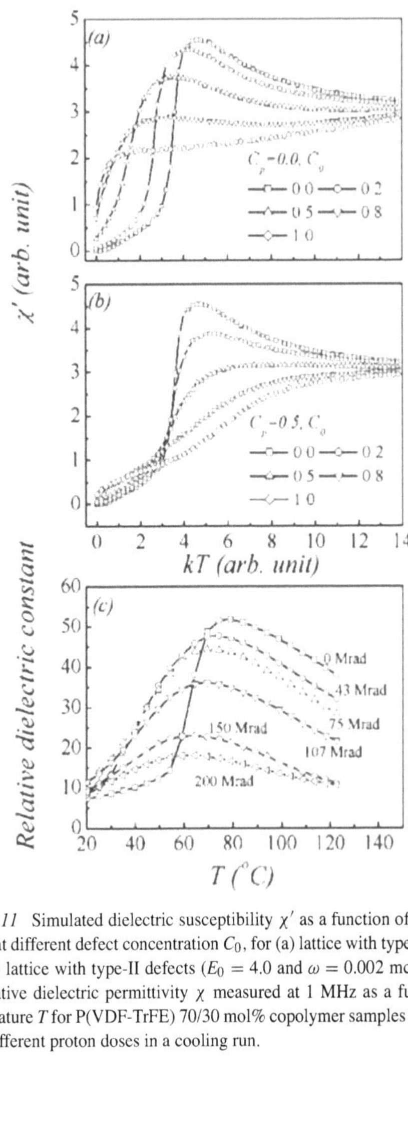

Figure 11 Simulated dielectric susceptibility χ' as a function of temperature T at different defect concentration C_0, for (a) lattice with type-I defects and (b) lattice with type-II defects ($E_0 = 4.0$ and $\omega = 0.002$ mcs^{-1}), and (c) relative dielectric permittivity χ measured at 1 MHz as a function of temperature T for P(VDF-TrFE) 70/30 mol% copolymer samples irradiated with different proton doses in a cooling run.

different dose levels. While the non-irradiated sample shows the typical first-order FE phase transitions at 70°C, the irradiated samples exhibit a broader transition peak and the peak height becomes smaller and the peak position shifts to a lower T, as the irradiation dose increases. These features are well reproduced in our simulations on the lattice with type-I defects, as shown in Fig. 11a. In fact, hysteresis measurements indicated that a relaxor-like two-phase microstructure is formed in irradiated ferroelectric copolymers, and the measured hysteresis loops are quite similar to the simulated ones shown in Fig. 9. What should be mentioned here is that the shift of the peak position becomes quite small when the dose is higher than 150Mrad and this is not consistent with our simulations where a significant shift continues at a very high defect concentration ($C_0 > 0.6$).

6. Conclusions

In conclusion, we have presented a Monte Carlo simulation on the dielectric and ferroelectric properties of ferroelectric lattices with two-types of randomly distributed point-like defects and compared the simulated results with the properties of proton-irradiated copolymer (P(VDF-TrFE)). The algorithm of simulation is based on the Ginzburg-Landau theory for first-order ferroelectric phase transition with the inclusion of the contributions from dipole-dipole interaction, gradient energy and electrostatic energy. The simulation has revealed that the introduction of the two types of lattice defects results in an evolution of the dipole configuration from a normal multi-domain ferroelectric lattice to a relaxor-like two-phase coexisted microstructure consisting of ferroelectric regions embedded in the matrix of a paraelectric phase. The dielectric susceptibility as a function of defect concentration has been simulated and the simulated results are very similar to those observed for relaxor ferroelectrics.

Acknowledgment

This work is supported by the 973 Project of China (2002CB613303) and NSFC Project of China (10021001, 50332020, 10474039), the Centre for Smart Materials of the Hong Kong Polytechnic University and the Hong Kong Research Grants Council (PolyU 5147/02E).

References

1. M. E. LINES and A. M. GLASSS, "Principles and applications of ferroelectrics and related materials" (Gordon and Breach, New York, 1977).
2. I. S. ZHELUDEV, in "Solid State Physics," edited by H. Ehrenreich, F. Seitz and D. Turnbull (Academic Press, New York, 1971) Vol. 26, p. 429.
3. L. E. CROSS, *Ferroelectrics* **76** (1987) 241.
4. A. P. LEVANYUK and A. S. SIGOV, *Defects and structural phase transitions* (Gordon and Breach, New York, 1988).
5. N. ICHINOSE, *Ferroelectrics* **203** (1997) 187.
6. Q. M. ZHANG, J. ZHAO, T. R. SHROUT and L. E. CROSS, *J. Mater. Res.* **12** (1997) 1777.
7. C. H. PARK and D. J. CHADI, *Phys. Rev. B.* **57** (1998) 13961.
8. Z. WU, W. DUAN, Y. WANG, B. L. GU and X. W. ZHANG, *ibid.* **67** (2003) 052101.
9. B. E. VUGMEISTER and M. D. GLINCHUK, *Rev. Mod. Phys.* **62** (1990) 993.
10. E. COURTENS, *Phys. Rev. Lett.* **52** (1984) 69.
11. A. K. TAGANTSEV, *ibid.* **72** (1994) 1100.
12. C. C. SU, B. VUGMEISTER and A. G. KHACHATURYAN, *J. Appl. Phys.* **90** (2001) 6345.
13. R. FISCH, *Phys. Rev. B.* **67** (2003) 094110.
14. D. VIEHLAND, S. J. JANG and L. E. CROSS, *J. Appl. Phys.* **68** (1990) 2916.
15. C. RANDALL, D. J. BARBER, R. W. WHATMORE and P. GROVES, *J. Mater. Sci.* **21** (1987) 4456.
16. X. H. DAI, Z. XU and D. VIEHLAND, *Philos. Mag. B.* **70** (1994) 33.
17. G. Smolenski and A. Agranovska, *Sov. Phys. Solid State* **1** (1960) 1429.
18. A. J. LOVINGER, *Macromolecules* **18** (1985) 910.
19. B. DAUDIN, M. DUBUS, F. MACCHI and L. F. LEGRAND, *Nucl. Inst. Meth. Phys. Res. B* **32** (1988) 177.
20. Q. M. ZHANG, V. BHARTI and X. ZHAO, *Science* **280** (1998) 2101.
21. S. SEMENOVSKAYA and A. G. KHACHATURYAN, *J. Appl. Phys.* **83** (1998) 5125.
22. J.-M. LIU, X. WANG, H. L. W. CHAN and C. L. CHOY, *Phys. Rev. B.***69** (2004) 094114.
23. X. WANG, J.-M. LIU, H. L. W. CHAN and C. L. CHOY, *J. Appl. Phys.* **95** (2004) 4282.
24. J.-M. LIU, K. F. WANG, S. T. LAU, H. L. W. CHAN and C. L. CHOY, *Comput. Mater. Sci.* **33** (2005) 66.
25. S. T. LAU, H. L. W. CHAN and C. L. CHOY, *Appl. Phys. A* **80** (2005) 289–294.
26. H. L. HU and L. Q. CHEN, *Mater. Sci. & Eng. A* **238** (1997) 182; *J. Am. Ceram. Soc.* **81** (1998) 492.

JOURNAL OF MATERIALS SCIENCE 41 (2006) 177–198

Solid freeform fabrication of piezoelectric sensors and actuators

A. SAFARI*, M. ALLAHVERDI, E. K. AKDOGAN
Department of Materials Science and Engineering, Rutgers-The State University of New Jersey, Piscataway, New Jersey 08854, U.S.A.
E-mail: safari@rci.rutgers.edu

The last two decades have witnessed the proliferation piezoelectric composite transducers for an array of sensor and actuator applications. In this article, a concise summary of the major methods used in composite making, with special emphasis on Solid Freeform Fabrication (SFF), is provided. Fused Deposition of Ceramics (FDC) and Sanders Prototyping (SP) are two SFF techniques that have been utilized to make a variety of novel piezocomposites with connectivity patterns including (1-3), (3-2), (3-1), (2-2) and (3-3). The FDC technique has also been used to prototype a number of actuators such as tube arrays, spiral, oval, telescoping, and monomorph multi-material bending actuators. It has been demonstrated that SFF technology is a viable option for fabricating piezocomposite sensors and actuators with intricate geometry, unorthodox internal architecture, and complex symmetry. The salient aspects of processing of such composite sensors and actuators are summarized, and structure-processing-property relations are elaborated on.

1. Introduction

In today's transducer technology, the use of advanced functional materials (AFM) is indispensable. Such advanced functional materials are required to perform several tasks that produce a useful correlation or feedback mechanism in a transducer system. Piezoelectrics and electrostrictors inherently possess both direct (sensing) and converse (actuation) effects. A piezoelectric/electrostrictive sensor converts a mechanical input (displacement or force) into a measurable electrical quantity by means of piezoelectric/electrostrictive effect. Alternatively, the actuator converts the input electrical signal into displacement or force. As a consequence, sensing and actuation functions can exist in tandem in a given transducer design by the intelligent use of such materials. Piezoelectrics and electrostrictors, therefore, are AFM that constitute the backbone of modern transducer technology [1].

The generic definition of an electromechanical transducer is "A device that converts electrical energy into mechanical vibration and vice versa by utilizing piezoelectricity or electrostriction" [2]. The direct piezoelectric effect enables a transducer to function as a passive sound receiver or as a pickup by the conversion of acoustic energy into an electrical signal. Applications in which such electromechanical transducers are used include hydrophones for under water low frequency noise detection and microphones. On the other hand, the converse piezoelectric effect permits a transducer to act as an active sound transmitter or a loudspeaker. A transducer can also perform both active and passive functions quasi-simultaneously in the so-called pulse-echo mode. In this mode of operation, the transducer element emits an acoustic wave into the medium of interest, and also senses the echoes reflected back onto it. Such echoes are produced when sound waves are reflected back from an interface (boundary) between two substances of different acoustic impedances. The magnitude of the echo is proportional to the acoustic impedance mismatch between the two materials forming the interface. Electrostrictive materials, on the other hand, can be used as an actuator since the strain is coupled to the applied field quadratically. Such materials have also been used as sensors where the change in permittity with applied stress constitutes the mechanism for sensing.

*Author to whom all correspondence should be addressed.

0022-2461
DOI: 10.1007/s10853-005-6062-x

The design and fabrication of composite materials opens new avenues to optimizing the electrical, magnetic and mechanical properties for specific application. Properties, which are otherwise impossible to achieve with constituent materials in single-phase fashion, can be synthesized by the use of several connectivity patterns. Piezoelectric composites originated from this line of reasoning so as to maximize the input and output responses of an electromechanical transducer. For instance, in the context of electromechanical transducers, one may wish to maximize the piezoelectric sensitivity, minimize the density to obtain a good acoustic matching with water, and also make the transducer mechanically flexible to conform to a curved surface. These properties are mutually exclusive; and Mother Nature does not provide us with single-phase materials that simultaneously satisfy such requirements. Thus, in many applications, one might meet conflicting design requirements by combining the most useful properties of two or more phases through the effective use of translational symmetry, i.e. the composite approach.

This review article focuses on the development of piezoelectric ceramic-polymer composites, with special emphasis on solid freeform fabrication (SFF). This advanced fabrication method brings unique degrees of freedom to composite-making, which is likely to set the stage for the prototyping of next generation piezoelectric composites in the twenty-first century.

2. Piezoelectric ceramic/polymer composites

2.1. Piezoelectric ceramics

The perovskite family comprises the largest class of piezoelectric ceramics that are used in transducer technology today [2]. Typical examples of piezoelectric ceramics with the perovskite structure include barium titanate ($BaTiO_3$), lead titanate ($PbTiO_3$), lead zirconate titanate ($Pb(Zr_xTi_{1-x})O_3$, or PZT), lead lanthanum zirconate titanate ($Pb_{1-x}La_x(Zr_yT_{1-y})_{1-x/4}O_3$, or PLZT), and lead magnesium niobate-lead titanate ($Pb(Mg_{1/3}Nb_{2/3}O_3)$-$PbTiO_3$, or PMN-PT), among others. After the independent discovery of ferroelectricity in $BaTiO_3$ ceramics in the 1940's [2] in various parts of the world, a large array of ceramic ferroelectric compounds and solid solutions have been synthesized [2–4]. Particularly, lead zirconate titanate (PZT), a binary solid solution of $PbZrO_3$ (PZ -an antiferroelectric), and PT (PT -a ferroelectric), has found widespread use in transducer technology because of its remarkable piezoelectric properties [2–4].

In addition to the monolithic ceramics, composites of piezoelectric ceramics with polymers have also been fabricated. Ceramics are less expensive and easier to fabricate than polymers or single crystals. Ceramics also have relatively high dielectric constants as compared to polymers and good electromechanical coupling coefficients. On the other hand, ceramics are disadvantageous due to their high acoustic impedance, which results in poor acoustic matching with media such as water and human tissue—the media through which it is typically transmitting or receiving a signal. The notable exception to this problem is non-destructive testing of materials (NDT), where the impedance of the ceramic can be matched to that of a metal such as steel with relative ease. In addition, ceramics exhibit high stiffness and brittleness; and cannot be formed onto curved surfaces, which contributes to limited design flexibility in a given transducer. Finally, the electromechanical resonances of piezoelectric ceramics give rise to a high degree of noise, which is an unwanted artifact in the context of transducer engineering. The best compromise in properties can be obtained by the judicious combination of piezoelectric ceramics and polymers. Piezoelectric ceramic/polymer composites possess high electromechanical coupling, low acoustic impedance, fewer spurious modes, and an intermediate dielectric constant. Moreover, composites are flexible, and are moderately priced.

Recently, there has been a growing interest in high Curie temperature systems, with special emphasis on single crystals. Materials such as lead metaniobate ($PbNb_2O_6$) [5, 6], or relaxor systems such as $Pb(Sc_{1/2}Nb_{1/2})O_3$-$PbTiO_3$ [7–11], $Pb(In_{1/2}Nb_{1/2})O_3$-$PbTiO_3$ [12], and $Pb(Yb_{1/2}Nb_{1/2})O_3$-$PbTiO_3$ [12–15] have been investigated, among others. Also, compositions in the $(1-2\times)BiScO_3-\times PbTiO_3$ ferroelectric family were studied [16–20]. The major driver for this line of research in ferroelectric materials is the sensor/actuator needs of the automotive industry.

Of very current interest is the development of non lead-based piezoelectric materials with properties comparable to that of PZT. New regulations appertaining to the use of lead-based materials in regard to environmental pollution as well as health concerns that will go in effect as early as 2007 have been the major driving force behind research activity for such piezoelectric materials in Europe and Japan. Very recently, new lead-free morphotropic phase boundary compositions of alkaline-based niobate piezoelectric solid solutions such as the $(K,Na)NbO_3$–$LiTaO_3$–$LiSbO_3$) system have been reported, whose properties are as good as PZT, especially when the material has a $\langle 100 \rangle$ texture [21].

Therefore, it is to be expected that the research in ferroelectric materials will be geared towards non-lead based materials in the future in an effort to obtain properties exceeding that of PZT along with a high transition temperature.

2.2. Connectivity patterns of piezocomposites

The arrangement of the phases comprising a composite dictate the field patterns inside the composite, which in turn, govern the electromechanical properties. A system that allows one to codify the manner in which the individual phases are self-connected (continuous) was first proposed by Newnham *et al.* [22, 23]. There are ten connectivity patterns for a two-phase (diphasic) system, where each phase could be continuous in zero, one, two or three-dimensions as illustrated in Fig. 1.

The internationally accepted nomenclature to describe such composites is (0-0), (0-1), (0-2), (0-3), (1-1), (1-2), (2-2), (1-3), (2-3) and (3-3). The first digit within the parentheses refers to the number of dimensions of connectivity for the piezoelectrically active phase, while the latter is used for the inactive polymer phase. The polymer phase is virtually always chosen to be inactive since the purpose here is to reduce the acoustic impedance and dielectric permittivity, and not to contribute to the emission of acoustic energy. An array of piezoelectric ceramic/polymer composites have been developed based on the connectivity concept, as shown in Fig. 2 [22, 23].

Previous research has unequivocally shown that these composites exhibit superior piezoelectric properties in comparison to single-phase piezoelectric ceramics for transducer applications. In the following section, we will discuss the means by which such composites can be brought to life.

2.3. Fabrication techniques of piezoelectric composites

Various processing techniques have been utilized to fabricate piezoelectric composites. The processing techniques that found widespread acceptance include dice and fill [24, 25], extrusion [24], injection molding [26–29], lost mold [30–39], tape lamination [40–44], dielectrophoresis [45, 46], relic processing [41–43], laser or ultrasonic cutting [47–49], jet machining [33–37], reticulation [50, 51], and co-extrusion [52, 53] etc. Among these, the dice and fill, injection molding and lost mold techniques are the

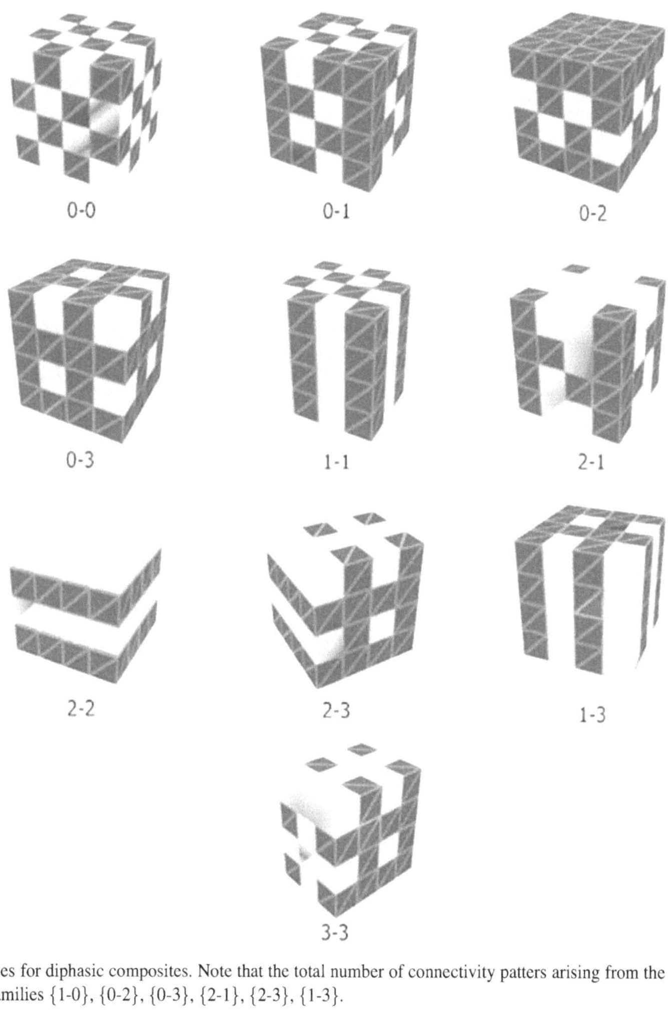

Figure 1 Connectivity families for diphasic composites. Note that the total number of connectivity patters arising from the 10 families depicted is 16 due to permutations involved in families {1-0}, {0-2}, {0-3}, {2-1}, {2-3}, {1-3}.

popular methods used for making composites on a commercial basis. Moreover, in the 1990's solid freeform fabrication techniques have also been proposed to make various piezocomposites with complex design and hierarchy, and a series of composites have been successfully prototyped.

Dice and fill method

The dice and fill method involves making a series of parallel cuts on a sintered piezoelectric block whereby a (2-2) connectivity pattern is created. If the diced sintered piezoelectric block (2-2 connectivity) is rotated by 90° and then further cut, (1-3) connectivity can be imparted to the composite. The diced ceramic block, which is still attached to the ceramic base, is backfilled with an epoxy, and then the base ceramic support is removed by polishing. The grain size of the sintered block to be cut has a strong impact on the ceramic fineness one can achieve in the dicing process—the smaller the grain size, the better the machinability. The simplicity of this method, in conjunction with readily available CAD-based wafer dicing systems, has made it very popular for fabricating (2-2) and (1-3) piezocomposite arrays for transducer applications, in general, and medical imaging applications in particular [54, 55].

Lost mold method

Skinner *et al.* and Rittenmeyer *et al.* were the first who proposed and implemented the so-called lost mold technique to fabricate (3-3) composites [56]. Siemens Inc. of Germany has also carried out pioneering work on the lost mold technique for fabricating piezoelectric composites by using a plastic mold manufactured by the LIGA (lithography, galvano-forming and plastic molding) process [31]. In the Siemens process, the plastic mold of the desired structure is filled with piezoelectric slurry. After drying, the mold is burned out and the structure sintered to >98% of the X-ray density. Composites incorporating rods with various sizes, shape and spacing could be straightforwardly made with the lost mold method, making it a very attractive process. However, the inability to rapidly prototype samples is a major drawback of this technique. Furthermore, the LIGA process is also expensive and time consuming, albeit highly accurate and of high resolution [30, 36].

Injection molding

The injection molding process is very suitable for making fine scale (2-2) and (1-3) ceramic structures with relative ease. The production of preforms with (2-2) sintered sheet composites as fine as 25-m, and (1-3) PZT rods as fine as 30–40-m in diameter has been reported by this technique [28]. Injection molding can also be used to make composites with a variety of rod sizes, shapes and spacing. The major advantages of this technique are the low material waste, flexibility with respect to the transducer design and a low cost per part. The limitation of this processing

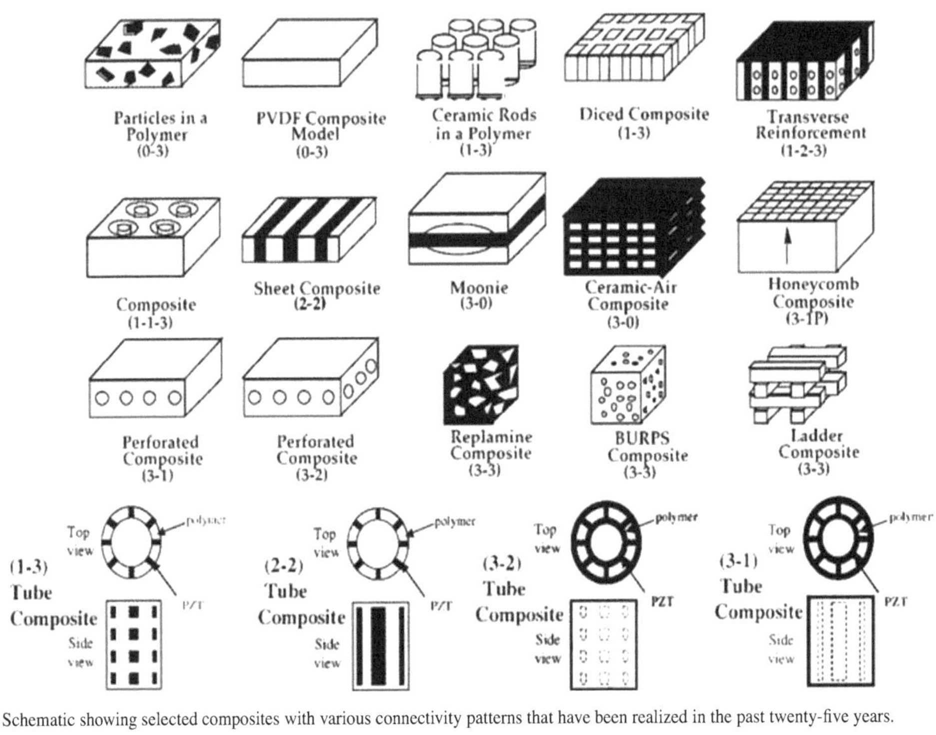

Figure 2 Schematic showing selected composites with various connectivity patterns that have been realized in the past twenty-five years.

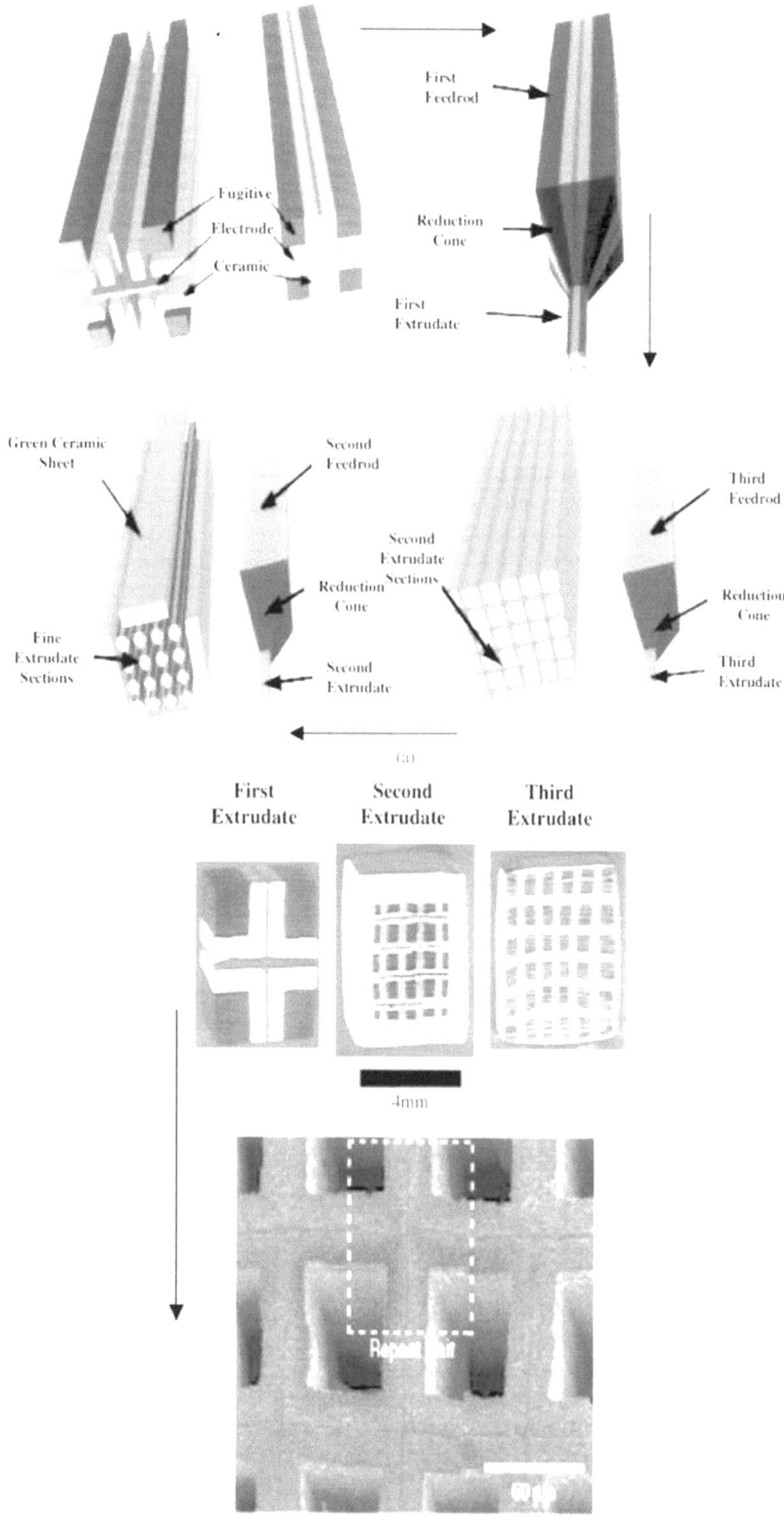

Figure 3 (a) Stepwise reduction in size by Micro-Fabrication by Co-Extrusion (MFCX), (b) cross-sectional view of the extrudates after each reduction step, and the resultant sintered microstructure.

method is the time and cost associated with making the mold. However, significant strides have been made to circumvent such problems associated with the lost mold method [26–29]. Materials Systems Incorporated (MSI) has adapted injection molding to produce net shape piezoceramics, greatly facilitating the manufacturing of large volumes of complex ceramic parts for underwater transducer applications, among others [57].

Tape casting

Tape casting has been effectively used for fabricating (2-2) composites. High frequency ultrasound transducers (>20 MHz) have been demonstrated using stacked tapes of PZT and polymer [54, 55]. With today's tape casting technology, very thin tapes (down to a several microns) can be cast, allowing manufacture of transducers for high frequency operation in high-resolution medical imaging. In addition to (2-2) composites, composites with (2-0-2) connectivity were also fabricated using PZT tapes and polymer mixed with different types of ceramic powders [59]. Other piezocomposites such as (1-3) can be made using stacked tapes and dicing. Multilayer composites with (2-2) and (1-3) connectivity patterns have also been fabricated using piezoelectric and conductive tapes [60]. Superior signal to noise ratios is within the realm of possibility in such multi-layered composites, and therefore offer high resolution capability to ultrasonic transducers [61].

Microfabrication by coextrusion

Recently, Halloran *et al.* [62, 63] developed a microfabrication by coextursion (MFCX) process, which involves forcing a thermoplastic ceramic extrusion compound through a die of a given reduction ratio. By assembling an extrusion feedrod from a shaped ceramic compound with space-filling fugitive compound, objects with complex shapes can be fabricated as shown in Fig. 3a. After each reduction state, extrudates are assembled into a feedrod and extruded again, reducing the size and multiplying the number of shaped objects. In this process, the size of the feature of interest can significantly be reduced with consecutive extrusions.

In Fig. 3b, a piezoelectric (PMN-PT) array with a final spatial resolution of 60 μm, as fabricated by MFCX, is depicted. The process has the potential for fabricating objects in the 10 μm size range [62, 63]. Therefore, the size of the features produced by MFCX is comparable to those produced using the "lost mold" technique, synchrotron radiation lithography, and also approaches the resolution realized by micromolding using photolithography [62, 63].

Solid freeform fabrication (SFF)

While the traditional ceramic processing techniques described thus far have proven to be effective for making composites with simple connectivity patterns, none of them permit the fabrication of composites with complex internal hierarchy and symmetry. The advent of a new manufacturing/prototyping technique to fabricate polymer, ceramic and metal components with very high design flexibility for novel structures, and fast prototyping is, therefore, needed. Solid Freeform Fabrication (SFF), an emerging technology that provides an integrated way of manufacturing 3-D components from computer aided design (CAD) files, arose from the concept of additive processing—a processes by which an arbitrary 3-D structure is made by cumulative deposition of material, without using any hard tooling, dies, molds, or machining operations [64–69].

In the mid-1990's, several SFF methods have been developed to fabricate polymer, metal or ceramic structures on a fixtureless platform, directly from a CAD file. Some of these techniques are designed to produce large parts with a fast output rate, and with modest surface finish. On the other hand, some of SFF techniques target markets where a very high resolution and a good surface finish are very critical. SFF or rapid prototyping methods that have found commercial success include Stereolithography (SLA, 3-D Systems Inc.) [68, 70, 71], Fused Deposition Modeling (FDM, Stratasys Inc.) [64–66], Selective Laser Sintering (SLS, DTM Corp.) [68], Laminated Object Manufacturing (LOM, Helisys Inc.) [72, 73], 3-D Printing (3-DP, Soligen Inc.) [74], Robocasting (Sandia National Labs) [74, 75], and Sanders Prototyping (SP, Sanders Prototyping Inc.) [76]. Most of these techniques are designed to manufacture net shape polymer parts for form-fitting applications and design verification. However, some are also capable of manufacturing metal or ceramic parts as exemplified by the Rutgers Fused Deposition of Ceramics (FDC) and Fused Deposition of Multi-Materials (FDMM) processes [77].

All SFF techniques begin with a common approach. Firstly, a CAD data description of the desired component is prepared. Secondly, a surface file (a.k.a. stl file) is created from the CAD file, which is later input to the manufacturing system. Thirdly, the stl file is converted into cross sectional slices, or a slice file, in which each slice can be uniquely defined about its build strategy by varying the tool path. The slices collectively define the shape of the part. And finally, the information appertaining to each slice (i.e., toolpath) is then transmitted in a layer-by-layer fashion to the SFF machine. As shown in Fig. 4, where an example from the FDC process is shown, the lowering of the fixtureless platform by one slice thickness in the Z-direction follows the completion of each layer. Subsequent layers are built on top of the preceding one in a sequential manner, and the process is repeated until the whole part is finished. The sequential or layered approach

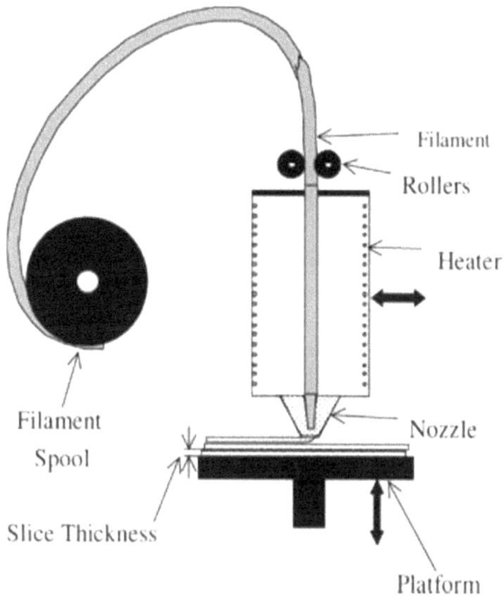

Figure 4 Schematic showing the layer-by-layer fabrication in Fused Deposition of Ceramics (FDC).

of manufacturing a 3-D object of arbitrary shape is indeed the very quintessential feature of SFF. In the Fused Deposition of Multiple Materials (FDMM) process, the same approach is pursued as in the FDC process; however, a multi-head assembly is used instead of a single head assembly. The multi-head assembly enables one to build 3D structures comprising two or more materials in a sequential fashion. While the FDMM process offers great flexibility in the design and fabrication of multi-material preforms, co-densification is a major challenge due to the differential shrinkage of in each layer. However, judicious materials selection as well as the fine-tuning of the solids loading in the filament feedstock can circumvent the aforementioned difficulty [77].

Ceserano *et al.* [78] developed the so-called Robocasting process –another near-net-shape processing method that is member of the greater SFF family of processes. Computer controlled layer-by-layer extrusion of colloidal slurries is accomplished in Robocasting. The liquid carrier is usually water, with some minor additions of dispersants and organic binders to form pastes with a plastic consistency suitable for extrusion. It has been recently shown that Robocasting could effectively be used to create complex piezoelectric lattices, which could later filled with an inactive material to make composites as shown in Fig. 5.

SFF techniques provide many advantages for the manufacturing of advanced functional components. A part normally takes weeks or months to fabricate because of the time spent in fabrication of the mold, tools and machining operations etc., in a typical conventional process. The same structure, on the other hand, can be prototyped in a few days time using SFF, thereby providing reduced lead times and costs in the development of new ceramic parts. This technology also provides the ability to do cost-effective iterative designing for producing components with optimum shape and desired properties.

3. Piezoelectric composites fabricated by SFF

The feasibility of the FDC process in making piezoelectric composites was demonstrated in the fabrication of various PZT/polymer composites with different connectivity patterns. The fused deposition process was used to fabricate piezoelectric ceramic-polymer composites via two distinct processing routes: (i) Direct fused deposition of ceramics (FDC) and (ii) Indirect technique or lost mold process where polymeric molds were manufactured by FDM^{TM} as shown in Fig. 6.

In the direct processing route, FDC was used with a PZT powder loaded polymer filament as the feed material for a direct layered manufacturing of the 3D green

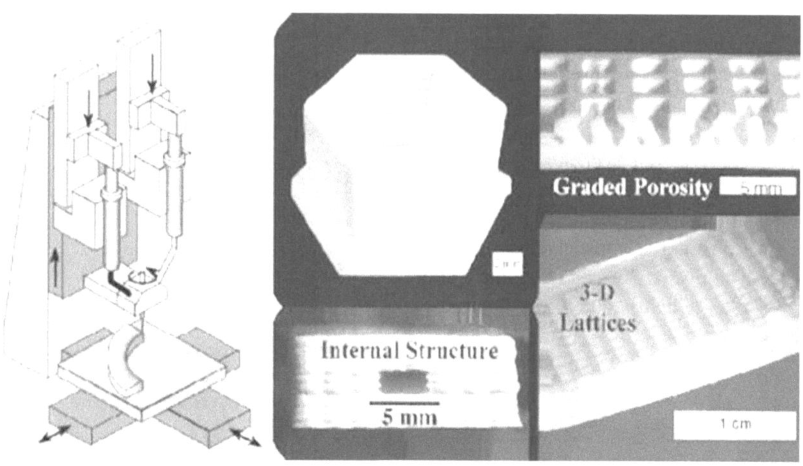

Figure 5 Three dimensional PZT lattices with (3-3) connectivity built by Robocasting.

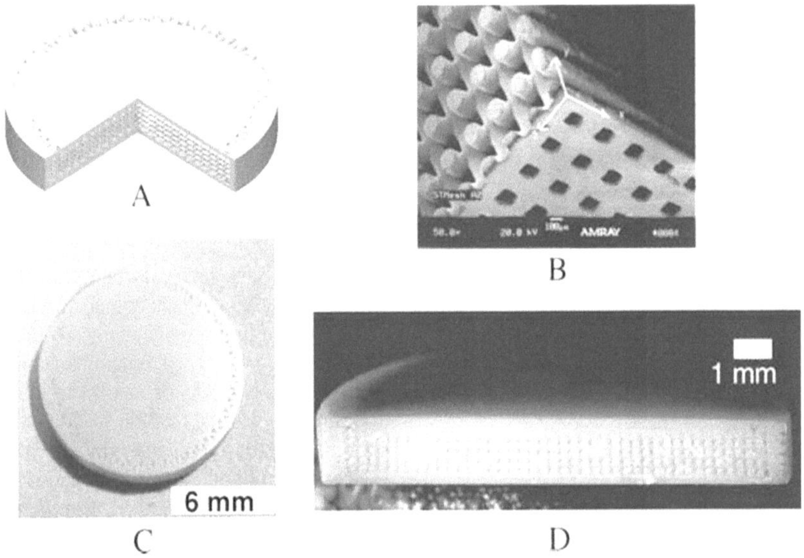

Figure 6 Piezoelectric ceramic-polymer composites via Direct fused deposition of ceramics (FDC) shown (see A & B) and Indirect technique or lost mold (C & D). Polymeric molds fabricated by FDM™.

ceramic structure. In the indirect process, Stratasys™ (Eden Prairie, MN) commercialized polymer filament was used to fabricate a mold via FDM™. A lost mold technique was then implemented to create the final structure. Embedding the ceramic structure in epoxy, followed by electroding and poling in both the indirect and direct processes obtained the composite.

Direct fused deposition of ceramics

In Fig. 7, a schematic of the direct fabrication technique is depicted on the upper path of the processing map. The optimum solids loading used in the production of filaments are typically 50–55 volume percent in a four-component thermoplastic binder system containing polymer, tackifier, wax, and plasticizer. It was found that powder pretreatment with an organic surfactant before compounding was critical in obtaining extrudable feedstock materials. Filaments of 1.78 mm in diameter are used in the process. Fused deposition of these filaments was accomplished using a 3D Modeler™ by Stratasys™, Inc. The CAD data description of the object was used to create the input file (.sml) for the 3-D Modeler™. The .sml file controls the build strategy for the part in 3D Modeler™. Samples with different volume fractions and various spacing between ceramic phases were fabricated by changing the build strategy (or the .sml file), which is due to the flexibility offered by this technique.

Indirect technique

In the indirect process, a polymer mold having the negative of the desired structure is formed via the FDM™ process using Stratasys™ commercial filament (ICW-04) and Sanders Prototyping. As shown in Fig. 7, the mold is infiltrated with the PZT slurry and then placed in a vacuum oven to ensure complete filling of the voids. The samples are then dried in an ambient atmosphere for 2 h and transferred to an oven that is maintained at 70°C. The initial slow drying step has to be included to further reduce the possibility of cracking in the slip. The thermoplastic polymer mold is evaporated during the early stages of a specifically designed binder burnout cycle. Details of each step in the process are described in the sections below.

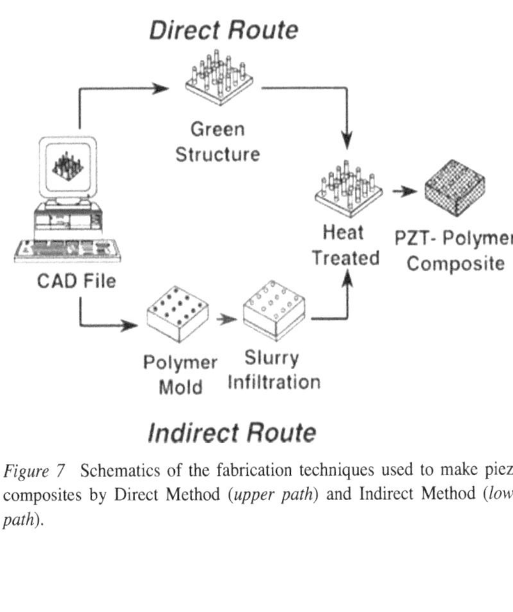

Figure 7 Schematics of the fabrication techniques used to make piezocomposites by Direct Method (*upper path*) and Indirect Method (*lower path*).

3.1. Examples of piezocomposites by SFF

Composites with (2-2) connectivity and volume fraction gradient

Regular (2-2) piezocomposites can be readily made using the dicing technique and SFF technology. However, the traditional dicing technique poses severe difficulties when one wishes to introduce a volume fraction gradient in the elevation direction of a (2-2) composite [79, 80]. That is so because the so-called kerf width has to be increased from the center to the edges, while a cutting blade with a single thickness is used in the dicing machine [81]. Furthermore, the resolution of the traditional techniques cannot be pushed under 50 μm [82]. To overcome these limitations, Panda [77] studied volume fraction gradient (VFG) (2-2) that were fabricated by the indirect lost mold route using Sanders Prototyping—a CAD-based prototyping technology. Many mathematical functions including regular, Gaussian, linear and exponential gradients were designed in order to study the effect of introducing different types of gradients along the elevation direction. The electromechanical and acoustic properties of such VFG composites were compared to those having no VFGs.

In Panda's work [77], all the VFG distributions were designed to have ~60 vol.% ceramic in that center region of the composite to obtain a high acoustic pressure output at the center. However, no further increase in the ceramic content at the center was pursued, as that would also increase the acoustic impedance difference between the central region and the edges albeit further improving the output. As a consequence, very complex matching layers would be needed for good energy transfer across the whole composite. The ceramic content in such VFG composites was gradually decreased to approximately 20% at the edges for all the types of distributions studied to ensure that the edges would act as a good receiver of ultrasonic waves due to the low dielectric constant of those regions. It should be noted that a further decrease in the total ceramic amount at the edges could further enhance the receiving sensitivity. However, this would concomitantly cause a fall in the thickness-coupling coefficient due to interference from the lateral modes, for a given ceramic element width and spacing of interest.

Fig. 8a and b show the sacrificial molds for two different functional distributions, Gaussian and linear VFG, as obtained using the Sanders prototyping technique. Throughout the structure, an air gap of 135 μm between the walls was created. Such a design enables one to ascertain that ceramic walls maintain a uniform width after slurry infiltration, as well as after subsequent heat treatment. In that particular design, the mold wall thickness was increased from 95 μm at the center to 575 μm at the edges, following the respective mathematical function. The positioning and thickness of the mold walls were very accurate with an X-Y deposition error of only $\pm$5 μm. After infiltrating PZT slurry into the molds and subsequent heat treatment, the sintered structures were embedded in Spurr epoxy. The measurement of the dielectric properties of such linear, exponential, and regular (2-2) composites has shown that the dielectric constant was as high as 950 for the Gaussian, and 430 for the exponential design, following the total volume percent of PZT in the composite. All structures had well-discernable thickness mode resonances with k_t on the order of 66%, and moderately low values of the k_p. The high k_t in such composites was attributed to the high aspect ratio of the ceramic elements (>6), and to the soundness of the VFG design by Sanders prototyping. The longitudinal piezocharge coefficient (d_{33}) of all these composites was found to be greater than 400 pC/N.

(a)

(b)

Figure 8 (a) SEM photograph showing the Sanders-built wax sacrificial polymer molds for the gaussian, and (b) linear volume fraction gradient distribution.

The vibration profiles of the (2-2) composites with and without VFGs are shown in Fig. 8 as a function of the distance from the center. The pressure outputs of the VFG designs were compared with a regular (2-2) composite, which was a diced PZT-5H ceramic structure having a wall width of 220 μm and a kerf of 440 μm. The regular

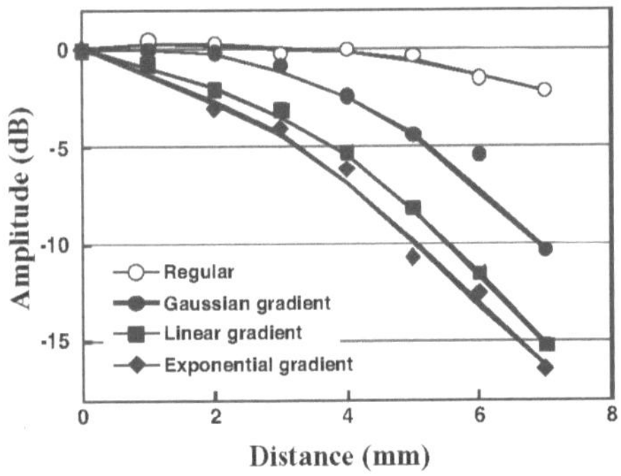

Figure 9 Graph showing the vibration amplitude profiles for 2-2 sheet composites with regular, gaussian, linear and exponential VFG distributions, as a function of the distance from the center of the structure.

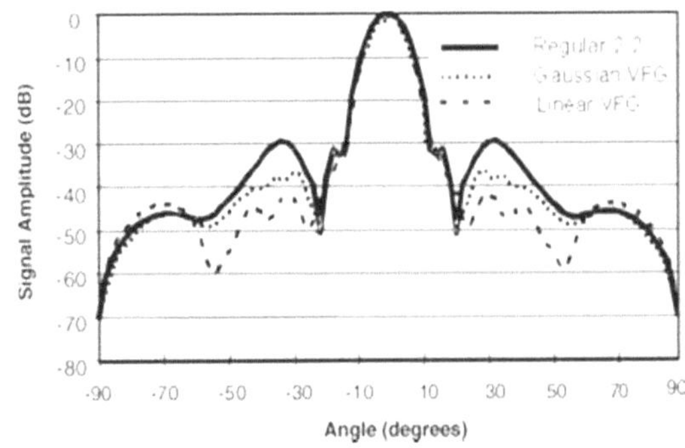

Figure 10 Predicted far field beam pattern plots for a regular, gaussian and linear VFG distribution.

design had ~33 vol.% ceramic with a d_{33} and k_t of 340 pC/N and 63%, respectively. As expected, the amplitude of the vibration for this composite did not vary much with the distance from the center, as this structure was the same everywhere. The small fall in the amplitude at the edges could be due to the edge clamping effect. On the other hand, all the VFG distributions showed a decrease in the vibration amplitude as distance from the center increased. The Gaussian VFG was designed to have a very slow decrease in ceramic volume fraction with distance from the center. The vibration amplitude observed in Fig. 9 followed a similar trend, initially remaining constant and the dropping by only −4 dB at a distance of 5 mm. Thus, this distribution was not very different from the regular (2-2) composite out to 5 mm from the center. The linear gradient showed a very smooth decline in the vibration output with distance, linearly falling to about −8 dB at a distance of 5 mm. The exponential distribution had a vibration profile that was slightly lower in amplitude than the linear distribution. It should be noted that although the ratio of the total ceramic content in the center and the edges is constant for all VFG distributions, the pressure output ratios are not similar. This could be because the 2.3 mm diameter photonic sensor probe used in the measurements of such composites shows an average reading over the probed area. Hence, the readings encompass not only the edge, but also a slightly larger region. The precise manipulation of the volume fraction within the same composite allows the control of the vibration amplitude profile to obtain the desired pressure output as seen in Fig. 10. The Gaussian vibration amplitude profile did not fall down fast and was very similar to the regular distribution. The linear or the exponential gradient showed a very smooth decrease in the vibration from the center to the edges.

Turcu *et al.* [83, 84] studied oriented (2-2) soft PZT-Epoxy composites, which were fabricated by FDC (Fig. 11). The inspiration for this study came from the modeling work of Nan et al., who investigated the effects of orientation of the ceramic phase relative to the poling directions in (0-3) and (1-3) piezocomposites [85, 86]. Nan et al.'s modeling results showed that the dielectric and piezoelectric properties of the piezocomposites with (1-3) connectivity decreased when the poling direction deviated from the orientation of the ceramic phase. In Turcu's work, the orientation angle of the ceramic phase relative to the poling direction varied in the range of 0° to 75° with 15° increments (Fig. 11) [83, 84]. The volume fraction of the ceramic phase in both composites was fixed at 0.30. It was found that the dielectric constant and piezoelectric d_{33} coefficients of both composites decreased with increasing orientation angle. However, the piezoelectric d_{31} coefficient became positive at 45° reaching a value of 150 pC/N. The relevant hydrostatic piezocharge and voltage coefficients, and the FOM of the oriented composites were calculated as shown in Figs 12 and 13, respectively [83, 84]. The data exhibit $d_h g_h$ maxima at an orientation angle of ~45°.

Composites with (3-3) Connectivity

A composite with (3-3) connectivity is comprised of two phases that mutually penetrate, and thereby form two three dimensionally self-connected networks in intimate contact. The feasibility of the FDC process for making piezoelectric composites was demonstrated in the fabrication of PZT/polymer 3-D honeycomb and ladder composites with (3-3) connectivity patterns [77].

Two scanning electron micrographs of a typical heat-treated ceramic ladder structure prepared by the direct FDC method prior to epoxy infiltration are shown in Fig. 14. The ladder structures were built by using a raster fill strategy with a fixed inter-road spacing. The consecutive layers were built 90° to each other. The volume fraction of the ceramic phase in the structure shown in Fig. 14 is approximately 70 percent, and can be varied by varying the width and spacing between the ceramic

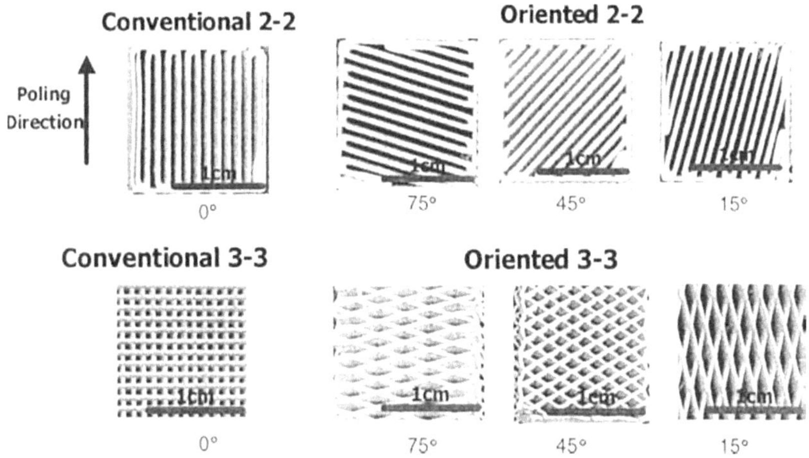

Figure 11 Oriented (2-2) and (3-3) composites fabricated by the Fused Deposition of Ceramics process.

roads. In Fig. 14, the ceramic roads are ~300 μm wide with ~800 μm center-to-center spacing. The mesoscopic structures created by FDC are very uniform with excellent unit cell repeatability, unequivocally demonstrating the superiority of the FDC process over traditional methods of composite making [77].

The indirect method was also used to fabricate 3-D honeycomb composites. The scanning electron micrographs (SEM) in Fig. 15 depict the various orientations of a typical heat-treated ceramic structure prior to epoxy infiltration from. Fig. 15a shows the top, front and side faces, while Fig. 15b shows the top and front faces. The structure consists of a 3-D lattice of "air pipes" defining the 3-D piezoelectric skeleton (matrix). The "air pipes" were created upon burning out the sacrificial polymer. By virtue of the flexibility offered by the FDC process, the diameter of the holes, the spacing and the volume fraction of the ceramic phase can be varied between the holes. In the example given in Fig. 15, the diameter of the holes is ~200 μm, with ~350 μm center-to-center uniform spacing, and the PZT content is ~25% by volume [77, 87, 88].

The dielectric constant (K_{33}) and piezocharge coefficient (d_{33}) of several 3-D honeycomb composites fabricated by the indirect method with 10–35 volume percent PZT loading is are shown in Fig. 16. An increase in K as the volume percent of PZT increased was observed, which is in conformity with mixing laws to a first approximation. It is also seen that there is an increase in d_{33} as the volume percent of PZT increases from 10 to 30. The d_{33} value then levels off at about 35 volume percent PZT, where pseudo-single phase behavior is reached [88].

Very recently in another application of SFF, namely Robocasting, it has been demonstrated that one can successfully fabricate piezoelectric composites [89]. As shown earlier in Fig. 5, composites with (3-3) connectivity have been built out of PZT, and were found to exhibit reasonable hydrostatic FOM. It was also shown that the incorporation of a faceplate in these (3-3) composites increased the FOM by approximately one order of magnitude, reaching a value of approximately 8000×10^{-15} m^2/N.

Turcu *et al.* [83, 84] studied oriented (3-3) soft PZT-Epoxy composites by FDC (Fig. 11). This CAD-based approach allowed the design of the piezoelectric "road" thickness, line spacing, composite thickness, as well as the orientation of the "roads". Testing of the said composites revealed that the piezoelectric voltage coefficient (g_{33}) showed a significant increase (about 85%) in (3-3) composites at an orientation angle of 45°. The FOM of the oriented (3-3) composite reached an extremely high value of ~$50{,}000\times10^{-15}$ m^2/N, while a more modest increase was seen for the (2-2) composites ($14{,}000\times10^{-15}$ m^2/N) as shown in Figs 12 and 13, respectively.

Radial composites

Monolithic ceramic tubes, like their planar counterparts, suffer from similar deficiencies such as high weight, high capacitance, and low figure of merit $d_h g_h$. Until recently, the fabrication of functional piezoelectric composites with radially-oriented connectivity has been difficult or impossible using conventional processing. With the advent of Rapid Prototyping, these types of composites can be fabricated with relative ease.

Radial composites with four different connectivity patterns have been fabricated using FDC and Sanders Prototyping (SP) processes. These consist of (1-3), (2-2), (3-1), and (3-2) type composites. For the purpose of simplicity, cylindrical coordinate systems are being used to designate the connectivity of each composite design. For the (1-3) type composite, the ceramic phase is continuous in only the r direction, while the polymer phase is continuous in the diameter (r), height (z), and sweep angle (θ) directions (Fig. 2). For (2-2) type composites,

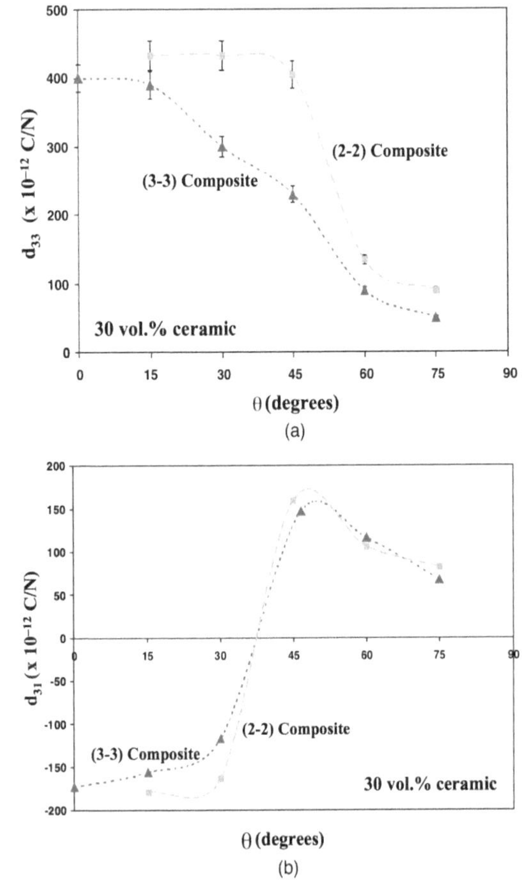

Figure 12 Piezoelectric properties of oriented (2-2) and (3-3) composites fabricated by FDC: (a) longitudinal piezostrain response, (b) transverse piezostrain response.

both the ceramic phase and the polymer phases are continuous in the r and z directions. In the case of (3-1) type composites, the ceramic phase is continuous in all three directions, while the polymer is continuous in only the z direction. Finally, for the (3-2) type composites, the ceramic phase is continuous in all three directions, while the polymer phase is continuous in the z and θ directions.

Fig. 17a and b shows the photograph of a tubular (3-1) PZT ceramic preform made using Fused Deposition of Ceramics (FDC). In this structure, the inner and outer ceramic rings are approximately 1 mm in thickness, and the ceramic sheets connecting the inner and outer rings are approximately 1 mm by 1 mm. Fig. 17c shows the picture of a (3-3) sintered ceramic made using a SP rapid prototyping system.

Curved transducers

Curved ceramic skeletons for (2-2) composite were built with 10 and 20 cm radii of curvature (Fig. 18a). Flat

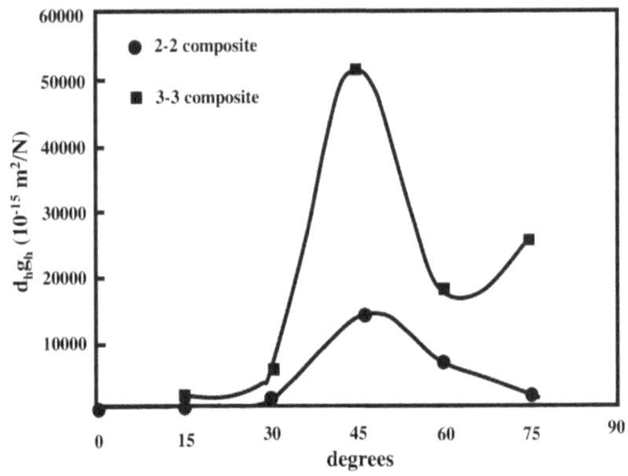

Figure 13 Effect of piezoelectric phase orientation on the figure of merit (d_hg_h) in (2-2) and (3-3) oriented composites made by FDC.

samples were also built for comparison of the electromechanical properties. In the FDC built green parts, the ceramic plates were 330 μm thick and the spacing between the plates was 760 μm. Physical characterization of the samples revealed that 95% of the theoretical density was achieved upon sintering. Optical micrographs of the parts in the green and sintered states were taken. Fig. 18b shows the surface roughness associated with the layer-by-layer building of the FDC technique. The slice thickness was found to be constant; however, the road width showed a significant spread in values. SEM of fracture surfaces of the sintered samples showed that good bonding occurred between the roads; no delamination was visible.

4. Piezoelectric actuators fabricated by SFF

Several piezoelectric actuators with novel designs have been made and characterized in the last few years [90–92]. It has been demonstrated that improved electromechanical properties can be achieved by tailoring the geometry and dimensions of an actuator. Ceramic processing techniques such as injection molding is costly and time-consuming to develop and prototype new actuators. Solid freeform fabrication has brought a high degree of flexibility, cost-effectiveness and speed to the field of piezoelectric actuator prototyping.

In the last ten years, SFF methods in particular FDC have been utilized to fabricate a series of piezoelectric actuators [91, 93–98]. Processing and properties of a few rapidly prototyped actuators will be reviewed with emphasis on the recent results on unimorph and multimaterial spirals and bending actuators, as well as tube actuators with curved tube walls.

4.1. Prototyped actuators

Piezoelectric actuators generate field-induced displacements that can be used in motion systems such as micro-

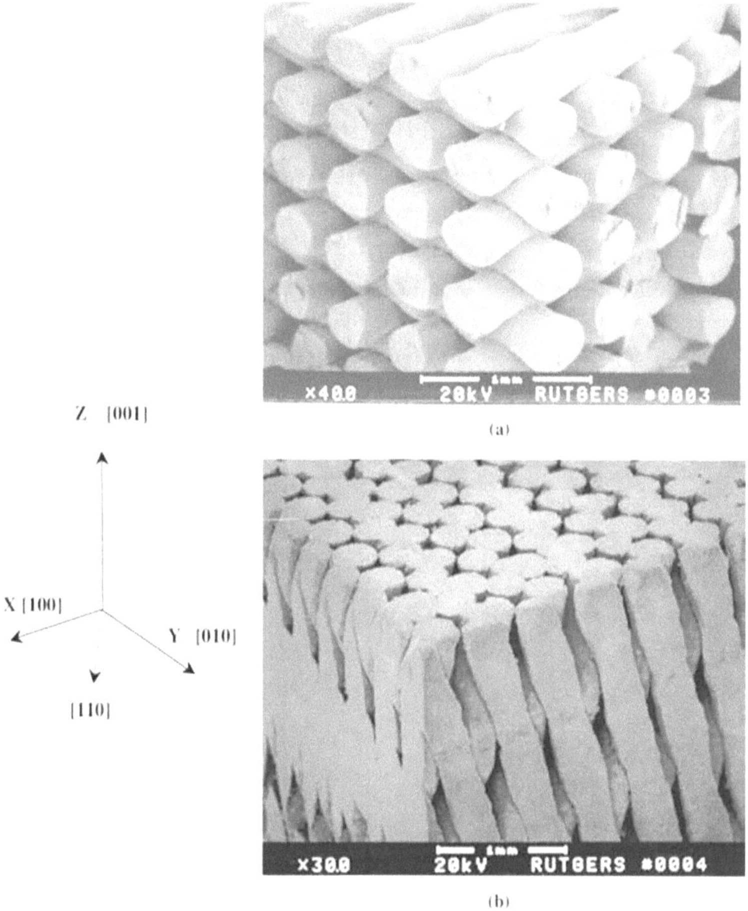

Figure 14 Ladder structure with (3-3) connectivity as obtained by FDC. Upon impregnation with polymer, a (3-3) piezoelectric ceramic/polymer is formed.

positioners, ultrasonic motors, and vibration damping. It is desirable to enhance the displacement while maintaining a reasonable amount of load bearing capability. This has been achieved by manipulating the geometry of simple shape (rod or strip) actuators into more complex geometries [1, 2, 99, 100]. For example, the axial displacement of a solid piezoelectric cylinder (rod) can be improved by replacing it with a tube of similar height and diameter. Other techniques such as the addition of a stiff inactive layer (a metal shim in unimorph and bimorph actuators) have also been utilized to improve displacement [92, 101].

Using the FDC process, a series of novel actuators have been designed and prototyped to study the effect of design/geometry on the field-induced displacement. The flexibility of FDC has allowed for the rapid fabrication of actuators with complex geometry that cannot easily be made by conventional manufacturing techniques. These actuators include spiral [2], telescoping [97], dome [98], tube [102], curved and multilayered tubes, bellows, oval, and monomorph benders [7]. Spiral and tube actuators will be briefly reviewed with the emphasis on the recent results.

Spiral actuators

Spiral actuators with a compact geometry show large field-induced strains (Fig. 19a). The application of an electric field across the spiral width induces both tangential and radial displacements, the former being much larger than the latter. It has been shown that for a PZT spiral actuator with 32 mm diameter, the tangential movement reaches 1.9 mm at ~11.6 kV/cm [97]. This is about twelve times higher than the displacement of a PZT strip of the same dimensions, which clearly shows that displacement can profoundly be enhanced by tailoring the geometry of the actuator.

In a curved geometry like spiral, there exists a slight gradient in the displacement of small elements across the thickness (Fig. 19b). This means that the displacement of an element in the outer diameter (r_o) is larger than that of

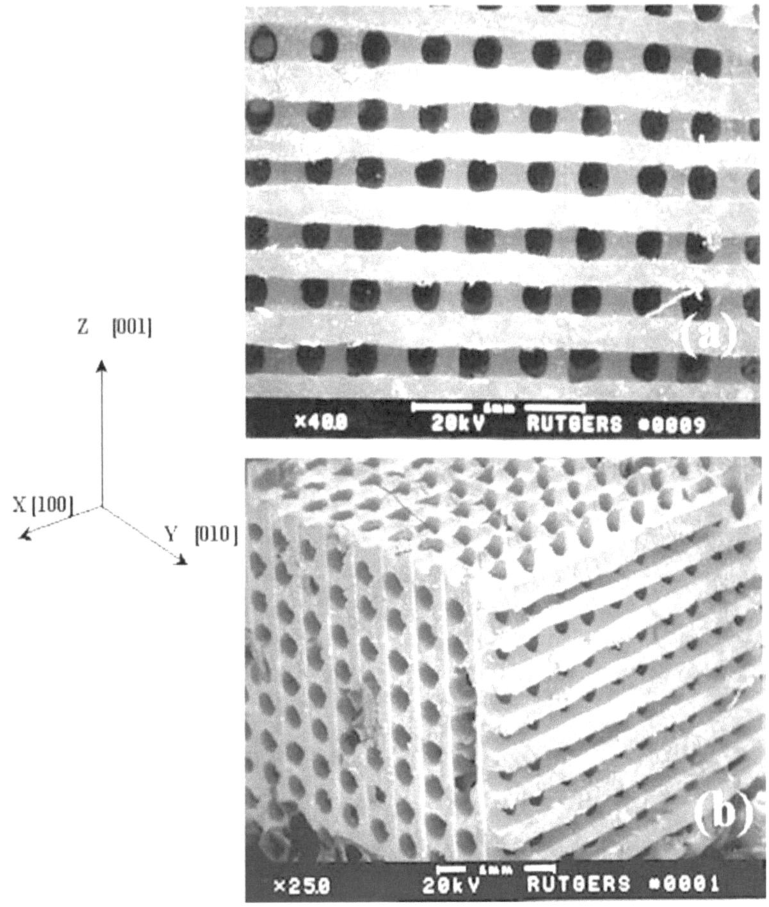

Figure 15 Honeycomb structure made out of PZT using FDC (a) front view, (b) oblique view.

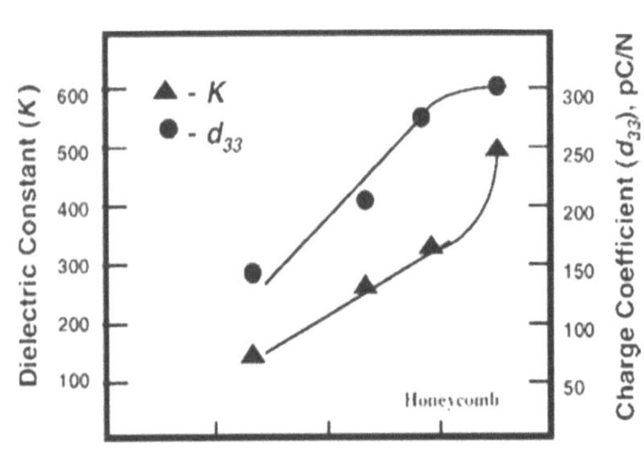

Figure 16 Variation of dielectric constant and piezocharge coefficient for a honeycomb composite with volume percent of PZT loading.

an element in the inner diameter (r_i). Such a phenomenon creates a bending moment, which leads to rotation or large tangential displacement in spirals. The large actuation of spirals is obtained at the cost of load bearing capability since the two are inversely related. For example, the blocking force for the above-mentioned spiral actuator, measured at various electric fields (1.3 to 3.3 kV/cm), is ~0.8 N [8]. Despite low blocking force, spiral actuators are of interest due to large displacement and compact geometry.

To further enhance the actuation, unimorph PZT spiral actuators have been investigated. Modeling results of spirals have suggested that asymmetric electrodes/shims can enhance the tangential displacement significantly [103]. Unimorph spirals were fabricated by attaching a stainless steel shim on the inner side of the spiral wall. In a conventional unimorph, the relative difference between the length of ceramic and metal strips under an electric field results in bending and therefore, large tip displacement. A PZT spiral actuator (width = 1.05 mm, diameter = 28.85 mm, height = 2.55 mm) with a 37.5 μm stainless steel shim was fabricated and tested. The addition of the stainless steel shim increased the tangential displacement by 200% at 850 V, as depicted in Fig. 20. This increase in displacement is analogous to the bending of a

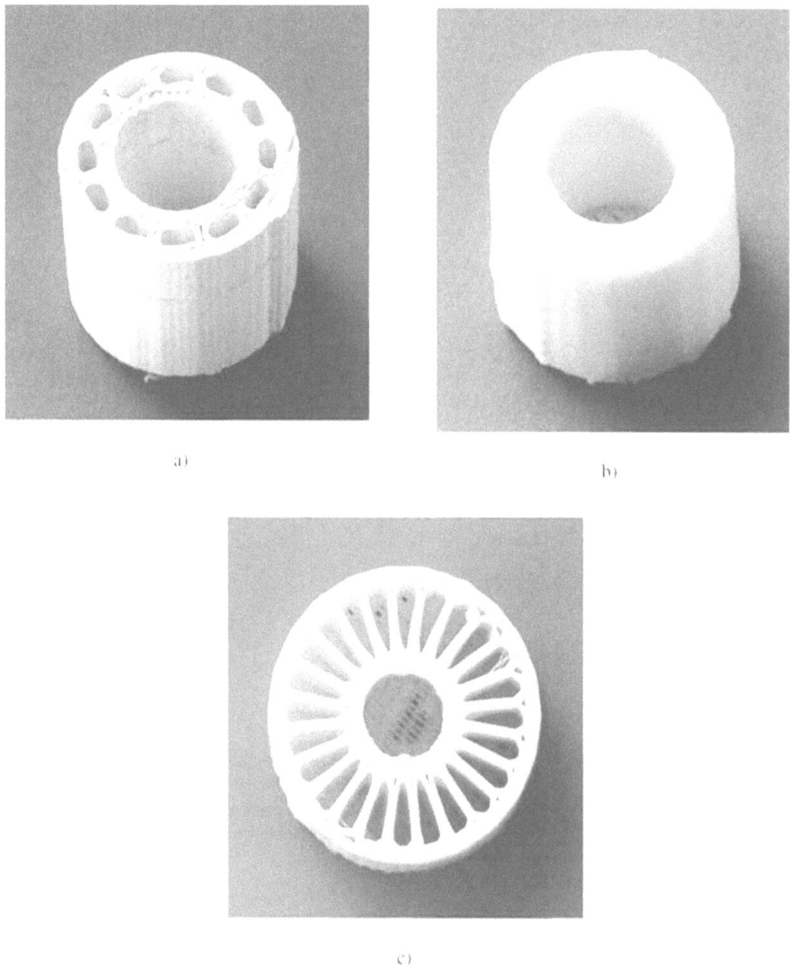

Figure 17 (a) Photograph of radial 3-1 type (design 2) hydrophone preform made via FDC, and (b) radial 3-2 type hydrophone, and (c) radial 2-2 type hydrophone.

conventional unimorph. It is also seen that the hysteresis of the unimorph spiral was much higher (e.g., about six times higher at 400 V) than the regular spiral actuator.

In another series of experiments, the tangential displacement of spirals with dissimilar ceramic strips was investigated. For this purpose, multimaterial spiral actuators consisting of one piezoelectric and one electrostrictive PMN-PT layer with the same thickness were fabricated and tested. These actuators can be driven in such a way that there is a strain gradient in the actuation between the piezoelectric and electrostrictive layers. This would further enhance the bending moment of the spiral. The multimaterial spiral actuator was designed such that the outer side of the spiral wall was the piezoelectric 0.65PMN-0.35PT layer, while the inner side of the wall was the electrostrictive 0.90PMN-0.10PT layer. Such spirals were prototyped by FDC and co-fired successfully at 1250°C for 1 h. Measurements have shown that the combined piezoelectric/electrostrictive spiral actuator yielded higher displacements than a piezoelectric actuator of similar dimensions. For example, the piezoelectric/electrostrictive actuator displaced to ~510 μm at 500 V as compared to ~200 μm for the piezoelectric actuator at the same DC biased field. The enhancement of tangential displacement was accompanied by large hysteresis of the multimaterial actuators similar to the unimorph spirals.

Tube actuators

Piezoelectric tubes are utilized for actuator (Ink Jet Printing) and sensor (1-3 piezocomposites) applications, among others [102, 103]. Tube actuators have demonstrated promising properties and have been successfully used to shift mirrors of laser resonators in scanning tunnel microscopy and atomic force microscopes.

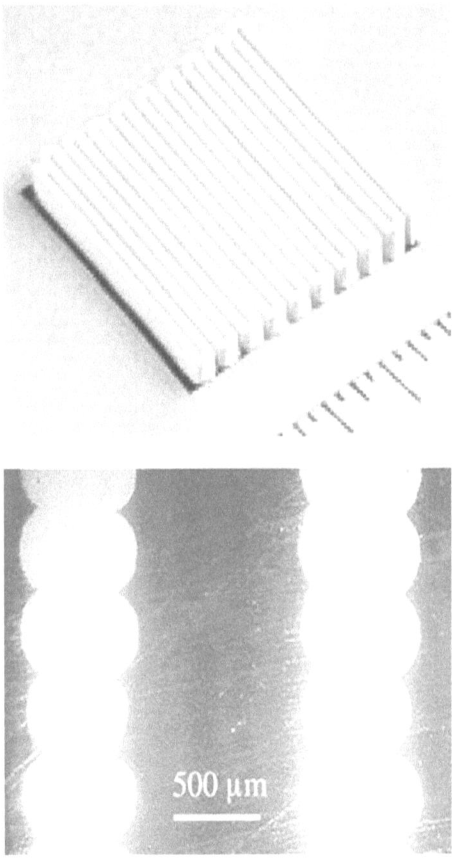

Figure 18 (a) Photograph of a curved green 2-2 PZT-5H preform. (b) Optical micrograph of the cross section of the walls of the structure.

Single tube and tube arrays (Fig. 21a) were fabricated with various diameter, height, and wall thickness via the FDC process [102]. Tube geometry has been modified to tune properties (e.g., resonance frequency and axial displacement) for a given application. Tube actuators with zigzag wall design (bellows, Fig. 21b) have also been prototyped to study the effect of wall configuration (bellows angle) on the axial and radial displacements [104]. Comparison of bellows with the straight tubes revealed that the bellows possess up to 50% larger radial displacements, whereas their axial displacements were about 20–30% lower.

Another type of tube actuator with helical electrodes has also been investigated (Fig. 21c). Normally, the inner and outer surfaces of the tubes are electroded and the tubes are poled radially. In this manner, the actuation or sensing is based on the d_{31} coefficient. However, in the new design the electrodes are laid side by side in helical form along the tube height [105]. For such piezoelectric tubes, displacement is derived from the d_{33} coefficient, resulting in a potential improvement of more than two times in axial displacement ($d_{33} / d_{31} > 2$). In addition

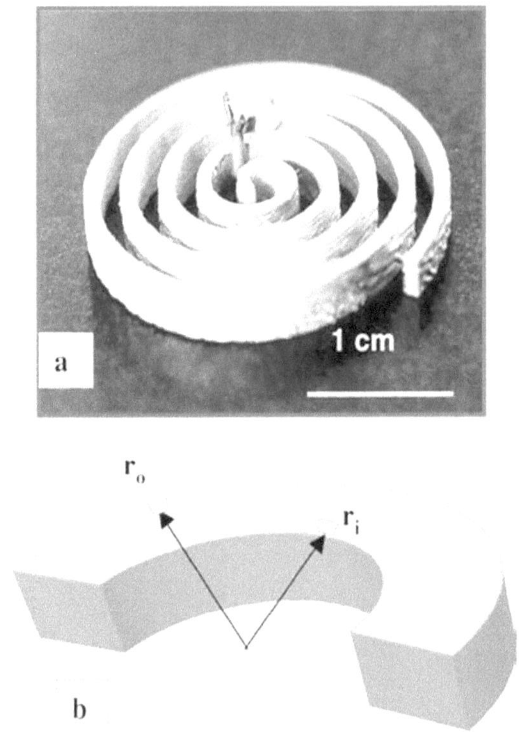

Figure 19 (a) Spiral actuator and (b) schematic of a curved piezoelectric segment, illustrating elements in the outer and inner diameters. A slight displacement gradient exist across the thickness, resulting in bending stress and segment rotation.

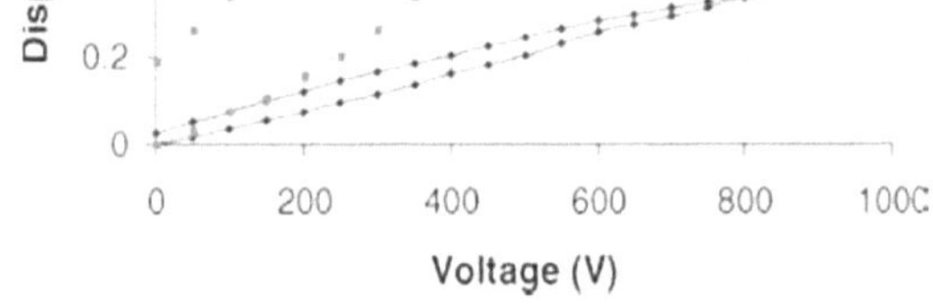

Figure 20 Displacement enhancement in PZT spiral actuator as a result of adding a 37.5 μm shim.

to tubes with helical electrodes, multilayer tube actuators (Fig. 21d) also benefit from the larger piezoelectric coefficient of d_{33}. Successful fabrication of these actuators depends upon the electrode materials, its thickness, and sintering conditions/atmosphere [104, 105].

It has been shown that the actuation of a piezoelectric element can be improved if a given structure is curved. For example, dome [101], RAINBOW (Reduced and

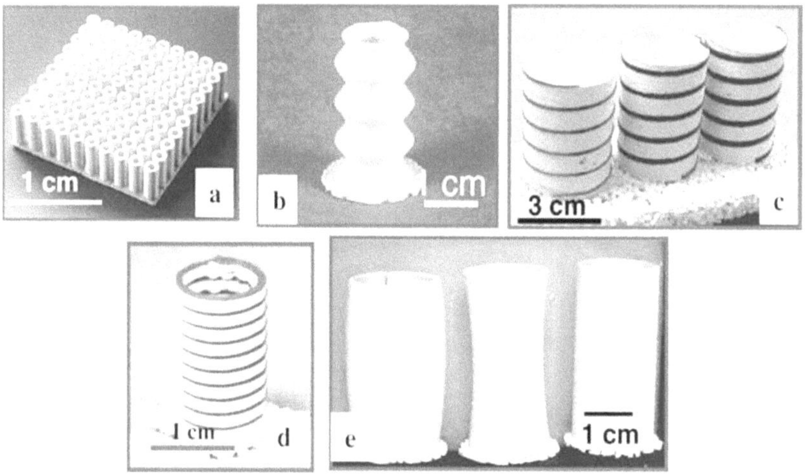

Figure 21 Various tube actuators made by the FDC process. (a) tube array, (b) bellows, (c) tubes with helical electrodes, (d) multilayer tube, and (e) tubes with curved walls.

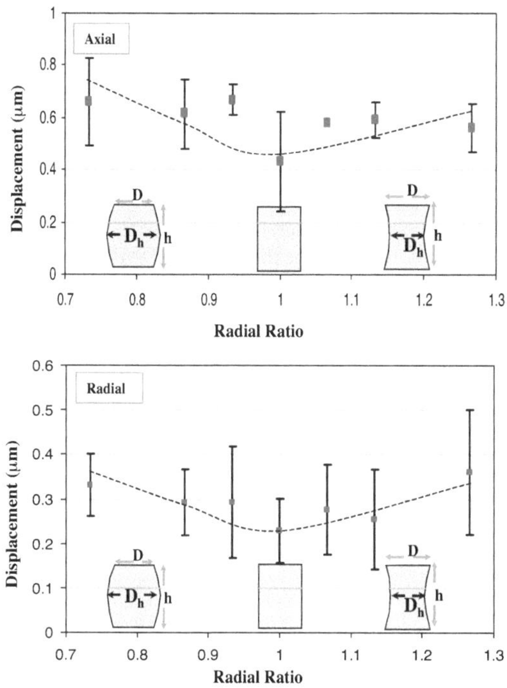

Figure 22 The effect of tube curvature on the axial (top) and radial (bottom) displacements of the tubes (radial ratio is D/Dh).

Internally Biased Oxide Wafer [1]), and THUNDER (Thin Layer Unimorph Driver [99] actuators illustrate displacement enhancement as a result of curved geometry and internal stresses. In order to determine if curvature would have a positive impact on tube actuators, PZT tubes with curved walls (both with inward and outward curvatures, as shown in Fig. 21e, were designed, prototyped, and characterized. All the green tubes were of the same height, end diameter and wall thickness (20 , 15 , and 1 mm, respectively). The axial and radial displacements of these tubes, measured by a photonic sensor, are plotted in Fig. 22. Despite the large variation of data points, it appears that the regular (straight-walled or radial ratio of 1) tubes showed minimum displacement whereas the curved-wall tubes showed slightly larger actuations, likely due to the wall curvature. It is speculated that the constrained geometry of these actuators (tubular shape) limits the contribution of the curvature, and therefore, only slight improvement can be gained in radial and axial displacements.

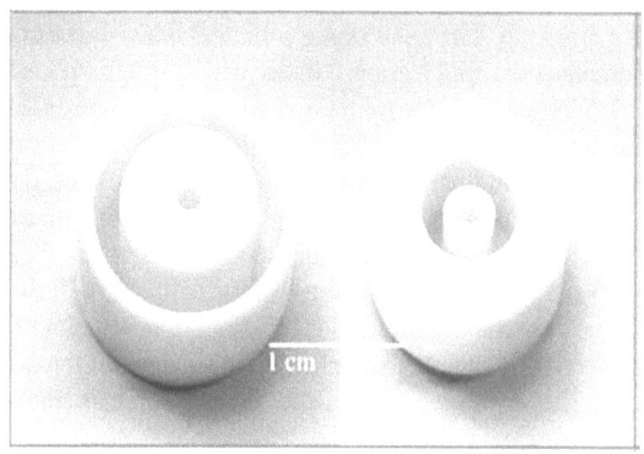

Figure 23 Photograph showing the top and bottom faces of a telescoping actuator structure.

Telescoping actuators

Another monolithic ceramic actuator with telescoping actuation was also fabricated by the FDC method. The device consists of interconnected PZT ceramic tubes bearing a common center with internal displacement amplification under electric field (Fig. 23). Upon the application of an

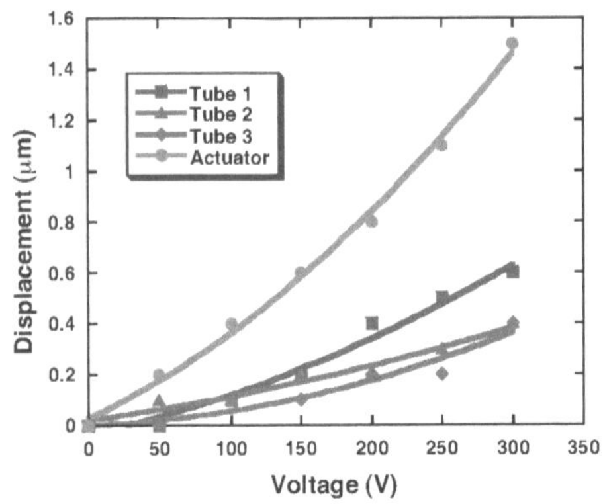

Figure 24 Field induced displacement of a telescoping actuator.

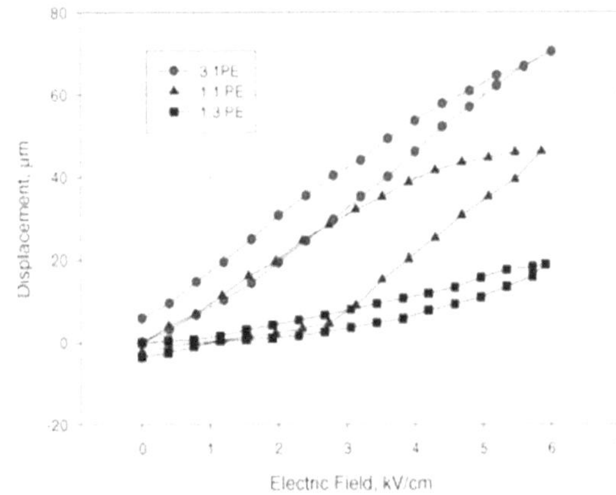

Figure 26 Displacement vs. electric field for 1:1, 1:3, and 3:1 PE monomorph (Thickness=1.26 mm).

electric field, the tubes elongate and shrink alternatively in the z direction, yielding a net additive displacement in the z direction. The telescoping actuators consist of three interconnected PZT ceramic tubes with a common center. Under applied electric field, the inner and outer tubes elongate and the middle tube shrinks in the z direction. The telescoping actuation of the device is the summation of actuation of all individual tubes in the actuator (Fig. 24). Therefore, increasing the number of the tubes, which are the driving component of the actuator, will further increase the displacement. Tailoring of the thickness, height, and the number of driving components of the actuator, the displacement and the exerted force of these actuators, can be altered.

Monomorph bending actuators

Monolithic multi-material monomorphs comprised of varying ratios of piezoelectric $0.65Pb(Mg_{2/3}Nb_{1/3})O_3$-$0.35PbTiO_3$ to electrostrictive $0.90Pb(Mg_{2/3}Nb_{1/3})O_3$-$0.10PbTiO_3$, have been designed and fabricated. The strain gradient between the piezoelectric and electrostrictive layers causes a bending moment to develop and therefore, the tip deflects. The schematic of piezoelectric/electrostrictive actuators is shown in Fig. 25 The relative permittivity, displacement, and polarization hysteresis were investigated for varying ratios of piezoelectric to electrostrictive material. The P-E hysteresis loop of the 1:1 sample exhibited saturation and remnant polarization slightly less than the piezoelectric PMN-PT 65/35, but higher than the electrostrictive PMN-PT 90/10.

Fig. 26 depicts the dc-displacement of 1:1, 3:1, and 1:1 monomorph actuators, respectively. The maximum tip deflection of the 1:1 monomorph was found to be 44 mm at 6.0 kV/cm bias field strength (actuator dimensions: 30 mm in length, 8 mm in width and 1.3 mm in thickness). The maximum tip displacement of the 3:1

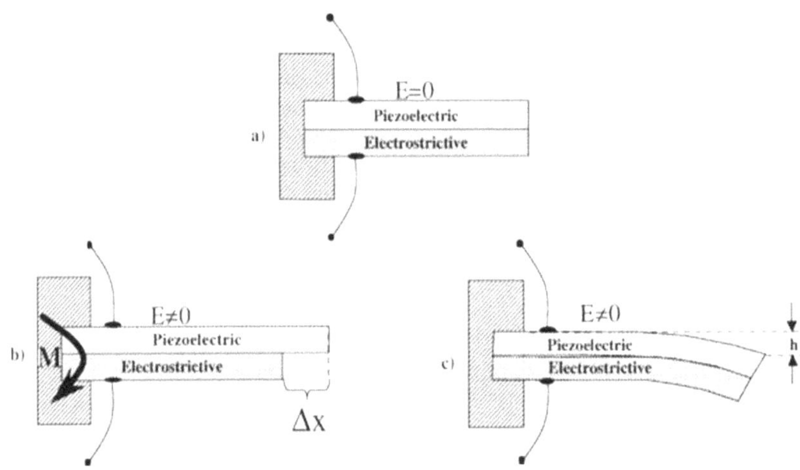

Figure 25 (a) Typical bender configuration used in this study, (b) Application of an electric field causes a transverse differential strain Δx (c) Transverse differential strain caused by tip deflection.

monomorph actuator was 76 mm at 6 kV/cm, which corresponds to 50% increase with respect to the 1:1 monomorph. It is interesting to note that the hysteresis of the 3:1 monomorph is also much less in comparison to the 1:1 despite its higher piezoelectric volume fraction, indicating that it would provide for a better positioning accuracy. The tip displacement of the 1:3 monomorph is the smallest of the three, with a maximum at ~20 mm at 6 kV/cm. The aforementioned comparison indicates that the tip displacement in such monomorph actuators can substantially be increased by clamping the transverse displacement of the piezoelectric with an electrostrictive layer that is lower in volume fraction. The electrostrictive layer also imparts lower hysteresis to the actuators, which is desirable in regard to positioning precision. In principle, one should be able to further increase the tip displacement, while at the same time reducing the hysteresis by decreasing the thickness of the electrostrictive layer.

Oval actuators

Monolithic piezoelectric actuators with oval geometry have been prototyped by FDC and characterized for applications such as fluid delivery systems as depicted in Fig. 27 [107]. The minor diameter of the ovals varied between 2 and 14 mm, and their major diameter, wall thickness, and width were 20, 0.85, and 7 mm, respectively. When driven under electric field, the actuators expanded along their minor diameter. The static and dynamic displacements of 7 and 5.6 mm were observed at 850 V DC and 100 V AC. The static displacement of the ovals varied almost linearly with voltage and did not change under the application of external load in the range of 1–15 N [107]. Fig. 27 displays the effect of minor diameter on the displacement of ovals at their resonance frequency driven under a 100 V AC sinusoidal excitation at 50 Hz [107]. As seen, the ovals with 4 mm minor diameter had the maximum displacement of 5.6 mm. The displacement at 100 V is notably amplified ~7 times at the resonance frequency. Such amplification is important for applications in piezoelectric motors where actuators usually operate in their resonance mode. As the minor diameter increases above 4 mm, the displacement decreases. From the geometric point of view, an oval gradually transforms into a tube when the minor diameter approaches the major diameter. In a tube, the actuation is attributed to the expansion and contraction of the wall thickness due to the contribution of the d_{31} and d_{33} coefficients. Thus similar to a tube, the displacement of the oval actuators is attributed to the contribution of the d_{31} and d_{33} coefficients. This denotes that when the displacement is in the range 4–14 mm, a non-uniform stress develops due to the effect of d_{31} coefficient

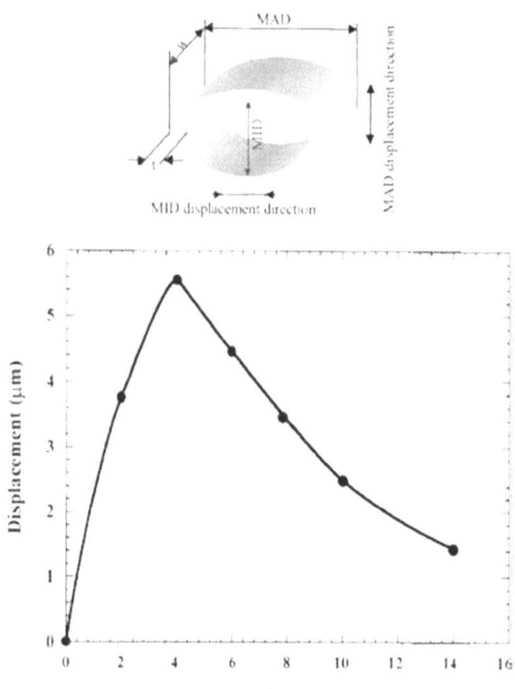

Figure 27 Schematic of Oval Actuators with minor and major diameter (MID and MAD), and the effect of effect of minor diameter on the displacement of ovals at their resonance frequency driven under 100 V ~ac sinusoidal wave at 50 Hz.

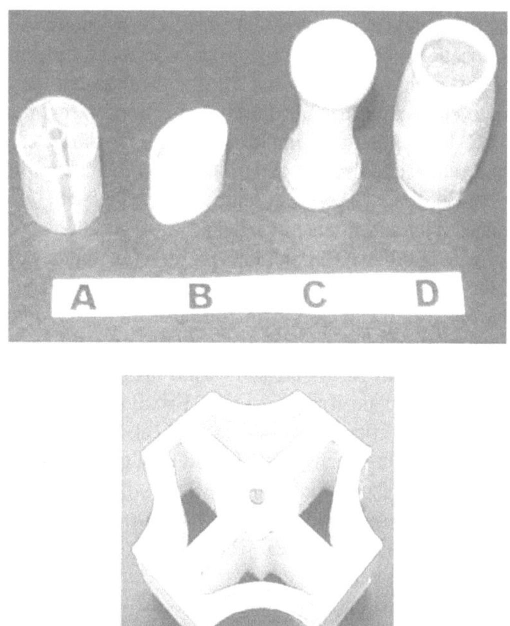

Figure 28 Monolithic multimode transducers with different designs: A. Wagon-Wheel (extruded), B. Class IV (FDC), C. Concave Class I (FDC), D. Convex Class I (FDC) and a Class VII design (bottom).

that creates a bending moment and actuation. When the minor diameter decreases below 4 mm, the displacement again decreases. This suggests that the oval structure behaves similar to a flat ceramic and therefore the bending contribution, as a result of the d_{31}, coefficient becomes less pronounced [107].

New monolithic multimode transducers

Recently, an array of novel multi-mode monolithic piezoelectric transducers has been fabricated using the FDC process [108]. Fig. 28 shows the photographs of some of the prototyped multimode transducers, which include flextensional transducers such as the Wagon Wheel, Class IV, Concave Class I, and Convex Class I, and vector sensors (projectors) [108]. These transducer designs possess internal displacement and/or stress amplification mechanisms that allow tuning of the resonance frequencies and optimization of the vibration modes. In addition, the monolithic nature of the transducers and the flexibility to driving each segment of the transducers separately or in tandem allow one to tune the bandwidth, acoustic radiation beam pattern, and directional response in an unprecedented fashion [108].

5. Summary

There have been great developments in fabrication of piezoelectric ceramic-polymer composites in the last two decades, which were summarized in the paper. The dice and fill method as well as recent fabrication methods, such as injection molding, the relic process, jet machining and the lost mold techniques, are methods with much to offer to composite making. Fused deposition of ceramics (FDC) is an effective technique to further develop piezocomposite technology as demonstrated by the fabrication of "ladder" and "3-D Honeycomb" composites with (3-3) connectivity and oriented (2-2) and (3-3) piezocomposites among others. The indirect and direct methods were both shown to be suitable for making composites. These processes start with a CAD file for the desired part and/or mold. In the indirect method, a CAD file for the negative of the final part is created, while the direct method uses directly deposit the final part. Excellent electromechanical properties for transducer applications have been obtained by these techniques. In addition, the FDC technique clearly shows the ability to form composites with controlled phase periodicity to vary volume fractions and micro and macro-structures, which are not possible with the traditional techniques.

Fused deposition of ceramics has also been used to make an array of novel actuators in the last decade. The flexibility of the FDC system allows for rapid fabrication of parts with frequent change of dimensions for optimization purposes. Actuators such as tubes with straight, bellows, and curved wall geometry, spirals, telescoping, ovals, and monomorphs have been prototyped and characterized. Spiral actuators show large tangential displacements with moderate blocking force. Monomorph piezoelectric/electrostrictive PMN-PT bender is a novel design with large tip deflection. By varying the ratio of the piezoelectric to electrostrictive layer, it is feasible to optimize the field-induced strain of the actuator for a given application. SFF technology is a great tool for rapid prototyping of new designs and verification of modeling.

Acknowledgments

We wish to thank for the generous financial support provided by the Office of Naval Research (ONR), the Defense Advanced Research Projects Agency (DARPA), and New Jersey Commission on Science and Technology (NJCTS). The contributions of S. C. Danforth, N. Langrana, M. Jafari, S. Guceri, and T.-W. Chou to the SFF work presented herein is gratefully acknowledged. We also wish to thank the past members of the Electroceramics Research Group at Rutgers University, A. Bandhophadhyay, R. K. Panda, T. F. McNulty, S. Turcu, G. Lous, F. Mohammadi, and B. Jadidian, whose work is summarized in this article.

References

1. K. UCHINO, in "Piezoelectric Actuators and Ultrasonic Motors" (Kluwer Academic Publisher, New York, 1996).
2. B. JAFFE, W. R. COOK and H. JAFFE, in "Piezoelectric Ceramics" (R.A.N. Publishers, Marietta, Ohio, 1971).
3. Y. XU, in "Ferroelectric Materials and Their Applications" (North-Holland, Amsterdam, 1991).
4. L. E. CROSS, in "Ferroelectric ceramics: Tailoring properties for specific applications," edited by N. Setter and E. L. Colla. (Birkhauser, Basel, 1993).
5. G. GOODMAN, *J. Amer. Ceram. Soc.* **36** (1960) 368.
6. E. C. SUBBARAO, *ibid.* **43** (1960) 439.
7. G. Y. XU, Z. ZHONG, Y. BING, Z. G. YE, C. STOCK and G. SHIRANE, *Phys. Rev.* **B 67** (2003) 104102.
8. Z. G. YE, Y. BING, J. GAO, A. A. BOKOV, P. STEPHENS, B. NOHEDA and G. SHIRANE, *Phys. Rev.* **B 67** (2003) 104104.
9. H. M. JI, Z. BING and M. X. XU, *Rare Metal Mater. Eng.* **31** (2002) 292.
10. Y. H. BING, R. GUO and A. S. BHALLA, *Ferroelectrics* **242** (2000) 1.
11. Y. H. BING and Z. G. YE, *J. Cryst. Growth* **250** (2003) 118.
12. N. YASUDA, N. UEMURA, H. OHWA, Y. YAMASHITA, M. IWATA, M. MAEDA, I. SUZUKI and Y. ISHIBASHI, *J. Korean Physical Soc.* **42** (2003) S1261.
13. M. IWATA, N. TOMISATO, H. ORIHARA, H. OHWA, N. YASUDA and Y. ISHIBASHI, *Ferroelec.* **261** (2001) 747.
14. M. IWATA, N. TOMISATO, H. ORIHARA, N. ARAI, N. TANAKA, H. OHWA, N. YASUDA and Y. ISHIBASHI, *Jpn. J. Appl. Phys. Part 1* **40** (2001) 5819.
15. H. OHWA, M. IWATA, H. ORIHARA, N. YASUDA and Y. ISHIBASHI, *J. Phys. Soc. Jpn.* **70** (2001) 3149.
16. R. E. EITEL, C. A. RANDALL, T. R. SHROUT and S. E. PARK, *Jpn. J. Appl. Phys. Part 1* **41** (2002) 2099.
17. S. J. ZHANG, L. LEBRUN, S. RHEE, R. E. EITEL, C. A. RANDALL and T. R. SHROUT, *J. Crys. Growth* **236** (2002) 210.

18. S. J. ZHANG, P. W. REHRIG, C. RANDALL and T. R. SHROUT, *ibid.* **234** (2002) 415.
19. S. J. ZHANG, S. RHEE, C. A. RANDALL and T. R. SHROUT, *Jpn. J. Appl. Phys. Part 1* **41** (2002) 722.
20. S. J. ZHANG, C. A. RANDALL and T. R. SHROUT, *Appl. Phys. Lett.* **83** (2003) 3150.
21. L. E. CROSS, *Nature* **432**(4) (2004) 84 (and references cited therein).
22. D. P. SKINNER, R. E. NEWNHAM and L. E. CROSS, *Mat. Res. Bull.* **13** (1978) 599.
23. R. E. NEWNHAM, *Ferroelec.* (1986) 1.
24. H. P. SAVAKUS, K. A. KLICKER and R. E. NEWNHAM, *Mater. Res. Bull.* **16** (1981); KLICKER, Ph.D. Thesis. (The Pennsylvania State University, 1980).
25. J. W. SLIWA, S. AYTER and J. P. MOHR, U. S. Pat. No. 5239736, 1993.
26. R. L. GENTILMAN, D. F. FIORE, H. T. PHAM, K. W. FRENCH and L. J. BOWEN, in "Ceramic Transactions, Ferroic Materials: Design, Preparation, and Characteristics", edited by A. S. Bhalla, K. M. Nair, I. K. Lloyd, H. Yanagida and D. A. Payne. (American Ceramic Society, Westerville, OH, 1994), Vol. 43: 239.
27. L. J. BOWEN and K. W. FRENCH, IEEE *Proceedings of the Int'l Symp. on the Appl. of Ferroelec.* (1992), 160.
28. L. J. BOWEN, R. L. GENTILMAN, H. T. PHAM, D. F. FIORE and K. W. FRENCH, in Proceedings of the IEEE Ultrasonics Symposium (1993), 499.
29. R. L. GENTILMAN, D. FIORE, H. PHAM, W. SERWATKA and L. BOWEN, in Proceedings of the SPIE: Industrial and Commercial Applications of Smart Structures Technologies, Vol. 2447, (1995) p. 274.
30. W. WERSING, in Proceedings IEEE International Symposium on Applications of Ferroelectrics (1986) p. 212.
31. E. W. BECKER, W. EHRFELD, P. HAGMANN, A. MANER and D. MUNCHMEYER, *Microelec. Eng.* **4** (1986) 35.
32. U. BAST, D. CRAMER and A. Wolff, in Ceramics Today-Tomorrow's Ceramics, edited by. P. Vincenzini. (Elsevier Science Publishers, Amsterdam, 1991), Vol. 66C.
33. U. BAST, H. KAARMANN, K. LUBITZ, M. VOGT, W. WERSING and D. CRAMER, U.S. Patent No. 5164920 (1992).
34. K. LUBITZ, A. WOLFF and G. PREU, *Ferroelec.* **133** (1992).
35. K. LUBITZ, A. WOLFF and G. PREU, in IEEE Ultrasonics Symposium (Piscataway, New Jersey, 1993).
36. K. LUBITZ, A. WOLFF and B. SCHULMEYER, *Ferroelec.* **133** (1992) 21.
37. K. A. LUBITZ, A. WOLFF and G. PREU, Proc. 1993 IEEE Ultrasonics Symposium (Piscataway, New Jersey, 1993).
38. D. J. WALLER, M. S. Thesis, Rutgers University, New Brunswick, NJ (1991).
39. D. J. WALLER, A. SAFARI and R. J. CARD, Proc. of the 7th IEEE Int'l Symp. on the Appl. of Ferroelectr., IEEE, (Piscataway, NJ, USA, 1990) p. 82.
40. J. ZOLA, U.S. Patent No. 4572981 (1986).
41. S. LIVNEH, V. F. JANAS and A. SAFARI, *J. Am. Ceram. Soc.* **78** (1995) 1900.
42. D. J. WALLER, T. IQBAL and A. SAFARI, *ibid.* **72** (1989) 322.
43. S. S. LIVNEH, S. M. TING and A. SAFARI, *Ferroelec.* **157** (1994) 421.
44. J. W. STEVENSON, M. R. REIDMEYER and W. HEUBNER, *J. Am. Ceram. Soc.* **77** (1994) 2491.
45. C. A. RANDALL, D. V. MILLER, J. H. ADAIR and A. S. BHALLA, *J. Mater. Res.* **8** (1993) 899.
46. C. A. RANDALL, C. P. BOWEN, T. R. SHROUT, A. S. BHALLA and R. E. NEWNHAM, in Proceedings of the 6th U.S. Japan Seminar on Dielectric and Piezoelectric Ceramics (1993) p. 152.
47. M. EYETT, D. BAUERLE and W. WERSING, *J. Appl. Phys.* **62** (1987) 1511.
48. Y. OHARA, M. MIYAYAMA, K. KOUMOTO and H. YANAGIDA, *Sensors and Actuators A* **40** (1994) 345.
49. *Idem., J. Mater. Sci. Lett.*, **12** (1993) 1279.
50. M. J. CREEDON, S. GOPALAKRISHAN and W. A. SCHULZE, IEEE International Symposium on the Application of Ferroelectrics (State College, PA, 1994).
51. M. J. CREEDON and W. A. SCHULZE, *Ferroelec.* **153** (1994) 333.
52. J. D. ERVIN, D. BREI, C. A. V. HOY, J. R. MAWDSLEY and J. W. HALLORAN, in Proceedings of the ASME Aerospace Division (1996), Vol. 52, pp. 695–697.
53. C. A. V. HOY, A. BARDA, M. GRIFFITH and J. W. HALLORAN, *J. Am. Ceram. Soc.* **81** (1998) 152.
54. D. P. SKINNER, R. E. NEWNHAM and L. E. CROSS, *Mat. Res. Bull.* (1978) 1553.
55. K. A. KLICKER, W. A. SHULTZE and J. V. BIGGERS, *J. Am. Ceram. Soc.* **65** (1982) C208.
56. K. RITTENMYER, T. SHROUT, W. A. SCHULZE and R. E. NEWNHAM, *Ferroelec.* **41** (1982) 323.
57. Materials System Inc., www.matsysinc.com, (2004).
58. W. HACKENBERGER, P. MING-JEN, D. KUBAN, T. RITTER and T. SHROUT, Novel method for producing high frequency 2-2 composites from PZT ceramic, in IEEE Proceedings of the Ultrasonics Symposium, (2000), Vol. 2, pp. 969.
59. R. P. SCHAEFFER, V. F. JANAS and A. SAFARI, Proceedings of the 10th IEEE International Symposium on Applications of Ferroelectrics, Part 2 of 2 (1996).
60. W. HUEBNER, M. R. REIDMEYER, J. W. STEVENSON and L. BUSSE, Proceedings of the 9th IEEE International Symposium on Applications of Ferroelectrics (University Park, PA, USA, 1994).
61. D. M. MILLS and S. W. SMITH, *IEEE Transactions of the Ultrasonics, Ferroelectrics and Frequency Control.* **49** (2002) 1005.
62. C. VAN HOY, A. BARDA, M. GRIFFITH and J. W. HALLORAN, *J. Amer. Ceramic Soc.* **81** (1998) 152.
63. J. W. HALLORAN, *Brit. Ceramic Trans.* **98** (1999) 299.
64. H. L. MARCUS, J. J. BEAMAN, J. W. BARLOW, D. L. BOURELL and R. H. CRAWFORD, Solid Freeform Fabrication Symposium, (Austin, TX, 1992).
65. H. L. MARCUS, J. J. BEAMEN, J. W. BARLOW, D. L. BOURELL and R. H. CRAWFORD, Solid Freeform Fabrication Proceedings (Austin, TX, 1992).
66. H. L. MARCUS, J. J. BEAMEN, J. W. BARLOW, D. L. BOURELL and R. H. CRAWFORD, Solid Freeform Fabrication Proceedings, (University of Texas at Austin, Austin, TX, 1991).
67. H. L. MARCUS and D. L. BOURELL, *Adv. Mat. Proc.* **9** (1993) 677.
68. H. L. MARCUS, *Mech. Eng.* **117** (1995) 62.
69. D. L. BOURELL, J. J. BEAMEN, H. L. MARCUS and J. W. BARLOW, Solid Freeform Fabrication Proceedings (Austin, TX, 1990).
70. P. F. JACOBS, "Stereolithography and Other RP & M Technologies from Rapid Prototyping to Rapid Tooling," Dearborn, MI: Society of Manufacturing Engineers (1995).
71. P. F. JACOBS, "Rapid Prototyping & Manufacturing: Fundamentals of StereoLithography," Dearborn, MI: Society of Manufacturing Engineers (1992).
72. M. FEYGIN and B. HSIEH, in Proceedings of the Solid Freeform Fabrication Symposium. (1991), p. 123.
73. M. FEYGIN, U. S. Patent No. 5354414 (1994).
74. J. CESARANO, T. A. BAER and P. CALVERT, in Proceedings of the Solid Freeform Fabrication Symposium. (1997) Vol. 8, pp. 25.

75. J. CESARANO, B. H. KING and H. B. DENHAM, in Proceedings of Solid Freeform Fabrication Symposium, (1998) Vol. 9, pp. 697.
76. E. M. SACHS, M. J. CIMA, P. WILLIAMS, D. BRANCAZIO and J. CORNIE, *J. Eng. Ind.* **114** (1992) 481.
77. R. K. PANDA, Ph.D. Thesis, Rutgers University, New Brunswick, New Jersey (1998).
78. J. CESARANO and P. CALVERT, U.S. Patent No. 6027326 (2000).
79. A. SAFARI, Ph.D. Thesis, The Pennsylvania State University, State College, PA (1983).
80. W. SCHULZE, *Ferroelectrics* **50** (1983) 33.
81. T. R. SHROUT, L. J. BOWEN and W. A. SCHULZE, *Mater. Res. Bulletin* **15** (1980) 1371.
82. A. SAFARI, *ibid.* **17** (1982) 301.
83. S. TURCU, M.S. Thesis, Rutgers University, New Brunswick, NJ (2002).
84. S. TURCU, B. JADIDIAN, S. C. DANFORTH and A. SAFARI, *J. Electroc.* **9** (2002) 165.
85. C. W. NAN, L. LIU and L. LI, *J. Phys. D* **33** (2000) 2988.
86. C. W. NAN and G. W. WENG, *J. Appl. Phys.* **88** (2000) 416.
87. A. BANDYOPADHYAY, R. K. PANDA, V. E. JANAS, M. K. AGARWALA, S. C. DANFORTH and A. SAFARI, *J. Amer. Ceramic Society* **80** (1997) 1366.
88. A. SAFARI, V. F. JANAS and R. K. PANDA, SPIE Symposium on Smart Structures and Materials, (San Diego, CA, 1996).
89. A. SAFARI, J. CESARANO, P. G. CLEM and B. BENDER, Proceedings of the 13th IEEE International Symposium on Applications of Ferroelectronics, (Nara, Japan, 2002).
90. G. H. HAERTLING, *Am. Ceram. Soc. Bull.* **73** (1994) 93.
91. F. MOHAMMADI, A. KHOLKIN, B. JADIDIAN and A. SAFARI, *Appl. Phys. Lett.* **75** (1999) 2488.
92. Q. M. WANG, Q. ZHANG, B. XU, R. LIU and L. E. CROSS, *J. Appl. Phys.* **86** (1999) 3352.
93. A. SAFARI, S. C. DANFORTH, M. A. JAFARI, M. ALLAHVERDI, B. JADIDIAN, F. MOHAMMADI, N. VANKATARAMAN and S. RANGARAJAN, in Proceedings of the 9th European Conference on Rapid Prototyping and Manufacturing, (2000) p. 247.
94. A. SAFARI and S. C. DANFORTH, in Proceedings of the 11th IEEE International Symposium on the Applications of Ferroelectrics, (1998) p. 229.
95. A. BANDYOPADHYAY, R. K. PANDA, T. F. MCNULTY, F. MOHAMMADI, S. C. DANFORTH and A. SAFARI, *Rapid Prototyping Journal* **4** (1998) 37.
96. A. SAFARI and M. ALLAHVERDI, *Ceram. Eng. Sci. Proc.* **22** (2001) 473.
97. F. MOHAMMADI, Ph.D. Thesis, Department of Ceramic and Materials Engineering, Rutgers University, New Jersey, USA (2001).
98. F. MOHAMMADI, A. KHOLKIN, S. C. DANFORTH and A. SAFARI, in Proceedings of the 11th IEEE International Symposium on the Applications of Ferroelectrics, pp. 273–275 (1998).
99. S. A. WISE, *Sensors and Actuators A* **69** (1998) 33.
100. Q. M. ZHANG, H. WANG and L. E. CROSS, *J. Mater. Sci.* **28** (1993) 3962.
101. Y. SUGAWARA, K. ONITSUKA, S. YOSHIKAWA, Q. C. XU, R. E. NEWNHAM and K. UCHINO, *J. Am. Ceram. Soc.* **75** (1992) 996.
102. A. M. UMARJI, A. L. KHOLKIN, T. F. MCNULTY, S. C. DANFORTH and A. SAFARI, in Proceedings of the 11th IEEE International Symposium on the Applications of Ferroelectrics, (1998) p. 269.
103. T. W. CHOU, B. A. CHEESEMAN, A. SAFARI and S. C. DANFORTH, *Ceramic Eng. Sci. Proc.* **22** (2001) 497.
104. J. CHEN, Q. M. ZHANG, L. E. CROSS and C. M. TROTTIER, in Proceedings of International Conference on Intelligent Materials (ICIM), Williamsburg, VA, pp. 316–318 (1994).
105. M. ALLAHVERDI and A. SAFARI, *J. Eur. Ceram. Soc.* **21** (2001) 1485.
106. M. ALLAHVERDI, B. JADIDIAN, B. HARPER, S. RANGARAJAN, M. JAFARI, S. C. DANFORTH and A. SAFARI, in Proceedings of 12th IEEE International Symposium on Applications of Ferroelectrics, edited by S. K. Streiffer, B. J. Gibbons and T. Tsurumi, (IEEE-UFFC, New Jersey, 2001) p. 381.
107. B. JADIDIAN, M. ALLAHVERDI, F. MOHAMMADI and A. SAFARI, *Appl. Phys. Lett.* **80** (2002) 1981.
108. R. E. NEWNHAM, D. C. MARKLEY, R. J. MEYER JR, W. J. HUGHES, A.-C. HLADKY-HENNION and J. K. COCHRAN JR, *Ceramic Transactions* **150** (2004) 427.

JOURNAL OF MATERIALS SCIENCE **41** (2006) 199–210

Kinetics of ferroelectric domains: Application of general approach to $LiNbO_3$ and $LiTaO_3$

VLADIMIR YA. SHUR
Ferroelectric Laboratory IPAM, Ural State University, Ekaterinburg, Russia
E-mail: vladimir.shur@usu.ru

We review the most interesting aspects of the domain structure kinetics in ferroelectrics important for "domain engineering" and discuss them in the framework of a unified nucleation approach. In our approach the nucleation rate is determined by the local value of electric field produced not only by bound charges and voltage applied to the electrodes, but also by screening charges. As a result, any kinetically produced domain pattern, even being far from the equilibrium, can be stabilized by bulk screening. The domain evolution represents a self-organizing process in which the screening of polarization plays the role of feedback. The general approach was applied for the description of the domain kinetics in lithium niobate and lithium tantalate as the most versatile materials for applications. The revealed original scenarios of the domain structure evolution are attributed to the retardation of the screening processes. The decisive role of screening effectiveness for shapes of individual domains and scenarios of the sideways domain wall motion is demonstrated both experimentally and by computer simulation. The possibility to produce a self-assembled nano-scale domain structures with controlled periods has been shown.

1. Introduction

A new branch of science and technology directed to the creation of periodic and quasi-periodic domain structures with desired parameters in commercially available ferroelectrics denoted as "domain engineering" is rapidly developing nowadays. Domain engineering in ferroelectric crystals, such as lithium niobate $LiNbO_3$ (LN) and lithium tantalate $LiTaO_3$ (LT), has revolutionized their use in nonlinear optical applications [1, 2]. The performance of LN and LT as an electro-optic, photorefractive, piezoelectric, and nonlinear optical crystals, make them useful for many different applications. It has been shown that LN and LT with periodical 1*D*- and 2*D*-domain structures possessing an efficient quasi-phase-matching open up a wide range of possibilities for bulk and waveguide nonlinear optical devices [2–4]. During ten years after the first electrical poling of bulk LN samples [5], research on periodically poled LN and LT is under intense interest around the world resulting in production of photonic devices. Breaking the micron-period barrier for periodical domain patterning in LN and LT is very desirable for several new electro-optic applications such as tunable cavity mirrors, which need a periodicity of about 350 nm. The most efficient exploitation of engineered sub-micron domain gratings in LN and LT is related with the waveguide photonic devices.

The switching by application of the external field using the electrode patterns produced by photolithography is the most popular method of periodical poling. Fabrication of the precisely governed regular domain patterns with periods about several microns needs the solution of key problems hindering the improvement of the characteristics of electro-optical and nonlinear optical devices [1]. The broadening of the domains out of electrode area leads to violation of duty-cycle and period due to domain merging. Loss of stability is a great problem for production of sub-micron domain patterns. The uniformity of the domain patterns is destroyed in the bulk. It is impossible to produce the individual domains with desirable shapes which is important for 2*D* patterning. It is clear that the understanding of the physical basis of the domain engineering is the only way to overcome the listed problems. The polarization reversal phenomenon demonstrating essential dependence on the experimental conditions and material still waits for a systematic explanation based on a universal approach.

0022-2461
DOI: 10.1007/s10853-005-6065-7

We present the systematic explanation of the experimental investigations of the domain kinetics in a wide range of domain growth velocities in uniaxial ferroelectrics with optically distinguished domains, LN and LT. We investigate in detail the domain evolution using *in situ* optical observation of the instantaneous domain patterns under application of an electric field. "Slow", "fast" and "super-fast" domain boundary motion regimes have been revealed and investigated. It was claimed that the key property for proposed classification is the screening effectiveness. The crucial role of the intrinsic or artificial dielectric surface layer is clearly exhibited as well. Computer simulation has been used for a verification of the proposed models.

In this paper we introduce the experimental evidence and theoretical considerations which reveal the kinetic nature of the observed domain configurations. In other words, the real domain configurations are determined by the domain kinetics prehistory. We demonstrate the crucial role of the bulk screening processes in the stabilization of the metastable domain structures [6–8]. This approach allows one to choose the proper experimental conditions for stabilization of almost any domain pattern.

2. Domain structure evolution during polarization reversal

Direct experimental visualization of domain kinetics has revealed four main stages of domain evolution [9].

The first stage, "*nucleation of new domains*", develops in perfect crystals at the surface only. The nucleation probability $p \sim \exp(-E_{ac}/E_{loc})$ is determined by activation field E_{ac}, which depends on intrinsic material properties, shape of nuclei and temperature, and electric field averaged over the volume of the nucleus E_{loc}. Experimental observation reveals the nucleation sites at intrinsic or artificial surface defects [8]. The facilitation of the nucleation in real crystals at the given sites is caused by: (1) frozen-in spatial inhomogeneity of E_{ac} due to presence of structural defects, (2) singularities of E_{loc} along the edges of finite electrodes, and (3) E_{loc} created by field concentrators resulting from pits at the sample surface. Moreover it is an open question weather the initial domain state is single-domain or contains the nano-scale residual domains [8, 9]. In the latter case the nucleation of new domains represents a field-induced transition of the "invisible" residual nano-domains into "visible" ones.

The second stage, "*forward growth*", represents an expansion of the formed "nuclei" in polar direction through the sample. All domains at this stage have the charged domain walls and needle-like shape is the most typical. For example, the ratio of the longitudinal to transversal domain sizes in *LN* reaches one hundred [10].

The third stage, "*sideways domain growth*", represents the domain expansion by wall motion in the direction transversal to the polar axis. This stage is the best studied experimentally by *in situ* optical observation of domains. The wall motion anisotropy results in formation of the polygon domains with sides oriented along crystallographic directions.

The fourth stage, "*domain coalescence*", prevails at the completion of the switching process. At this stage the sideways wall motion decelerates. The walls halt and the residual region between walls disappears very rapidly after certain rest time. This process causes the jump-like switching behavior. It is displayed as a noise component of the switching current, known as Barkhausen noise [11–13].

3. General consideration

We discuss all experimentally observed stages of domain kinetics from a unified point of view. According to this approach, the domains with different orientation of spontaneous polarization are considered as the regions of different phases divided by domain walls which represent the phase boundaries. Thus the domain structure evolution during switching is an example of a first-order phase transformation.

The theory of a first-order phase transformation based on kinetic theory of nucleation in its classical version is 80 years old [14, 15]. Nevertheless up to now it is still an active field of research [16].

This approach has been used for description of the growth of solid fraction during crystallization of metals in famous Kolmogorov–Mehl–Johnson–Avrami (KMJA) theory [17, 18]. Ishibashi and Takagi have applied KMJA theory for explanation of the time dependence of the ferroelectric switching current behavior [19].

We will demonstrate how the consistent application of the concepts of the first-order phase transformation theory can be used for explanation of the main features of the domain kinetics during polarization reversal. In this approach all above mentioned stages of the domain kinetics are governed by evolution of thermally activated nuclei with a preferred orientation of the spontaneous polarization.

The nucleation processes: The domain kinetics under application of electric field is governed by competition between nucleation processes of different dimensionalities. Each nucleus represents the minimum domain with preferential orientation of the spontaneous polarization determined by the direction of the local electric field. The appearance of new domains is due to formation of three-dimensional (*3D*) nuclei. The domain growth by motion of the domain walls is a result of *2D*-nucleation (generation of the steps at the wall) and *1D*-nucleation (step growth along the wall). The nucleation probability depends on E_{loc}, being the driving force of all nucleation processes during polarization reversal [8].

Local value of electric field: The internal electric field E_{loc} representing a superposition of electric fields produced by different sources is essentially inhomogeneous. In a ferroelectric capacitor $E_{loc}(r, t)$ is determined mainly by the sum of (1) the external field $E_{ex}(r)$, produced by the voltage applied to the electrodes, (2) the depolarization field $E_{dep}(r, t)$ produced by bound charges developing as a result of spatial inhomogeneity of the spontaneous polarization, (3) the external screening field $E_{scr}(r, t)$ originating from the redistribution of the charges at the electrodes, and (4) the bulk screening field $E_b(r, t)$ governed by bulk screening processes [6–8].

$$E_{loc}(r, t) = E_{ex}(r) + E_{dep}(r, t) + E_{scr}(r, t) + E_b(r, t) \quad (1)$$

The external field $E_{ex}(r)$ strongly depends on the electrode shape. It demonstrates the singularities in the surface layer at the electrode edges due to the fringe effect. This effect is the most pronounced in the vicinity of the ends and corners of the stripe electrodes. The field singularities lead to the dominance of primary nucleation at the electrode boundary. Thus the start of the switching process at the threshold field is determined by field singularities at the electrode edges. It is clear that the "real value" of the threshold field is essentially higher than the value obtained under the common assumption that the switching field is equal to applied potential difference divided by sample thickness.

The depolarization field $E_{dep}(r, t)$ is produced by bound charges existing at the polar surfaces of the sample and at the charged domain walls ("head-to-head" or "tail-to-tail"). For typical ferroelectrics E_{dep} in single-domain plate can reach 10^8–10^9 V/m. This enormous value is essentially reduced for narrow domains and especially for the needle-like ones. That is the reason why triangular and spindle nuclei shapes are the most favorable ones. E_{dep} decelerates the sideways wall motion and limits the wall shift from the initial state. The screening processes diminish the impact of E_{dep}. Thus for low switching rate E_{dep} can be compensated almost totally and complete switching can be observed. For high switching rate the screening retards and applied field leads only to a small wall shift. In this case the initial domain state recovers after switch-off of the external field ("backswitching"). The screening processes can be divided in external and bulk screening.

External screening in ferroelectric capacitor is caused by a current in external circuit. Its rate is defined by the characteristic time of the external circuit τ_{ex}. This time constant is determined by the product of resistance and capacity of the circuit, and usually ranges from nanoseconds to microseconds. The experimentally observed switching time t_s cannot exceed τ_{ex}.

The fast external screening never compensates E_{dep} completely due to existence of the intrinsic dielectric surface layer ("dielectric gap" or "dead layer") [8, 20]. For ferroelectric capacitor of the thickness d the bulk residual depolarization field E_{rd} remains in the area, freshly switched from one single-domain state to another one, even after complete external screening due to existence of the dielectric layer of thickness L and dielectric permittivity ε_L [8, 20]:

$$E_{rd} = E_{dep} - E_{scr} = (2L/d)(P_S/(\varepsilon_L \varepsilon_o)) \quad (2)$$

where P_S is spontaneous polarization.

The existence of the residual depolarization field is the reason why "slow" bulk screening processes are very important.

Bulk screening is the only way to compensate E_{rd}. Three bulk screening mechanisms are considered usually: (1) redistribution of the bulk charges [8, 20], (2) reorientation of the defect dipoles [21], and (3) injection of carriers from the electrode through the dielectric gap [22]. The time constants of all considered mechanisms τ_b range from milliseconds up to months. While E_{rd} is several orders of magnitude less than E_{dep}, nevertheless it is of the same order as experimentally observed threshold fields. For short field pulse the bulk screening of the new state lags behind. Thus such switching is ineffective for irreversible change of the domain structure.

The cooperative action of E_{rd} and E_b leads to the backswitching after switch-off of the external field. This effect could be observed in the areas where $E_{loc}(r, t)$ exceeds the threshold field E_{th}, i.e.

$$\begin{aligned} E_{loc}(r, t) &= -[E_{dep}(r, t) - E_{scr}(r, t) + E_b(r, t)] \\ &= -[E_{rd}(r, t) + E_b(r, t)] > E_{th} \end{aligned} \quad (3)$$

The limiting values of screening effectiveness leads to different variants of domain kinetics after field switch-off: (1) for ineffective bulk screening the initial domain state can be reconstructed completely; (2) for effective bulk screening almost any field-induced domain pattern can be stabilized.

It has been shown experimentally that each of domain kinetics stages can develop according to different scenarios which are strongly dependent on bulk screening effectiveness. The critical values of the applied fields and switching rates corresponding to replacement of domain kinetic scenarios are dependent on material and experimental conditions. The crucial parameter determining the selection of the particular scenario of the domain kinetics is the ratio R between switching rate ($1/t_s$) and bulk screening rate ($1/\tau_{scr}$). Here t_s is the switching time and τ_{scr} is the screening time constant [23]. Three ranges of the bulk screening effectiveness are considered: (1) $R \ll 1$—"complete screening", (2) $R > 1$—"incomplete screening", and (3) $R \gg 1$ — "ineffective screening". In the following Sections we discuss the domain kinetic scenarios for sideways domain wall motion and growth of individual domain, which is crucial for manufacturing of engineered $1D$ and $2D$ periodical domain structures.

It must be stressed that static ferroelectric domain structure configuration obtained under application of the proposed approach differs from the equilibrium domain structures obtained in classical theory [11]. The equilibrium multi-domain state is determined in this theory by the minimum of the total free energy. Only two limiting types of the equilibrium domain structures can exist according to this approach. First, the periodical laminar or maze domain structure with neutral domain walls forms if the screening effects are neglected. The domain period is determined by the values of spontaneous polarization, dielectric permittivity, density of the domain wall energy, and sample thickness [9, 11]. Second type demonstrates the absence of any domain structure (single-domain state without domain walls) due to complete screening of the depolarization field. It is well-known that both classical predictions contradict experimentally observed domain patterns. In contrast our approach predicts that practically any domain pattern can be produced and stabilized. These metastable domain structures remain for a long time (about months and years) and present a static one for any application.

It is clear that the domain kinetics depends also on the mechanical properties of the material due to electromechanical coupling. In proper ferroelectrics the first order corrections to the driving force due to piezoelectric effect leads to renormalization of the local value of the switching field [11]. Gopalan *et al.* [24] explained the hexagonal shape of the isolated domains in LN taking into account the anisotropy of the domain wall energy induced by piezoelectric effect. In this paper these effects are accounted for implicitly. We assume that the electromechanical coupling leads to observed anisotropy of the nucleation processes. This fact allows us to explain the experimentally observed continuous transformation of the of domain shape from regular hexagon to three-ray stars (see Section 6).

4. Materials and experimental conditions

The discussed general consideration has been applied for explanation of the domain kinetics in lithium niobate $LiNbO_3$(LN) and lithium tantalate $LiTaO_3$(*LT*). These materials are widely used for creation of periodically poled structures for nonlinear optical applications. The production of the precise tailored periodical domain structures requires the deep understanding of the domain kinetic processes in an inhomogeneous electric field produced by electrode patterns.

LN and LT are the favorite objects of the domain engineering, in spite of the fact that the coercive fields in the most popular congruent compositions (CLN and CLT) are enormously high (about 210 kV/cm). For many years both materials were classified as "frozen ferroelectrics". Quite recently new LN and LT families of crystals closer to stoichiometric composition (SLN and SLT) demonstrating essentially lower coercive fields have become available [25–28].

These crystals are uniaxial with C_3 symmetry in the ferroelectric phase and domain structure with 180° domain walls only. The domain walls are visualized due to pronounced electro-optical effects and direct optical methods can be used for *in situ* observation of the domain kinetics [29, 30]. In the framework of our approach the experimentally observed optical contrast of the domain walls and "domain wall prints" (see Section 5) can be attributed to change of the refractive index induced by incompletely compensated depolarization field in the vicinity of the domain wall [23].

Our switching experiments were held in optical-grade single-domain wafers cut perpendicular to the polar axis and carefully polished. The thickness for CLN and CLT ranges from 0.2 to 0.5 mm, and for SLN and SLT – from 1 to 2 mm.

For the *in situ* investigation of the domain kinetics in an uniform electric field we prepared 1-mm-diameter circular transparent electrodes of: (1) liquid electrolyte (water solution of LiCl) in a special fixture and (2) In_2O_3: Sn (ITO) films deposited by magnetron sputtering. The used electrode sizes allow visualizing the domain kinetics over the whole switched area with enough spatial resolution. The direct observations of the domain evolution were carried out using a polarizing microscope with simultaneous TV-recording and subsequent processing of the image series.

For the periodical poling the wafers were lithographically patterned with periodic stripe metal electrodes deposited on Z^+-surface only and oriented along one of Y-directions. The patterned surface was covered by a photoresist layer. High voltage pulses producing an electric field greater than the coercive one were applied through the fixture containing a saturated water solution of LiCl (Fig. 1). The domain patterns produced by partial poling were revealed on both Z-surfaces by etching using pure hydrofluoric acid (HF) at room temperature. The obtained surface relief was visualized by optical microscopy, scanning electron microscopy (SEM) and atomic force microscopy (AFM). Moreover we have used the methods of domain visualization without etching by optical microscopy with phase contrast and by piezoelectric force microscopy (PFM) [31].

5. Sideways domain wall motion

5.1. Slow domain growth

"Slow domain growth" is obtained for complete bulk screening ($R \ll 1$). In this case the switching process in

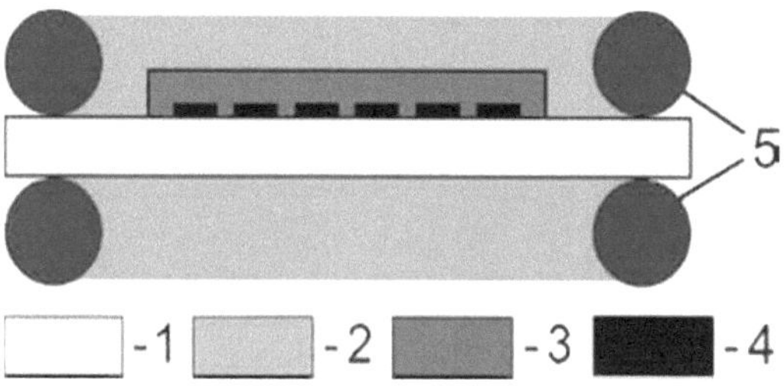

Figure 1 Experimental setup used for periodical poling: 1—wafer, 2—liquid electrolyte, 3—insulating layer, 4—periodic electrodes, 5—O-rings.

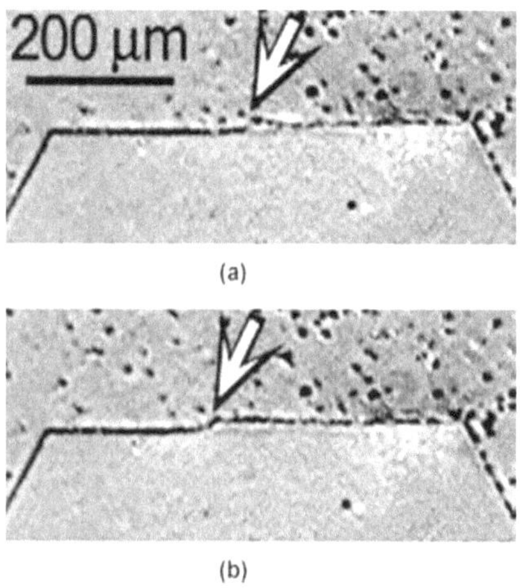

Figure 2 *In situ* optical visualization of the layer-by-layer domain growth by step propagation along the wall in CLN. Time intervals from field switch-on: (a) 1.20 s, (b) 1.28 s. Liquid electrodes. E_{ex}=220 kV/cm.

SLT, CLN and SLN is achieved through the sideways motion of the walls strictly oriented along the Y-directions.

In situ observation of the domain kinetics shows that the switching always starts with nucleation at the electrode edges or at the artificially produced surface defects in the center of the electrode area in accordance with our approach [32, 33].

Domain wall motion usually proceeds via the propagation along the wall of an optically distinguished micro-scale domain steps (bunches of the elementary nano-scale steps) (Fig. 2) [32–34], thus confirming above discussed approach. The wall decelerates while shifting from the initial position and partial backswitching is observed after field switch-off.

For quantitative description of the deceleration effect let us consider the shift of the single plane domain wall from the initial state with accomplished bulk screening (E_{rd} completely compensated by the bulk screening). For slow bulk screening the spatial distribution of E_b remains fixed during field-induced wall shift and becomes asymmetric relatively to the shifted wall. In this case E_b is codirectional with E_{rd} in the switched area. The total field at the wall averaged over the sample thickness ΔE_{loc} coincides with the field produced by a stripe capacitor with the width equal to the wall shift Δx and the surface charge determined by the doubled bulk screening charge density $2\sigma_b$ [34–36]

$$\Delta E_{loc}(\Delta x) = (2\sigma_b/\varepsilon\varepsilon_o)F(\Delta x/d) \tag{4}$$

where $F(\Delta x/d) = 1/\pi[2\text{arctg}(\Delta x/d) + (\Delta x/d)\ln(1 + d^2/\Delta x^2)]$, and $\sigma_b = 2P_s(\varepsilon/\varepsilon_L)L/d$.

The decelerating field at the wall increases with Δx thus suppressing the step generation ($2D$-nucleation). As a result the wall motion velocity diminishes and the wall stops at some distance from the initial position. After external field switch-off the action of $E_{rd}+E_b$ leads to return of the domain wall to the initial state ("complete backswitching"). The experimentally obtained field dependence of the wall shift measured during step-by-step increase of E_{ex} amplitude is fairly well described by Equation 4 [36].

When the generation of the individual steps is suppressed due to the incomplete screening ($R>1$), two unusual scenarios of sideways wall motion can be realized: (1) loss of the domain wall stability through formation of the finger structure, and (2) acceleration of the wall motion due to domain merging.

5.2. Loss of the domain wall shape stability

The perturbation of the planar wall shape leads to a loss of the domain wall shape stability through formation of the self-assembled domain structure with sub-micron fingers (Fig. 3). The effect is the most pronounced while switching with artificial dielectric layer. The perturbation of the domain wall shape can be induced by (1) inhomogeneity of E_{loc} due to electrode shape irregularity or (2) local decreasing of the threshold field. The perturbation of the domain wall shape evolves forming a ledge ("finger") strictly oriented along Y-direction (Fig. 3a). The correlated nucleation, leading to peculiarities of E_{loc} spatial distribution, result in the appearance of the neighboring fingers and propagation of the finger structure along the wall (Fig. 3b). The decelerating effect for shifted wall, characterized by $E_{loc}(\Delta x)$, is suppressed for a finger tip due to diminishing (comparatively with the plane wall) of the neighboring switched area, which is the source of ΔE_{loc} (see Equation 3). Finally the correlated finger structure forms along the wall (Fig. 3c). During periodical domain poling this mechanism provides the abnormal "domain broadening" (large shift of the domain wall from the electrode edge), which is extremely undesirable for periodical poling. The "fingering" leads to the domain merging, thus destroying the periodicity of the domain pattern [37–39].

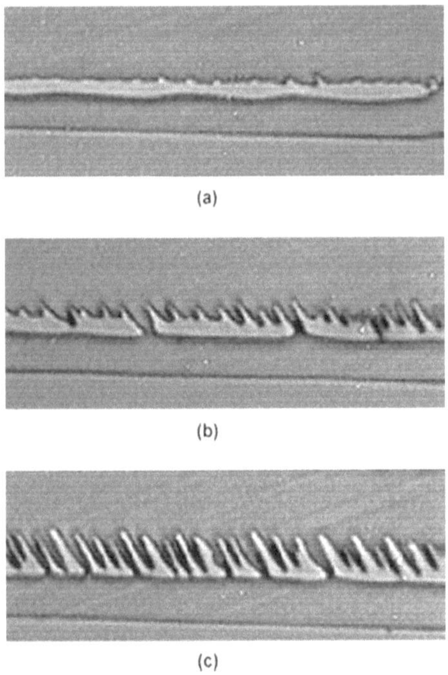

Figure 3 Loss of the wall shape stability during switching in CLN in the area covered by artificial dielectric layer. Stages of the structure evolution. Optical observation of domains revealed by etching.

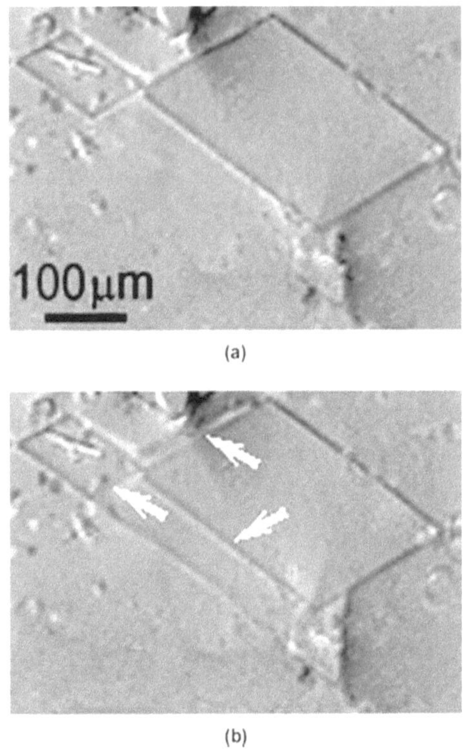

Figure 4 *In situ* optical visualization of "domain gulping" during reversal poling in uniform field in *CLN*. Arrows indicate the domain wall prints. Time interval between the frames 0.04 s. Liquid electrodes. E_{ex}=153 kV/cm.

5.3. Acceleration of the domain wall motion

Although the step generation probability becomes negligible for incomplete screening ($R > 1$), the applied field is strong enough for the step growth. That is the reason why the domain wall propagation can be accelerated by domain merging (Fig. 4), which leads to an effective generation of a great number of steps at the wall. These steps rapidly propagate along existing walls [32–34]. As a result the local deviation of the wall orientation from the allowed crystallographic directions disappears (Fig. 4b). The obtained abnormally fast domain growth ("domain gulping") is followed by easily visualized memory effect caused by retardation of the bulk screening.

"Domain wall prints" remain at the places, where the walls stay for a comparatively long-time before jump to a new position induced by merging (Fig. 4b). Such effect can be understood if we take into consideration the electro-optical nature of the observed contrast of the domain walls. The calculated bulk distribution of E_{loc} in the ferroelectric capacitor with dielectric surface layer demonstrates the field increase in the vicinity of the static domain wall [38]. The field induced variation of the refractive index leads to observed optical contrast of the domain wall. It is clear that E_b also demonstrates a spatial anomaly in the vicinity of the wall (see Equation 1). After the wall jump the domain wall prints fade gradually due to comparatively slow bulk screening process. The print life-time is defined by the bulk screening time constant. This effect has been experimentally observed in CLT by Gopalan and Mitchell [40].

The above discussed acceleration effect prevails for switching from multi-domain initial state with high concentration of small isolated domains. The switching rate for "step generation by merging only" mechanism is determined by only one parameter—the concentration of the individual domains in the initial state. The complete switching can be achieved if the relative concentration exceeds the critical value, which is about 0.02 according to our computer simulations on triangular lattice [41]. The discussed switching scenario prevails for the first switching in CLT [42, 43], which starts with a formation of sub-micron-diameter domains with a density about 1000 mm^{-2} (Fig. 5). It leads to an acceleration of the domain wall motion velocity by more than two orders of magnitude from 1 μm/s for isolated domains to 130 μm/s (for E_{ex}=190 kV/cm).

5.4. Super-fast motion of the switching front

In strong enough fields the switching process can continue even for absolutely ineffective bulk screening ($R \gg 1$). In this case the continuous motion of the wall is absolutely

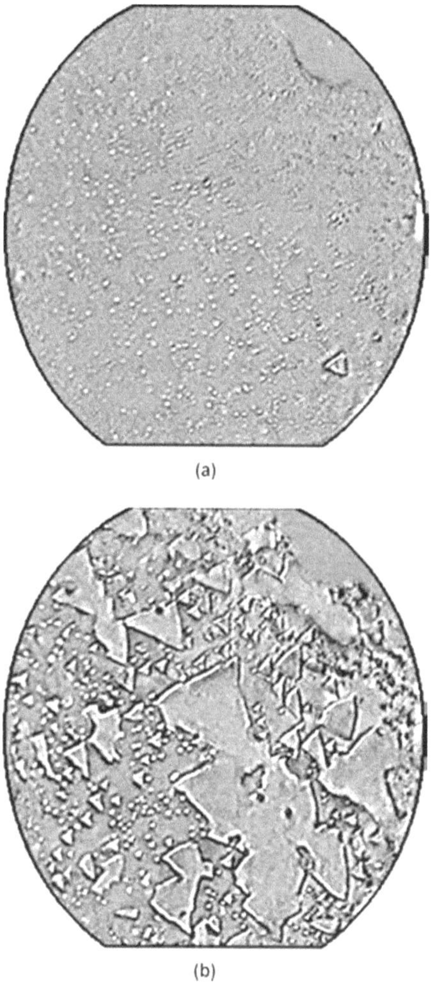

Figure 5 *In situ* optical visualization of the domain kinetics during switching in *CLT*. Time intervals from the field switch-on: (a) 0.3 s, (b) 0.9 s. 1-mm-diameter liquid electrodes. $E_{ex}=190$ kV/cm.

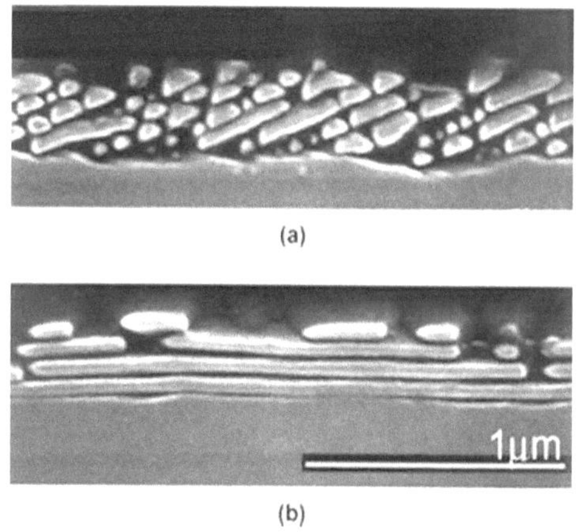

Figure 6 Quasi-periodic nano-scale domain structures: (a) arrays and (b) stripes. Backswitching near the edge of the stripe electrode under the artificial dielectric layer in CLN. SEM images. Domains revealed by etching.

forbidden and domain growth is achieved through propagation of the boundary of ensemble of isolated needle-like domains. Such "discrete switching" results in super-fast motion of the switching front demonstrating pronounce self-assembling behavior.

This effect has been observed during spontaneous backswitching after abrupt removing of the external switching field in LN and LT crystals with an artificial dielectric surface layer [44–46]. Various self-organized nanoscale domain structures develop (1) arrays strictly oriented along crystallographic directions (Fig. 6a) and (2) stripes (Fig. 6b). Similar effect has been observed during "super-fast switching" in PGO [47, 48].

Formation of all obtained self-assembled structures can be considered as manifestation of the correlated nucleation effect in the vicinity of the shifted domain wall which is stopped by incomplete screening. The correlated nucleation is caused by a pronounced maximum of E_{loc}, existing in this case in front of the wall at the distance nearly equal to the thickness of the surface dielectric layer L, shown by computer simulation [38]. This field maximum essentially increases the 3*D*-nucleation probability in front of the wall thus leading to appearance of isolated domains at the distance L along the boundary. Any arising isolated domain cannot spread out due to suppression of 2*D*-nucleation at its wall by the uncompensated E_{dep}. The arising domains repel each other due to electrostatic and electro-mechanical interaction. As a result the quasi-regular domain chain consisting of needle-like domains with submicro- or nano-scale transversal sizes aligns along the wall (Fig. 7). It has been shown by simulation that the new field maximum appears at approximately same distance from the formed domain chain thus initiating formation of the second one [38]. Thus, self-maintained enlarging quasi-regular domain structure can cover the areas of about square millimeters.

The role of the correlated nucleation can be intensified by: (1) increasing of R through rising of the switching field or by hindering of the screening processes, (2) deposition of the artificial surface dielectric layer, thus increasing E_{rd}. The later allows us to control the period of the quasi-regular structure, which is important for domain engineering. It has been revealed experimentally that the velocity of such process exceeds by orders of magnitude the usual wall motion velocity. That is the reason why the process can be named as "super-fast domain growth."

Correlated nucleation plays the most important role during backswitching in CLN after an abrupt removing of the external field. The record value of P_s in LN leads to abnormally high value of E_{rd}, which induces backswitching. The backward motion of the domain wall is achieved through propagation of the highly organized quasi-periodical structure of domain arrays strictly ori-

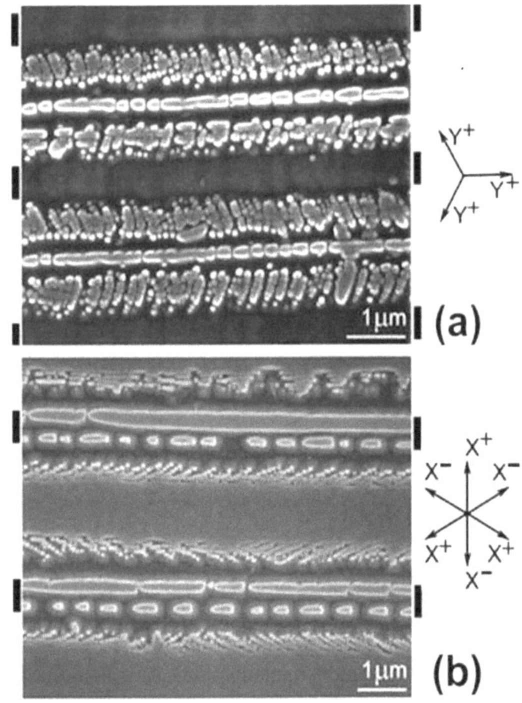

Figure 7 Nanodomain arrays oriented along (a) Y^+ and (b) X-directions. Backswitching near the edge of the stripe electrode under the artificial dielectric layer in CLN. SEM images. Domains revealed by etching. Black rectangles show electrode positions [33].

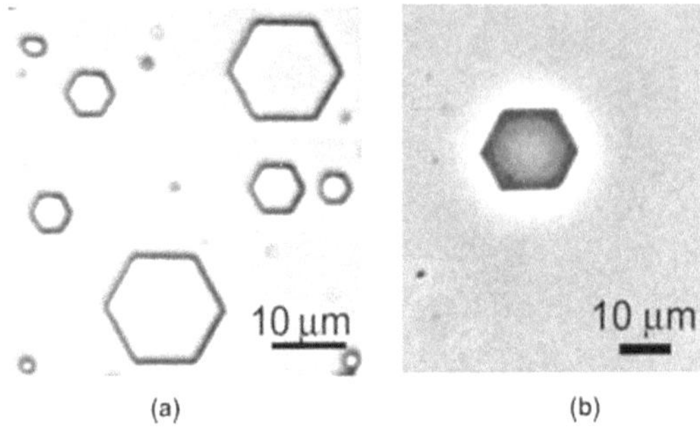

Figure 8 Hexagon domains: (a) CLN, (b) SLT. Optical images: (a) domains revealed by etching, (b) phase-contrast-microscopy [21].

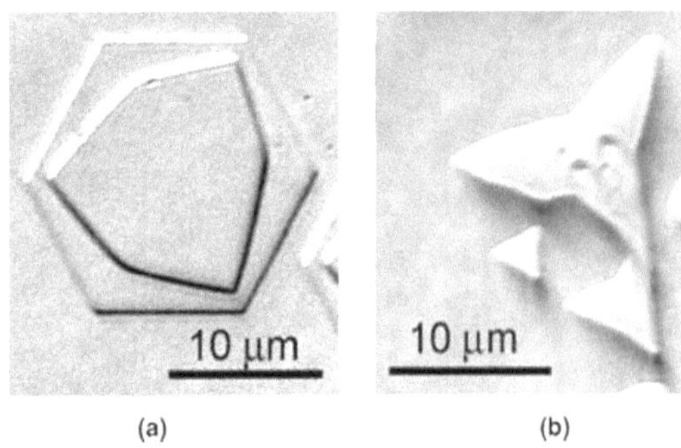

Figure 9 Exotic polygon domains formed during switching in CLN under artificial dielectric layer: (a) convex hexagons, (b) hexagon with concave angles ("Mercedes star") and regular triangle. Domains revealed by etching. (a) Switching by two pulses and etching of Z^- surface after each pulse [21].

ented along crystallographic directions (Fig. 7) [38, 44–46, 49]. Each quasi-regular array is comprised of nano-domains with a diameter 30–100 nm and an average linear density exceeding 10^4 mm^{-1}. Two variants of array orientation are obtained: (1) along the Y^+-direction at 60° to the electrode edges (Fig. 7a) and (2) along the X-directions (Fig. 7b). In the later case the fast growth of nano-domains along electrodes can lead to the formation of a periodic set of nano-scale stripe domains with period about 100 nm (Fig. 6b).

6. Growth of isolated domains

It has been shown experimentally for several ferroelectrics that the shapes of isolated domains growing in uniform electric field essentially depend on the field value [42, 50–53].

6.1. Hexagonal domain shape

The perfect hexagon domains with sides strictly oriented along Y-directions are formed for switching under complete screening ($R<1$) in any LN (Fig. 8a) and SLT (Fig. 8b).

For incomplete screening ($R>1$) E_{rd} suppresses the step propagation along the walls, thus leading to the essential deviation from the hexagonal shape. The domain shapes essentially deviating from the equiangular hexagons forms in LN during very fast switching. Similar effect is observed also in the samples covered by an artificial dielectric layer (Fig. 9).

6.2. Triangular domain shape

Individual domains of triangular shape with the sides strictly oriented along the X-directions were observed in CLN and CLT (Fig. 10). In CLT, the regular triangular domains were always obtained for any switching conditions (Fig. 10a), whereas in CLN such domains form for switching with an artificial dielectric layer or as a result of fast spontaneous backswitching (Fig. 10b).

The observed quantitative difference between domain shapes in CLT (triangles) and SLT or any LN (hexagons) for switching in ordinary conditions can be attributed to the large difference of the screening times. It has been shown experimentally that the screening process in CLT is essentially slower ($\tau_{scr} \sim 1$ s) as compared with LN and SLT ($\tau_{scr} \sim 50$–100 ms). The continuous transformation from hexagons to triangles in CLN induced by incomplete screening is a clear demonstration that the domain shape

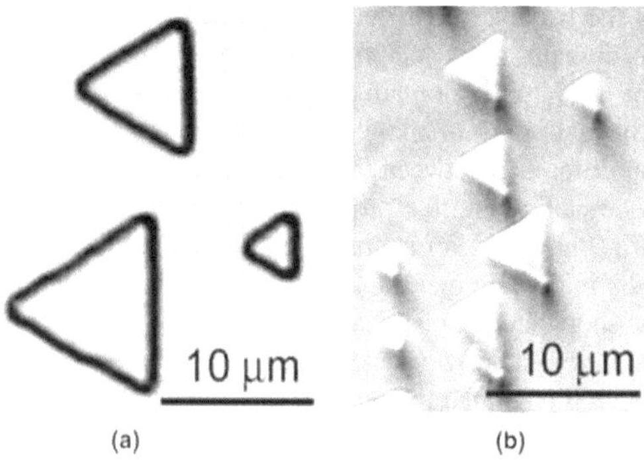

Figure 10 Triangle domains: (a) CLT, (b) CLN (formed during switching under artificial dielectric layer). Optical images. Etch-revealed domains [21].

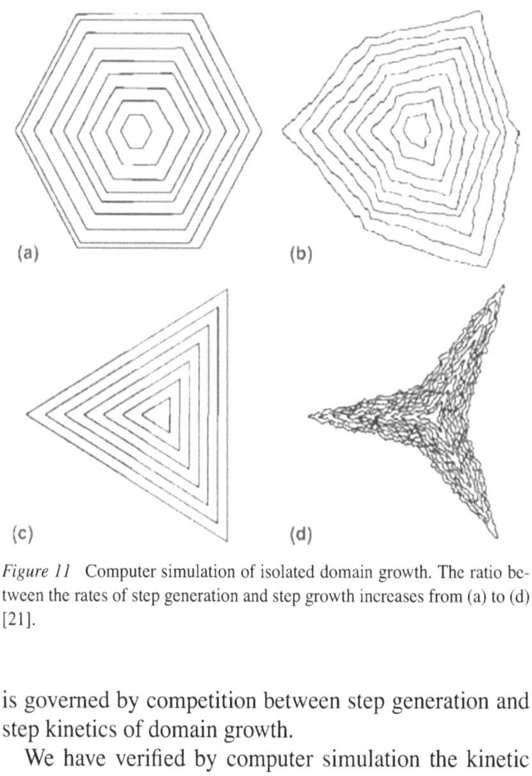

Figure 11 Computer simulation of isolated domain growth. The ratio between the rates of step generation and step growth increases from (a) to (d) [21].

is governed by competition between step generation and step kinetics of domain growth.

We have verified by computer simulation the kinetic nature of the domain shape. The model is based on the experimentally observed main features of the domain growth in crystals with C_3 symmetry. First, the generation of new steps occurs at three vertices of the regular hexagonal isolated domain (Y^+-crystallographic directions). Second, the steps grow along Y^+-directions. The wall orientation is determined by the step concentration similar to formation of vicinal faces during crystal growth. The hexagons are formed for high step growth velocity and low step generation rate (low step concentration) (Fig. 11a).

The variation of the ratios between the step generation and step growth rates due to retardation of the bulk screening changes the domain shape. Existence of a trail

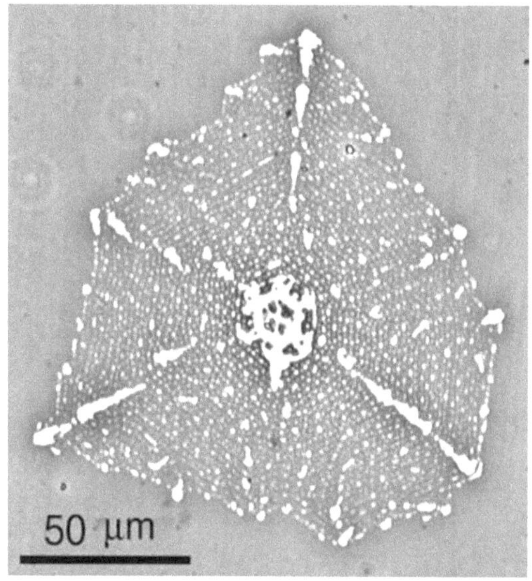

Figure 12 Web-type domain structure for switching with artificial dielectric layer in *SLT*. Optical image. Domains revealed by etching [31].

of uncompensated field following the moving step decelerates the step growth thus increasing the step concentration at the wall. The simulation of the isolated domain growth predicts existence of hexagons, triangles and even polygons with concave angles similar to "Mercedes star" (Fig. 11d). All shapes have been experimentally observed during switching in *CLN* covered by artificial dielectric layer (Fig. 9).

6.3. Web-like domain shape

It is interesting that the similar laws of the domain growth can be observed even during above discussed discrete switching for completely ineffective screening ($R \gg 1$). The effect has been studied during switching in *SLT* completely covered by artificial dielectric layer (photoresist). The switching starts with the formation of a conventional hexagon domain around the pin-hole in the dielectric layer. The subsequent growth is achieved through spreading of the quasi-regular ensemble of micro-scale isolated domains following the same mechanisms as the continuous growth of the isolated domain. The shape of the switched area is the same regular polygon as in the case of growth of isolated domains. The "steps" forms and propagates along the boundary of the ensemble playing the role of the domain wall (Fig. 12). The averaged distance between the nuclei is very close to the thickness of the artificial dielectric gap.

7. Domain engineering

The understanding of the field induced domain kinetics in LN and LT allows us to propose recently an original pol-

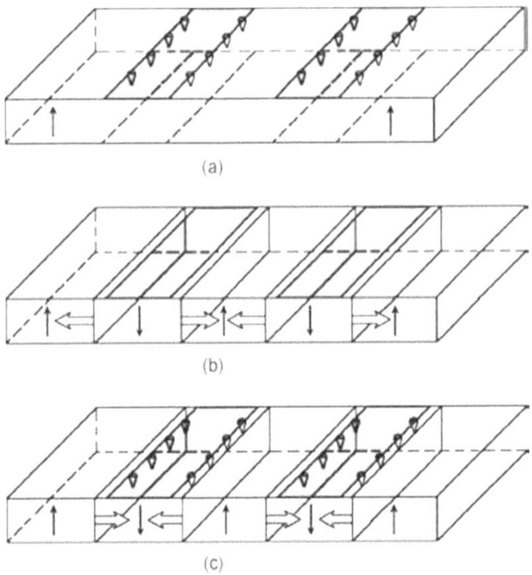

Figure 13 Main stages of the domain evolution during backswitched poling. (a) nucleation, (b) broadening, (c) backswitching after field switch off. White arrows show the directions of the domain wall motion [33].

ing method for creation of short-pitch periodical domain structures, so-called "backswitched poling" [44, 45, 54].

We exploit the unique abilities of the backswitching process, which was always considered as undesirable one, because it destroys the tailored structure. During backswitched poling, several distinguishable stages of domain evolution can be revealed (Fig. 13). The process starts with formation of new domains at the z^+- polar surface along the electrode edges due to the field singularities caused by the fringe effect (Fig. 13a). During the second stage, the domains grow and propagate through the wafer forming the laminar domains. The always observed domain broadening out of the electrode area leads to an essential difference between the lithographically defined electrode pattern and the produced periodical domain structure (Fig. 13b). For short periods domain broadening results in domain merging, thus limiting the production of the short-pitch domain patterns. After rapid decreasing of the poling field, the backswitching starts through shrinkage of the laminar domains by the backward wall motion and formation of the domains with the initial orientation of P_s along the electrode edges (Fig. 13c).

The application of this improved poling method to LN demonstrates the spatial frequency multiplication of the domain patterns as compared to the spatial frequency of the electrodes and self-maintained formation of the oriented domain arrays consisting of individual nano-scale domains [44, 46]. The mechanism of frequency multiplication is based on domain formation along the electrode edges during backswitching. For "frequency tripling" (Fig. 14c), the subsequent growth and merging of the domains lead to formation of a couple of strictly oriented sub-micron-width domain stripes with depth about 20–50 μm under the edges of wide electrodes (Fig. 14d). For narrow electrodes only the "frequency doubling" can be obtained with the depth of the backswitched domain stripes about 50–100 μm (Fig. 14a). The cross sections of the backswitched domain reveal two distinct variants of domain evolution: "erasing" and "splitting". During "erasing" the backswitched domains are formed in the earlier switched area without any disturbance of the external shape of the laminar domain (Fig. 14e). During

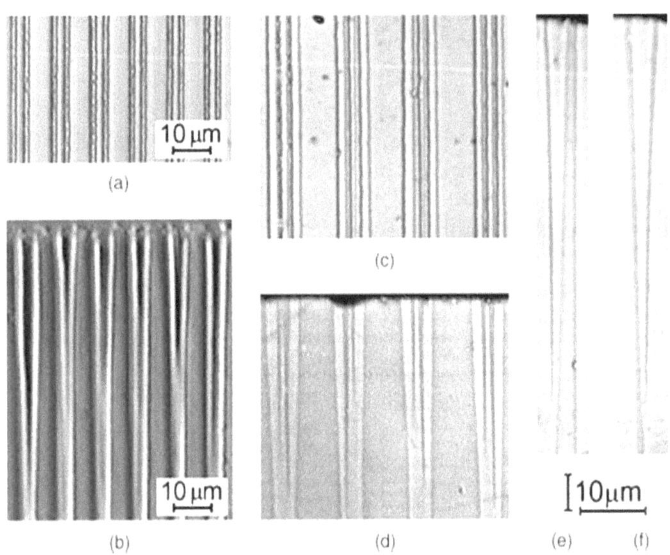

Figure 14 Domain frequency multiplication (a), (b) "doubling," (c), (d) "tripling," (e) "erasing," (f) "splitting". (a), (c) z^+ view, (b), (d), (e), (f) y cross-sections. Optical images. Domains revealed by etching. CLN [33].

"splitting" the backswitched domains cut the switched one conserving its volume and varying the shape (Fig. 14f).

Backswitched poling in *CLN* enables higher fidelity and shorter period domain patterning of thick substrates than can be achieved with conventional poling. 4 μm-period 5-cm-length devices were characterized for continuous-wave (cw) single-pass second harmonic generation (*SHG*) of blue light [45]. First-order single-pass cw *SHG* at 460–465 nm produced 61 mW of power at 6.1%/W efficiency from a Ti:sapphire laser source and 60 mW at 2.8%/W efficiency for a laser-diode source was achieved [55].

8. Conclusion

In the present paper, we have formulated the unified approach to the domain kinetics based on the nucleation mechanism of the polarization reversal and demonstrated its validity for understanding the variety of experimentally observed domain evolution scenarios. We prove that the kinetics of the ferroelectric domain structure essentially depends upon effectiveness of the screening processes. Original scenarios of the domain structure evolution were revealed experimentally and discussed within unified approach accounting for the decisive role of the retardation of the screening process. We demonstrate that the evolution of a domain structure in ferroelectrics during decay of the highly-nonequilibrium state presents the self-organizing process, in which the screening of polarization reversal plays the role of feedback. The discussed results of the fundamental investigations can be used as a physical basis of "domain engineering". We have proposed and realized several new techniques, which allow to produce the short-pitch regular domain patterns with record periods and nano-scale quasi-regular domain structures in lithium niobate and lithium tantalate single crystals. The crystals possessing such regular structures demonstrate new non-linear optical properties required for modern coherent light frequency conversion devices.

Acknowledgements

This work was supported in part by INTAS (Grant 03-51-6562), by RFBR-DFG (Grant 04-02-04007), by RFBR-NNSF (Grant 03-02-39004), by Program "Development of the Scientific Potential of High Education" of Federal Agency of Education (Grant 48859), by Program BRHE of U.S. CRDF and Federal Agency of Education (Award No.EK-005-X1). It is a pleasure to acknowledge the many helpful stimulating discussions with A. Alexandrovski, L.E. Cross, M.M. Fejer, K. Kitamura, A.L. Korzhenevskii, E.L. Rumyantsev, and J.F. Scott.

References

1. R. L. BYER, *J. Nonlinear Opt. Phys. & Mater.* **6** (1997) 549.
2. L. E. MYERS, R. C. ECKHARDT, M. M. FEJER, R. L. BYER, W. R. BOSENBERG and J. W. PIERCE, *J. Opt. Soc. Am. B* **12** (1995) 2102.
3. G. W. ROSS, M. POLLNAU, P. G. R. SMITH, W. A. CLARKSON, P. E. BRITTON and D. C. HANNA, *Opt. Lett.* **23** (1998) 171.
4. N. G. R. BRODERICK, G. W. ROSS, H. L. OFFERHAUS, D. J. RICHARDSON and D. C. HANNA, *Phys. Rev. Lett.* **84** (2000) 4345.
5. M. YAMADA, N. NADA, M. SAITOH and K. WATANABE, *Appl. Phys. Lett.* **62** (1993) 435.
6. V. YA. SHUR and E. L. RUMYANTSEV, *Ferroelectrics* **191** (1997) 319.
7. V. YA SHUR, *Phase Transitions* **65** (1998) 49.
8. V. YA. SHUR, in Ferroelectric Thin Films: Synthesis and Basic Properties, edited by C. A. Paz de Araujo, J. F. Scott and G. W. Taylor (Gordon and Breach, New York, 1996) p. 153.
9. E. FATUZZO and W. J. MERZ, "Ferroelectricity" (North-Holland Publishing Company, Amsterdam, 1967).
10. V. YA. SHUR, E. L. RUMYANTSEV, E. V. NIKOLAEVA, E. I. SHISHKIN, R. G. BATCHKO, G. D. MILLER, M. M. FEJER and R. L. BYER, *Ferroelectrics* **236** (2000) 129.
11. M. E. LINES and A. M. GLASS, "Principles and Application of Ferroelectrics and Related Materials" (Clarendon Press, Oxford, 1977).
12. R. C. MILLER, *J. Phys. Chem. Solids* **17** (1960) 93.
13. V. YA. SHUR, E. L. RUMYANTSEV, D. V. PELEGOV, V. L. KOZHEVNIKOV, E. V. NIKOLAEVA, E. I. SHISHKIN, A. P. CHERNYKH and R. K. IVANOV, *Ferroelectrics* **267** (2002) 347.
14. M. VOLMER, "Kinetik der Phasenbildung" (Steinkopff, Dresden-Leipzig, 1939).
15. YA. B. ZELDOVICH, *Zh. Eksp. Theor. Fiz.* **12** (1942) 525 (in Russian).
16. J. W. P. Schmelzer (ed.) Nucleation Theory and Applications, (Wiley-VCH, Weinheim, 2005).
17. A. N. KOLMOGOROV, *Izv. Acad. Nauk USSR., Ser. Math.* **3** (1937) 355 (in Russian).
18. M. AVRAMI, *J. Chem. Phys.* **7** (1939) 1103.
19. Y. ISHIBASHI and Y. TAKAGI, *J. Phys. Soc. Jap.* **31** (1971) 506.
20. V. M. FRIDKIN, "Ferroelectrics Semiconductors" (Consult. Bureau, New York and London, 1980).
21. P. V. LAMBECK and G. H. JONKER, *J. Phys. Chem. Solids* **47** (1986) 453.
22. A. K. TAGANTSEV, I. STOLICHNOV, E. L. COLLA and N. SETTER, *J. Appl. Phys.* **90** (2001) 1387.
23. V. YA. SHUR, in Nucleation Theory and Applications, edited by J. W. P. Schmelzer (Wiley-VCH, Weinheim, 2005) p. 178.
24. Y. FURUKAWA, K. KITAMURA, S. TAKEKAWA, K. NIWA and H. HATANO, *Opt. Lett.* **23** (1998) 1892.
25. V. GOPALAN, N. A. SANFORD, J. A. AUST, K. KITAMURA and Y. FURUKAWA, in Handbook of Advanced Electronic and Photonic Materials and Devices, edited by H. S. Nalwa (Academic Press, 2001) Ch. 2, p. 57.
26. K. KITAMURA, Y. FURUKAWA, K. NIWA, V. GOPALAN and T. MITCHELL, *Appl. Phys. Lett.* **73** (1998) 3073.
27. K. NIWA, Y. FURUKAWA, S. TAKEKAWA and K. KITAMURA, *J. Crystal Growth* **208** (2000) 493.
28. L. HUANG, D. HUI, D. J. BAMFORD, S. J. FIELD, I. MNUSHKINA, L. E. MYERS and J. V. KAYSER, *Appl. Phys. B* **72** (2001) 301.
29. V. GOPALAN, Q. JIA and T. MITCHELL, *Appl. Phys. Lett.* **75** (1999) 2482.
30. M. MULLER, E. SOERGEL and K. BUSE, *Opt. Lett.* **28** (2003) 2515.
31. A. GRUVERMAN, O. KOLOSOV, J. HATANO, K. TAKAHASHI and H. TOKUMOTO, *J. Vac. Sci. Technol. B* **13** (1995) 1095.
32. V. YA. SHUR, E. L. RUMYANTSEV, E. V. NIKOLAEVA, E. I. SHISHKIN, R. G. BATCHKO, M. M. FEJER, R. L.

BYER and I. MNUSHKINA, *Ferroelectrics* **269** (2002) 189.
33. V. YA. SHUR, E. L. RUMYANTSEV, E. V. NIKOLAEVA, E. I. SHISHKIN, R. G. BATCHKO, M. M. FEJER and R. L. BYER, *ibid.* **257** (2001) 191.
34. M. E. DROUGARD and R. LANDAUER, *J. Appl. Phys.* **30** (1959) 1663.
35. V. YA. SHUR, A. L. GRUVERMAN, V. V. LETUCHEV, E. L. RUMYANTSEV and A. L. SUBBOTIN, *Ferroelectrics* **98** (1989) 29.
36. V. YA. SHUR, E. L. RUMYANTSEV, V. P. KUMINOV, A. L. SUBBOTIN and E.V. NIKOLAEVA, *Phys. Solid State* **41** (1999) 112.
37. G. ROSENMAN, A. SKLIAR and A. ARIE, *Ferroelectrics Review* **1** (1999) 263.
38. V. YA. SHUR, E. L. RUMYANTSEV, E. V. NIKOLAEVA, E. I. SHISHKIN, R. G. BATCHKO, G. D. MILLER, M. M. FEJER and R. L. BYER, *SPIE Proceedings on Smart Structures and Materials* **3992** (2000) 143.
39. V. SHUR, E. RUMYANTSEV, R. BATCHKO, G. MILLER, M. FEJER and R. BYER, *Ferroelectrics* **221** (1999) 157.
40. V. GOPALAN and T. MITCHELL, *J. Appl. Phys.* **85** (1999) 2304.
41. A. P. CHERNYKH, V. YA. SHUR, E. V. NIKOLAEVA, E. I. SHISHKIN, A. G. SHUR, K. TERABE, S. KURIMURA, K. KITAMURA and K. GALLO, *Material Science & Engineering B* **120** (2005) 109.
42. V. YA. SHUR, E. V. NIKOLAEVA, E. I. SHISHKIN, A. P. CHERNYKH, K. TERABE, K. KITAMURA, H. ITO and K. NAKAMURA, *Ferroelectrics* **269** (2002) 195.
43. V. YA. SHUR, E. V. NIKOLAEVA, E. I. SHISHKIN, V. L. KOZHEVNIKOV, A. P. CHERNYKH, K. TERABE and K. KITAMURA, *Appl. Phys. Lett.* **79** (2001) 3146.
44. V. YA. SHUR, E. L. RUMYANTSEV, E. V. NIKOLAEVA, E. I. SHISHKIN, D. V. FURSOV, R. G. BATCHKO, L. A. EYRES, M. M. FEJER and R. L. BYER, *ibid.* **76** (2000) 143.
45. R. G. BATCHKO, V. Y. SHUR, M. M. FEJER and R. L. BYER, *ibid.* **75** (1999) 1673.
46. V. YA. SHUR, E. L. RUMYANTSEV, E. V. NIKOLAEVA, E. I. SHISHKIN, D. V. FURSOV, R. G. BATCHKO, L. A. EYRES, M. M. FEJER, R. L. BYER and J. SINDEL, *Ferroelectrics* **253** (2001) 105.
47. V. YA. SHUR, A. L. GRUVERMAN, N. YU. PONOMAREV, and N. A. TONKACHYOVA, *ibid.* **126** (1992) 371.
48. V. YA. SHUR, A. L. GRUVERMAN, N. YU. PONOMAREV, E. L. RUMYANTSEV and N. A. TONKACHYOVA, *Integrated Ferroelectrics* **2** (1992) 51.
49. V. YA. SHUR, E. V. NIKOLAEVA and E. I. SHISHKIN, *Physics of Low-Dimensional Structures* **3/4** (2003) 139.
50. R. C. MILLER and A. SAVAGE, *Phys. Rev.* **115** (1959) 1176.
51. J. HATANO, F. SUDA and H. FUTAMA, *J. Phys. Soc. Jap.* **45** (1978) 244.
52. V. YA. SHUR, V. V. LETUCHEV and E. L. RUMYANTSEV, *Sov. Phys. Solid State* **26** (1984) 1521.
53. V. YA. SHUR, V. V. LETUCHEV, E. L. RUMYANTSEV and I. V. OVECHKINA, *ibid.* **27** (1985) 959.
54. R. BATCHKO, G. MILLER, R. BYER, V. SHUR and M. FEJER, *United States Patent* No. 6, 542,285 B1, April 1, 2003.
55. R. G. BATCHKO, M. M. FEJER, R. L. BYER, D. WOLL, R. WALLENSTEIN, V. YA. SHUR and L. ERMAN, *Optics Letters* **24/18** (1999) 1293.

JOURNAL OF MATERIALS SCIENCE 41 (2006) 211–216

Ferroelectric transducer arrays for transdermal insulin delivery

BENJAMIN SNYDER
Department of Bioengineering, The Pennsylvania State University, University Park, PA 16802, USA; Material Research Laboratory, The Pennsylvania State University, University Park, PA 16802, USA

SEUNGJUN LEE
Department of Bioengineering, The Pennsylvania State University, University Park, PA 16802, USA

NADINE BARRIE SMITH*
Department of Bioengineering, The Pennsylvania State University, University Park, PA 16802, USA; Material Research Laboratory, The Pennsylvania State University, University Park, PA 16802, USA; Graduate Program in Acoustics, The Pennsylvania State University, University Park, PA 16802, USA
E-mail: nbs@engr.psu.edu

ROBERT E. NEWNHAM
Material Research Laboratory, The Pennsylvania State University, University Park, PA 16802, USA

The goal of this research was the development of a portable therapeutic device for the treatment of diabetes. A 3×3 cymbal array has been successfully tested in noninvasive insulin delivery experiments. The cymbal is a composite flextensional transducer constructed from a poled PZT ceramic and shaped metal endcaps that amplify the transducer motions by two orders of magnitude. Nine cymbals are wired in parallel and potted in polyurethane to form the flat panel arrays. When driven near resonance (20 kHz), the array generated a low intensity acoustic beam of 100 mW/cm^2. Animal experiments on hyperglycemic rats and rabbits demonstrated the ultrasonic enhancement of transdermal insulin delivery.

1. Introduction

Flextensional transducers are mechanical amplifiers coupling the small longitudinal strains in piezoelectric drive elements to large flexural motions in a metal shell. By means of converting the high impedance of a stiff ceramic into a low acoustic impedance, the shell acts as a mechanical amplifier [1, 2]. The cymbal transducer is a miniature Class V flextensional transducer first developed at Penn State a decade ago [3–5]. Originally developed as a small actuator capable of generating moderate force and a sizable displacement, the cymbal was later utilized as an underwater acoustic projector [6]. The cymbal transducer consists of a piezoelectric ceramic disk poled in the thickness direction and sandwiched between two metal endcaps shaped similar to musical cymbals, hence the name cymbal transducer. By incorporating both flexural and rotational motions that convert the small radial displacements of the disk into large axial motions normal to the endcap surface, the endcaps act as mechanical amplifiers (Fig. 1). The simple manufacturing process of the cymbal is well-suited to inexpensive mass production and incorporation in a low-cost therapeutic device.

According to a recent epidemiological study, there are approximately 12 million diabetics in the United States [7]. Daily insulin injections are the usual treatment but a number of alternative therapies are under investigation.

*Author to whom all correspondence should be addressed.

0022-2461
DOI: 10.1007/s10853-006-6080-3

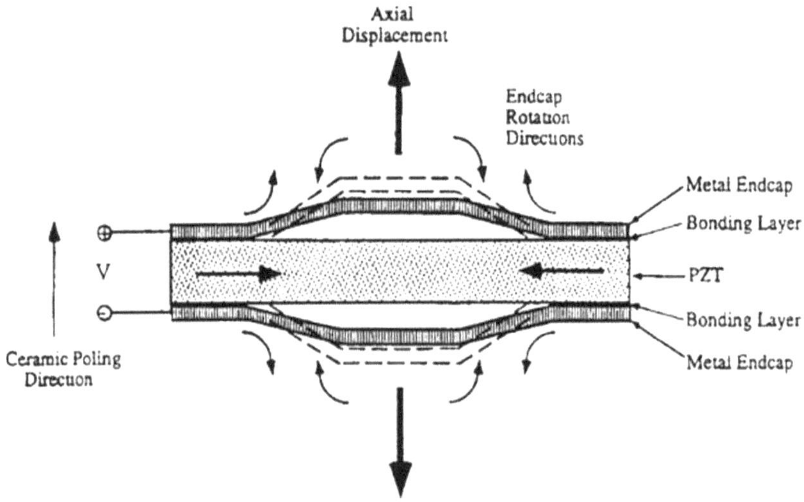

Figure 1 Cross-sectional view of a cymbal flextensional transducer where the hatched areas represent the metal endcaps, and the meshed area the poled PZT disk. Arrows incorporated into the schematic delineate the radial displacement of the disk and rotational motion of the endcaps, which both combine to produce the amplified axial displacement of the transducer [6].

One alternative involves the use of ultrasound to enhance the painless transdermal delivery of insulin. Sonophoresis, as it is called, possesses many potential advantages, but the human skin is a stubborn obstacle to large molecules like insulin [8]. Low frequency ultrasound (20–100 kHz) is found to be about a thousand times more effective than high frequencies (1–3 MHz) [9]. Most of these sonophoresis experiments were carried out with immobile commercial sonicators, which are impractical for a small portable device. The low profile cymbal arrays described in this paper appear to satisfy this requirement. Earlier experiments using a 2×2 array demonstrated the transport of insulin across *in vitro* human skin and with live *in vivo* animal studies [10, 11].

This paper describes the design, fabrication, and testing of a 3×3 cymbal array for use in transdermal drug delivery. Due to an increase in spatial area of ultrasound output in the 3×3 array as compared to the 2×2 array, a comparison between the respective *in vivo* results of the two arrays will elucidate the effect of spatial area on insulin delivery and subsequent decrease in blood glucose levels.

2. Design and construction of a 3 × 3 array

U.S. Patents 5,729,077 and 6,232,702 describe the fabrication and performance of cymbal actuators and transducers [12, 13]. The cymbal transducers were 12.7 mm in diameter and 2.2 mm thick. Hard lead zironate-titanate (PZT 4, Piezokinetics, Bellefonte, PA) disks were used as the driving elements, and poled in the thickness direction. Circular titanium caps were punched from sheet metal 0.25 mm thick with a diameter of 12.7 mm and a cavity depth of 0.32 mm. Eccobond epoxy 0.02 mm thick was used to bond thecaps to the ceramic.

Individual cymbals resonate at 39±2 kHz in air and 20.5 kHz in water with an omni-directional beam pattern. The cymbals are wired into arrays to increase efficiency, source level, and transmitting voltage response (TVR). Nine cymbals were electrically wired in parallel and potted in polyurethane to lower the acoustic impedance and improve the coupling to human tissue. As shown in Fig. 2, the dimensions of the 3×3 array were 57×57×7 mm^3 and it weighed less than 35 g. Electrical leads were encased in heat-shrink tubing and assembled into a BNC connector. A water-tight square-shaped Plexiglass® standoff 2.25 mm thick was attached to the face of the array and provided a reservoir for insulin or for saline solution.

3. Ultrasound exposimetry

An electrical control system consisting of a radiofrequency (RF) waveform generator, digital oscilloscope, RF amplifier, and matching circuit was used to drive the 3×3 array. The voltage to the array was monitored in real-time using an oscilloscope probe. Conditions of the insulin delivery experiments required the waveform generator operating at 20 kHz, a pulse duration of 200 msec, and a pulse repetition period of 1 sec (20% duty cycle). The waveform generator output voltage and RF amplifier were set so that the maximum spatial peak-temporal peak acoustic intensity (I_{SPTP}) was 100 mW/cm^2 at an offset distance of 1 mm below the bottom surface of the array, simulating the thickness between the surface of the array and the skin.

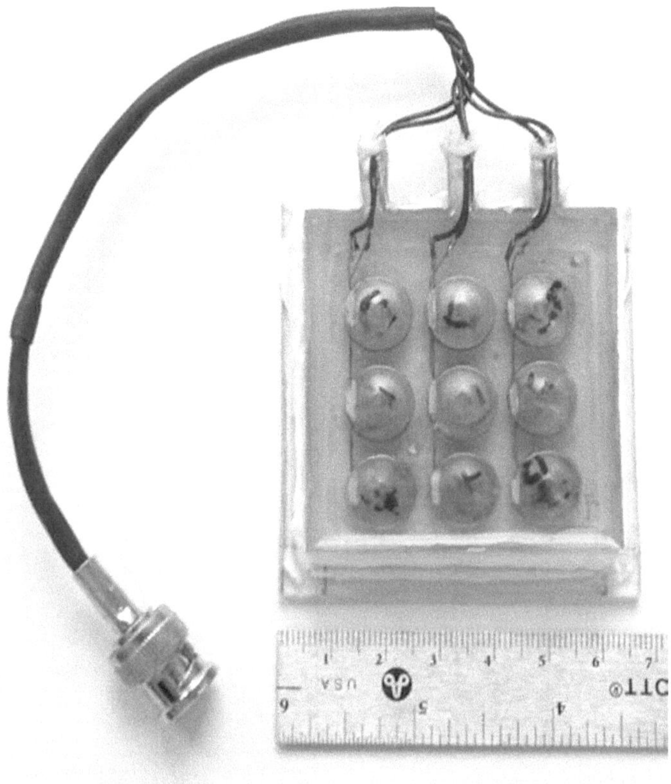

Figure 2 An assembled 3×3 cymbal array with BNC connector, heat shrink tubing, and Plexiglass ® standoff. Nine cymbal transducers are wired in parallel and potted in a polyurethane polymer. The overall dimensions of the low profile array are $57\times57\times7$ mm^3 and weighing less than 35 g.

A calibrated, miniature omnidirectional reference hydrophone and a computer-controlled exposimetry system were used to make the intensity measurements. First, the hydrophone (Model TC4013, RESON, Inc., Goleta, CA) was stepped incrementally across the bottom surface of the array in 1 mm steps over a cross sectional area of 40×40 mm^2. Two- and three-dimensional maps were generated to ensure that all cymbals were working individually and in conjunction, and that the I_{SPTP} output cone was centered over the face of the array. Second, by placing the hydrophone at an offset distance of 1 mm from the surface of the array over the central cymbal position and spanning the peak-to-peak input voltage from the waveform generator (at a constant input frequency of 20 kHz), a graph of maximum I_{SPTP} versus input voltage was obtained. By interpolating the results of these graphs, the input peak-to-peak voltage could be adjusted to produce the maximum I_{SPTP} output of 100 mW/cm^2 required for the animal experiments.

4. Animal experiments

The New Zealand White rabbits were anesthetized by procedures approved by the Institutional Animal Care and Use Committee (IACUC) at the Pennsylvania State University. A total of 17 rabbits (3.0–4.5 kg) were divided into three groups with five rabbits in "ultrasound only" control, five in "insulin only" control, and seven with "ultrasound & insulin". The rabbits were anesthetized with a combination of ketamine hydrochloride (40 mg/kg administered subcutaneously, Ketaject ®, Phoenix, St. Joseph, MO) and sodium xylazine (10 mg/kg administered subcutaneously, Xyla-ject®, Phoenix, St. Joseph, MO). In addition to anesthetizing the rabbits, the combination of ketamine and xylazine caused temporary but sustained (up to 12 hours) hyperglycemia in the test rabbits.

While anesthetized, the hair in the abdominal region of the rabbits was shaved with clippers and a depilatory cream was briefly applied to remove any remnant hair; the area was then thoroughly cleaned with water and alcohol. With the rabbit placed in the dorsal recumbent position, the 3×3 array with standoff was affixed to the exposed abdominal skin using double-sided industrial carpet tape (3M, St. Paul, MN) cut to the imprint shape of the standoff (Fig. 3).

The reservoir created by the standoff was filled with saline solution for the ultrasound only control experiments and about 5 mL of insulin solution for the insulin only control and insulin combined with ultrasound experiments (Humilin®R, rDNA U-100, Eli Lilly and Co., Indianapo-

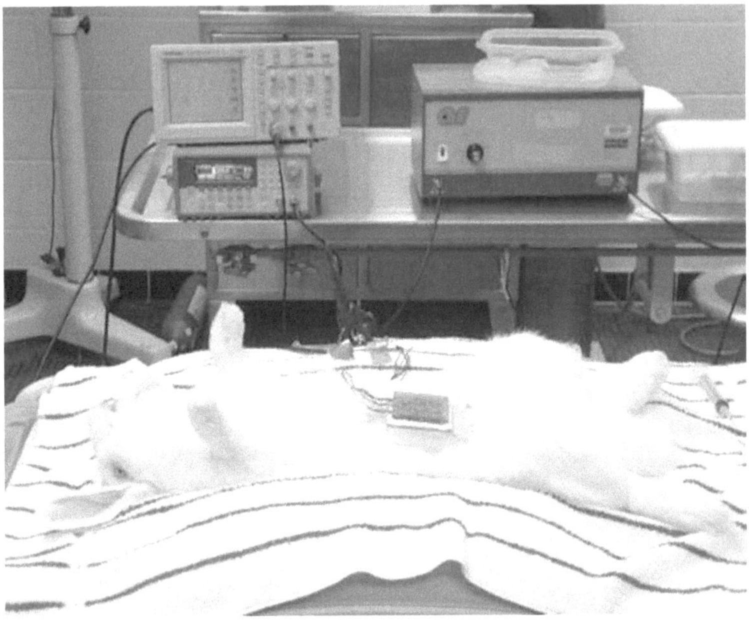

Figure 3 Experimental set-up of the rabbit, 3×3 cymbal array, and electronic drive system during the noninvasive transdermal insulin delivery experiments. With the rabbit anesthetized and lying in the dorsal recumbent position, the waveform generator provided a pulsed signal at 20 kHz, pulse duration of 200 msec, and a pulse repetition period of 1 sec (20% duty cycle). The input peak-to-peak voltage and amplifier gain were adjusted to supply an acoustic I_{SPTP} output of 100 mW/cm^2 through the 60 min exposure.

lis, IN). The fluids were added through small holes drilled through the top of the array. Care was taken to remove all bubbles from the solution within the reservoir. The insulin solution consisted of 10 mL insulin (100 U/mL) diluted with 10 mL saline to produce a total volume of 20 mL and a final insulin concentration of 50 U/mL. Previous research has shown that dilution of the insulin does not affect the transdermal delivery of insulin or subsequent decrement in blood glucose [11].

Prior to beginning the experiment, 0.03 mL of blood was collected from the ear vein of each rabbit for a baseline glucose comparison. The blood glucose level (mg/dL) was determined using an ACCU-CHEK™ Instant® (Roche Diagnostics Co., Indianapolis, IN) blood glucose monitor; multiple (3–6) blood glucose readings were recorded for the baseline value and every 15 min for 90 min after experiment initiation. The time elapsed between induction of anesthesia and the baseline glucose reading was about 30 min. In order to compare decrements in blood glucose between successive rabbit experiments, the blood glucose level was normalized to the baseline recording in each experiment.

For each rabbit, the entire experiment lasted a total of 90 min. The first control group ($n = 5$) was exposed to insulin solution in the reservoir but no ultrasound (designated "insulin only") to study the effects of passive diffusion. The second control group ($n = 5$) involved the application of ultrasound (100 mW/cm^2 for 60 min) with saline in the reservoir (designated "ultrasound only") to study the effects of ultrasound alone. The final group ($n = 7$) combined the application of ultrasound (100 mW/cm^2 for 60 min) with insulin solution in the reservoir (designated "insulin & ultrasound"). For all three groups, the reservoir with saline or insulin and the array were removed after 60 min exposure, although the blood glucose continued to be monitored at 15 min intervals for an additional 30 min.

5. Results

An initial exposimetry of the 3×3 ultrasound array indicated the integrity of the array design and assembly. As shown in Fig. 4, the initial I_{SPTP} acoustic intensity output resembles a relatively smooth, centered cone. Each 3×3 array was constructed with the cymbal having the lowest fundamental resonance peak positioned in the center location. This minimized detrimental acoustic interactions resulting in a more controlled, rounded peak, rather than an exaggerated spike. Once the array integrity was verified, a more rounded 100 mW/cm^2 I_{SPTP} output was achieved through the addition of a RF matching circuit, Plexiglass® standoff, and interpolation of input peak-to-peak voltage graphs.

The anesthetic combination of ketamine and xylazine induced a hyperglycemic state in all rabbits by increasing their normal blood glucose levels of 100–130 mg/dL to 197.6 ± 26.9 mg/dL immediately before the start of the transdermal insulin delivery experiments. This was the initial blood glucose reading at 0 min and was thus

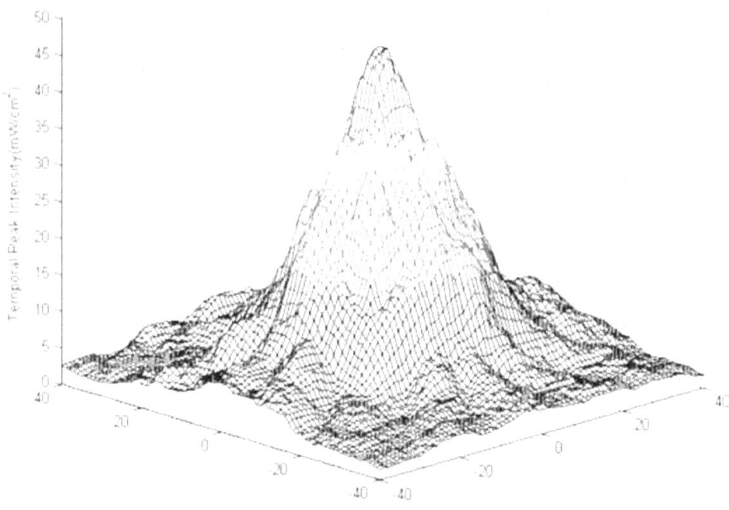

Figure 4 Exposimetry integrity test results of a correctly fabricated 3×3 cymbal array: 3D I_{SPTP} contour plot with reference bar. This initial exposimetry measurement was performed without the enhancement of the RF matching circuit or Plexiglass ® standoff, and was used to validate that the transducers were working in concert to produce a smooth and centered I_{SPTP} output cone.

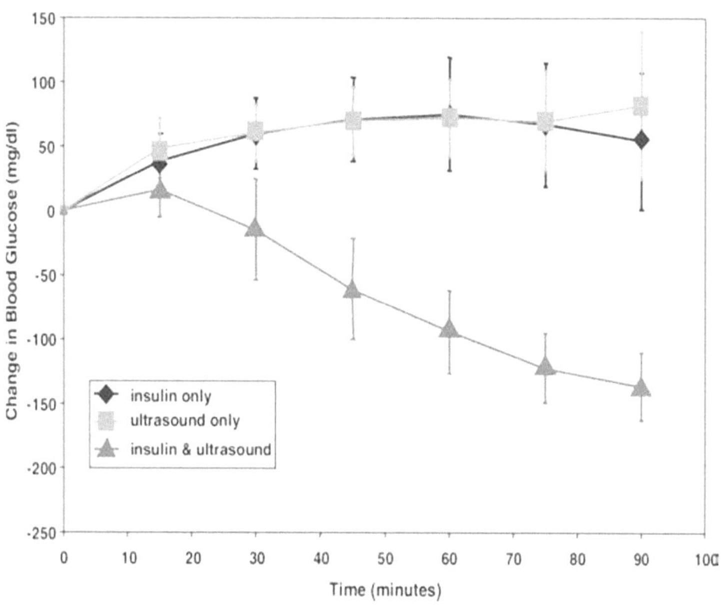

Figure 5 Graph of the change in blood glucose over the 90 min transdermal insulin delivery experiment duration, comparing both controls ("insulin only" and "ultrasound only") to the "insulin & ultrasound" results generated by the 3×3 array. The two controls showed an increase in the normalized blood glucose level to ~70 mg/dL after 60 min. The collaboration of insulin with ultrasound resulted in the normalized blood glucose level decreasing to −94.1 ± 32.3 mg/dL after 60 min exposure and to 136.1 ± 26.1 mg/dL at 90 min.

considered the baseline value. For comparison between rabbits in the three groups, the change in the blood glucose level was normalized to the baseline value. The normalized blood glucose levels for the three rabbit groups in the experiments are graphed in Fig. 5 as a mean and standard deviation (mean ± SD) for each group at each particular time interval. The results illustrate the change in blood glucose levels during the 90 min (15 min increments) experiment. For the "ultrasound only" control group, the blood glucose increased to 69.7 ± 26.8 mg/dL after 45 min and maintained approximately at that level until 90 min, where it increased to 82.5 ± 57.7 mg/dL. For the "insulin only" control group, the blood glucose continually increased to 75.2 ± 44.0 mg/dL after 60 min and then decreased to 54.6 ± 52.8 mg/dL at 90 min. For the "insulin & ultrasound" group, the blood glucose decreased to −94.1 ± 32.3 mg/dL after 60 min exposure, at which time the insulin and array were removed, and continued to decrease to −136.1 ± 26.1 mg/dL at 90 min. Results from the "ultrasound & insulin" rabbit group indicate that the ultrasound enhanced the delivery of insulin. The initial hyperglycemic blood glucose levels for the rabbits in this group returned to the normal blood glucose range via the ultrasonic enhancement of transdermal insulin delivery. A careful examination of the rabbit skin was performed after ultrasound exposure to detect any visible lesions on the skin surface. Visual examination of the ultrasound exposed skin did not reveal any palpable damage or significant change in the skin.

A statistical analysis between the blood glucose decrements achieved utilizing the earlier 2×2 array and 3×3 array calculated that there exists no significant statistical difference [14].

6. Conclusions

An exposimetry system was employed to quantify the spatial and temporal acoustic intensity (I_{SPTP} [mW/cm^2]) and was used to validate the performance of the 3×3 cymbal arrays utilized in the insulin delivery experiments: the I_{SPTP} output was centered with a smooth three-dimensional cone in all the 3×3 cymbal arrays.

Enhanced transdermal insulin delivery via ultrasound was demonstrated using the 3×3 cymbal arrays. Following an approved IACUC protocol, five New Zealand White rabbits were utilized in rotation and were anesthetized using a combination of ketamine and xylazine, which also induced a hyperglycemic state in the rabbits (blood glucose levels increased from 100–135 mg/dL to 197.6 ± 26.9 mg/dL at the start of the experiments). A total of 17 rabbit experiments were conducted, and were divided into three categories: ultrasound only control ("ultrasound only"); insulin only control ("insulin only"); and, the combination of insulin with ultrasound ("insulin & ultrasound"). The two controls, determined to be not significantly different ($p > 0.5$), showed an increase in the normalized blood glucose level to ~70 mg/dL after 60 min. Conversely, by applying ultrasound with an insulin solution contained within the reservoir, the normalized blood glucose level decreased to −94.1 ± 32.3 mg/dL after 60 min exposure and to 136.1 ± 26.1 mg/dL at 90 minutes. The decrement in blood glucose due to the 3×3 array was determined to be not significantly different ($p > 0.5$) to the decrement in blood glucose level for the 2×2 array under similar experimental conditions [14]. Therefore, the increase in spatial area of ultrasound output appears to have a negligible effect on insulin delivery. This result is beneficial in that a small, portable insulin delivery device can potentially deliver the same amount of insulin and cause a similar decrease in blood glucose levels as a large, immobile device. The transdermal delivery of insulin may only be dependent upon the ultrasound dosage. Feasibility studies that independently vary the ultrasound dosage - quantified by exposure time, acoustic intensity, or ultrasonic frequency — may elucidate a direct relationship.

Transdermal insulin delivery experiments on pigs are now being performed as a prelude to therapeutic use on human beings. A greater correlation between ultrasound dosage and transdermal insulin delivery is a notable objective. The transdermal delivery of other drugs such as steroids and antibiotics are also worthy goals for future work.

Acknowledgements

The support provided by Life Sciences Greenhouse of Central Pennsylvania and the Penn State Tuition Grant-in-Aid Program is gratefully acknowledged.

References

1. K. D. ROLT, *J. Acoust. Soc. Am.* **87**(3) (1990) 1340.
2. W. J. TOULIS, Us Patent 3,277,433 (1966).
3. Y. K. SUGAWARA, S. YOSHIKAWA, Q. C. XU and R. E. NEWNHAM, *J. Amer. Ceramic Soc.* **75** (1992) 996.
4. J. F. TRESSLER, Q. C. XU, S. YOSHIKAWA, K. UCHINO and R. E. NEWHAM, *Ferroelectrics* **156** (1994) 67.
5. A. DOGAN, K. UCHINO and R. E. NEWHAM, *IEEE UFFC* **44** (1997) 597.
6. J. F. TRESSLER and R. E. NEWHAM , *J. Acoust. Soc. Am.* **105(2)** (1999) 591.
7. American Diabetes Association, *Diabetes Care* **26**(Supp. 1) (2003) s121.
8. S. MITRAGOTRI, *J. Pharm. Sci.* **84** (1995) 697.
9. *Idem., Pharm. Res.* **13** (1996) 411.
10. N. B. SMITH, S. LEE, E. MAIONCE, R. B. ROY, S. MCELLIGOTT and K. K. SHUNG, *Ultrasound Med. Biol.* **29** (2003) 311.
11. N. B. SMITH, S. LEE and K. K. SHUNG, *ibid.* **29** (2003) 1205.
12. R.E. NEWNHAM and A. DOGAN, US Patent 5,729,077 (1995).
13. R.E. NEWNHAM and J. ZHANG, US Patent 6,232,702 (1999).
14. S. LEE, "Ultrasound-Mediated Transdermal Insulin Delivery and Glucose Measurement Using the Cymbal Array," Ph.D. Thesis, Penn State University (2004).

JOURNAL OF MATERIALS SCIENCE 41 (2006) 217–228

Loss mechanisms and high power piezoelectrics

K. UCHINO, J. H. ZHENG, Y. H. CHEN, X. H. DU, J. RYU, Y. GAO, S. URAL, S. PRIYA
International Center for Actuators and Transducers, Penn State University, University Park, PA, USA

S. HIROSE
Faculty of Eng., Yamagata University, Yonezawa, Japan

Heat generation is one of the significant problems in piezoelectrics for high power density applications. In this paper, we review the loss mechanisms in piezoelectrics first, followed by the heat generation processes for various drive conditions. Heat generation at off-resonance is caused mainly by dielectric loss tan δ' (i.e., P-E hysteresis loss), not by mechanical loss, while the heat generation at resonance is mainly attributed to mechanical loss tan ϕ'. Then, practical high power materials developed at Penn State is introduced, which exhibit the vibration velocity more than 1 m/s, leading to the power density capability 10 times of the commercially available "hard" PZTs. We propose a internal bias field model to explain the low loss and high power origin of these materials. Finally, using a low temperature sinterable "hard" PZT, we demonstrated a high power multilayer piezoelectric transformers.

1. Intoduction

Loss or hysteresis in piezoelectrics exhibits both merits and demerits. For positioning actuator applications, hysteresis in the field-induced strain provides a serious problem, and for resonance actuation such as ultrasonic motors, loss generates significant heat in the piezoelectric materials. Further, in consideration of the resonant strain amplified in proportion to a mechanical quality factor, low (intensive) mechanical loss materials are preferred for ultrasonic motors. To the contrary, for force sensors and acoustic transducers, a low mechanical quality factor Q_m (which corresponds to high mechanical loss) is essential to widen a frequency range for receiving signals.

K. H. Haerdtl wrote a review article on electrical and mechanical losses in ferroelectric ceramics [1]. Losses are considered to consist of four portions: (1) domain wall motion, (2) fundamental lattice portion, which should also occur in domain-free monocrystals, (3) microstructure portion, which occurs typically in polycrystalline samples, and (4) conductivity portion in highly-ohmic samples. However, in the typical piezoelectric ceramic case, the loss due to the domain wall motion exceeds the other three contributions significantly. They reported interesting experimental results on the relationship between electrical and mechanical losses in piezoceramics, $Pb_{0.9}La_{0.1}(Zr_{0.5}Ti_{0.5})_{1-x}Me_xO_3$, where Me represents the doped ions Mn, Fe or Al and x varied between 0 and 0.09. However, they measured the mechanical losses on poled ceramic samples, while the electrical losses on unpoled samples, i.e., in a different polarization state, which lead big ambiguity in the discussion.

Authors are aware little systematic studies of the loss mechanisms in piezoelectrics, particularly in high electric field and high power density ranges. Although T. Ikeda described part of the formulas of this paper in his textbook [2], he totally neglected the piezoelectric losses, which have been found not to be neglected in our investigations. In this paper, we review the loss mechanisms in piezoelectrics first, followed by the heat generation processes for various drive conditions. Then, practical high power materials developed at Penn State are introduced, exhibiting the power density capability 10 times of the commercially available "hard" PZTs. We propose a model to explain the low loss and high power origin of these materials.

0022-2461 © *2006 Springer Science + Business Media, Inc.*
DOI: 10.1007/s10853-005-7201-0

Finally, using a low temperature sinterable "hard" PZT, we demonstrated a high power multilayer piezoelectric transformers.

The terminologies, "intensive" and "extensive" losses are introduced in the relation with "intensive" and "extensive" parameters in the phenomenology. These are not directly relevant with "intrinsic" and "extrinsic" losses which were introduced to explain the loss contribution from the mono-domain single crystal status and from the others [3]. In this paper, our discussion is focused on the "extrinsic" losses, in particular, domain-reorientation originated losses.

2. General consideration of loss and hysteresis in piezoelectrics

2.1. Theoretical formulas

Since we have described the detailed mathematics in the previous paper [4], we just summarize the results in this section. We start from the following two piezoelectric equations:

$$x = s^{\mathrm{E}} X + d\,E, \tag{1}$$

$$D = d\,X + \varepsilon^{\mathrm{X}} \varepsilon_0 E. \tag{2}$$

Here, x is strain, X, stress, D, electric displacement, E, electric field. Equations 1 and 2 are the expression in terms of intensive (i.e., externally controllable) physical parameters X and E. The elastic compliance s^{E}, the dielectric constant ε^{X} and the piezoelectric constant d are temperature-dependent. Note that the piezoelectric equations cannot yield a delay-time related loss, without taking into account irreversible thermodynamic equations or dissipation functions, in general. However, the latter considerations are mathematically equivalent to the introduction of complex physical constants into the phenomenological equations, if the loss is small and can be treated as a perturbation.

Therefore, we will introduce complex parameters $\varepsilon^{\mathrm{X}^*}$, s^{E^*} and d^* in order to consider the hysteresis losses in dielectric, elastic and piezoelectric coupling energy:

$$\varepsilon^{\mathrm{X}^*} = \varepsilon^{\mathrm{X}}(1 - j \tan \delta'), \tag{3}$$

$$s^{\mathrm{E}^*} = s^{\mathrm{E}}(1 - j \tan \phi'), \tag{4}$$

$$d^* = d(1 - j \tan \theta'). \tag{5}$$

θ' is the phase delay of the strain under an applied electric field, or the phase delay of the electric displacement under an applied stress. Both delay phases should be exactly the same if we introduce the same complex piezoelectric constant d^* into Equations 1 and 2. δ' is the phase delay of the electric displacement to an applied electric field

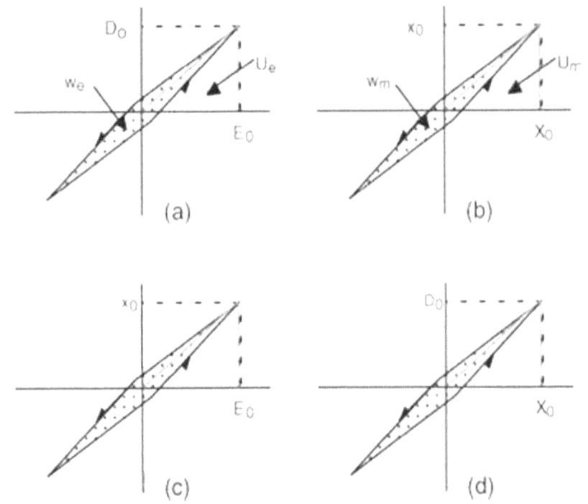

Figure 1 (a) D vs. E (stress free), (b) x vs. X (short-circuit), (c) x vs. E (stress free) and (d) D vs. X (short-circuit) curves with a slight hysteresis in each relation.

under a constant stress (e.g., zero stress) condition, and ϕ' is the phase delay of the strain to an applied stress under a constant electric field (e.g., short-circuit) condition. We will consider these phase delays as "intensive" losses.

Fig. 1a–d correspond to the model hysteresis curves for practical experiments: D vs. E curve under a stress-free condition, x vs. X under a short-circuit condition, x vs. E under a stress-free condition and D vs. X under a short-circuit condition for measuring current, respectively. Notice that these measurements are easily conducted in practice.

The stored energies and hysteresis losses for pure dielectric and elastic energies can be calculated as:

$$U_{\mathrm{e}} = (1/2)\varepsilon^{\mathrm{X}} \varepsilon_0 E_0^2, \tag{6}$$

$$w_{\mathrm{e}} = \pi \varepsilon^{\mathrm{X}} \varepsilon_0 E_0^2 \tan \delta', \tag{7}$$

$$U_{\mathrm{m}} = (1/2) s^{\mathrm{E}} X_0^2, \tag{8}$$

$$w_{\mathrm{m}} = \pi s^{\mathrm{E}} X_0^2 \tan \phi'. \tag{9}$$

The electromechanical hysteresis losses are more complicated, which can be calculated as follows, depending on the measuring ways; when measuring the induced strain under an electric field,

$$U_{\mathrm{em}} = (1/2)(d^2/s^{\mathrm{E}}) E_0^2, \tag{10}$$

and

$$w_{\mathrm{em}} = \pi (d^2/s^{\mathrm{E}}) E_0^2 (2 \tan \theta' - \tan \phi'). \tag{11}$$

Note that the strain vs. electric field measurement should provide the combination of piezoelectric loss $\tan \theta'$ and elastic loss $\tan\phi'$. When we measure the induced charge under stress, the stored energy U_{me} and the hysteresis loss W_{me} during a quarter and a full stress cycle, respectively, are obtained as

$$U_{me} = (1/2)(d^2/\varepsilon_0\varepsilon^X)X_0^2, \quad (12)$$

$$w_{me} = \pi(d^2/\varepsilon_0\varepsilon^X)X_0^2(2\tan\theta' - \tan\delta'). \quad (13)$$

Hence, from the measurements of D vs. E and x vs. X, we obtain $\tan\delta'$ and $\tan\phi'$, respectively, and either the piezoelectric (D vs. X) or converse piezoelectric measurement (x vs. E) provides $\tan\theta'$ through a numerical subtraction.

So far, we discussed the "intensive" dielectric, mechanical and piezoelectric losses in terms of "intensive" parameters X and E. In order to consider real physical meanings of the losses in the material, we will introduce the "extensive" losses [4] in terms of "extensive" parameters x and D. In practice, intensive losses are easily measurable, but extensive losses are not, but obtainable from the intensive losses. When we start from the piezoelectric equations in terms of extensive physical parameters x and D:

$$X = c^D x - h\,D, \quad (14)$$

$$E = -hx + \kappa^x\kappa_0 D, \quad (15)$$

we introduce the extensive dielectric, elastic and piezoelectric losses as

$$\kappa^{x*} = \kappa^x(1 + j\tan\delta), \quad (16)$$

$$c^{D*} = c^D(1 + j\tan\phi), \quad (17)$$

$$h^* = h(1 + j\tan\theta). \quad (18)$$

It is notable that the permittivity under a constant strain (e.g., zero strain or completely clamped) condition, ε^{x*} and the elastic compliance under a constant electric displacement (e.g., open-circuit) condition, s^{D*} can be provided as an inverse value of κ^{x*} and c^{D*}, respectively, in this simplest one dimensional expression (in the case of a general 3-D expression, this part must be translated as "inverse matrix components of κ^{x*} and c^{D*} tensors."). Thus, using the exactly the same losses in Equations 16 and 17,

$$\varepsilon^{x*} = E^x(1 - j\tan\delta), \quad (19)$$

$$s^{D*} = s^D(1 - j\tan\phi), \quad (20)$$

We will consider these phase delays again as "extensive" losses.

Here, we consider the physical property difference between the boundary conditions; E constant and D constant, or X constant and x constant in a simplest 1-D model. When an electric field is applied on a piezoelectric sample as illustrated in the top of Fig. 2, this state will be equivalent to the superposition of the following two steps: first, the sample is completely clamped and the field E_0 is applied (pure electrical energy (1/2) $\varepsilon^x\varepsilon_0\, E_0^2$ is input); second, keeping the field at E_0, the mechanical constraint is released (additional mechanical energy (1/2) $(d^2/s^E)E_0^2$ is necessary). The total energy should correspond to the total input electrical energy (1/2) $\varepsilon^X\varepsilon_0\, E_0^2$. Similar energy calculation can be obtained from the bottom of Fig. 2, leading to the following equations:

$$\varepsilon^x/\varepsilon^X = (1 - k^2), \quad (21)$$

$$s^D/s^E = (1 - k^2), \quad (22)$$

$$\kappa^X/\kappa^x = (1 - k^2), \quad (23)$$

$$c^E/c^D = (1 - k^2), \quad (24)$$

where

$$k^2 = d^2/(s^E\varepsilon_0\varepsilon^X) = h^2/(c^D\kappa^x\kappa_0). \quad (25)$$

This k is called the *electromechanical coupling factor*, which is defined as a real number in this manuscript.

In order to obtain the relationships between the intensive and extensive losses, the following three equations are essential:

$$\varepsilon_0\varepsilon^X = [\kappa^x\kappa_0(1 - h^2/(c^D\kappa^x\kappa_0))]^{-1}, \quad (26)$$

$$s^E = [c^D(1 - h^2/(c^D\kappa^x\kappa_0))]^{-1}, \quad (27)$$

$$d = [h^2/(c^D\kappa^x\kappa_0)][h(1 - h^2/(c^D\kappa^x\kappa_0))]^{-1}. \quad (28)$$

Replacing the parameters in Equations 26 and 27 by the complex parameters in Equations 3, 5), 16–18, we obtain the relationships between the intensive and extensive losses:

$$\tan\delta' = (1/(1 - k^2))[\tan\delta + k^2(\tan\phi - 2\tan\theta)], \quad (29)$$

$$\tan\phi' = (1/(1 - k^2))[\tan\phi + k^2(\tan\delta - 2\tan\theta)], \quad (30)$$

$$\tan\theta' = (1/(1 - k^2))[\tan\delta + \tan\phi + (1 + k^2)\tan\theta], \quad (31)$$

where k is the electromechanical coupling factor defined by Equation 25, and here as a real number. It is important

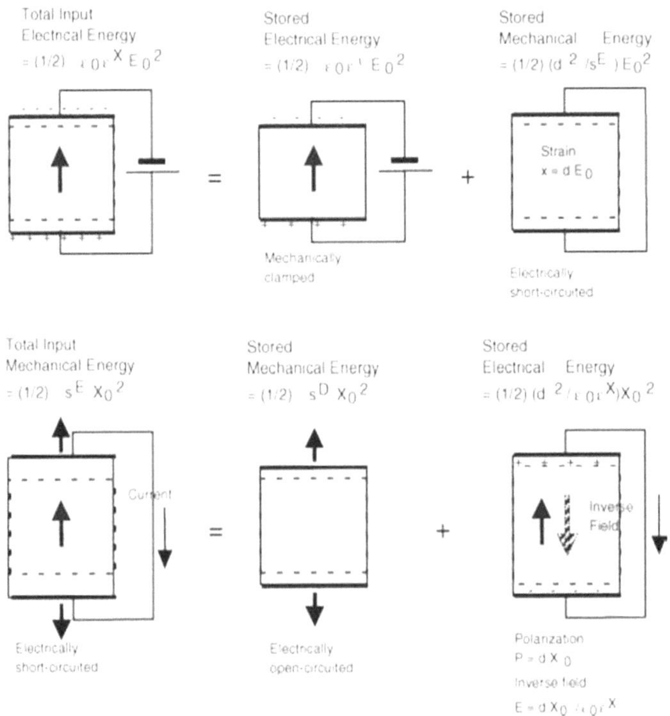

Figure 2 Conceptual figure for explaining the relation between ε^X and ε^x, s^E and s^D

that the intensive dielectric and elastic losses are mutually correlated with the extensive dielectric, elastic and piezoelectric losses through the electromechanical coupling k^2, and that the denominator $(1-k^2)$ comes basically from the ratios, $\varepsilon^x/\varepsilon^X = (1-k^2)$ and $s^D/s^E = (1-k^2)$, and this real part reflects to the dissipation factor when the imaginary part is divided by the real part.

2.2. Experimental example

We determined "intensive" dissipation factors first from (a) D vs. E (stress free), (b) x vs. X (short-circuit), (c) x vs. E (stress free) and (d) D vs. X (short-circuit) curves for a soft PZT based multilayer actuator [5]. Then, we calculated the "extensive" losses as shown in Fig. 3. Note that the piezoelectric losses tan θ' and tan θ are not so small as previously believed, but comparable to the dielectric and elastic losses, and increase gradually with the field or stress. Also it is noteworthy that the extensive dielectric loss tan δ increases significantly with an increase of the intensive parameter, i.e., the applied electric field, while the extensive elastic loss tan ϕ is rather insensitive to the intensive parameter, i.e., the applied compressive stress. When similar measurements to Fig.1a and b, but under constrained conditions; that is, D vs. E under a completely clamped state, and x vs. X under an open-circuit state, respectively, we can expect smaller hystereses; that is, extensive losses, tan δ and tan ϕ. These measurements seem to be alternative methods to determine the three losses separately, however, they are rather difficult in practice.

2.3. Physical meaning of extensive losses

For making the situation simplest, we consider here only the domain wall motion-related losses. Taking into account the fact that the polarization change is primarily attributed to 180 degree domain wall motion, while the strain is attributed to 90 degree (or non-180°) domain wall motion, we suppose that the extensive dielectric and mechanical losses are originated from 180° and 90° domain wall motions, respectively, as illustrated in Fig. 4. The dielectric loss comes from the hysteresis during the 180° polarization reversal under E, while the elastic loss comes from the hysteresis during the 90° polarization reorientation under X. In this model, the intensive (observable) piezoelectric loss is explained by the 90° polarization reorientation under E, which can be realized by superimposing the 90° polarization reorientation under X and the 180° polarization reversal under E. This is the primarily reason why Equation 11 includes a combination term as (2 tan θ' − tan ϕ').

If we adopt the Uchida-Ikeda polarization reversal/reorientation model [6], we can explain the loss change with intensive parameter (externally controllable parameter). By finding the polarization P and the field-induced

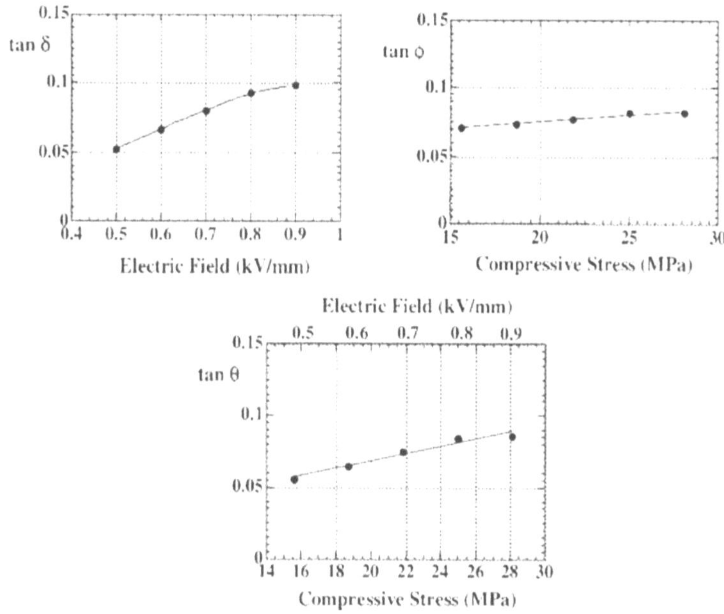

Figure 3 Extensive loss factors, tan δ, tan ϕ and tan θ as a function of electric field or compressive stress, measured for a PZT based actuator.

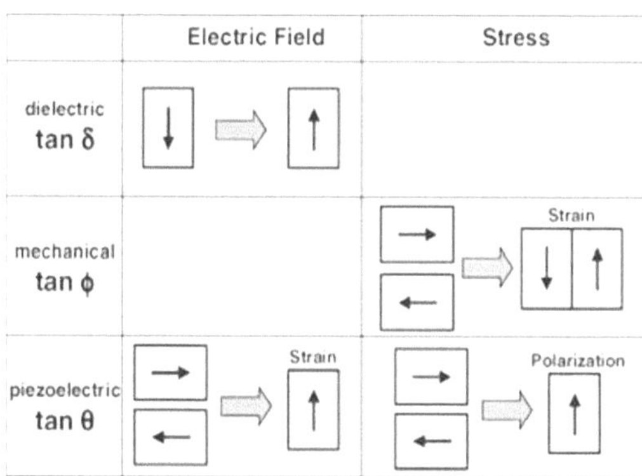

Figure 4 Polarization reversal/reorientation model for explaining dielectric, elastic and piezoelectric losses.

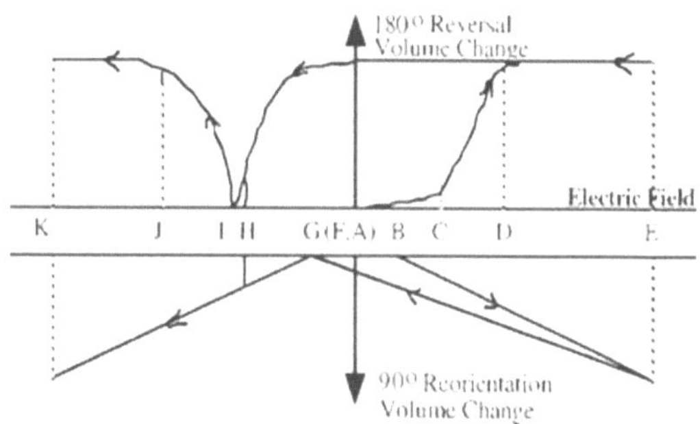

Figure 5 Polarization reversal/reorientation model for explaining the loss change with electric field.

strain x as a function of the electric field E, it is possible to estimate the volume in which a 180° reversal or a 90° rotation occurred. This is because the 180° domain reversal does not contribute to the induced strain, only the 90° rotation does, whereas the 180° domain reversal contributes mainly to the polarization. The volume change of the domains with external electric field is shown schematically in Fig. 5 that with the application of an electric field the 180° reversal occurs rapidly whereas the 90° rotation occurs slowly. It is notable that at G in the figure, there remains some polarization while the induced strain is zero, at H the polarizations from the 180° and 90° reorientations cancel each other and become zero, but the strain is not at its minimum. Due to a sudden change in 180° reversal above a certain electric field, we can expect a sudden increase in the polarization hysteresis and in the loss [this may reflect to the extensive dielectric loss measurement in Fig. 3 top left]; while the slope of 90° reorientation is almost constant, we can expect a constant loss or a mechanical quality factor Q_m with changing the external parameter, E or X [extensive elastic loss in Fig. 3 top right]. This situation will be discuss again in the later section.

3. Losses at a piezoelectric resonance

So far, we have considered the losses for a quasi-static or off-resonance state. Problems in ultrasonic motors which are driven at the resonance frequency include significant distortion of the admittance frequency spectrum due to nonlinear behavior of elastic compliance at a high vibration amplitude, and heat generation which causes a serious degradation of the motor

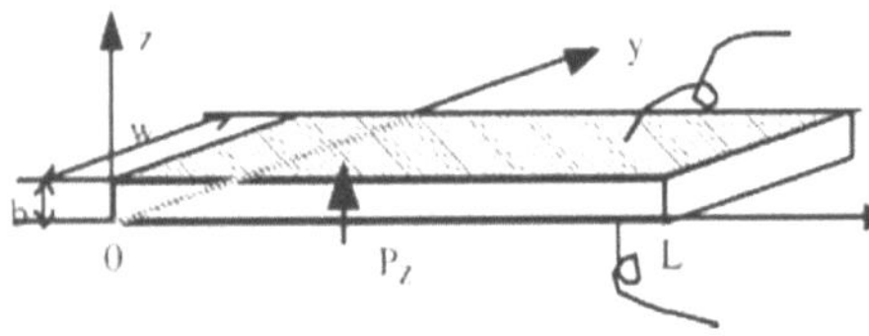

Figure 6 Longitudinal vibration through the transverse piezoelectric effect (d_{31}) in a rectangular plate.

characteristics through depoling of the piezoceramic. Therefore, the ultrasonic motor requires a very hard type piezoelectric with a high mechanical quality factor Q_m, leading to the suppression of heat generation. It is also notable that the actual mechanical vibration amplitude at the resonance frequency is directly proportional to this Q_m value.

3.1. Vibration at a piezoelectric resonance

Let us review the longitudinal mechanical vibration of a piezo-ceramic plate through the transverse piezoelectric effect (d_{31}) as shown in Fig. 6 [7]. Assuming that the polarization is in the z-direction and the x-y planes are the planes of the electrodes, the extensional vibration in the x direction is represented by the following dynamic equation:

$$(\partial^2 u/\partial t^2) = F = (\partial X_{11}/\partial x) + (\partial X_{12}/\partial y) + (\partial X_{13}/\partial z), \quad (32)$$

where u is the displacement of the small volume element in the ceramic plate in the x-direction. When the plate is very long and thin, X_2 and X_3 may be set equal to zero through the plate, and the following solutions can be obtained:

$$\text{(strain)} \quad \partial u/\partial x = x_1 = d_{31} E_Z [\sin \omega(L-x)/v + \sin(\omega x/v)]/\sin(\omega L/v), \quad (33)$$

$$\text{(total displacement)} \quad \Delta L = \int_0^L x_1 dx = d_{31} E_z L(2v/\omega L)\tan(\omega L/2v). \quad (34)$$

Here, v is the sound velocity in the piezoceramic which is given by

$$v = 1/\sqrt{\rho s_{11}^E}. \quad (35)$$

The admittance for the mechanically free sample is calculated to be:

$$\begin{aligned} Y &= (1/Z) = (i/V) = (i/E_z t) \\ &= (j\omega w L/t)\varepsilon_0 \varepsilon_3^{LC}[1 + (d_{31}^2/\varepsilon_0 \varepsilon_3^{LC} s_{11}^E) \\ &\quad \times (\tan(\omega L/2v)/(\omega L/2v)], \end{aligned} \quad (36)$$

where w is the width, L the length, t the thickness of the sample, and V the applied voltage. ε_3^{LC} is the permittivity in a longitudinally clamped sample, which is given by

$$\varepsilon_0 \varepsilon_3^{LC} = \varepsilon_0 \varepsilon_3^X - \left(d_{31}^2/s_{11}^E\right) = \varepsilon_0 \varepsilon_3^X\left(1 - k_{31}^2\right). \quad (37)$$

Now, we will introduce the complex parameters into the admittance curve around the resonance frequency, in a similar way to the previous section: $\varepsilon_3^{X*} = \varepsilon_3^X(1 - j \tan \delta')$, $s_{11}^{E*} = s_{11}^E(1 - j \tan \phi\prime)$, and $d_{31}^* = d(1 - j \tan \theta')$ into Equation 36:

$$\begin{aligned} Y = Y_d + Y_m &= j\omega C_d(1 - j \tan \delta) \\ &+ j\omega C_d K_{31}^2[(1 - j(2 \tan \theta' - \tan \phi')] \\ &\times [(\tan(\omega L/2v^*)/(\omega L/2v^*)], \end{aligned} \quad (38)$$

where

$$C_0 = (wL/t)\varepsilon_0 \varepsilon_3^X, \quad (39)$$

,

$$C_d = \left(1 - k_{31}^2\right)C_0. \quad (40)$$

Note that the loss for the first term (damped conductance) is represented by the "extensive" dielectric loss tan δ, not by the intensive loss tan δ'. We further calculate $1/(\tan(\omega L/2v^*)$ with an expansion-series approximation around $(\omega L/2v) = \pi/2$, taking into account that the resonance state is defined in this case for the maximum admittance point.

Using new frequency parameters,

$$\Omega = \omega L/2v, \quad \Delta\Omega = \Omega - \pi/2(\ll 1), \quad (41)$$

and $K_{31}^2 = k_{31}^2/(1 - k_{31}^2)$, the motional admittance Y_m is approximated around the first resonance frequency by

$$\begin{aligned} Y_m = j(8/\pi^2)\omega_0 C_d K_{31}^2[(1 + j((3/2)\tan \phi' \\ - 2 \tan \theta')]/[(-(4/\pi)\Delta\Omega + j \tan \phi'). \end{aligned} \quad (42)$$

The maximum Y_m is obtained at $\Delta\,\Omega = 0$:

$$\begin{aligned} Y_m^{max} &= (8/\pi^2)\omega_0 C_d K_{31}^2(\tan \phi')^{-1} \\ &= (8/\pi^2)\omega_0 C_d K_{31}^2 Q_m, \end{aligned} \quad (43)$$

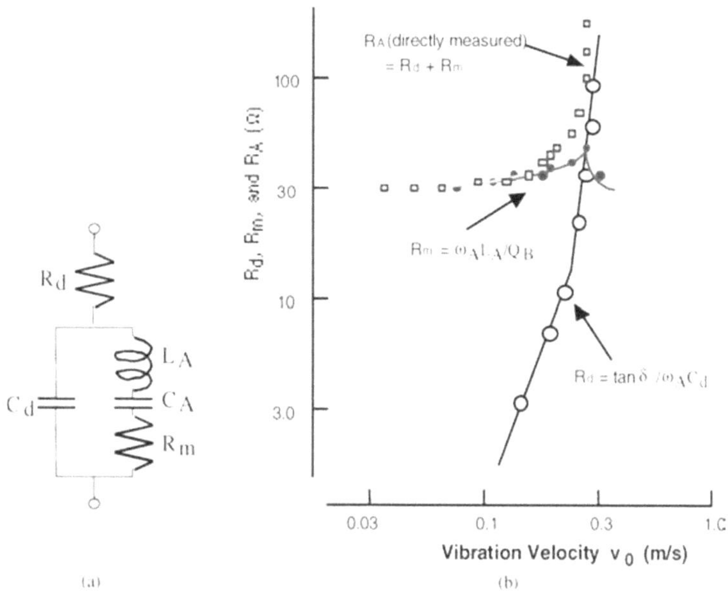

Figure 7 (a) Equivalent circuit of a piezoelectric device for the resonance under high power drive. (b) Vibration velocity dependence of the resistances R_d and R_m in the equivalent electric circuit for a longitudinally vibrating PZT ceramic transducer through the transverse piezoelectric effect d_{31}. Notice a dramatic change in R_d above a certain threshold vibration velocity.

where $Q_m = (\tan\phi')^{-1}$. Similarly, the maximum displacement u_{max} is obtained at $\Delta\Omega$=0:

$$u_{max} = (8/\pi^2)d_{31}E_Z L\, Q_m. \quad (44)$$

The maximum displacement at the resonance frequency is $(8/\pi^2)Q_m$ times larger than that at a non-resonance frequency, $d_{31}E_Z L$.

In a brief summary, when we observe the admittance or displacement spectrum as a function of drive frequency, and obtain the mechanical quality factor Q_m estimated from $Q_m = \omega_0/2\Delta\omega$, where $2\Delta\omega$ is a full width of the 3 dB down (i.e., $1/\sqrt{2}$) of the maximum value at $\omega = \omega_0$, we can obtain the intensive mechanical loss $\tan\phi'$.

3.2. Equivalent circuit under high power drive

The equivalent circuit for the piezoelectric actuator is represented by a combination of L, C and R. Fig.7a shows an equivalent circuit for the resonance state, which has very low impedance. Taking into account Equation 42, we can understand that C_d and R_d correspond to the electrostatic capacitance (for a longitudinally clamped sample in the previous case, not a free sample) and the clamped (or "extensive") dielectric loss $\tan\delta$, respectively, and the components L_A and C_A in a series resonance circuit are related to the piezoelectric motion. For example, in the case of the longitudinal vibration of the above rectangular plate through d_{31}, these components are represented approximately by

$$L_A = (\rho/8)(Lb/w)\left(s_{11}^{E2}/d_{31}^2\right), \quad (45)$$

$$C_A = (8/\pi^2)(Lw/b)\left(d_{31}^2/s_{11}^E\right). \quad (46)$$

The total resistance R_A ($=R_d + R_m$) should correspond to the loss $\tan\phi'$, which is composed of the extensive mechanical loss $\tan\phi$ and dielectric/piezoelectric coupled loss ($\tan\delta - 2\tan\theta$) (see Equation 30). Thus, intuitively speaking, R_d and R_m correspond to the extensive dielectric and mechanical losses, respectively. Note that we introduced an additional resistance R_d to explain a large contribution of the dielectric loss when a vibration velocity is relatively large. There are, of course, different ways to introduce R_d in an equivalent circuit [8].

3.3. Losses as a function of vibration velocity

Let us consider here the degradation mechanism of the mechanical quality factor Q_m with increasing electric field and vibration velocity. Fig. 8 shows the change in mechanical Q_m with vibration velocity. Q_m is almost constant for a small electric field/vibration velocity, but above a certain vibration level Q_m degrades drastically, where temperature rise starts to be observed [9].

Fig. 7b depicts an important notion on heat generation from the piezoelectric material, where the damped and motional resistances, R_d and R_m, in the equivalent electrical circuit of a PZT sample (Fig. 7a are separately plotted

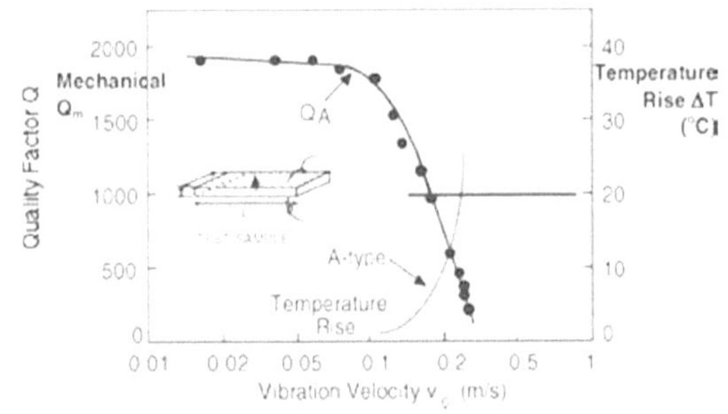

Figure 8 Vibration velocity dependence of the quality factor Q_A and temperature rise for A (resonance) type resonance of a longitudinally vibrating PZT ceramic transducer through the transverse piezoelectric effect d_{31}. The maximum vibration velocity is defined at the velocity where 20°C temperature rise from room temperature is exhibited.

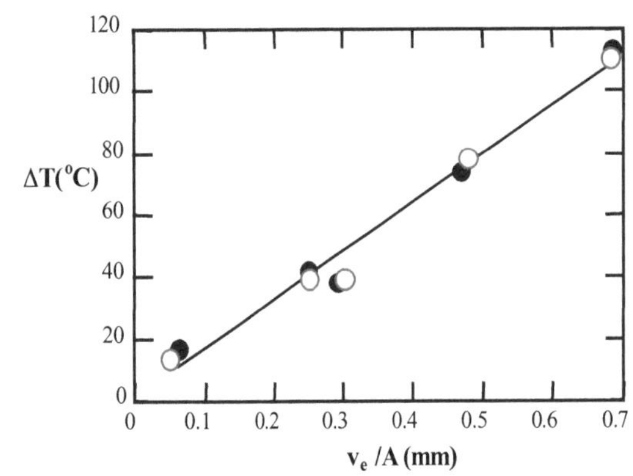

Figure 9 Temperature rise at off-resonance versus V_e/A (3 kV/mm, 300 Hz) in various size soft PZT multilayer actuators, where V_e is the effective volume generating the heat and A is the surface area dissipating the heat.

as a function of vibration velocity. Note that R_m, mainly related to the extensive mechanical loss (90° domain wall motion), is insensitive to the vibration velocity, while R_d, related to the extensive dielectric loss (180° domain wall motion), increases significantly around a certain critical vibration velocity. Thus, the resonance loss at a small vibration velocity is mainly determined by the extensive mechanical loss which provides a high mechanical quality factor Q_m, and with increasing vibration velocity, the extensive dielectric loss contribution significantly increases. This is consistent with the discussion made on Fig. 5. After R_d exceeds R_m, we started to observe heat generation.

4. Heat generation in piezoelectrics

Heat generation in various types of PZT-based actuators has been studied under a large electric field applied (1 kV/mm or more) at an off-resonance frequency and under a relatively small electric field applied (100 V/mm) at a resonance frequency.

Zheng *et al.* reported the heat generation at an off-resonance frequency from various sizes of multilayer type piezoelectric ceramic actuators [6]. The temperature change with time in the actuators was monitored when driven at 3 kV/mm and 300 Hz, and Fig. 9 plots the saturated temperature as a function of V_e/A, where V_e is the effective volume (electrode overlapped part) and A is the surface area. This linear relation is reasonable because the volume V_e generates the heat and this heat is dissipated through the area A. Thus, if we need to suppress the temperature rise, a small V_e/A design is preferred. From these experimental results, we calculated the total loss u of the piezoelectric, which is summarized in Table I. The experimental data of *P-E* hysteresis losses under a stress-free condition is also listed for comparison. It is very important that the *P-E* hysteresis intensive loss agrees well with the total loss contributing to the heat generation under an off-resonance drive.

Tashiro *et. al.* observed the heat generation in a rectangular piezoelectric plate during a resonating drive [10]. Even though the maximum electric filed is not very large, heat is generated due to the large induced strain/stress at the resonance. Fig. 10 depicted an infrared image taken for a resonating rectangular PZT plate in our laboratory. The maximum heat generation was observed at the nodal point of the resonance vibration, where the maximum strain/stress are generated. This observation supports that the heat generation in a resonating sample is attributed to the intensive elastic loss tan ϕ'. This is not contradictory to the result in the previous paragraph, where a high-voltage was applied at an off-resonance frequency. We concluded there that the heat is originated from the intensive dielectric loss tan δ'. In consideration that both the "intensive" dielectric and mechanical losses are composed of the "extensive" dielectric and mechanical losses, and that the extensive dielectric loss tan δ changes significantly with the external electric field and stress, the major contribution to the heat generation seems to come from the "extensive" dielectric loss (i.e., 180° domain wall motion). Since this is just our model, there can be different domain reorienta-

TABLE I Loss and overall heat transfer coefficient for PZT multilayer samples ($E = 3$ kV/mm, $f = 300$ Hz). The effective heat transfer coefficient here is the sum of the rates of heat flow by radiation and by convection, neglecting the conduction effect

Actuator	4.5×3.5×2 mm	7×7×2 mm	17×3.5×1 mm
Total loss ($\times 10^3$ J/m^3) $u = \frac{\rho c v}{f v_e}\left(\frac{dT}{dt}\right)_{t-} > 0$	19.2	19.9	19.7
P-E hysteresis loss ($\times 10^3$ J/m^3)	18.5	17.8	17.4
$k(T)$ (W/m^{2}K)	38.4	39.2	34.1

Figure 10 An infrared image of a "hard" PZT rectangular plate driven at the second resonance mode. Note three hot points, which correspond to the nodal points for this vibration mode.

tion models, and further investigations are waited for the microscopic observation of this phenomenon.

5. High power piezoelectrics

5.1. Practical PZT based ceramics

"High Power" in this paper stands for high power density in mechanical output energy converted from the maximum input electrical energy under the drive condition with 20°C temperature rise. The mechanical power density can be evaluated by the square of the maximum vibration velocity (v_0^2), which is a sort of material's constant (Remember that there exists the maximum mechanical energy density, above which level, the piezoelectric material becomes a ceramic heater.). Fig. 11 shows the mechanical Q_m versus basic composition x at two effective vibration velocities $v_0 = 0.05$ m/s and 0.5 m/s for $Pb(Zr_xTi_{1-x})O_3$ doped with 2.1 at % of Fe [11]. The decrease in mechanical Q_m with an increase of vibration level is minimum around the rhombohedral-tetragonal morphotropic phase boundary (52/48). In other words, the smallest Q_m material under a small vibration level becomes the highest Q_m material under a large vibration level, and the data obtained by a conventional impedance analyzer with a small voltage/power does not provide data relevant to high power characteristics. Thus, we developed various measuring techniques of high power piezoelectricity, including "Constant Current" and "pulse drive" methods [12].

The conventional piezo-ceramics have the limitation in the maximum vibration velocity (v_{max}), since the additional input electrical energy is converted into heat, rather than into mechanical energy. The typical rms value of v_{max} for commercially available materials, defined by the temperature rise of 20°C from room temperature, is around 0.3 m/sec for rectangular samples operating in the k_{31} mode (like a Rosen-type transformer) [9]. $Pb(Mn,Sb)O_3$ (PMS) – lead zirconate tatanate (PZT) ceramics with the v_{max} of 0.62 m/sec are currently used for NEC transformers [11]. By doping the PMS-PZT or $Pb(Mn,Nb)O_3$-PZT with rare-earth ions such as Yb, Eu and Ce, we recently developed the high power piezoelectrics, which can operate with v_{max} up to 1.0 m/sec [13, 14]. Compared with commercially available piezoelectrics, 10 times higher input electrical energy and output mechanical energy can be expected from these new materials without generating significant temperature rise, which corresponds to 100 W/cm^2. Fig. 12 shows the dependence of the maximum vibration velocity v_0 (20°C temperature rise) on the atomic % of rare-earth ion, Yb, Eu or Ce in the $Pb(Mn,Sb)O_3$ (PMS) – PZT based ceramics. Enhancement in the v_0 value is significant by adding a small amount of the rare-earth ion [14].

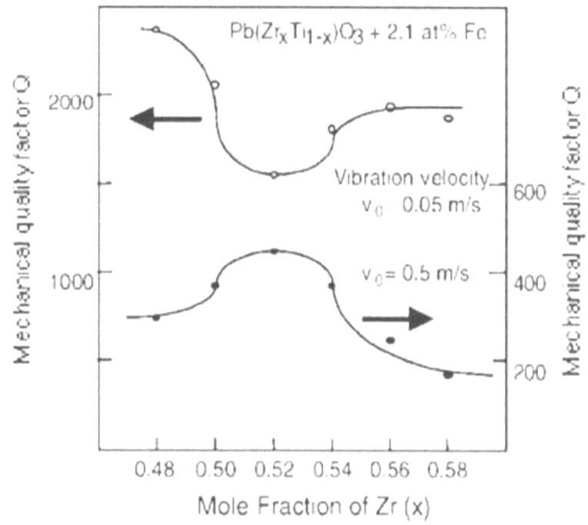

Figure 11 Mechanical Q_m versus basic composition x at two effective vibration velocities $v_0 = 0.05$ m/s and 0.5 m/s for $Pb(Zr_xTi_{1-x})O_3$ doped with 2.1 at % of Fe.

5.2. Origin of the high power piezoelectrics

"Hard" PZT is usually used for high power piezoelectric applications, because of its high coercive field; in other words, the stability of the domain walls. Acceptor ions, such as Fe^{3+}, introduce oxygen deficiencies in the PZT crystal (In the case of donor ions, such as Nb^{5+}, Pb deficiency is introduced). Thus, in the conventional model, the acceptor doping causes domain pinning through the easy reorientation of deficiency-related dipoles, leading to "hard" characteristics (Domain Wall Pinning Model

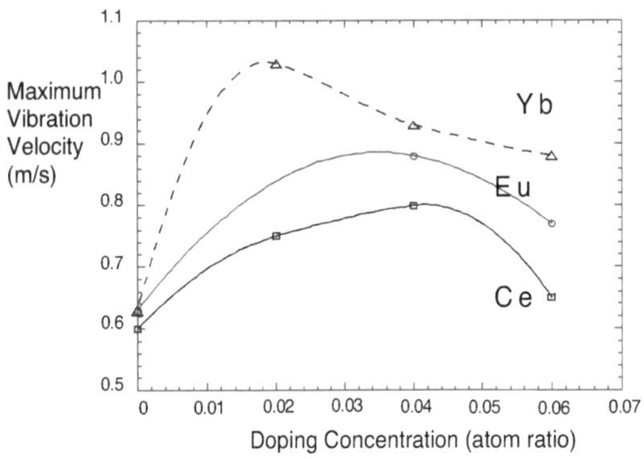

Figure 12 Dependence of the maximum vibration velocity v_0 (20°C temperature rise) on the atomic % of rare-earth ion, Yb, Eu or Ce in the $Pb(Mn,Sb)O_3$ (PMS) – PZT based ceramics.

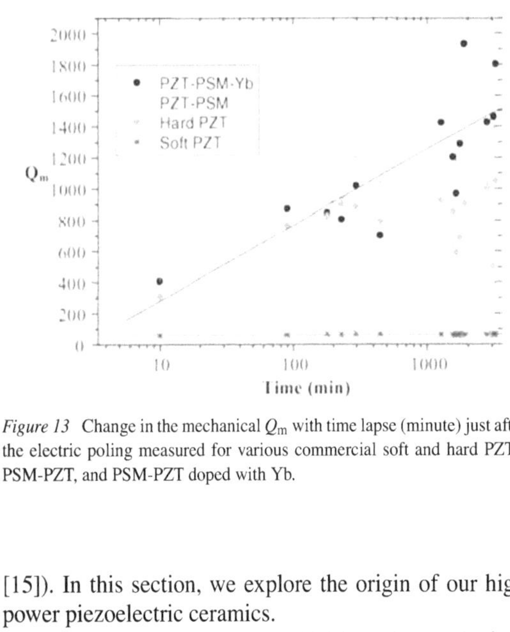

Figure 13 Change in the mechanical Q_m with time lapse (minute) just after the electric poling measured for various commercial soft and hard PZTs, PSM-PZT, and PSM-PZT doped with Yb.

[15]). In this section, we explore the origin of our high power piezoelectric ceramics.

High mechanical Q_m is essential in order to obtain a high power material with a large maximum vibration velocity. Fig. 13 exhibits suggestive results in the mechanical Q_m increase with time lapse (minute) after the electric poling measured for various commercial soft and hard PZTs, PSM-PZT, and PSM-PZT doped with Yb. It is notable that the Q_m values for commercial hard PZT and our high power piezoelectrics were almost the same, slightly higher than soft PZTs, and around 200–300 immediately after the poling. After a couple of hours passed, the Q_m values increased more than 1000 for the "hard" materials, while no change was observed in the "soft" material. The increasing slope is the maximum for the Yb-doped PSM-PZT. We also found a contradiction that this gradual increase (in a couple of hours) in the Q_m cannot be explained by the above-mentioned "domain wall pinning" model, but more likely by some ionic diffusion model.

Fig. 14 shows the polarization vs. electric field hysteresis curves measured for the Yb-doped $Pb(Mn,Sb)O_3$- PZT sample immediately after poling (fresh), 48 hr after, and a week after (aged). Remarkable aging effects could be observed; (a) in the decrease in the magnitude of the remnant polarization, and (b) in the positive internal bias electric field growth (i.e., the hysteresis curve shift leftwards in terms of the external electric field axis). The phenomenon (a) can be explained by the local domain wall pinning effect, but the large internal bias (more than 1 kV/mm) growth (b) seems to be the origin of the high power characteristics. Suppose that the vertical axis in Fig. 5 shifts rightwards, the inverse electric field required for realizing the 180° polarization reversal is increased, leading to the resistance enhancement against generating the hysteresis or heat.

Finally, let us propose the origin of this internal bias field growth. Based on the necessity of the oxygen deficiencies and the relatively slow (a couple of hours) growth rate, we assume here the oxygen deficiency diffusion model, which is illustrated in Fig.15. Under the electric poling process, the defect dipole P_{defect} (a pair of acceptor ion and oxygen deficiency) will be arranged parallel to the external electric field. After removing the field, the oxygen diffusion occurs, which can be estimated in a scale of hour at room temperature. Taking into account slightly different atomic distances between the A and B ions in the perovskite crystal in a ferroelectric (asymmetric) phase, the oxygen diffusion probability will be slightly higher for the downward, as shown in the figure, leading to the increase in the defect dipole with time. This may be the origin of the internal bias electric field.

6. High power piezoelectric components

Though we have developed "high power density" piezoelectric ceramics, the multilayer structure is a key to develop actual "high power" components from the device designing viewpoint. However, the present Ag-Pd electrode structure includes two-fold problems; (1) expensive Pd, and (2) Ag migration under a high field. In order to solve the problems, pure Cu electrode will be a key. But, the multilayer samples need to be sintered at a relatively low temperature (900°C or lower), when utilizing Cu embedded electrodes. Thus, low temperature sintering of "hard" type PZTs is a necessary technology to be developed. Different from soft PZTs, most of the conventional dopants to decrease the sintering temperature failed to be used, because these dopants also degrade the Q_m value significantly.

Our recent development of multilayered piezoelectric transformers with a hard piezoelectric ceramic is introduced here. In this sort of electric-mechanical-electric conversion application (combination of the converse and direct piezoelectric effects), since the fig. of merit is v_0k, the composition needs to sustain high d

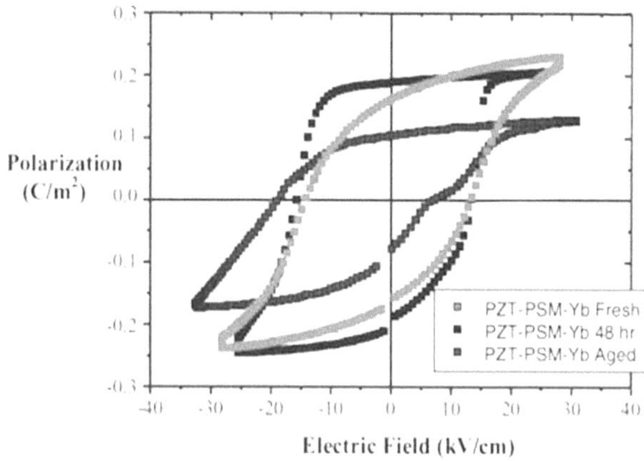

Figure 14 Polarization vs. electric field hysteresis curves measured for the Yb-doped $Pb(Mn,Sb)O_3$- PZT sample just after poling (fresh), 48 hours after, and a week after (aged).

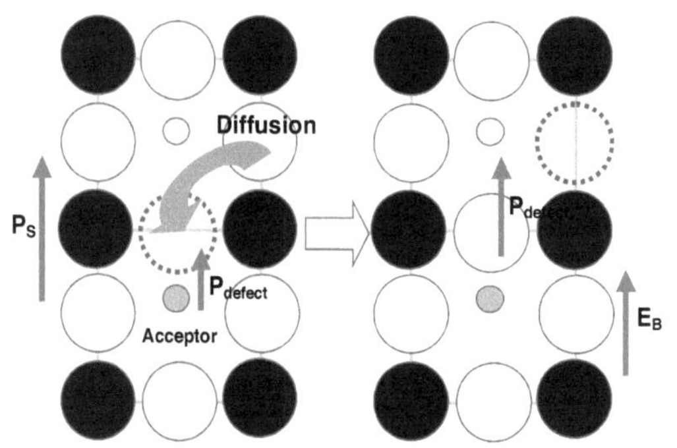

Figure 15 Oxygen deficiency diffusion model for explaining the internal bias electric field growth.

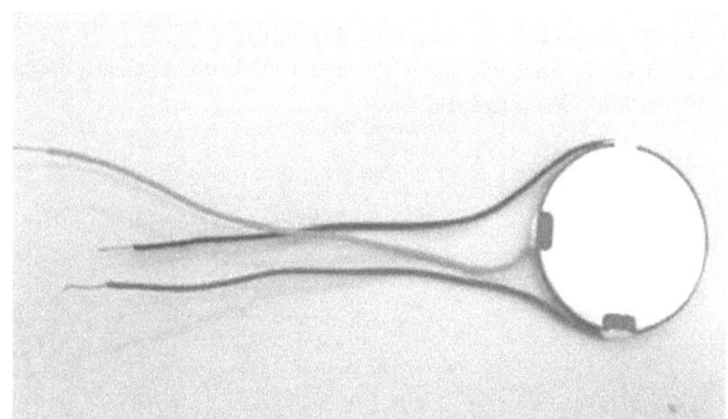

Figure 16 Multilayer co-fired transformer (35 W) with "hard" PZT and pure Ag electrodes, sintered at 880°C [Penn State trial product].

and k by keeping reasonable maximum vibration velocity. From the above reason, a new hard PZT composition system has been tried, starting from the originally soft $Pb(Zn_{1/3}Nb_{2/3})_x(Zr_{0.5}Ti_{0.5})_{1-x}O_3$. The composition corresponding to 0.2 PZN−0.8 PZT+0.5 wt% MnO_2 was found to have high piezoelectric properties along with a reasonable mechanical quality factor (d_{33}=277 pC/N, $\varepsilon_{33}/\varepsilon_o$=946, tan δ'=0.35%, k_p=0.58, Q_m=1402, $v_{0,rms}$=0.55 m/s). Furthermore, we found that modification of 0.2 PZN−0.8 PZT+0.5 wt% MnO_3 with $PbTiO_3$ and CuO lowered the sintering temperature to 880°C [See Table II]. The low field piezoelectric properties of the sample sintered at 880°C for 2 hr were found to be as: d_{33}=300 pC/N, $\varepsilon_{33}/\varepsilon_o$=1100, tan δ'=1.2%, k_p=0.52 and Q_m=900. Cu-embedded multilayer transformers are now being developed in trial as depicted in Fig. 16.

7. Conclusions

(1) There are three loss origins in piezoelectrics: dielectric, elastic and piezoelectric losses. 180° and non-180° domain wall motions contribute primarily to the extensive dielectric and elastic losses, respectively.

(2) Heat generation occurs in the sample uniformly under an off-resonance mainly due to the intensive dielectric loss, while heat is generated primarily at the vibration nodal points via the intensive elastic loss under a resonance. In both cases, the loss increase is originated from the extensive dielectric loss change with electric field and/or stress.

(3) Doping rare-earth ions into PZT-$Pb(Mn,X)O_3$ (X = Sb, Nb) ceramics increases the maximum vibration velocity up to 1 m/s, which corresponds to one order of magni-

TABLE II Electromechanical coupling parameters for low temperature sintering "hard" PZT based ceramics: 0.2 PZN−0.8 PZT + 0.5 wt% MnO_2 with $PbTiO_3$ and CuO

wt% PT extra	Sintering Temp.(°C)	k_p	Q_m	d_{33} (pC/N)	$\varepsilon_{33}/\varepsilon_0$	tan δ (%)
0	1200	0.55	1502	265	1011	0.56
6.5	1000	0.47	1308	260	1300	1
6.3 + 0.3 wt% CuO	880	0.52	900	300	1100	1.2

tude higher energy density than conventionally commercialized piezo-ceramics.

(4) To obtain high power density/high vibration velocity materials, domain wall immobility/stabilization via the positive internal bias field seems to be essential, rather than the local domain wall pinning effect.

Since the above conclusions were derived only from a limited number of PZT-based soft and hard piezoelectrics, it is too early to generalize these conclusions. Further investigations are highly required.

Acknowledgments

Part of this research was supported by the Office of Naval Research through the grant no. N00014-96-1-1173 and N00014-99-1-0754.

References

1. K. H. HAERDTL, *Ceram. Int'l.* **8** (1982) 121.
2. T. IKEDA, "Fundamentals of Piezoelectric Materials Science", (Ohm Publication Co., Tokyo, 1984) 83.
3. N. SETTER, (ed.) *Piezoelectric Materails in Devices*, (2002).
4. K. UCHINO and S. HIROSE, *IEEE-UFFC Trans.* **48** (2001) 307.
5. J. ZHENG, S. TAKAHASHI, S. YOSHIKAWA, K. UCHINO and J. W. C. DE VRIES, *J. Amer. Ceram. Soc.* **79** (1996) 3193.
6. N. UCHIDA and T. IKEDA, *Jpn. J. Appl. Phys.* **6** (1967) 1079.
7. K. UCHINO, "Piezoelectric Actuators and Ultrasonic Motors", (Kluwer Academic Publ., Boston, 1997) p. 197.
8. M. UMEDA, K. NAKAMURA and S. UEHA, *Jpn. J. Appl. Phys.* **38** (1999) 3327.
9. S. HIROSE, M. AOYAGI, Y. TOMIKAWA, S. TAKAHASHI and K. UCHINO, *Proc. Ultrasonics Int'l '95*, Edinburgh, (1995) 184.
10. S. TASHIRO, M. IKEHIRO and H. IGARASHI, *Jpn. J. Appl. Phys.* **36** (1997) 3004.
11. S. TAKAHASHI and S. HIROSE, *ibid.* **32** (1993) 2422.
12. K. UCHINO, J. ZHENG, A. JOSHI, Y. H. CHEN, S. YOSHIKAWA, S. HIROSE, S. TAKAHASHI and J. W. C. DE VRIES, *J. Electroceramics* **2** (1998) 33.
13. J. RYU, H. W. KIM, K. UCHINO and J. LEE, *Jpn. J. Appl. Phys.* **42**(3) (2003) 1.
14. Y. GAO, K. UCHINO and D. VIEHLAND, *J. Appl. Phys.* **92** (2002) 2094.
15. K. UCHINO, "Ferroelectric Devices" (Marcel Dekker, Inc., New York, 2000), p. 63.

JOURNAL OF MATERIALS SCIENCE **41** (2006) 229–249

Effect of electrical conductivity on poling and the dielectric, pyroelectric and piezoelectric properties of ferroelectric 0-3 composites

C. K. WONG
Department of Applied Physics, The Hong Kong Polytechnic University, Hong Kong, China

F. G. SHIN
Department of Applied Physics, Materials Research Center and Center for Smart Materials, The Hong Kong Polytechnic University, Hong Kong, China

We have investigated the effects of electrical conductivity of the constituents on the poling behavior, dielectric, pyroelectric and piezoelectric properties of ferroelectric 0-3 composites. Modeling of polarization behavior is explored for both dc and ac poling procedures. Simulated results show that, in addition to the poling schedule, conductivity plays an important role in the poling process. Calculations are carried out for the time dependent internal electric fields induced by an ac field in dielectric measurements, by modulated temperature in pyroelectric measurement or by stress in piezoelectric measurement. Our previously developed models for dielectricity, pyroelectricity and piezoelectricity have been extended to include the additional contribution from the electrical conductivities. These can be significant for ceramic/polymer composites possessing high conductivity in the matrix phase. Calculated values based on the new model are discussed and compared with the previous models, and in particular the pyroelectric activities reported in the literature for a graphite doped lead zirconate titanate / polyurethane composite. Explicit expressions for the transient and steady state responses are given and the effective permittivity, pyroelectric and piezoelectric coefficients are derived in this paper. Remarkable enhancement in these coefficients is obtained when higher conductivity is introduced in the matrix phase.

1. Introduction

One of the primary goals of embedding ferroelectric ceramic particles within a polymer matrix (i.e. to form a 0-3 composite) is to combine the better properties of ceramic and polymer. Ferroelectric ceramics have high piezoelectric and pyroelectric properties, but their poor mechanical properties and the large difference in acoustic impedance with water and human tissues tend to restrict their usefulness in many applications. Ferroelectric polymers, on the other hand, have good mechanical flexibility, but their piezoelectric and pyroelectric coefficients are low. The use of ferroelectric composites seems capable of overcoming these deficiencies and their properties can be tailored for specific situations. One may in principle design a 0-3 composite with high ceramic volume fraction for applications necessitating high piezoelectric and pyroelectric values. However, it is difficult to fabricate a 0-3 composite sample of high ceramic concentration and the high ceramic content will also lower the flexibility of the composite.

In an interesting paper [1], Chen *et al.* reported very high piezoelectric coefficients of a solvent treated ferroelectric 0-3 composite. They immersed the composite sample of lead zirconate titanate/polyvinylidene fluoride (PZT/PVDF) in a solvent [e.g., *N*-dimethylacetamide] for a period of time. The piezoelectric and dielectric properties were measured after the sample was removed from the solvent environment. They observed that both the permittivity and d_{33} coefficient were significantly larger than those of the virgin specimen. Such high values are quite likely to be out of reach of existing model predictions for "normal" dielectric and piezoelectric 0-3 composites. We have suggested

0022-2461
DOI: 10.1007/s10853-005-7243-3

in a previous paper [2] that these enhancements could very well be related to an increased of electrical conductivity in the matrix material brought about by the solvent.

Indeed, the effect of higher conductivity in the polymer matrix is not limited only to enhancing the pyroelectric and piezoelectric properties. Recently we find that, in the poling process of ferroelectric composites, it shortens the time development of the electric field acting on the ceramic to reach saturation, therefore poling of the ceramic phase in such composite systems can be more efficient [3, 4], which is of practical significance since the pyroelectric and piezoelectric activities of the constituents are generally proportional to their degree of poling [4, 5]. Sakamoto *et al.* [6] also attempted to improve the poling efficiency of lead zirconate titanate/polyurethane (PZT/PU) 0-3 composites by doping with graphite filler. By adding about 1% by volume of graphite to increase the electrical conductivity of PU, the poling of the PZT phase becomes easier. In addition, they also observed that the pyroelectric and piezoelectric d_{33} coefficients of the graphite doped composite show superior properties over their undoped samples.

The study of the poling process is highly important since an understanding of the physical processes behind poling will be useful for the selection of materials and the poling method of composite systems for particular applications. In one poling format of a ferroelectric ceramic/polymer composite, the ceramic phase is first polarized under a dc field for a certain duration at temperatures higher than the Curie transition temperature of the polymer phase, then the polymer phase is polarized under a constant field at decreasing temperature. The large deviation between the permittivities of the ferroelectric ceramic and the polymer will lead to a large difference in the electric field acting on the two phases. In this context, it is difficult to simultaneously obtain a high poling degree for both phases. Recently, it was suggested that the polymer phase could be polarized under an ac field; the ac field does not polarize the ceramic phase if the poling duration is substantially shorter than the dielectric relaxation time of charge in the composite [7, 8]. In other words, a two stage poling process can be employed, i.e., a dc poling at elevated temperature followed by an ac poling at a lower temperature. In this article, theoretical simulations of dc poling, ac poling and their combinations are performed for different poling formats/schedules. The effect of electrical conductivity on the poling results of composites with low and medium high ceramic volume fractions will be discussed.

This paper also attempts to investigate the effect of electrical conductivity on the dielectric, pyroelectric and piezoelectric properties of a ferroelectric composite of a dispersion of spherical inclusions in a continuous matrix, assuming the constituents possess finite conductivity. The sample is excited by an ac electric field of small amplitude or stress for dielectric and piezoelectric measurement, whilst for pyroelectric measurement the sample is excited by an ac modulated or linearly ramped temperature. An analytical model is given and explicit expressions are derived for the dynamic behavior of the electric fields in the constituents. The steady state solutions are then used to obtain explicit expressions for the effective permittivity, pyroelectric and piezoelectric coefficients. Compared to our previously derived analytical expressions for permittivity [9], pyroelectricity [10], and piezoelectricity [11] which assume perfectly insulating constituents, the new set of expressions contains a new factor describing the coupled effects of permittivity, conductivity and the measuring frequency. A generalization of the effective permittivity expression for the concentrated suspension regime is also given by the present model. To illustrate the various new results, theoretical evaluations based on typical ferroelectric composites (e.g. PZT/PVDF) are discussed. Sakamoto *et al.* reported a monotonic increasing profile of pyroelectric coefficients when their PZT/PU sample was heated at a constant rate [6]. This can also be qualitatively simulated by assuming an increase in the conductivity of PU. We will demonstrate that this enhancement is significant for samples with a moderately conductive matrix phase.

2. Theory

To model the polarization behavior and to find the effective permittivity, pyroelectric and piezoelectric coefficients of a 0-3 composite of two ferroelectric phases, we first obtain the time development of the internal electric fields within the individual phases, given the external sinusoidal electric field, modulating temperature or stress. Then the poling process can be modeled and the steady state solution of the in-phase component of internal fields will be used to obtain expressions for the permittivity, pyroelectric and piezoelectric coefficients.

2.1. A formulation for calculating the dynamic behavior of internal electric fields

Suppose the composite is initially polarized in the z direction, in which case we only need to be concerned with the electric field and polarization in the "3" direction. We first write the volumetric average electric displacement D and conduction current density j for the ferroelectric constituent materials in the composite as [12]

$$\begin{cases} \langle D_{3i} \rangle = \varepsilon_i \langle E_{3i} \rangle + \langle P_{3i} \rangle \\ \langle D_{3m} \rangle = \varepsilon_m \langle E_{3m} \rangle + \langle P_{3m} \rangle \end{cases}, \tag{1}$$

$$\begin{cases} \langle j_{3i} \rangle = \sigma_i \langle E_{3i} \rangle \\ \langle j_{3m} \rangle = \sigma_m \langle E_{3m} \rangle \end{cases}, \tag{2}$$

where the angular brackets denote volume-averaged fields enclosed. P is polarization, ε and σ denote permittivity and electrical conductivity respectively, and E is electric field. Subscripts i and m denote "inclusion" and "matrix" respectively.

Consider the single inclusion problem of a ferroelectric sphere surrounded by a ferroelectric matrix medium with a uniform electric field applied along the 3-direction far away from the inclusion. The boundary value problem gives the following equations (see ref. [12]):

$$\langle D_{3i}\rangle + 2\varepsilon_m\left(\langle E_{3i}\rangle - \langle E_{3m}\rangle\right) = \langle D_{3m}\rangle - q_0, \quad (3)$$

$$\langle j_{3i}\rangle + 2\sigma_m\left(\langle E_{3i}\rangle - \langle E_{3m}\rangle\right) = \langle j_{3m}\rangle + \partial q_0/\partial t. \quad (4)$$

In Equations 3 and 4, we have assumed both constituent materials are uniformly polarized and the homogeneously polarized sphere is covered with surface charge of density q_0 at the pole along the polarizing direction ($\theta = 0$) with a distribution given by $q_0 \cos\theta$.

In a previous paper [9], Poon and Shin considered nonferroelectric composites of zero conductivities (where $\langle P_{3i}\rangle = \langle P_{3m}\rangle = 0$ and $q_0 = 0$). They suggested Equation 3 can be written more accurately as

$$\langle D_{3i}\rangle + 2\varepsilon_m\left(\langle E_{3i}\rangle - \langle E_{3m}\rangle\right) = \langle \tilde{D}_{3m}\rangle \quad (5)$$

at higher volume fraction ϕ of inclusions, where $\langle \tilde{D}_3 m\rangle$ is the electric displacement of the surrounding matrix material as seen by a particular single inclusion in the composite. It has two contributions (the pure medium and the polarization from other inclusions):

$$\langle D_{3m}\rangle = \langle D_{3m}\rangle + \phi\,\langle P_{3i}\rangle, \quad (6)$$

where $\langle \tilde{P}_{3i}\rangle = (\varepsilon_i - \varepsilon_m)\langle E_{3i}\rangle$. For a ferroelectric composite with interfacial charge accumulation, Equation 5 may be rewritten, following the same rationale, as

$$\langle D_{3i}\rangle + 2\varepsilon_m\left(\langle E_{3i}\rangle - \langle E_{3m}\rangle\right) = \langle D_{3m}\rangle - q_0, \quad (7)$$

where $\langle \tilde{D}_{3m}\rangle = \langle D_{3m}\rangle + \phi\langle \tilde{P}^f_{3i}\rangle$ and $\langle \tilde{P}^f_{3i}\rangle = \langle P_{3i}\rangle - \langle P_{3m}\rangle + q_0 + (\varepsilon_i - \varepsilon_m)\langle E_{3i}\rangle$. Hence, Equation 4 may also be rewritten for higher volume fraction as

$$\langle j_{3i}\rangle + 2\sigma_m\left(\langle E_{3i}\rangle - \langle E_{3m}\rangle\right) = \langle j_{3m}\rangle + \phi\left(\sigma_i - \sigma_m\right)\langle E_{3i}\rangle + (1-\phi)\,\partial q_0/\partial t. \quad (8)$$

For a composite comprising spherical particles uniformly distributed in the matrix material, the volumetric averages of the electric fields satisfy [9, 11]

$$\langle E_3\rangle = \phi\langle E_{3i}\rangle + (1-\phi)\,\langle E_{3m}\rangle. \quad (9)$$

We can then obtain from Equations 1, 2, 7–9

$$\frac{\partial\langle E_{3i}\rangle}{\partial t} + \frac{\langle E_{3i}\rangle}{\tau} = \frac{3[\sigma_m\langle E_3\rangle + \varepsilon_m\partial\langle E_3\rangle/\partial t] + (1-\phi)^2\,\partial\left[\langle P_{3m}\rangle - \langle P_{3i}\rangle\right]/\partial t}{\phi 3\varepsilon_m + (1-\phi)\{\varepsilon_i + 2\varepsilon_m - \phi(\varepsilon_i - \varepsilon_m)\}}, \quad (10)$$

where

$$\tau = \frac{\phi 3\varepsilon_m + (1-\phi)\{\varepsilon_i + 2\varepsilon_m - \phi(\varepsilon_i - \varepsilon_m)\}}{\phi 3\sigma_m + (1-\phi)\{\sigma_i + 2\sigma_m - \phi(\sigma_i - \sigma_m)\}}. \quad (11)$$

2.2. Modeling the hysteresis behavior of a ferroelectric composite

Equation 10 is a first order differential equation. For a given external field, which may be equated to $\langle E_3\rangle$, we may obtain $\langle E_{3i}\rangle$ as a function of time t when the P-E relations for the individual constituents ($\langle P_{3i}\rangle$ versus $\langle E_{3i}\rangle$ and $\langle P_{3m}\rangle$ versus $\langle E_{3m}\rangle$) are known. The P-E relationship of a ferroelectric material is very complex and it has been the subject of many investigations. The ferroelectric polarization is generally not a simple function of the electric field. The change of polarization depends on a change of electric field as well as on the polarization already present. Here we use the model of Miller *et al.* to describe P-E relations of the constituent materials [13, 14]:

$$\frac{\partial P}{\partial E} = \left(1 - \tanh\sqrt{\frac{P - P_{sat}}{\xi P_s - P}}\right)\left(\frac{\partial P_{sat}}{\partial E}\right), \quad (12)$$

where P_{sat} is the polarization of the saturated hysteresis loop at the field of interest, and P and P_s are the magnitudes of the ferroelectric polarization and saturation polarization, respectively. In this model, ξ takes $+1$ and -1 for increasing E and decreasing E respectively. The polarization of the saturated hysteresis loop is written as a function of the electric field as

$$P_{sat} = \xi P_s \tanh\left\{\frac{\xi E - E_c}{2E_c}\ln\left(\frac{1 + P_r/P_s}{1 - P_r/P_s}\right)\right\}, \quad (13)$$

where P_r and E_c are taken as positive quantities representing the magnitude of remanent polarization and coercive field, respectively. Since different kinds of ferroelectric materials may have different hysteresis loop shapes (corresponding to different $\partial P/\partial E$ for a given field), even though they may have the same values of P_r, P_s and E_c, we have therefore slightly modified Equation 13 to include a factor $n \equiv n_1/n_2$ to extend the usage of Miller *et al.*'s model for a broader range of ferroelectrics, and as a result the saturated hysteresis loop may be written as

$$P_{sat} = \xi P_s \tanh\left\{\frac{\xi E - E_c}{2E_c} \ln\left[\frac{1-(-P_r/P_s)^{1/n}}{1+(-P_r/P_s)^{1/n}}\right]\right\}^n. \quad (14)$$

In general, it is found that odd-numbered n_1 and n_2 with n_1/n_2 ranging from 1/3 to 1 will give results close to hysteresis loops of realistic materials. When Equation 12 is solved with Equation 10, the relations $\partial\langle P_{3i}\rangle/\partial t = [\partial\langle P_{3i}\rangle/\partial\langle E_{3i}\rangle][\partial\langle E_{3i}\rangle/\partial t]$ and $\partial\langle P_{3m}\rangle/\partial t = [\partial\langle P_{3m}\rangle/\partial\langle E_{3m}\rangle][\partial\langle E_{3m}\rangle/\partial t]$ should be employed. Concerning the initial condition of internal fields, our theoretical calculations assume electric fields in the constituents are both initially zero.

2.3. Effective permittivity of a ferroelectric 0-3 composite

For dielectric measurement, assume $\langle E_3\rangle = E_0 \sin \omega t$ where $\omega = 2\pi f$ and f is the frequency of the applied field. We further assume the amplitude E_0 is small such that the contribution from the hysteresis behavior may be neglected. Thus, $\partial\langle P_{3m}\rangle/\partial t = \partial\langle P_{3i}\rangle/\partial t = 0$ in Equation 10 which becomes

$$\frac{\partial\langle E_{3i}\rangle}{\partial t} + \frac{\langle E_{3i}\rangle}{\tau} = L_E\left\{\frac{\partial\langle E_3\rangle}{\partial t} + \frac{\langle E_3\rangle}{\tau_m}\right\}, \quad (15)$$

where $\tau_m = \varepsilon_m/\sigma_m$ and

$$L_E = \frac{3\varepsilon_m}{\phi 3\varepsilon_m + (1-\phi)\{\varepsilon_i + 2\varepsilon_m - \phi(\varepsilon_i - \varepsilon_m)\}}. \quad (16)$$

Assuming $\langle E_{3i}\rangle$ is initially zero, the solution from solving Equation 15 is:

$$\langle E_{3i}\rangle = \left(\frac{E_0\tau/\tau_m}{1+\omega^2\tau^2}\right) L_E\{\omega(\tau-\tau_m)e^{-t/\tau} + (1+\omega^2\tau\tau_m)\sin\omega t - \omega(\tau-\tau_m)\cos\omega t\}. \quad (17)$$

In a dielectric measurement, the electric current flowing in the composite is used to determine the permittivity value. The volumetric average of the total current density is

$$\langle J_3\rangle = \langle j_3\rangle + \partial\langle D_3\rangle/\partial t, \quad (18)$$

where

$$\begin{cases} \langle j_3\rangle = \phi\langle j_{3i}\rangle + (1-\phi)\langle j_{3m}\rangle \\ \langle D_3\rangle = \phi\langle D_{3i}\rangle + (1-\phi)\langle D_{3m}\rangle \end{cases}. \quad (19)$$

At sufficiently long time from the start of the dielectric measurement (i.e. at steady state), the term $\exp(-t/\tau)$ in Equation 17 may be omitted. The components of $\langle E_{3i}\rangle$ in phase and 90° out of phase with the applied field are

$$\langle E_{3i}\rangle|_{\text{in phase}} = \left(\frac{E_0\tau/\tau_m}{1+\omega^2\tau^2}\right) \times L_E\{(1+\omega^2\tau\tau_m)\sin\omega t\}, \quad (20)$$

$$\langle E_{3i}\rangle|_{\text{out of phase}} = \left(\frac{E_0\tau/\tau_m}{1+\omega^2\tau^2}\right) \times L_E\{\omega(\tau_m-\tau)\cos\omega t\}. \quad (21)$$

In this paper, we only focus on the in-phase component of the dielectric property. The permittivity is thus:

$$\varepsilon = \frac{\partial\left\{\int\langle J_3\rangle\, dt\right\}_{\text{in phase}}}{\partial\langle E_3\rangle}. \quad (22)$$

Substituting Equations 18–20 into Equation 22 and making use of Equations 1, 2, 9, the effective permittivity is obtained as:

$$\varepsilon = \varepsilon_m + \phi\Gamma_\varepsilon L_E(\varepsilon_i - \varepsilon_m), \quad (23)$$

where

$$\Gamma_\varepsilon = \frac{\tau\{\tau_m^{-1} + \tau_{neg}^{-1}(1-\tau/\tau_m)\} + \omega^2\tau^2}{1+\omega^2\tau^2}, \quad (24)$$

$$\tau_{neg} = \frac{\varepsilon_i - \varepsilon_m}{\sigma_i - \sigma_m}, \quad (25)$$

or in a more compact form:

$$\varepsilon = \varepsilon_m \frac{(\varepsilon_i + 2\varepsilon_m) + \{3\Gamma_\varepsilon - (2-\phi)\}\phi(\varepsilon_i - \varepsilon_m)}{(\varepsilon_i + 2\varepsilon_m) - \phi(2-\phi)(\varepsilon_i - \varepsilon_m)}. \quad (26)$$

Note that when $\sigma_i = \sigma_m = 0$, then $\Gamma = 1$ and Equation 26 reduces to

$$\varepsilon = \varepsilon_m \frac{(\varepsilon_i + 2\varepsilon_m) + \phi(1+\phi)(\varepsilon_i - \varepsilon_m)}{(\varepsilon_i + 2\varepsilon_m) - \phi(2-\phi)(\varepsilon_i - \varepsilon_m)}. \quad (27)$$

This equation is identical to the permittivity formula suggested by Poon and Shin [9].

2.4. Effective pyroelectric coefficient of a ferroelectric 0-3 composite

The polarization of a ferroelectric material will be influenced by temperature (pyroelectricity) and stress (piezoelectricity). We will first derive the pyroelectric coefficient which is measured by using an ac temperature modulation, then by a direct method using constant heating rate. In each case the pyroelectric current is determined.

2.4.1. Sinusoidally modulated temperature method

In the pyroelectric measurement, the polarizations $\langle P_{3i}\rangle$ and $\langle P_{3m}\rangle$ vary with the temperature modulation. Suppose the composite is subjected to a sinusoidal temperature Θ. The rates of change of polarizations in Equation 10 are related to the temperature due to pyroelectric effect:

$$\left\{ \frac{\partial \langle P_{3i}\rangle}{\partial t} = p_i \frac{\partial \Theta}{\partial t}, \quad \frac{\partial \langle P_{3m}\rangle}{\partial t} = p_m \frac{\partial \Theta}{\partial t} \right. , \tag{28}$$

where p denotes pyroelectric coefficient.

In short-circuit condition (i.e., $\langle E_3\rangle = 0$), Equation 10 becomes

$$\frac{\partial \langle E_{3i}\rangle}{\partial t} + \frac{\langle E_{3i}\rangle}{\tau} = \frac{(1-\phi)^2 (p_m - p_i)}{\phi 3\varepsilon_m + (1-\phi)\{\varepsilon_i + 2\varepsilon_m - \phi(\varepsilon_i - \varepsilon_m)\}} \frac{\partial \Theta}{\partial t}. \tag{29}$$

Assuming $\Theta = \Theta_0 \sin \omega t$ with Θ_0 being the amplitude of modulation temperature and $\langle E_{3i}\rangle$ initially zero, the solution from solving Equation 29 is:

$$\langle E_{3i}\rangle = \left(\frac{\Theta_0 \omega\tau}{1+\omega^2\tau^2}\right) \times \frac{(1-\phi)^2 (p_m - p_i)}{\phi 3\varepsilon_m + (1-\phi)\{\varepsilon_i + 2\varepsilon_m - \phi(\varepsilon_i - \varepsilon_m)\}} \times (\omega\tau \sin \omega t + \cos \omega t - e^{-t/\tau}), \tag{30}$$

where $\omega = 2\pi f$ and f is the frequency of modulated temperature. From Equation 9, $\langle E_{3m}\rangle = -\phi\langle E_{3i}\rangle/(1-\phi)$, since $\langle E_3\rangle = 0$ in pyroelectric measurement.

For a sufficiently long time in the pyroelectric measurement (i.e. at steady state), the term $\exp(-t/\tau)$ in Equation 30 may be neglected. The components of $\langle E_{3i}\rangle$ in phase and 90° out of phase with the modulated temperature are

$$\langle E_{3i}\rangle|_{\text{in phase}} = \left(\frac{\Theta_0 \omega^2\tau^2}{1+\omega^2\tau^2}\right) \times \frac{(1-\phi)^2 (p_m - p_i)\sin \omega t}{\phi 3\varepsilon_m + (1-\phi)\{\varepsilon_i + 2\varepsilon_m - \phi(\varepsilon_i - \varepsilon_m)\}}, \tag{31}$$

$$\langle E_{3i}\rangle|_{\text{out of phase}} = \left(\frac{\Theta_0 \omega\tau}{1+\omega^2\tau^2}\right) \times \frac{(1-\phi)^2 (p_m - p_i)\cos \omega t}{\phi 3\varepsilon_m + (1-\phi)\{\varepsilon_i + 2\varepsilon_m - \phi(\varepsilon_i - \varepsilon_m)\}}. \tag{32}$$

Again, we only focus on the in-phase component. The pyroelectric coefficient of the composite is calculated from:

$$p = \frac{\partial\{\int \langle J_3\rangle\, dt\}_{\text{in phase}}}{\partial \Theta}. \tag{33}$$

Note that this corresponds to the pyroelectric current which is out-of-phase with the temperature modulation, as measured in practice. Substituting Equation 31 into Equations 1, 2 and 18, we obtain from Equation 33

$$p = \phi p_i + (1-\phi) p_m + \phi(1-\phi)\Gamma_p^{\text{ac}} \times (L_E - \bar{L}_E)(p_i - p_m), \tag{34}$$

where

$$\Gamma_p^{\text{ac}} = \frac{\tau/\tau_{neg} + \omega^2\tau^2}{1+\omega^2\tau^2}, \tag{35}$$

$$\bar{L}_E = \frac{1-\phi L_E}{1-\phi} = \frac{(1-\phi)\varepsilon_i + (2+\phi)\varepsilon_m}{\phi 3\varepsilon_m + (1-\phi)\{\varepsilon_i + 2\varepsilon_m - \phi(\varepsilon_i - \varepsilon_m)\}}. \tag{36}$$

Alternatively, the effective pyroelectric coefficient can be re-expressed as

$$p = \phi[1 - \Gamma_p^{\text{ac}}(1 - L_E)]p_i + (1-\phi)[1 - \Gamma_p^{\text{ac}}(1 - \bar{L}_E)]p_m. \tag{37}$$

Note that when $\sigma_i = \sigma_m = 0$, we have $\Gamma_p^{\text{ac}} = 1$ and Equation 37 reduces to:

$$p = \phi L_E p_i + (1-\phi)\bar{L}_E p_m, \tag{38}$$

which is the familiar form for pyroelectric coefficient suggested in the literature [15–17]. If one follows the same footsteps to solve the pyroelectric coefficient but using Equations 3 and 4 rather than Equations 7 and 8 (i.e. for low ϕ), expressions for L_E and $\bar{L}_E$ become the pair derived previously [11, 12, 18, 19] and Equation 38 will be identical to the formula for primary pyroelectric coefficient suggested by Chew *et al.* [10].

Equation 37 is also a generalization of the expression suggested by Lam *et al.* who investigated the pyroelectric properties of dilute (small ϕ) composites with non-pyroelectric matrix [19].

2.4.2. Linear temperature ramp method

For the pyroelectric measurement with the sample being heated at a constant rate, Θ can be written as $\Theta = \Theta_{rm} + \Re_{\Theta} t$ with Θ_{rm} and $\Re_{\Theta}$ being the initial temperature and heating rate respectively.

Assuming $\langle E_{3i} \rangle$ is initially zero, the solution from solving Equation 29 now becomes:

$$\langle E_{3i} \rangle = \frac{(1-\phi)^2 (p_m - p_i) \Re_{\Theta} \tau}{\phi 3\varepsilon_m + (1-\phi)\{\varepsilon_i + 2\varepsilon_m - \phi(\varepsilon_i - \varepsilon_m)\}} \times (1 - e^{-t/\tau}), \tag{39}$$

and $\langle E_{3m} \rangle = -\phi \langle E_{3i} \rangle / (1-\phi)$, since $\langle E_3 \rangle = 0$ in pyroelectric measurement.

From Equations 1, 2, 18 and 39, the pyroelectric current of the composite obtained from measurement is derived as:

$$I_p = \Re_{\Theta} A \{\phi p_i + (1-\phi) p_m + \phi(1-\phi) \Gamma_p^{\text{ramp}} \times (L_E - \bar{L}_E)(p_i - p_m)\}, \tag{40}$$

where A is the electrode area and

$$\Gamma_p^{\text{ramp}} = e^{-t/\tau} + (1 - e^{-t/\tau}) \tau / \tau_{neg}. \tag{41}$$

The pyroelectric coefficient is then directly obtained from Equation 40

$$p = \phi p_i + (1-\phi) p_m + \phi(1-\phi) \times \Gamma_p^{\text{ramp}} (L_E - \bar{L}_E)(p_i - p_m). \tag{42}$$

Alternatively, the effective pyroelectric coefficient can be re-expressed as

$$p = \phi [1 - \Gamma_p^{\text{ramp}} (1 - L_E)] p_i + (1-\phi)[1 - \Gamma_p^{\text{ramp}} (1 - \bar{L}_E)] p_m. \tag{43}$$

2.5. Effective piezoelectric coefficients of a ferroelectric 0-3 composite

In the piezoelectric measurement, the polarizations $\langle P_{3i} \rangle$ and $\langle P_{3m} \rangle$ vary with the applied stress. Suppose the composite is subjected to an external tensile stress T. The rates of change of polarizations in Equation 10 can be related to the external stress due to the piezoelectric effect. Thus see ref. 2,

$$\begin{cases} \dfrac{\partial \langle P_{3i} \rangle}{\partial t} = d_i \dfrac{\partial T}{\partial t} \\ \dfrac{\partial \langle P_{3m} \rangle}{\partial t} = d_m \dfrac{\partial T}{\partial t} \end{cases}, \tag{44}$$

where $d_i = \partial \langle P_{3i} \rangle / \partial T$, $d_m = \partial \langle P_{3m} \rangle / \partial T$. When the stress is applied along the x direction, i.e. $T = T_{xx}$, then $d_i = d_i^{\perp}$ and $d_m = d_m^{\perp}$, where

$$d_i^{\perp} = \left(L_T^{//} + L_T^{\perp} \right) d_{31i} + L_T^{\perp} d_{33i}, \tag{45}$$

$$d_m^{\perp} = \left(\bar{L}_T^{//} + \bar{L}_T^{\perp} \right) d_{31m} + \bar{L}_T^{\perp} d_{33m}, \tag{46}$$

and

$$L_T^{\perp} = \frac{I_T}{1 - \phi(1 - 3I_T)} - \frac{J_T}{1 - \phi(1 - 3J_T)}, \tag{47}$$

$$L''_T = \frac{I_T}{1 - \phi(1 - 3I_T)} + \frac{2J_T}{1 - \phi(1 - 3J_T)}, \tag{48}$$

$$\bar{L}_T^{\perp} = \frac{-\phi L_T^{\perp}}{1-\phi} = \frac{1}{3} \left[\frac{1}{1 - \phi(1 - 3I_T)} - \frac{1}{1 - \phi(1 - 3J_T)} \right], \tag{49}$$

$$\bar{L}''_T = \frac{1 - \phi L''_T}{1-\phi} = \frac{1}{3} \left[\frac{1}{1 - \phi(1 - 3I_T)} + \frac{2}{1 - \phi(1 - 3J_T)} \right], \tag{50}$$

$$I_T = \frac{1}{3} \frac{k_i}{k_m} \frac{3k_m + 4\mu_m}{3k_i + 4\mu_m}, \tag{51}$$

$$J_T = \frac{5}{3} \frac{(3k_m + 4\mu_m)\mu_i}{6(k_m + 2\mu_m)\mu_i + (9k_m + 8\mu_m)\mu_m}. \tag{52}$$

Here d_{31} and d_{33} are the piezoelectric coefficients. k and μ denote bulk modulus and shear modulus respectively.

When the stress acts along the z direction, i.e. $T = T_{zz}$, then $d_i = d_i^{//}$ and $d_m = d_m^{//}$, where

$$d_i^{//} = L_T^{//} d_{33i} + 2L_T^{\perp} d_{31i}, \tag{53}$$

$$d_m^{//} = \bar{L}_T^{//} d_{33m} + 2\bar{L}_T^{\perp} d_{31m}. \tag{54}$$

In short-circuit condition (i.e., $\langle E_3 \rangle = 0$), Equation 10 becomes

$$\frac{\partial \langle E_{3i} \rangle}{\partial t} + \frac{\langle E_{3i} \rangle}{\tau} = \frac{(1-\phi)^2 (d_m - d_i)}{\phi 3\varepsilon_m + (1-\phi)\{\varepsilon_i + 2\varepsilon_m - \phi(\varepsilon_i - \varepsilon_m)\}} \frac{\partial T}{\partial t}. \tag{55}$$

Assuming $T = T_0 \sin \omega t$ with T_0 being the amplitude of applied stress and $\langle E_{3i} \rangle$ initially zero, the solution from solving Equation 55 is:

$$\langle E_{3i} \rangle = \left(\frac{T_0 \omega \tau}{1 + \omega^2 \tau^2} \right) \times \frac{(1-\phi)^2 (d_m - d_i)}{\phi 3\varepsilon_m + (1-\phi)\{\varepsilon_i + 2\varepsilon_m - \phi(\varepsilon_i - \varepsilon_m)\}} \times (\omega\tau \sin \omega t + \cos \omega t - e^{-t/\tau}), \tag{56}$$

where $\omega = 2\pi f$ and f is the frequency of applied stress. From Equation 9, $\langle E_{3m} \rangle = -\phi \langle E_{3i} \rangle / (1-\phi)$, since $\langle E_3 \rangle = 0$ in piezoelectric measurement.

For a sufficiently long time in the piezoelectric measurement, the term $\exp(-t/\tau)$ in Equation 56 drops out. The components of $\langle E_{3i} \rangle$ in phase and 90° out of phase with the applied stress are

$$\langle E_{3i} \rangle|_{\text{in phase}} = \left(\frac{T_0 \omega^2 \tau^2}{1 + \omega^2 \tau^2} \right) \times \frac{(1-\phi)^2 (d_m - d_i) \sin \omega t}{\phi 3\varepsilon_m + (1-\phi)\{\varepsilon_i + 2\varepsilon_m - \phi(\varepsilon_i - \varepsilon_m)\}}, \tag{57}$$

$$\langle E_{3i} \rangle|_{\text{out of phase}} = \left(\frac{T_0 \omega \tau}{1 + \omega^2 \tau^2} \right) \times \frac{(1-\phi)^2 (d_m - d_i) \cos \omega t}{\phi 3\varepsilon_m + (1-\phi)\{\varepsilon_i + 2\varepsilon_m - \phi(\varepsilon_i - \varepsilon_m)\}}. \tag{58}$$

Again, we only focus on the in-phase component of the piezoelectric responses. The piezoelectric d coefficient of the composite obtained from measurement is:

$$d = \frac{\partial \left\{ \int \langle J_3 \rangle \, dt \right\}_{\text{in phase}}}{\partial T}. \tag{59}$$

Substituting Equation 57 into Equations 1, 2 and 18, we obtain from Equation 59

$$d = \phi d_i + (1-\phi) d_m + \phi(1-\phi)\Gamma_d \left(L_E - \bar{L}_E \right) \times (d_i - d_m), \tag{60}$$

where

$$\Gamma_d = \frac{\tau/\tau_{neg} + \omega^2 \tau^2}{1 + \omega^2 \tau^2} = \Gamma_p^{\text{ac}}. \tag{61}$$

Further, using Equations 45, 46, 53 and 54, the effective d_{31} and d_{33} coefficients can be re-expressed as

$$\begin{aligned} d_{31} = {} & \phi[1 - \Gamma_d(1 - L_E)] \\ & \times \{(L_T^{//} + L_T^{\perp}) d_{31i} + L_T^{\perp} d_{33i}\} \\ & + (1-\phi)[1 - \Gamma_d(1 - \bar{L}_E)] \\ & \times \{(\bar{L}_T^{//} + \bar{L}_T^{\perp}) d_{31m} + \bar{L}_T^{\perp} d_{33m}\}, \end{aligned} \tag{62}$$

$$\begin{aligned} d_{33} = {} & \phi[1 - \Gamma_d(1 - L_E)] \left\{ L_T^{//} d_{33i} + 2L_T^{\perp} d_{31i} \right\} \\ & + (1-\phi)[1 - \Gamma_d(1 - \bar{L}_E)] \\ & \times \left\{ \bar{L}_T^{//} d_{33m} + 2\bar{L}_T^{\perp} d_{31m} \right\}. \end{aligned} \tag{63}$$

The effective hydrostatic piezoelectric d_h coefficient is defined by $d_h = d_{33} + 2d_{31}$ (and similarly for inclusion and matrix), thus:

$$d_h = \phi[1 - \Gamma_d(1 - L_E)] L_T^h d_{hi} + (1-\phi) \times [1 - \Gamma_d(1 - \bar{L}_E)] \bar{L}_T^h d_{hm}, \tag{64}$$

where

$$L_T^h = 2L_T^{\perp} + L_T^{//} = \frac{\left(1 + \frac{4}{3}\mu_m/k_m\right)}{\phi\left(1 + \frac{4}{3}\mu_m/k_m\right) + (1-\phi)\left(1 + \frac{4}{3}\mu_m/k_i\right)}, \tag{65}$$

$$\bar{L}_T^h = 2\bar{L}_T^\perp + \bar{L}_T^{//} = \frac{1-\phi L_T^h}{1-\phi}$$
$$= \frac{\left(1+\frac{4}{3}\mu_m/k_i\right)}{\phi\left(1+\frac{4}{3}\mu_m/k_m\right)+(1-\phi)\left(1+\frac{4}{3}\mu_m/k_i\right)}. \quad (66)$$

2.6. Effective piezoelectric coefficients of a composite in terms of effective ε, k and μ

The foregoing Equations 62–64 have considered a higher ϕ treatment for the electric problem [Equations 7 and 8], but the elasticity problem [Equations 45, 46, 53 and 54] is only valid for the dilute suspension regime. One can re-express the L_T's [Equations 47–50 and 65–66] in terms of the effective elastic properties of the composite as in refs. 11 and 20. We here adopt the same "transformed" expressions F_T's which have been demonstrated there to give results which are applicable to higher ϕ, provided that better estimates of effective properties are available. This scheme provides simple and tractable explicit expressions for higher ϕ. Thus Equations 62–64 are transformed to:

$$d_{31} = \phi[1-\Gamma_d(1-L_E)]\left\{\left(F_T'' + F_T^\perp\right)d_{31i} + F_T^\perp d_{33i}\right\} + (1-\phi)[1-\Gamma_d(1-\bar{L}_E)] \times \left\{\left(\bar{F}_T'' + \bar{F}_T^\perp\right)d_{31m} + \bar{F}_T^\perp d_{33m}\right\}' \quad (67)$$

$$d_{33} = \phi\,[1-\Gamma_d\,(1-L_E)]\left\{F''_T d_{33i} + 2F_T^\perp d_{31i}\right\} + (1-\phi)[1-\Gamma_d(1-\bar{L}_E)] \times \left\{\bar{F}''_T d_{33m} + 2\bar{F}_T^\perp d_{31m}\right\} \quad (68)$$

$$d_h = \phi\,[1-\Gamma_d\,(1-L_E)]\,F_T^h d_{hi} + (1-\phi) \times \left[1-\Gamma_d\left(1-\bar{L}_E\right)\right]\bar{F}_T^h d_{hm}, \quad (69)$$

where

$$F_T^\perp = \frac{1}{\phi}\left\{\frac{1}{3}\frac{k^{-1}-k_m^{-1}}{k_i^{-1}-k_m^{-1}} - \frac{1}{3}\frac{\mu^{-1}-\mu_m^{-1}}{\mu_i^{-1}-\mu_m^{-1}}\right\}, \quad (70)$$

$$F''_T = \frac{1}{\phi}\left\{\frac{1}{3}\frac{k^{-1}-k_m^{-1}}{k_i^{-1}-k_m^{-1}} + \frac{2}{3}\frac{\mu^{-1}-\mu_m^{-1}}{\mu_i^{-1}-\mu_m^{-1}}\right\}, \quad (71)$$

$$\bar{F}_T^\perp = \frac{-\phi F_T^\perp}{1-\phi} = \frac{1}{1-\phi} \times \left\{\frac{1}{3}\frac{k_i^{-1}-k^{-1}}{k_i^{-1}-k_m^{-1}} - \frac{1}{3}\frac{\mu_i^{-1}-\mu^{-1}}{\mu_i^{-1}-\mu_m^{-1}}\right\}, \quad (72)$$

$$\bar{F}''_T = \frac{1-\phi F''_T}{1-\phi} = \frac{1}{1-\phi} \times \left\{\frac{1}{3}\frac{k_i^{-1}-k^{-1}}{k_i^{-1}-k_m^{-1}} + \frac{2}{3}\frac{\mu_i^{-1}-\mu^{-1}}{\mu_i^{-1}-\mu_m^{-1}}\right\}, \quad (73)$$

$$F_T^h = \frac{1}{\phi}\frac{k^{-1}-k_m^{-1}}{k_i^{-1}-k_m^{-1}}, \quad (74)$$

$$\bar{F}_T^h = \frac{1-\phi F_T^h}{1-\phi} = \frac{1}{1-\phi}\frac{k_i^{-1}-k^{-1}}{k_i^{-1}-k_m^{-1}}. \quad (75)$$

Concerning the L_E's, they can also be expressed in terms of the effective dielectric property of the composite i.e. transformed to ℓ_E's following the same procedure as in ref. 2. The results are

$$\ell_E = \frac{1}{\phi\Gamma_\varepsilon}\frac{\varepsilon-\varepsilon_m}{\varepsilon_i-\varepsilon_m}, \quad (76)$$

$$\bar{\ell}_E = \frac{1}{(1-\phi)\bar{\Gamma}_\varepsilon}\frac{\varepsilon_i-\varepsilon}{\varepsilon_i-\varepsilon_m}, \quad (77)$$

where Γ_ε is given by Equation 24,

$$\bar{\Gamma}_\varepsilon = \frac{\tau\{\tau_{pos}^{-1}+\tau_{neg}^{-1}(1-\tau/\tau_{pos})\}+\omega^2\tau^2}{1+\omega^2\tau^2}, \quad (78)$$

and $\tau_{pos} = [(1-\phi)\varepsilon_i+(2+\phi)\varepsilon_m]/[(1-\phi)\sigma_i+(2+\phi)\sigma_m]$. Equations 76 and 77 can be used for analyzing the fractions of applied field distributed to the constituents. ℓ_E and $\bar{\ell}_E$ are essentially $\langle E_{3i}\rangle/\langle E_3\rangle$ and $\langle E_{3m}\rangle/\langle E_3\rangle$ respectively [11].

Note that when $\sigma_i = \sigma_m = 0$, $\Gamma_d = 1$ and Equations 67–69 reduce to:

$$d_{31} = \phi F_E\left\{\left(F_T^{//}+F_T^\perp\right)d_{31i} + F_T^\perp d_{33i}\right\} + (1-\phi)\,\bar{F}_E\left\{\left(\bar{F}_T^{//}+\bar{F}_T^\perp\right)d_{31m} + \bar{F}_T^\perp d_{33m}\right\}, \quad (79)$$

$$d_{33} = \phi F_E \left\{ F_T^{//} d_{33i} + 2F_T^{\perp} d_{31i} \right\} + (1-\phi)\bar{F}_E \left\{ \bar{F}_T^{//} d_{33m} + 2\bar{F}_T^{\perp} d_{31m} \right\}, \quad (80)$$

$$d_h = \phi F_E F_T^h d_{hi} + (1-\phi)\bar{F}_E \bar{F}_T^h d_{hm}, \quad (81)$$

where

$$F_E = \frac{1}{\phi}\frac{\varepsilon - \varepsilon_m}{\varepsilon_i - \varepsilon_m}, \quad (82)$$

$$\bar{F}_E = \frac{1-\phi F_E}{1-\phi} = \frac{1}{1-\phi}\frac{\varepsilon_i - \varepsilon}{\varepsilon_i - \varepsilon_m}, \quad (83)$$

which are the ℓ_E's with $\Gamma_\varepsilon = \bar{\Gamma}_\varepsilon = 1$. Equations 79–81 are identical to our previous model [11].

Theoretical predictions given by Equations 67–69 require the values of the bulk modulus and shear modulus of the composite. Sometimes the elastic properties of the composite are not measured together with the piezoelectric properties. In such cases, we can follow the same technique as in our previous articles [11, 20]. For the effective bulk modulus k,

$$k = k_m + \frac{\phi(k_i - k_m)}{1 + (1-\phi)(k_i - k_m)/(k_m + 4\mu_m/3)} \quad (84)$$

gives a good approximation for a composite with spherical inclusions [21]. For the effective shear modulus μ, explicit bounds may be employed. Following Christensen [22], the lower bound μ_l:

$$\mu_l = \mu_m \left\{ 1 + \frac{15(1-\nu_m)(\mu_i/\mu_m - 1)\phi}{7 - 5\nu_m + 2(4 - 5\nu_m)\left[\mu_i/\mu_m - (\mu_i/\mu_m - 1)\phi\right]} \right\} \quad (85)$$

given by Hashin and Shtrikman [23] for arbitrary phase geometry is adopted in our prediction. In Equation 85, ν_m is Poisson's ratio of the matrix phase. The upper bound μ_u given by Hashin for spherical inclusion geometry may also be adopted [21]. If both upper bound μ_u and lower bound μ_l are simultaneously adopted for the effective shear modulus, each prediction of d_{31} and d_{33} coefficients [Equations 67 and 68 respectively] then gives a pair of lines. In the next section, we will only adopt lower bound μ_l to evaluate the effective piezoelectric coefficients, since the resultant pair of lines given by the prediction of piezoelectric coefficients are narrow.

In summary, Equation 26 is used for the prediction of effective permittivity. For pyroelectric measurement using ac modulated temperature, Equation 37 is used. When the linear temperature ramp method is adopted, Equation 43 is to be used accordingly. For piezoelectric coefficients, Equations 67, 68 and 69 are used for the prediction of d_{31}, d_{33} and d_h respectively.

3. Results and discussion

In this section, we will first concentrate on the simulation of poling of ferroelectric 0-3 composites, then the theoretical predictions based on the foregoing formulas for dielectric, pyroelectric and piezoelectric properties are investigated. We will show that electrical conductivity in the matrix phase plays a significant role in influencing those properties.

3.1. Polarization behavior of ferroelectric 0-3 composites

In poling, normally a dc electric field is applied on the ferroelectric sample to force the dipoles to align. It is important to choose an appropriate poling field, temperature and time for a particular ferroelectric sample because the ferroelectric response varies among different materials. To obtain an optimal degree of poling, the poling field should be high (but below the breakdown field of the material) and the poling time should be long enough to get an appreciable stable polarization. For ferroelectric composites of ceramic inclusions dispersed in the polymer matrix, the selection of poling parameters will be more complicated since we cannot directly measure the internal field distribution to properly select the parameters. Theoretical simulation results of poling will be useful as a reference. The discussion of the effects of poling field and poling time have been systemically investigated before [4, 7]. The following discussions will focus on the effects of σ_m and the applied field schedule.

Fig. 1 shows the simulation results of lead zirconate titanate (PZT)/polyvinylidene fluoride-trifluoroethylene [P(VDF-TrFE)] 0-3 composites. The adopted properties for the constituents are shown in Table I. The composites are polarized by a dc electric field of 60 MV/m for an hour with the poled samples then short-circuited for 2 h. This process tends to polarize the two ferroelectric phases in the same direction. Two different conductivity values for the P(VDF-TrFE) copolymer have been adopted to demonstrate the effect of σ_m. σ_i is set to $10^{-12}\ \Omega^{-1}\text{m}^{-1}$, which is small, so that the effect of σ_i is effectively ignored (actually, the poling behavior to be shown below will not be significantly affected by a higher σ_i). In the figures, the solid and dashed lines are used to denote the results for higher and lower σ_m respectively. In Fig. 1a which takes $\phi = 0.1$, the poling in both phases is shown to behave quite differently for lower and higher σ_m. When $\sigma_m = 10^{-14}\ \Omega^{-1}\text{m}^{-1}$ which is smaller than

TABLE I Properties of constituents for PZT/P(VDF-TrFE) 0-3 composites

	$\varepsilon/\varepsilon_0$	σ ($\Omega^{-1}m^{-1}$)	P_r (C/m^2)	P_s (C/m^2)	E_c (MV/m)	n [in Equation 14]
PZT	1600	10^{-12}	0.33	0.35	1	1
P(VDF-TrFE)	20	varied	0.055	0.06	60	1

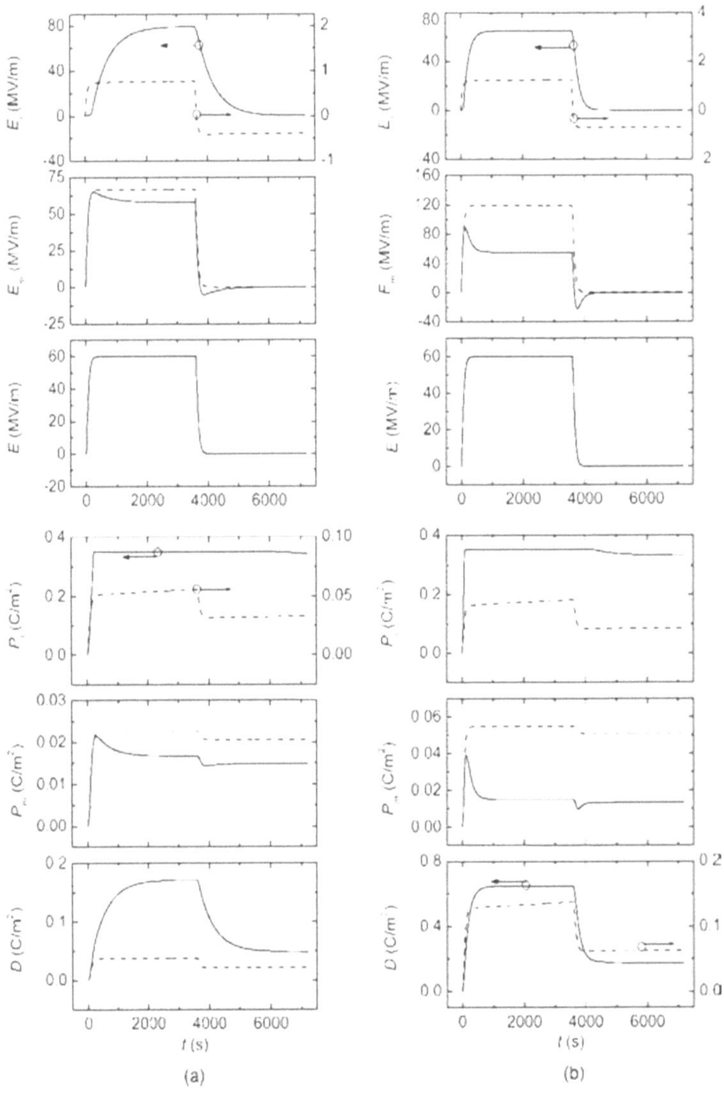

Figure 1 Simulated polarizations and electric fields of PZT/P(VDF-TrFE) composites with (a) ϕ =0.1, (b) $\phi = 0.5$, when the composites are polarized by a dc electric field. The dashed and solid lines denote $\sigma_m = 10^{-14}\Omega^{-1}m^{-1}$ and $\sigma_m = 10^{-11}\Omega^{-1}m^{-1}$ respectively.

σ_i, the degree of poling in PZT is negligibly small, but the copolymer phase is polarized to some extent. When $\sigma_m = 10^{-11}$ $\Omega^{-1}m^{-1}$, the situation is changed: the degree of poling in PZT is satisfactory, although the degree of poling in the copolymer phase is still restricted. For a higher ceramic volume fraction, Fig. 1b shows the simulation results of a composite ($\phi = 0.5$) with the same applied electric field. It is found that both the poling in the PZT and copolymer improves for the case $\sigma_m = 10^{-14}$ $\Omega^{-1}m^{-1}$. The simulated behaviors for P_i and P_m with $\sigma_m = 10^{-11}\Omega^{-1}m^{-1}$ are similar to the case for $\phi = 0.1$ (i.e., PZT is nearly fully poled, whilst the copolymer is poorly polarized). The comparison between Figs. 1a and 1b then clearly reveals that the PZT inclusion is difficult to pole if the polymer in the composite is highly insulating, especially for small ϕ. Thus, experimental techniques do exist to increase σ_m to facilitate poling [1, 6]. Also note from the figures that E_i and E_m take time to develop to a high value under an applied dc field. The poling time should therefore be sufficient to allow steady state to be reached. Our simulations show that an hour of poling time is about sufficient for both high or low σ_m samples. In addition, E_i and E_m also take time to relax back to zero under short circuit condition, hence the composite D also relaxes gradually. According to the general rule-of-thumb that the pyroelectric and piezoelectric strengths are roughly proportional to the polarization, the simulated results for D suggest the pyroelectricity and piezoelectricity may also gradually relax until $E_i = E_m = 0$.

In Fig. 2, we investigate the mechanism of poling by an ac electric field. The applied electric field is sinusoidal with 80 MV/m in magnitude and the frequency is 10 Hz. The duration of poling lasts for one and a half cycles. No matter whether $\sigma_m = 10^{-11}\Omega^{-1}\text{m}^{-1}$ or $10^{-14}\Omega^{-1}\text{m}^{-1}$, we find that the degree of poling in PZT is poor for both $\phi = 0.1$ [Fig. 2a] and $\phi = 0.5$ [Fig. 2b] (the simulated results for $\sigma_m = 10^{-11}\Omega^{-1}\text{m}^{-1}$ and $10^{-14}\Omega^{-1}\text{m}^{-1}$ overlap with each other). On the other hand, the degree of poling in the copolymer phase is better. Supposing σ_m increases

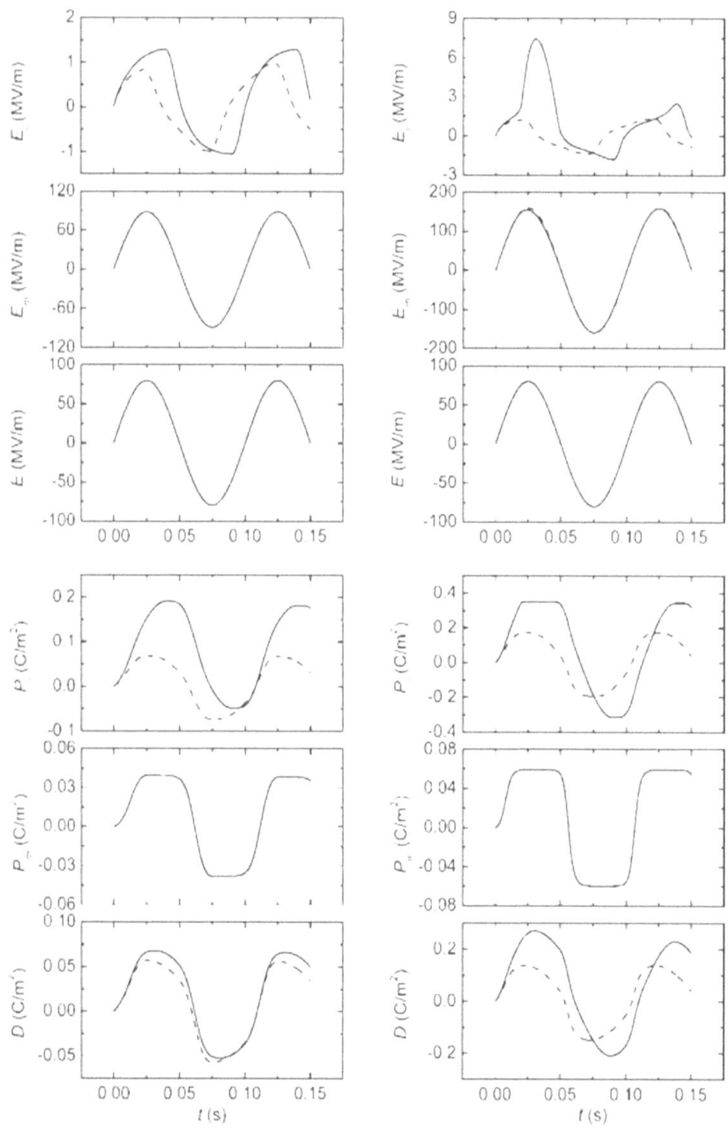

Figure 2 Simulated polarizations and electric fields of PZT/P(VDF-TrFE) composites with (a) $\phi = 0.1$, (b) $\phi = 0.5$, when the composites are polarized by an ac electric field. The dashed and solid lines denote $\sigma_m = 10^{-14}\sim10^{-11}\Omega^{-1}\text{m}^{-1}$ and $\sigma_m = 2\times10^{-8}\Omega^{-1}\text{m}^{-1}$ respectively.

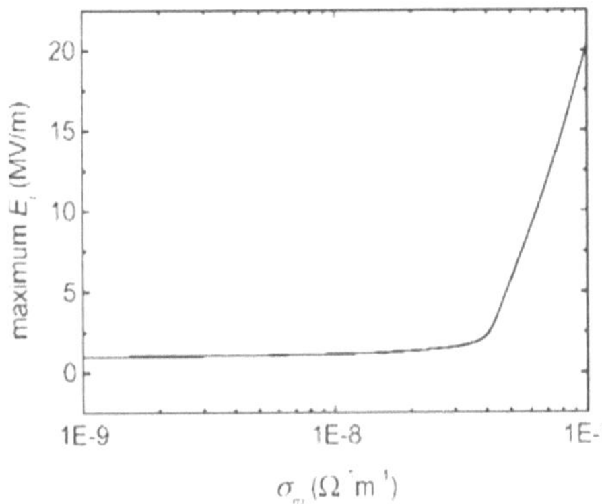

Figure 3 Variation of the maximum E_i with σ_m for a PZT/P(VDF-TrFE) composite with $\phi = 0.1$, when the composites are polarized by an ac electric field.

further to $2\times10^{-8}\Omega^{-1}\text{m}^{-1}$, the residual polarization (P_i at $t = 0.15$ s) in PZT will be high and the final P_m remains the same. Fig. 3 shows the variation of maximum E_i versus σ_m for $\phi = 0.1$, we find that there is a threshold value for σ_m which E_i is significantly increased. This confirms high σ_m facilitates poling, but such high σ_m value would not normally exist in polymeric materials at room temperature, or even at elevated temperature. It seems that ac poling is not an effective way to polarize PZT inclusions, but it has a strong effect for the polymer matrix. Also note from the figures that E_i is not zero at the end of the poling time ($t = 0.15$ s) but E_m is very close to zero, thus the relaxation in P_m after poling is expected to be not as large as in P_i.

From previous figures, we observe that dc poling is able to polarize the inclusions and matrix at different σ_m, whilst ac poling only has a strong effect in polarizing the copolymer matrix. One effective way to polarize both phases is to use a two-stage poling procedure [7, 8, 24]. Usually the sample is heated to above the Curie transition temperature of P(VDF-TrFE) and then subjected to a dc electric field for a certain duration. As the copolymer is in its paraelectric phase when the field is applied, only the ceramic phase has been polarized. This high temperature enhances σ_m and the poling in PZT is more complete [25]. After the PZT has been polarized, the composite is then cooled to below the Curie temperature of P(VDF-TrFE) and the copolymer in the composite is polarized by an ac field. This last procedure should not alter the polarization state of PZT significantly. Simulation of this poling process is shown in Figs 4 and 5.

Fig. 4 demonstrates the electric fields and polarization dynamics when both PZT and copolymer phases are polarized in the same direction. The adopted field parameters are also 60 MV/m and 80 MV/m for dc and ac poling fields. When $\sigma_m = 10^{-14}\Omega^{-1}\text{m}^{-1}$, it will not have sufficient E_i to polarize the PZT inclusion. This is similar to what we demonstrated in Fig. 1. However, we still note that the residual P_i in Fig. 1 is higher than the residual P_i (before ac poling) in Fig. 4. This suggests it is more difficult to polarize PZT inclusions in a nonferroelectric matrix. For $\sigma_m = 10^{-11}\Omega^{-1}\text{m}^{-1}$, it is shown that P_i is high after the dc poling for both $\phi = 0.1$ [Fig. 4a] and $\phi = 0.5$ [Fig. 4b]. P_m is zero because the copolymer is in the paraelectric phase. After the one and a half cycles of ac poling, the copolymer phase has been polarized to a high degree. The polarization in the PZT will be disturbed by the ac field, but its high poling degree will be restored at the end of the poling procedure. Hence, both phases are polarized to high degrees.

The above procedure may also be used to polarize the two phases in opposite directions. This has practical importance since it can reinforce piezoelectric activity and suppress pyroelectric activity to tailor for specific applications [26]. In contrast, pyroelectric activity is reinforced and piezoelectric activity is suppressed when the two phases are polarized in the same direction. Using the above two-stage poling procedure, the direction of the resulting polarization in the copolymer phase is determined by the electric field direction in the last half cycle of the ac poling field. Fig. 5 shows the simulation results for the same set of adopted σ_m and ϕ. Similar to the behavior demonstrated in Fig. 4, the degrees of poling are satisfactory for both phases for higher $\sigma_m (= 10^{-11}\Omega^{-1}\text{m}^{-1})$. One point worth noting is that, when σ_m is not sufficiently high (dashed line in Fig. 5, the residual P_i (after dc poling) could be easily switched by the ac field to the opposite direction. This may be an origin of the reduced electroactivity properties reported by Zeng *et al.* for their composite samples with oppositely poled constituents [24].

For comparison purposes, Fig. 6 shows the simulation results of composite samples subjected to a two-stage dc poling field with the two dc fields in opposite directions. It clearly shows that, for low ϕ or high ϕ and different σ_m's, the reduction of the initial polarization in the PZT phase due to the 2nd dc poling is significant and it is switched to the direction parallel to the 2nd poling field. This may be the origin of the diminished electroactivity properties reported by Ng *et al.* for their composite samples with oppositely poled constituents [17]. Indeed their composite samples with oppositely polarized phases were obtained by applying, as the final poling step, a constant poling field on the composite sample for half an hour in the reverse direction to the polarization of the pre-polarized ceramic phase, aimed at poling the copolymer phase.

Overall, the degree of poling of the constituent material depends on the poling procedure, and a higher σ_m is shown to facilitate poling. This is consistent with Sa-gong *et al.*'s conclusion [27]. A higher σ_m in composites can normally be obtained by adding conductive filler such as carbon black, or by poling at an elevated temperature. However, for a ferroelectric composite with a PTC (positive temperature coefficient of resistivity) matrix material such as carbon black loaded polyethylene [28], a higher σ_m is achievable at a lower temperature and thus in this case the composite is more effectively poled at lower temperatures.

3.2. Effect of σ_m and measuring frequency on ε, p and d_{33} coefficients of 0-3 composites

Here, we will demonstrate that high σ_m plays a significant role in the effective ε, p and d_{33}. The adopted properties of constituents for the calculations are listed in Table II (γ in the table denotes Young's modulus). For simplicity we are only concerned about the magnitude of the pyroelectric coefficient since the matrix materials of the 0-3 composites considered below are taken as nonferroelectric. Thus although p for PZT should formally be -330 μC/m^{2}K, it appears as $+330$ μC/m^{2}K in Table II. These adopted values are typical for a PZT/PVDF composite. Y and ν may be transformed to k and μ by using $k = Y/(3-6\nu)$ and $\mu = Y/(2+2\nu)$. The calculated results of ε [Equation 26], p [Equation 37] and d_{33} [Equation 68] for PZT/PVDF 0-3 composites are shown in Fig. 7. The measuring frequency for dielectric constant, pyroelectric coefficient and piezoelectric coefficient are 1 kHz, 5 mHz and 50 Hz respectively. For simplicity, only Equation 68 with μ_l is shown. In Figs 7a and c, the solid lines are based on our

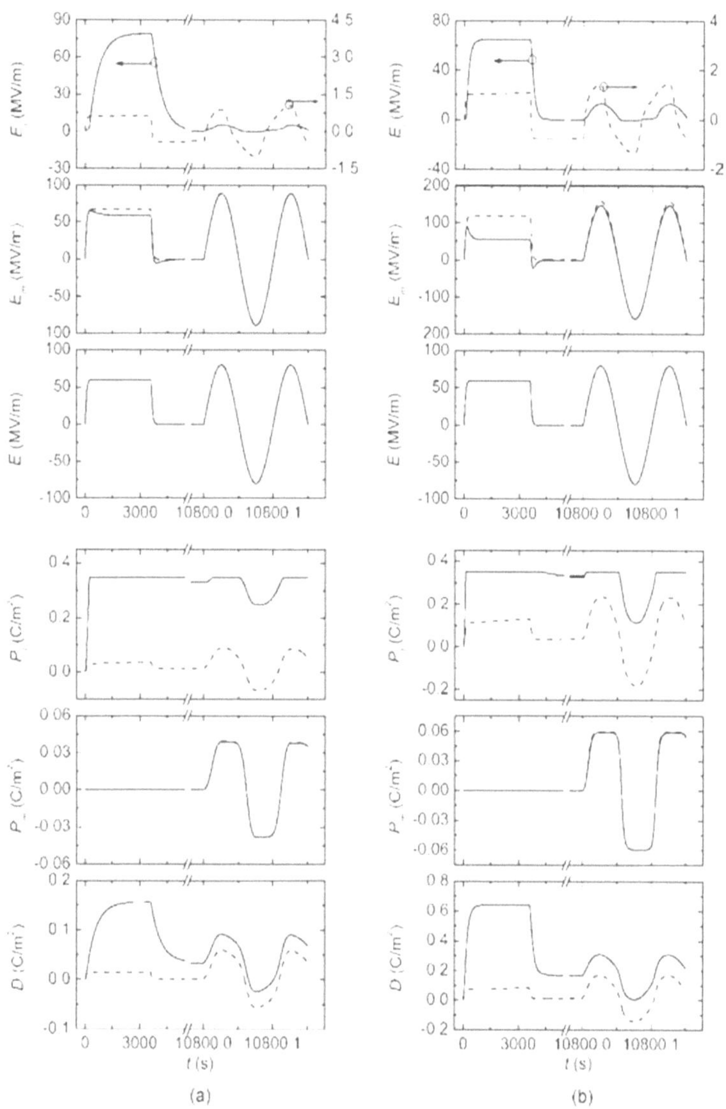

Figure 4 Simulated polarizations and electric fields of PZT/P(VDF-TrFE) composites with (a) $\phi = 0.1$, (b) $\phi = 0.5$, when the ceramic phase is first polarized by a dc electric field, then the composite is subject to an ac field to polarize the polymer phase in the same direction. The dashed and solid lines denote $\sigma_m = 10^{-14}\Omega^{-1}\text{m}^{-1}$ and $\sigma_m = 10^{-11}\Omega^{-1}\text{m}^{-1}$ respectively.

TABLE II Properties of constituents for PZT/PVDF 0-3 composites

	$\varepsilon/\varepsilon_0$	σ (Ω^{-1}m^{-1})	Y (GPa)	ν	p (μC/m^{2}K)	d_{33} (pC/N)	$-d_{31}$ (pC/N)
PZT	1200	10^{-14}	36	0.3	330	330	140
PVDF	11	varied	2.5	0.4	0	0	0

previous model [9, 11], which corresponds to $\Gamma_\varepsilon \varepsilon = 1$ in Equation 26 and $\Gamma_d = 1$ in Equation 68 of the present model (i.e., $\sigma_i = \sigma_m = 0$). The solid line in Fig. 7b corresponds to $\Gamma_p^{ac} = 1$ in Equation 37. A typical value of conductivity for PVDF at room temperature is about 10^{-12} Ω^{-1}m^{-1}. The prediction based on this σ_m value overlaps with our previous model. Thus, the effect of electrical conductivity on the effective ε, p and d_{33} of conventional samples can normally be neglected at room temperature. When σ_m is increased, no notable enhancement in ε, p and d_{33} is observed until $\sigma_m = 10^{-6}$ Ω^{-1}m^{-1}, $\sigma_m = 10^{-11}$ Ω^{-1}m^{-1} and $\sigma_m = 10^{-7}$ Ω^{-1}m^{-1} respectively. More enhancement of ε, p and d_{33} is observed at higher ceramic volume fraction. Further order of magnitude increment of

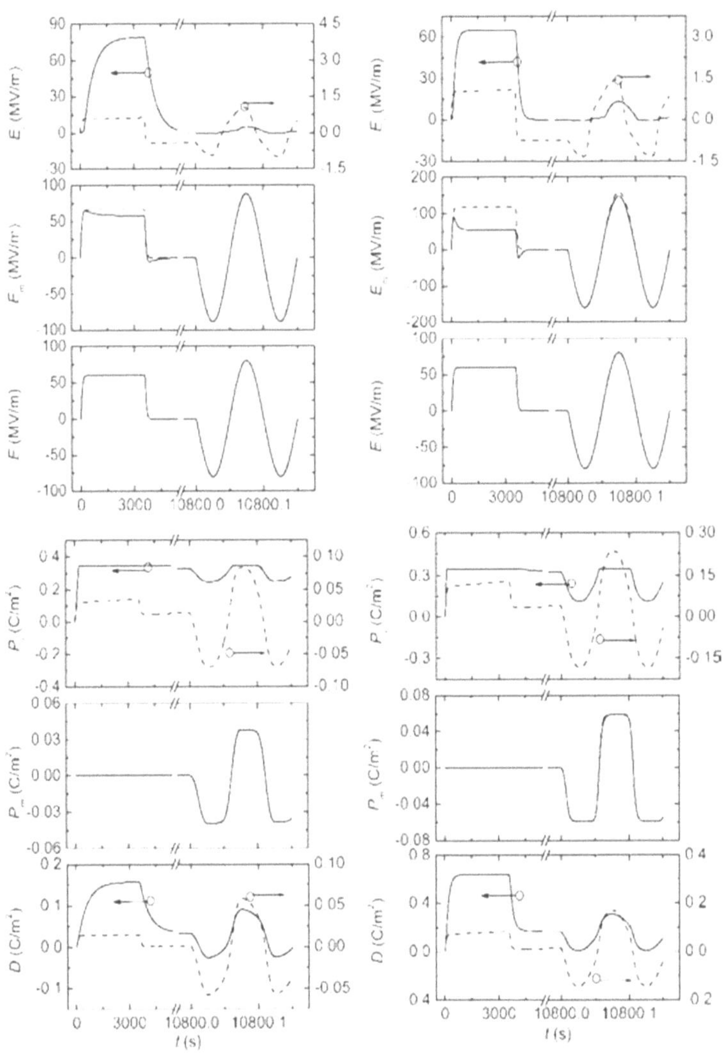

Figure 5 Simulated polarizations and electric fields of PZT/P(VDF-TrFE) composites with (a) $\phi = 0.1$, (b) $\phi = 0.5$, when the ceramic phase is first polarized by a dc electric field, then the composite is subjected to an ac field to polarize the polymer phase in an opposite direction. The dashed and solid lines denote $\sigma_m = 10^{-14}\Omega^{-1}$m^{-1} and $\sigma_m = 10^{-11}\Omega^{-1}$m^{-1} respectively.

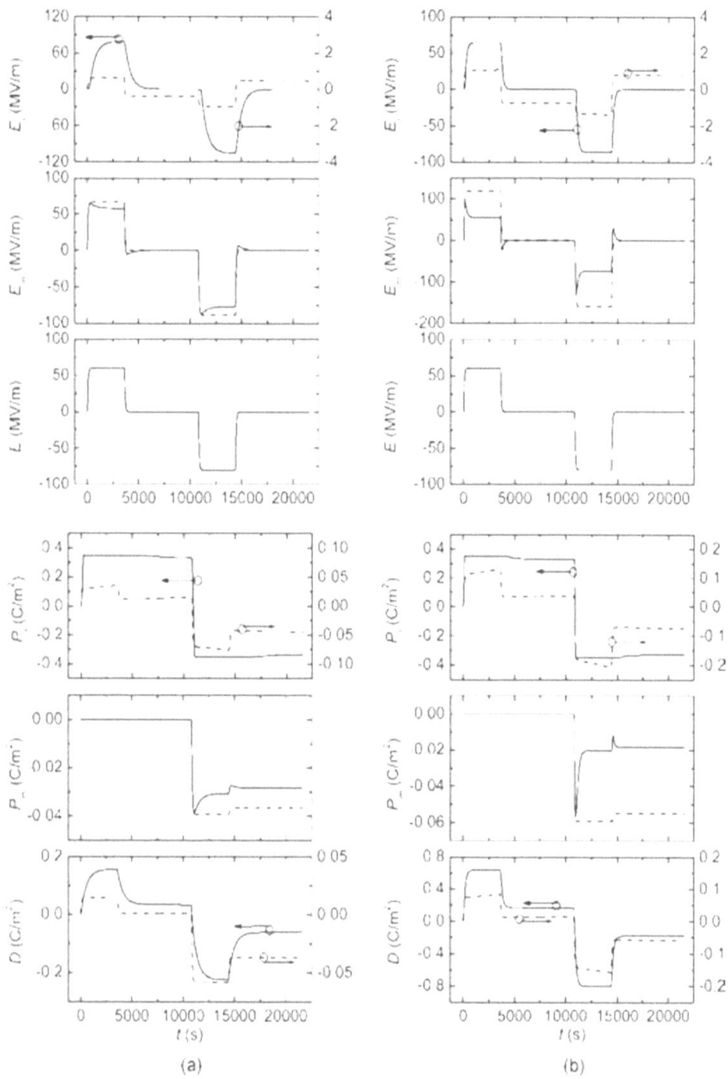

Figure 6 Simulated polarizations and electric fields of PZT/P(VDF-TrFE) composites with (a) $\phi = 0.1$, (b) $\phi = 0.5$, when the ceramic phase is first polarized by a dc electric field, then the composite is subjected to a dc field to polarize the polymer phase in an opposite direction. The dashed and solid lines denote $\sigma_m = 10^{-14}\Omega^{-1}\text{m}^{-1}$ and $\sigma_m = 10^{-11}\Omega^{-1}\text{m}^{-1}$ respectively.

σ_m will dramatically increase the values of ε, p and d_{33} for a given ϕ. The different σ_m values which start to "activate" the enhancement for ε, p and d_{33} are related to the different measuring frequencies. The smaller "threshold" σ_m value shown in Fig. 7b suggests electrical conductivity can affect pyroelectric measurement at lower values, whilst having least effect for permittivity measurement. This feature may be made use of in tailoring composites for higher than "normal" d_{33} constant, etc. In contrast, other approaches have been adopted in the literature to get an optimal piezoelectric response. For example, the well-known Piezo-Rubber manufactured by NTK-NGK [29–31] uses an extremely high content of PZT/$PbTiO_3$ particles (ϕ can be as large as 0.75) to obtain a resultant $d_{33} \approx 34\sim 56$ pC/ N [32, 33], which is quite difficult to fabricate.

In Fig. 7, we demonstrate the significant effect when $\sigma_m \geq 10^{-6}\ \Omega^{-1}\text{m}^{-1}$, $10^{-11}\ \Omega^{-1}\text{m}^{-1}$ and $10^{-7}\ \Omega^{-1}\text{m}^{-1}$ for ε, p and d_{33} respectively. It shows that all ε, p and d_{33} are very sensitive to a small change of σ_m in some region. In most of the calculations shown in Fig. 7, σ_i is taken as $10^{-14}\ \Omega^{-1}\text{m}^{-1}$. In Fig. 7, the predictions for $\sigma_i = \sigma_m = 5\times10^{-6}\ \Omega^{-1}\text{m}^{-1}$, $5\times10^{-11}\ \Omega^{-1}\text{m}^{-1}$ and $5\times10^{-7}\ \Omega^{-1}\text{m}^{-1}$ (for ε, p and d_{33} respectively) are also shown. It clearly reveals that the calculated ε, p and d_{33} values are only slightly different from the case of $\sigma_i = 10^{-14}\ \Omega^{-1}\text{m}^{-1}$ with the same $\sigma_m = 5\times10^{-6}\ \Omega^{-1}\text{m}^{-1}$ for ε, $5\times10^{-11}\ \Omega^{-1}\text{m}^{-1}$ for p, and $5\times10^{-7}\ \Omega^{-1}\text{m}^{-1}$ for d_{33}.

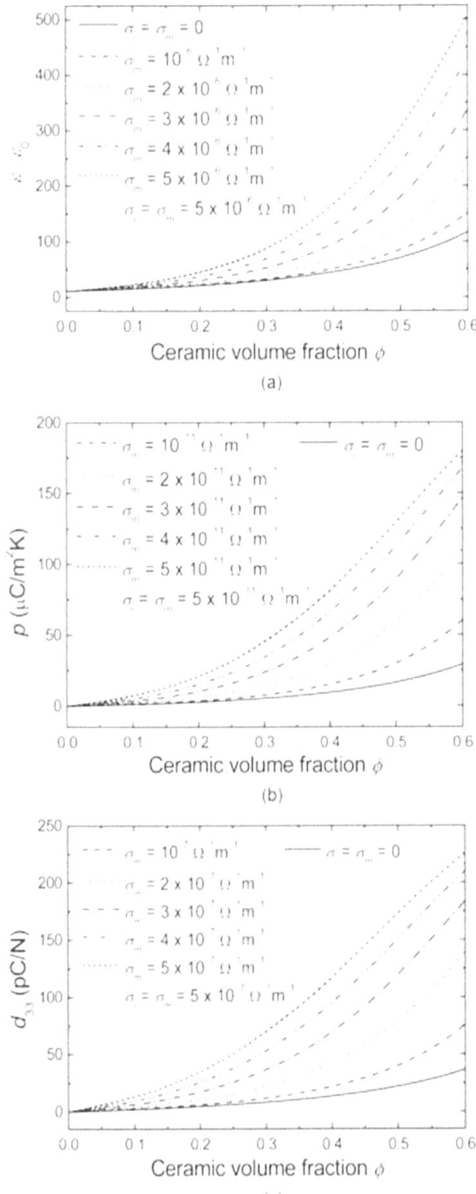

Figure 7 The variation of (a) permittivity ε [Equation 26], (b) pyroelectric coefficient [Equation 37] and (c) d_{33} constant [Equation 68] of PZT/PVDF composites with conductivity. Eqs 26, 37 and 68 with Γ's = 1 (solid lines) represent the constituents are perfectly insulating.

Actually, we generally find that ε, p and d_{33} are much less affected by σ_i than by σ_m.

In Equations 26, 37 and 68, τ is the governing factor which depends on the constituent permittivities, conductivities, and ϕ, rather than just σ_i or σ_m. Moreover, one can also appreciate from Equation 11 that σ_i is of lesser significance than σ_m. Fig. 8 shows the ceramic volume fraction dependence of τ with different σ_m's with $f = 50$ Hz. The figure shows that τ decreases monotonically with

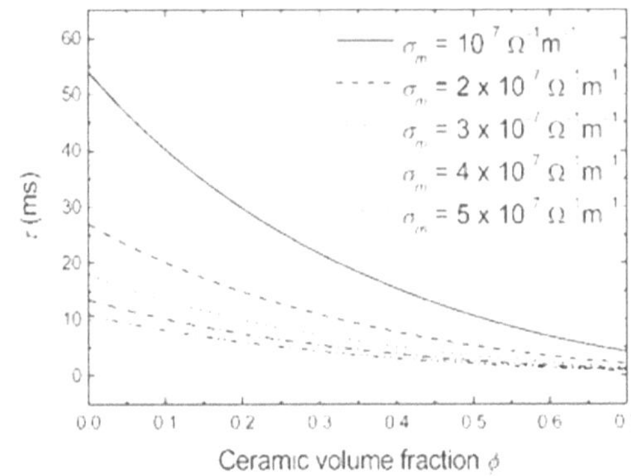

Figure 8 The variation of τ [Equation 11] of PZT/PVDF composites with conductivity in the matrix phase. The measuring frequency is 50 Hz.

ϕ for all σ_m's. Thus, ε, p and d_{33} increase with ceramic volume fraction, as noted in Fig. 7. From Fig. 8, it is found that a higher σ_m generally displaces the whole curve to lower τ values.

The effect of large σ_m essentially changes the magnitude of the induced internal electric fields as well as their phase difference with the applied electric field, modulating temperature or stress. Taking the piezoelectric measurement as an example, Fig. 9a shows a typical applied sinusoidal stress T of 50 Hz. The first three cycles of the induced electric field in the inclusion phase ($\langle E_{3i}\rangle$) are shown in Fig. 9b. Since $\langle E_3\rangle = 0$, $\langle E_{3m}\rangle$ is out-of-phase with $\langle E_{3i}\rangle$. When $\sigma_m = 10^{-14}\ \Omega^{-1}\text{m}^{-1}$ which is typically small for conventional sample, the magnitude of $\langle E_{3i}\rangle$ is almost the same for each cycle and $\langle E_{3i}\rangle$ is out-of-phase with T. When σ_m is large, the magnitude of $\langle E_{3i}\rangle$ and its phase difference with T changes with time until steady state is reached (not shown in Fig. 9).

In Fig. 7, we have demonstrated that substantial enhancement of ε, p and d_{33} starts from $\sigma_m = 10^{-6}\ \Omega^{-1}\text{m}^{-1}$, $10^{-11}\ \Omega^{-1}\text{m}^{-1}$ and $10^{-7}\ \Omega^{-1}\text{m}^{-1}$ respectively. This enhancement continues for increasing σ_m until steady state. Fig. 10 shows the σ_m dependence in a wide range for ε, p and d_{33}. For $\sigma_m < 10^{-6}\ \Omega^{-1}\text{m}^{-1}$, ε is small and nearly independent of σ_m. Similarly for $\sigma_m < 10^{-11}\ \Omega^{-1}\text{m}^{-1}$ and $\sigma_m < 10^{-7}\ \Omega^{-1}\text{m}^{-1}$ for p and d_{33}. For σ_m between $10^{-6}\ \Omega^{-1}\text{m}^{-1}$ and $10^{-4}\ \Omega^{-1}\text{m}^{-1}$, ε significantly increases to very high values. When $\sigma_m > 10^{-4}\ \Omega^{-1}\text{m}^{-1}$, steady state is reached and no additional change in ε is observed. Since the "threshold" σ_m value for ε is very high, permittivity measurement is not expected to be significantly affected by conductivity effect for many ferroelectrics. However, conductivity effect may be notable in pyroelectricity and piezoelectricity. Fig. 10b and c demonstrate the lower "threshold" σ_m ranges (for p and d_{33}) and their limits. These apparent limits will be smaller for a lower measuring frequency.

From Equations 26, 37 and 68, the Γ's should also be sensitive to changes in the measuring frequency. Fig. 11

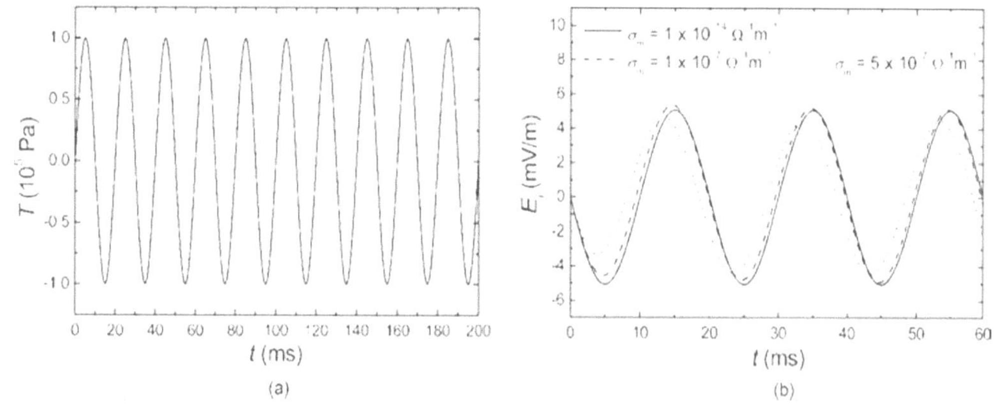

Figure 9 (a) A typical applied sinusoidal stress and (b) theoretical calculations of the induced electric field in the inclusion phase [Equation 56] for the d_{33} measurement of a PZT/PVDF composite with $\phi = 0.4$. The measuring frequency is 50 Hz.

shows the variation of ε, p and d_{33} with the measuring frequency f, assuming $\sigma_m = 5\times10^{-10}\ \Omega^{-1}\text{m}^{-1}$ for ε, $\sigma_m = 10^{-14}\ \Omega^{-1}\text{m}^{-1}$ for p, and $\sigma_m = 5\times10^{-10}\Omega^{-1}\text{m}^{-1}$ for d_{33}. At frequencies above 1 Hz, no notable enhancement is found for ε, and similarly for frequencies above 10 μHz and 1 Hz for p and d_{33} respectively. Further reduction in f will have an effect similar to increasing σ_m, resulting in significant enhancement of ε, p and d_{33}.

The above analyses for piezoelectricity have been confined to the d_{33} constant. Actually, the enhancement by conductivity also applies to the d_{31} [Equation 67] and d_h [Equation 69] constants. However, for the piezoelectric g_h coefficient ($\equiv d_h/\varepsilon$) which is important for hydrophone applications, piezoelectric enhancement may not be achieved by increasing σ_m, since both d_h and ε vary with σ_m and they have a common "threshold" σ_m since the "operating" frequencies are identical in this case [2]. Thus, the percentage increment of d_h must be larger than that for ε to achieve an enhanced g_h value. Summing up, with electrical conductivity considered, the new model suggested in this article yields significant effects for composite samples with high conductivity in the polymer matrix, but only minimal effects for normal composites possessing relatively low conductivity. Thus the Γ factors [Equations 26, 37 and 68] do not affect in a noticeable way the goodness of fit already obtained by many existing models [34] for ordinary composite samples. For samples possessing high σ, the effect of the Γ factors have to be included.

3.3. Comparison with experimental data for pyroelectric measurement

A previous article [6] reported that the pyroelectric coefficient of a PZT/PU 0-3 composite would be significantly increased by doping with graphite. The temperature dependence of the pyroelectric coefficient was measured for both conventional and graphite doped samples (1% by volume). The composite sample with $\phi = 0.5$ was heated at a constant rate of 1°C/min. Sakamoto *et al.* found that the pyroelectric coefficient at 303 K was 5.6 and 10.7 μC/m^2 for conventional and graphite doped samples respectively. However, the deviation of p between the two composite samples became very large at higher temperatures. Owing to the fact that the increment of p with temperature for the undoped sample is quite limited, the anomaly for the doped sample is thought to arise not solely from the increment of permittivities and pyroelectric coefficient of the constituents. The phenomenon is most likely the effect of incremental temperature on the electrical conductivity, hence the large p. Since the rate of temperature ramp is slow, the effect of conductivity of PU is thought to be significant, especially for the present situation of a graphite-doped sample.

Here we attempt to apply Equation 43 to investigate the discrepancy between the measured pyroelectric coefficients versus temperature for the foregoing PZT/PU composite system with and without graphite filler. Experimental data for pyroelectric measurement on a PZT/PU system at room temperature using the ac method can be found in ref. 19. As the dielectric and pyroelectric properties of the constituent materials provided in these articles have not been described in sufficient detail for our purposes, typical values have been adopted in our calculation, as will be explained in the next paragraph. In the present study, our theoretical calculation will only focus at the lower temperature region (303~330 K). At the high temperature region, variations of dielectric and pyroelectric properties of the constituents will be significant and complicates the problem.

Sakamoto *et al.* have measured the permittivity of PZT at 343 K which is 1200 ε_0, but the temperature variations are not given. They have also measured the ε against temperature for PZT/PU with $\phi = 0.19$ [35]. We assume the variation of ε_i versus Θ is similar to the measurement results of PZT by Furukawa *et al.* [18]. The results of

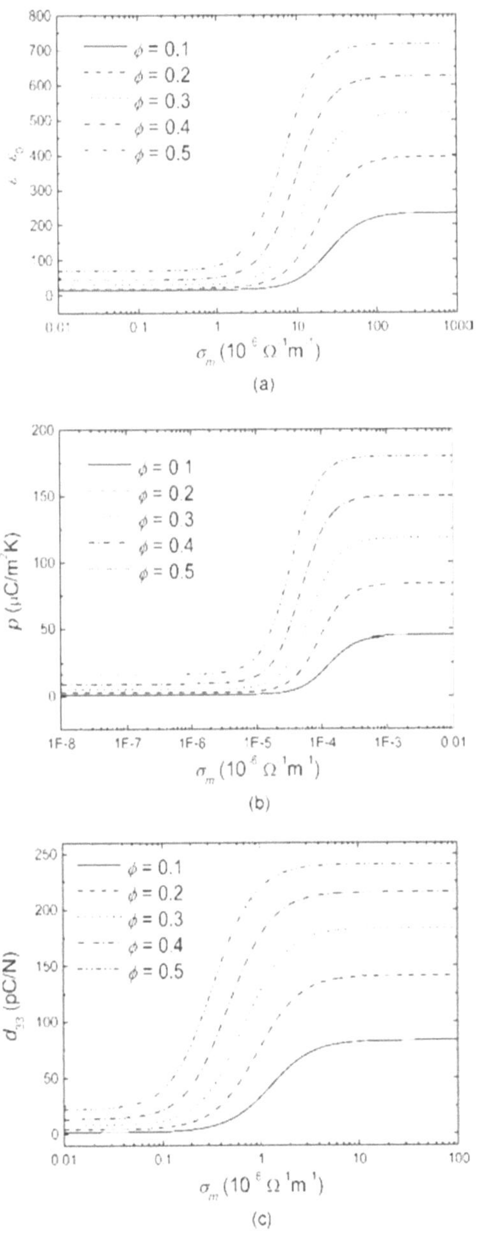

Figure 10 The variation of (a) permittivity ε [Equation 26], (b) pyroelectric coefficient [Equation 37] and (c) d_{33} constant [Equation 68] of PZT/PVDF composites with conductivity. The enhancement of ε, p and d_{33} with increasing σ_m becomes significant at $10^{-6}\ \Omega^{-1}m^{-1}$, $10^{-11}\ \Omega^{-1}m^{-1}$ and $10^{-7}\ \Omega^{-1}m^{-1}$ respectively. They are close to maximum value for (a) $\sigma_m > 10^{-4}\ \Omega^{-1}m^{-1}$, (b) $\sigma_m > 10^{-9}\ \Omega^{-1}m^{-1}$ and (c) $\sigma_m > 10^{-5}\ \Omega^{-1}m^{-1}$.

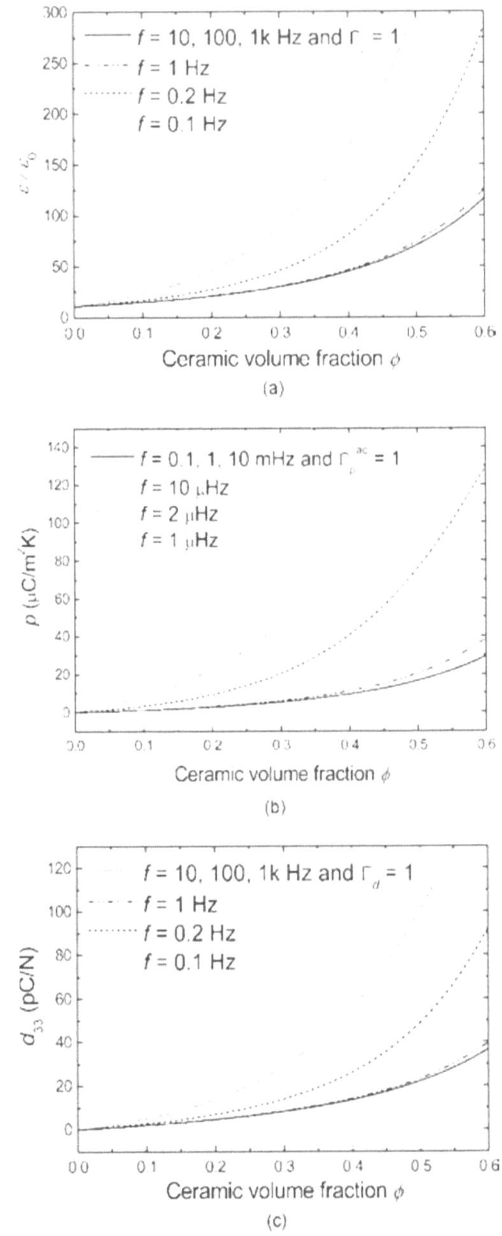

Figure 11 The variation of (a) permittivity ε [Equation 26], (b) pyroelectric coefficient [Equation 37] and (c) d_{33} constant [Equation 68] of PZT/PVDF composites with the measuring frequency. Equations 26, 37 and 68 with Γ's = 1 represent the constituents are perfectly insulating. Other lines with Γ's$\neq$1 assume the constituents possess small but finite conductivity.

permittivity against Θ by Furukawa *et al.* is then scaled in such a way that $\varepsilon_i = 1200\ \varepsilon_0$ at 343 K. For the permittivity of PU, we use Equation 27 to calculate the ε_m's from the experimental ε of PZT/PU given by Sakamoto *et al.* [35]. Fig. 12 shows our adopted temperature dependence of permittivity for the PZT and PU materials. Concerning the pyroelectric coefficients, we have $p_m = 0$ and p_i is calculated from the experimental p of undoped PZT/PU [6] with $\phi = 0.5$ [by Equation 38] and with permittivities taken from Fig. 12. In the above, we have assumed Sakamoto *et al.*'s measurement on PZT (permittivity measurement) and undoped PZT/PU (permittivity and pyroelectric measurement) is not affected by conductivity effects, so that Equations27and 38 may be employed. The temperature

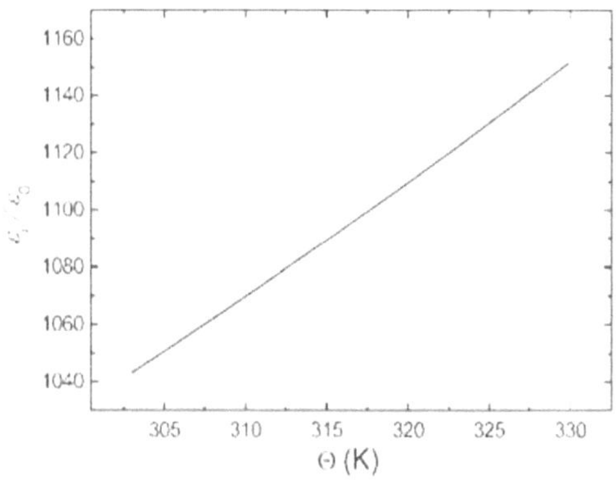

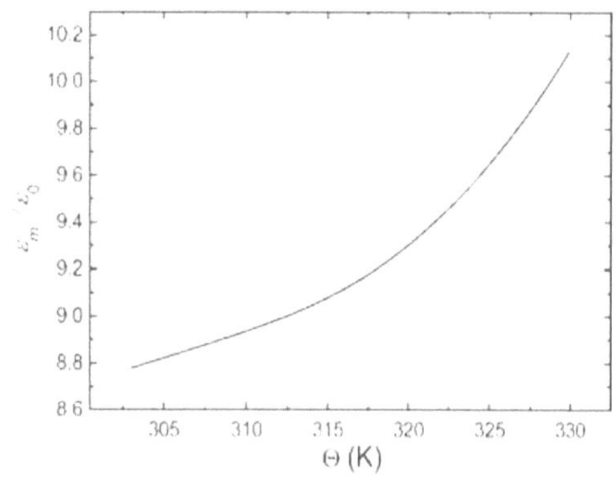

Figure 12 Temperature dependence of dielectric constant for (a) PZT ceramic, (b) PU polymer.

profile of the pyroelectric coefficient for PZT is shown in Fig. 13.

Theoretical prediction based on Equation 43 with the above constituent properties is shown in Fig. 14. When $\sigma_m = 10^{-14}\ \Omega^{-1}\text{m}^{-1}$ which is small, the predicted line is very close to the experimental data for undoped PZT/PU. When σ_m is increased to $3\times10^{-13}\ \Omega^{-1}\text{m}^{-1}$, we notice that, apart from the high temperatures ($\Theta > 320$ K), the general trend of p versus Θ for graphite doped PZT/PU has been reproduced by the present model incorporating conductivity. The discrepancy between the prediction and measured values may suggest the increment of σ_m with temperature should be included for higher temperature range. To investigate this, we adopt the following formula [36]

$$\sigma_m = C_1 + C_2 \exp\left(\frac{-U_{av}}{k_b\Theta}\right), \tag{86}$$

where

$$C_1 = \frac{m_b\sigma_m(\Theta_L) - m_a\sigma_m(\Theta_H)}{\sigma_m(\Theta_L) - \sigma_m(\Theta_H)}, \tag{87}$$

$$C_2 = \frac{m_b - m_a}{\sigma_m(\Theta_L) - \sigma_m(\Theta_H)}, \tag{88}$$

and $m_a = \exp(-U_{av}/k_b\Theta_L)$, $m_b = \exp(-U_{av}/k_b\Theta_H)$. U_{av} is the activation energy and k_b is the Boltzmann constant. Assume σ_m increases with temperature from $3\times10^{-13}\ \Omega^{-1}\text{m}^{-1}$ (at $\Theta_L = 303$) to $6\times10^{-13}\ \Omega^{-1}\text{m}^{-1}$ (at $\Theta_H = 330$) and $U_{av} = 2$ eV, the result shows excellent agreement between the prediction (solid line) and the experimental data for graphite doped PZT/PU. All in all, the consideration of electrical conductivity seems to provide a good understanding to the pyroelectric experimental results of Sakamoto *et al.*

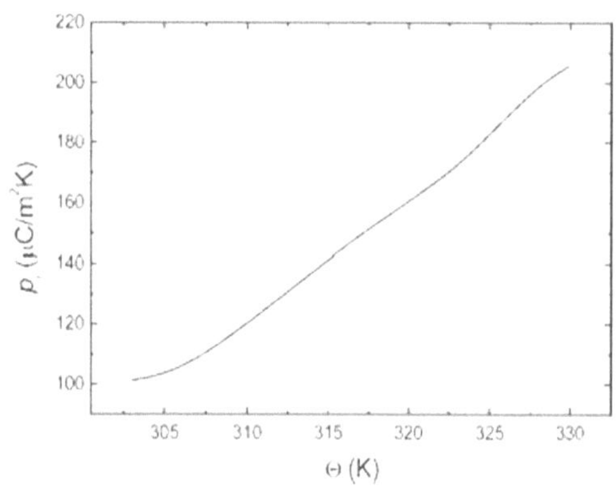

Figure 13 Temperature dependence of pyroelectric coefficient for PZT ceramic.

The present formulation neglects the loss components of the elastic and pyroelectric properties in the constituents. In the case that the loss components are also considered, all constituent parameters involved should take on complex values, and the final expressions would become much more complicated [37]. Nevertheless, an increased matrix conductivity will generally give higher, undesirable losses which need to be taken into account in considering applications. Thus, it is usually not worthwhile to make excessively conducting composites solely for rapid poling. This has been reported and discussed in the literature [27, 38]. On the other hand, manipulation of conductivity in piezoelectric and pyroelectric composites is useful for a number of devices. For instance, the damping characteristics of a mechanical composite damper can be tailored by changing the conductivity through doping the matrix [39]. For pyroelectric sensor applications, the gate voltage to the MOSFET used as a signal amplifier can be fed through the conductive composite to avoid the need for a high resistance gate resistor which is difficult and costly to realize in an integrated circuit [40]. The methodology and results of

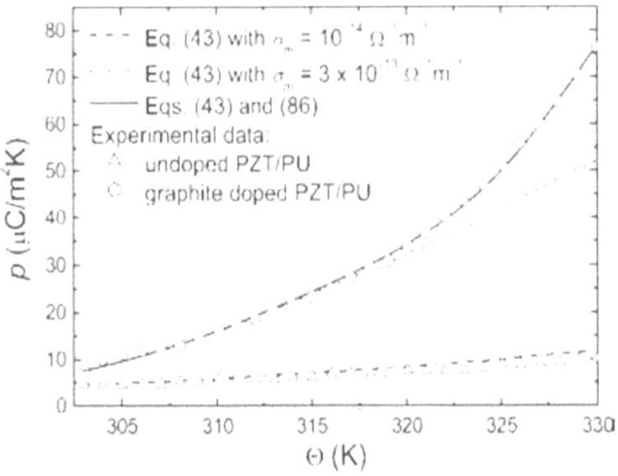

Figure 14 Theoretical prediction by Equation 43 is compared with the experimental data of Sakamoto *et al.* (see ref. 6) for the temperature dependence of *p* of a PZT/PU composite. The sample was heated at a constant rate of 1°C/min.

the present paper will certainly be of value for designing conducting composites for new applications.

4. Conclusions

The significance of electrical conductivities in ferroelectric 0-3 composites have been considered in this article. We have modeled the polarization behavior of ceramic/polymer 0-3 composites with different poling procedures. Our simulations suggest that higher σ_m facilitates poling. To obtain optimal poling of a ferroelectric-ferroelectric PZT/P(VDF-TrFE) composite, a two stage poling procedure with dc poling at evaluated temperature followed by an ac poling is highly advantageous. We have also calculated the effect of electrical conductivity on the dielectric, pyroelectric and piezoelectric properties. New explicit expressions have been derived for ε, p, and the d_{33}, d_{31} and d_h coefficients. A high conductivity in the matrix phase can alter the internal fields in a ferroelectric composite and allows the accumulation and dissipation of free charge at the matrix-inclusion interfaces. This can significantly enhance the permittivity, pyroelectric and piezoelectric coefficients of ferroelectric 0-3 composites. A high σ_m value may be induced by graphite doping and evaluated temperature. The permittivity ε, pyroelectric p and piezoelectric d constants will start to be noticeably enhanced when σ_m is larger than some threshold values and reach saturation beyond a certain σ_m. Comparison of our theoretical results with the experimental data of Sakamoto *et al.* show fairly good agreement, when σ_m is assumed to increase exponentially with temperature. The electrical conductivity effect is quite likely responsible for the reported anomalously high pyroelectric values in their work.

Acknowledgment

This work was partially supported by the Center for Smart Materials of The Hong Kong Polytechnic University.

References

1. X. D. CHEN, D. B. YANG, Y. D. JIANG, Z. M. WU, D. LI, F. J. GOU and J. D. YANG, *Sensor Actuat. A-Phys.* **65** (1998) 194.
2. C. K. WONG and F. G. SHIN, *J. Appl. Phys.* **97** (2005) 4111.
3. Y. T. OR, C. K. WONG, B. PLOSS and F. G. SHIN, *ibid.*, **93** (2003) 064112.
4. *Idem.*, *ibid.*, **94** (2003) 3319.
5. T. FURUKAWA, *IEEE Trans. Electr. Insul.* **24** (1989) 375.
6. W. K. SAKAMOTO, P. MARIN-FRANCH and D. K. DAS-GUPTA, *Sensor Actuat. A-Phys.* **100** (2002) 165.
7. K. W. KWOK, C. K. WONG, R. ZENG and F. G. SHIN, *Appl. Phys. A: Mater. Sci. Process* **81** (2005) 217.
8. B. PLOSS, B. PLOSS, F. G. SHIN, H. L. W. CHAN and C. L. CHOY, *Appl. Phys. Lett.* **76** (2000) 2776.
9. Y. M. POON and F. G. SHIN, *J. Mater. Sci.* **39** (2004) 1277.
10. K.-H. CHEW, F. G. SHIN, B. PLOSS, H. L. W. CHAN and C. L. CHOY, *J. Appl. Phys.* **94** (2003) 1134.
11. C. K. WONG, Y. M. POON and F. G. SHIN, *ibid.*, **90** (2001) 4690.
12. C. K. WONG, Y. M. WONG and F. G. SHIN, *ibid.*, **92** (2002) 3974.
13. S. L. MILLER, R. D. NASBY, J. R. SCHWANK, M. S. RODGERS and P. V. DRESSENDORFER, *ibid.* **68** (1990) 6463.
14. S. L. MILLER, J. R. SCHWANK, R. D. NASBY and M. S. RODGERS, *ibid.* **70** (1991) 2849.
15. C. J. DIAS and D. K. DAS-GUPTA, *IEEE Trans. Dielectr. Electr. Insul.* **3** (1996) 706.
16. B. PLOSS, B. PLOSS, F. G. SHIN, H. L. W. CHAN and C. L. CHOY, *ibid.* **7** (2000) 517.
17. K. L. NG, H. L. W. CHAN and C. L. CHOY, *IEEE Trans. Ultrason. Ferroelectr. Freq. Control* **47** (2000) 1308.
18. T. FURUKAWA, K. FUJINO and E. FUKADA, *Jpn. J. Appl. Phys.* **15** (1976) 2119.
19. K. S. LAM, Y. W. WONG, L. S. TAI, Y. M. POON and F. G. SHIN, *J. Appl. Phys.* **96** (2004) 3896.
20. C. K. WONG, Y. M. POON and F. G. SHIN, *ibid.* **93** (2003) 487.
21. Z. HASHIN, *J. Appl. Mech.* **29** (1962) 143.
22. R. M. CHRISTENSEN, in "Mechanics of Composite Materials" (Wiley, New York, 1979) Chap. 4.
23. Z. HASHIN and S. SHTRIKMAN, *J. Mech. Phys. Solids* **11** (1963) 127.
24. R. ZENG, K. W. KWOK, H. L. W. CHAN and C. L. CHOY, *J. Appl. Phys.* **92** (2002) 2674.
25. H. L. W. CHAN, Y. CHEN and C. L. CHOY, *Integr. Ferroelectr.* **9** (1995) 207.
26. H. L. W. CHAN, P. K. L. NG and C. L. CHOY, *Appl. Phys. Lett.* **74** (1999) 3029.
27. G. SA-GONG, A. SAFARI, S. J. JANG and R. E. NEWNHAM, *Ferroelectrics Lett.* **5** (1986) 131.
28. B. WANG, X. YI, Y. PAN and H. SHAN, *J. Mater. Sci. Lett.* **16** (1997) 2005.
29. H. BANNO, K. OGURA, H. SOBUE and K. OHYA, *Jpn. J. Appl. Phys.* **26**(suppl. 26-1) (1987) 153.
30. H. BANNO and K. OGURA, *ibid.* **30** (1991) 2247.
31. J. F. TRESSLER, S. ALKOY, A. DOGAN and R. E. NEWNHAM, *Compos. Part A—Appl. Sci.* **30** (1999) 477.

32. R. Y. TING, *Appl. Acoust.* **41** (1994) 325.
33. R. Y. TING and F. G. GEIL, in Proc. of the 7th IEEE International Symposium on Applications of Ferroelectrics, (1991), p. 14.
34. C. J. DIAS and D. K. DAS-GUPTA, *IEEE Trans. Dielectr. Electr. Insul.* **3** (1996) 706.
35. W. K. SAKAMOTO, S. KAGESAWA, D. H. KANDA and D. K. DAS-GUPTA, *J. Mater. Sci.* **33** (1998) 3325.
36. N. BRAITHWAITE and G. WEAVER, in "Electronic Materials" (Butterworth, Boston, 1990).
37. C. K. WONG, Y. M. POON and F. G. SHIN, *J. Appl. Phys.* **92** (2002) 3287.
38. A. SAFARI, G. SA-GONG, J. GINIEWICZ and R. E. NEWNHAM, in Proceedings of the 21st University Conference on Ceramic Science, (1986) Vol. 20 p. 445.
39. Y. SUZUKI, K. UCHINO, H. GOUDA, M. SUMITA, R. E. NEWNHAM and A. R. RAMACHANDRAN, *J. Ceram. Soc. Jpn.* **99** (1991) 1096.
40. B. PLOSS, Y. W. WONG and F. G. SHIN, Ferroelectrics [to appear in a special issue of Ferroelectrics for the Fourth Asian Meeting on Ferroelectrics (AMF-4)].

JOURNAL OF MATERIALS SCIENCE 41 (2006) 251–258

Properties of triglycine sulfate/poly(vinylidene fluoride-trifluoroethylene) 0-3 composites

Y. YANG, H. L. W. CHAN, C. L. CHOY
Department of Applied Physics and Materials Research Centre, The Hong Kong Polytechnic University, Hung Hom, Kowloon, Hong Kong, China
E-mail: apahlcha@polyu.edu.hk

Composites of triglycine sulfate (TGS) powder dispersed in a poly(vinylidene fluoride-trifluoroethylene) (P(VDF-TrFE), 70/30 mol%) copolymer matrix have been prepared using solvent casting followed by compression moulding. The composites have been characterized by means of scanning electron microscopy (SEM) and differential scanning calorimetry (DSC). Three groups of samples have been prepared: unpoled composites, composites with only the TGS phase poled and composites with both phases poled. The observed relative permittivity of these composites are consistent with the predictions of the Bruggeman model. When the TGS and copolymer phases are poled in the same direction, the piezoelectric activities of the two phases partially cancel each other while the pyroelectric activities reinforce. The pyroelectric coefficients exhibit good agreement with the effective-medium model. These composites have high pyroelectric figures of merit which increase as the volume fraction of TGS increases.

1. Introduction

Ferroelectric ceramic/polymer 0-3 composites have a considerable potential in pyroelectric sensor, ultrasonic transducer and underwater hydrophone applications as they combine high piezoelectric and pyroelectric activities of the ceramic with the mechanical flexibility and low acoustic impedance of the polymer [1–5]. If both the inclusion and the matrix are ferroelectric, the poling state of the matrix provides an additional feature and it is possible to produce composites with only the ceramic poled or with both the ceramic phase and the matrix poled in the same direction [4]. Hence, P(VDF-TrFE) copolymers (with TrFE content >20 mol%) have been chosen as the electroactive matrix in ceramic/polymer 0-3 composites in recent years [3, 4]. In selecting the inclusion, many conventional pyroelectric ceramic materials with high pyroelectric activity and dielectric permittivity have been used, e.g. $PbTiO_3$ [2] and lead zirconate titanate [5].

The main aim of this work is to produce 0-3 composites with high pyroelectric activity and high pyroelectric figures of merit. Based on the effective-medium approach [6], the pyroelectric coefficient p of a composite can be calculated by:

$$p = \left(\frac{\varepsilon - \varepsilon_m}{\varepsilon_i - \varepsilon_m}\right) \cdot p_i + \left(\frac{\varepsilon_i - \varepsilon}{\varepsilon_i - \varepsilon_m}\right) \cdot p_m \tag{1}$$

where ε, ε_i and ε_m are the relative permittivities of the composite, inclusion and matrix, respectively and p_i and p_m are the pyroelectric coefficients of the inclusion and matrix, respectively. Using this equation, it is found that high relative permittivity of the inclusion will inhibit the improvement of pyroelectric properties in 0-3 composites. Therefore, triglycine sulfate (TGS), which is a well-known pyroelectric crystal having a high pyroelectric coefficient and a low relative permittivity, would be an ideal material to be used as the inclusion phase. Although TGS single crystals have high pyroelectric figures of merit, they are very fragile and water-soluble. These make it difficult to use TGS crystals in pyroelectric devices. In previous publications, studies on TGS/PVDF 0-3 composites have been reported [7, 8]. However, there is no systematic study on TGS/P(VDF-TrFE) 0-3 composites. In this work, the dielectric, piezoelectric and pyroelectric properties of

0022-2461 © *2006 Springer Science + Business Media, Inc.*
DOI: 10.1007/s10853-005-7245-1

TGS/P(VDF-TrFE) 0-3 composites are studied in detail.

2. Experimental procedure

2.1. Fabrication of TGS/P(VDF-TrFE) 0-3 composites

The P(VDF-TrFE) 70/30 mol% copolymer used as the matrix phase in the present study was supplied by Piezotech, France. The TGS single crystal and powder were supplied by the Institute of Crystal Materials, Shandong University, China.

Powder of P(VDF-TrFE) 70/30 mol% copolymer was dissolved in dimethylformamide (DMF) and then predetermined amount of TGS powder (size $\sim$5 μm) was dispersed into the solution. After 20 min of ultrasonic agitation, the solution was stirred magnetically at 60°C until a gel-form composite was obtained. In order to remove the solvent completely, the thick composite film was dried at 80°C in a vacuum oven for 12 h. The thick film was then crushed into small pieces and compression moulded at 160°C under a pressure of 15 MPa. After 15 min, the mould was then removed and cooled to room temperature in air. In order to enhance the crystallinity, the composite films were annealed at 130°C for 2 h. The final thickness of the composite film was about 70 μm. Cr-Au electrodes were deposited on both sides of the film for subsequent property measurements. TGS/P(VDF-TrFE) 0-3 composites with various volume fractions of TGS $\phi = 0.05, 0.11, 0.22, 0.27, 0.33$ and 0.43 were fabricated.

2.2. Poling process

To impart piezoelectric and pyroelectric activities to the samples, they were subjected to a poling process so as to orient the dipoles in the TGS and P(VDF-TrFE) phases. In the present study, two groups of samples were poled in an oil bath. In group 1, both phases were poled and in group 2 only the TGS phase was poled. The poling procedures are described as follows:

Group 1: Samples were heated to 70°C and a stepwise poling method was used. Firstly, samples were poled at 70°C with a field of 60 V/μm for 30 min. At this temperature, only the copolymer phase was poled since TGS was in the paraelectric phase (the Curie temperature of TGS is 49°C). Secondly, the field was decreased to 10 V/μm and the sample was cooled from 70°C to ambient temperature in a period of about 1h with the electric field applied. During the cooling process, TGS went through a transition from the paraelectric phase to the ferroelectric phase with the field applied.

Group 2: Samples were heated to 70°C and a field of 10 V/μm was applied. Then the samples were cooled to room temperature with the electric field kept on. The copolymer matrix could not be poled with such a low field. To verify that only the TGS phase was poled, the same procedure was applied to a pure copolymer sample. The copolymer has no detectable piezoelectric or pyroelectric activities, thus showing that it was not polarized.

2.3. Characterization

Three series of composites were prepared for characterization: unpoled composites, composites with only the TGS phase poled and composites with both phases poled. The structure, phase transitions and dielectric, piezoelectric and pyroelectric properties of the composites were investigated. The microstructure of TGS/P(VDF-TrFE) 0-3 composites was observed using a scanning electron microscope (JEOL JSM-6335F FE-SEM). Thermal measurements were performed with a differential scanning calorimeter (Perkin-Elmer DSC7). The heating or cooling rate was 10°C/min in the temperature range of −20 to 130°C. The relative permittivity ε_r and loss factor tan δ as a function of temperature from −20 to 120°C at 10 kHz were determined using an impedance analyzer (HP 4194A) equipped with a temperature chamber (Delta 9023). The relative permittivity ε_r and loss factor tan δ as a function of frequency from 1 kHz to 10 MHz were also measured at room temperature (about 22°C). The piezoelectric coefficient d_{33} of the composites was measured using a piezo d_{33} meter (Model ZJ-3D, Beijing Institute of Acoustics, China). A sinusoidal mechanical stress with a frequency of 60 Hz was applied to the sample while the current signal was recorded. The pyroelectric coefficient p was measured using a dynamic method [9]. At a certain temperature T the sample temperature was sinusoidally modulated [$T(t) = T + T \sim \sin 2\pi ft$] with frequency $f =$ 55 mHz and amplitude $T_\sim$ = 1K using a Peltier element. The pyroelectric current signal was amplified with an electrometer and the 90° out of phase component of the current with respect to the temperature modulation was measured with a lock-in amplifier.

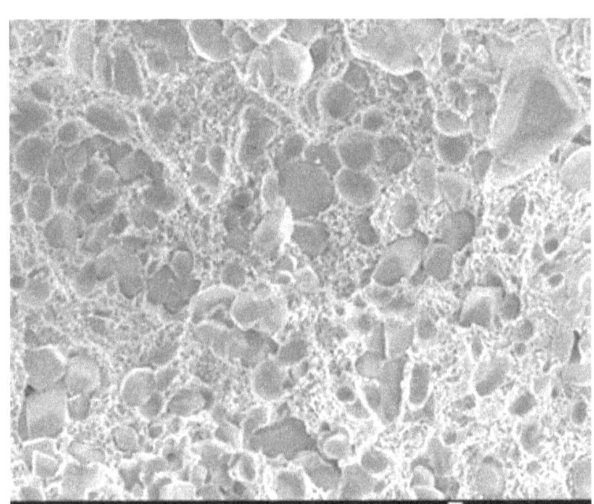

Figure 1 SEM micrograph of the fracture surface for TGS/P(VDF-TrFE) 0-3 composite with $\phi = 0.43$.

3. Results and discussion

3.1. Microstructures

The cross-sectional micrograph of the composite with $\phi = 0.43$ is shown in Fig. 1. It is seen that the TGS particles are dispersed rather uniformly in the copolymer matrix.

3.2. Phase transitions

Fig. 2 presents the DSC thermograms of the 0-3 composites during the heating and cooling processes and for comparison, the DSC curves of the pure copolymer and TGS powder are also plotted. Both the Curie temperature

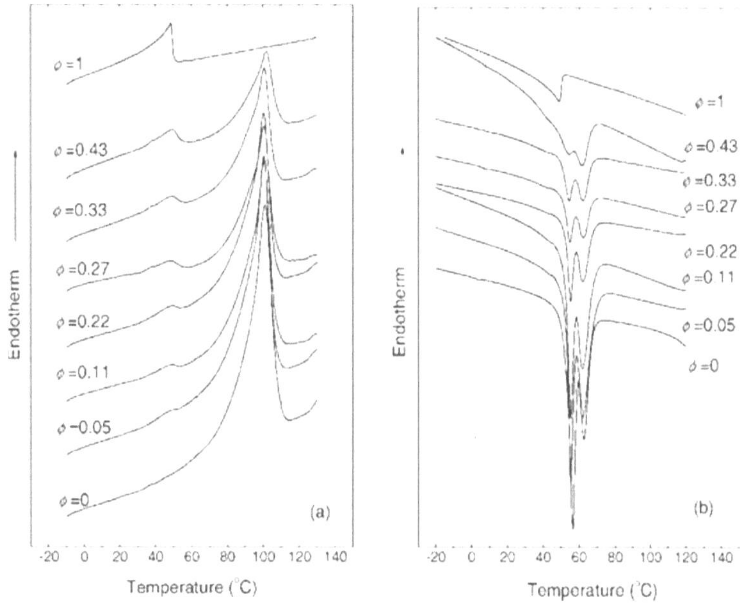

Figure 2 DSC thermograms of TGS/P(VDF-TrFE) 0-3 composites during the (a) heating and (b) cooling process.

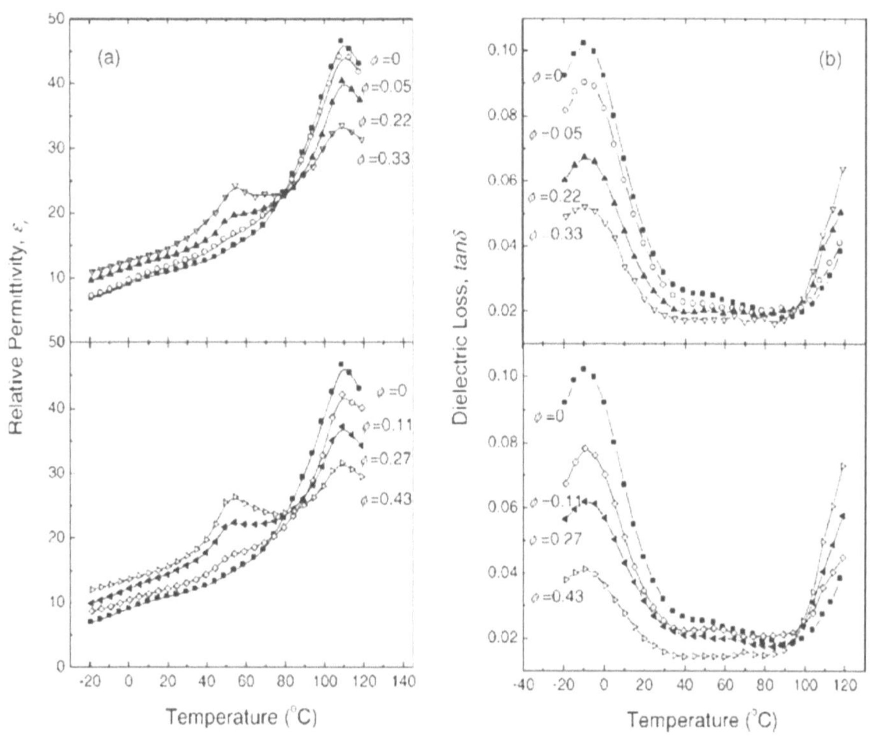

Figure 3 (a) Relative permittivity and (b) dielectric loss factor as a function of temperature for the unpoled TGS/P(VDF-TrFE) 0-3 composites measured at 10 kHz.

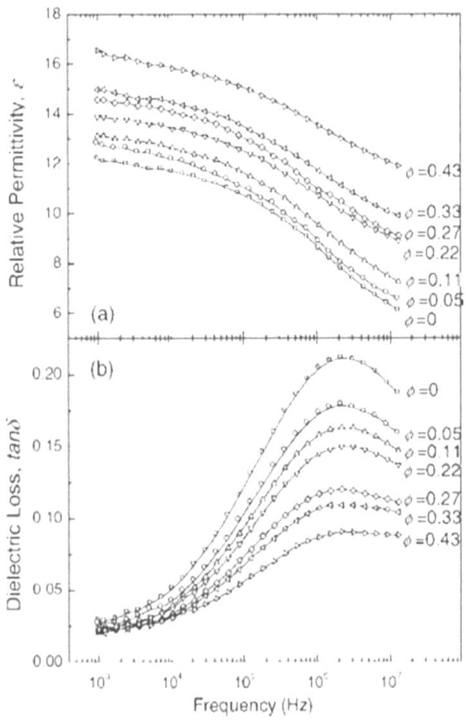

Figure 4 (a) Relative permittivity and (b) dielectric loss factor as a function of frequency for the unpoled TGS/P(VDF-TrFE) 0-3 composites measured at room temperature.

of TGS ($\phi = 1$) at 49°C and that of P(VDF-TrFE) ($\phi = 0$) at about 105°C (during heating) and 60°C (during cooling) appear in the DSC thermograms and the intensity of the endothermic/exothermic peaks follow the change in ϕ.

3.3. Dielectric permittivity

The temperature dependence of the relative permittivity ε_r and dielectric loss factor tan δ of the unpoled 0-3 composites measured at 10 kHz are presented in Fig. 3. The ferroelectric-paraelectric phase transition of TGS (at about 50°C) and P(VDF-TrFE) (at about 105°C) are both observed in the ε_r vs. temperature curves of the composites. The Curie point anomaly of TGS becomes more and more obvious as the TGS volume fraction increases. At the same time, the peak intensity corresponding to the phase transition of P(VDF-TrFE) copolymer decreases with increasing ϕ. In the tan δ vs temperature curves, a non-crystalline β relaxation of P(VDF-TrFE) is observed at about –10°C and the magnitude of the peak decreases gradually with increasing ϕ. In addition, there is a sharp rise of tan δ above 90°C in all samples presumably due to the Curie transition of the copolymer.

Fig. 4 shows the relative permittivity ε_r and dielectric loss factor tan δ as a function of frequency for the unpoled composites at room temperature. It is found that the dielectric behaviour of the 0-3 composites is dominated by the copolymer ($\phi = 0$). In the composites, the TGS phase possesses higher relative permittivity (the literature value of unpoled TGS single crystal along the b-axis [10] is $\varepsilon_r \sim$

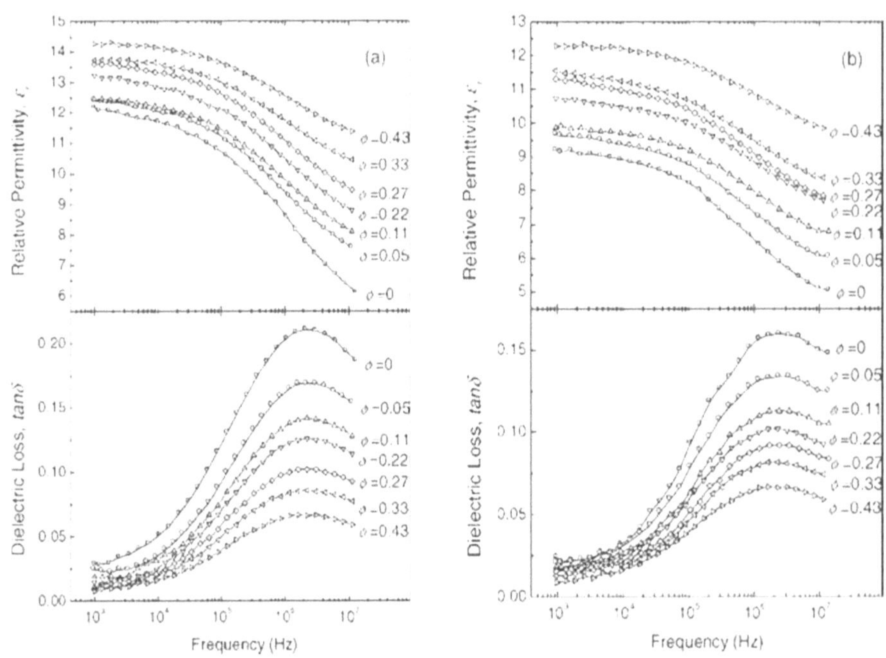

Figure 5 Relative permittivity and dielectric loss factor measured at room temperature as a function of frequency for (a) the composites with only the TGS phase poled and (b) with both phases poled.

38 at room temperature) than the P(VDF-TrFE) phase (ε_r $\sim$ 12 for unpoled P(VDF-TrFE)), therefore ε_r increases with increasing ϕ. In the dielectric loss, the peak related to the β relaxation (at 2 MHz) of the copolymer is suppressed as the volume fraction of TGS increases. Fig. 5 shows the dielectric properties of the 0-3 composites as a function of frequency with only the TGS phase poled and with both phases poled. The variation of ε_r and tan δ in the poled composites with ϕ is similar to that of the unpoled samples (Fig. 4). The values of ε_r and tan δ for the samples with both phases poled are lower than those for the samples with only TGS poled, which are in turn lower than those of the unpoled samples due to the dipole alignment.

The relative permittivity of the composites can be understood in terms of the Bruggeman model [11]:

$$\left(\frac{\varepsilon_m}{\varepsilon}\right)^{1/3} = (1-\phi)\cdot\frac{\varepsilon_i - \varepsilon_m}{\varepsilon_i - \varepsilon} \quad (2)$$

where ε, ε_i and ε_m represent the relative permittivity of the composite, the inclusion phase and the matrix phase, respectively. It should be noted that the relative permittivity of TGS powder is unknown and therefore the data for the copolymer and composites were least square fitted to the Bruggeman model. Fig. 6 shows the theoretical curves and experimental data for the unpoled samples, samples with only the TGS phase poled and samples with both phases poled. It can be seen that the theoretical values agree well with the experimental results with $\varepsilon_i \sim 20$ and $\varepsilon_i \sim 15$ for the unpoled and poled TGS powder, respectively. It is noted that $\varepsilon_r = 18.6$ as measured in a poled TGS single crystal supplied by the Institute of Crystal Materials, Shandong University, China.

3.4. Piezoelectric properties

Fig. 7 shows the piezoelectric coefficient d_{33} of the composites with only the TGS phase poled and the composites with both phases poled measured at room temperature. For the composites with only the TGS phase poled, the value of d_{33} increases from 0 to 15 pC/N when ϕ increases from 0 to 0.43. It should be noted that piezoelectric coefficients of P(VDF-TrFE) and TGS have opposite signs. Therefore, for the composites with both phases poled, d_{33} have neg-

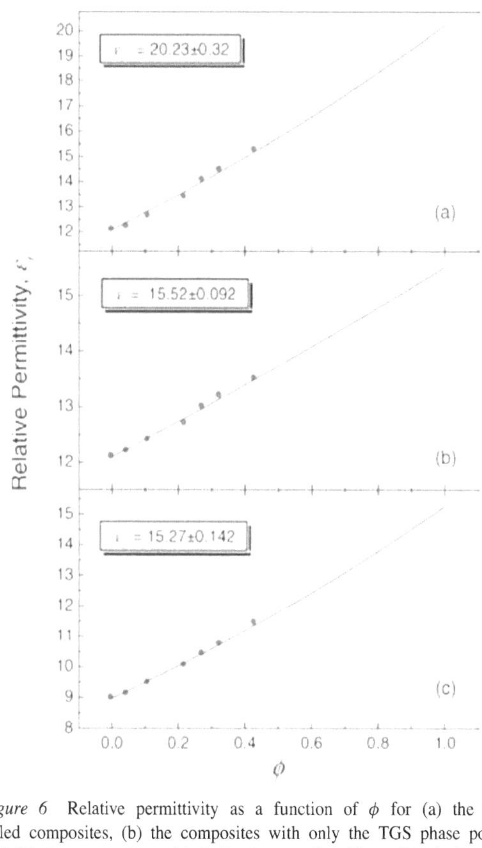

Figure 6 Relative permittivity as a function of ϕ for (a) the unpoled composites, (b) the composites with only the TGS phase poled and (c) the composites with both phases poled. The solid circles and lines represent the experimental data and Bruggeman model predictions, respectively.

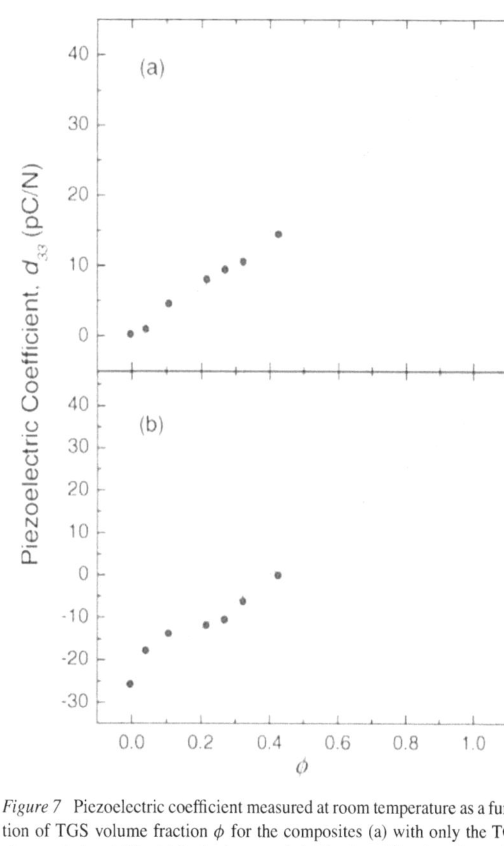

Figure 7 Piezoelectric coefficient measured at room temperature as a function of TGS volume fraction ϕ for the composites (a) with only the TGS phase poled and (b) with both phases poled. The dotted line is only a guide to the eye.

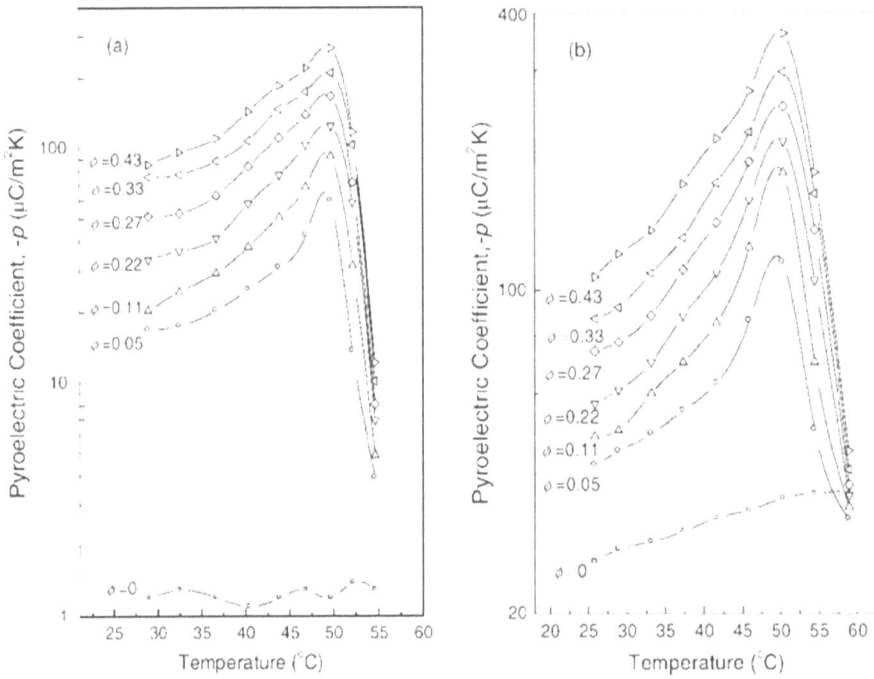

Figure 8 Pyroelectric coefficient as a function of temperature for the composites (a) with only the TGS phase poled and (b) with both phases poled.

ative values at low volume fraction of TGS mainly due to the contribution of the copolymer and as ϕ increases, d_{33} is reduced in magnitude and becomes nearly zero when $\phi = 0.43$. This $\phi = 0.43$ composite has the unique property that it is pyroelectric but non-piezoelectric. Vibration-induced noise can be minimized if this composite is used in fabricating pyroelectric sensors [12].

3.5. Pyroelectric properties

Fig. 8 shows the pyroelectric coefficients p of the 0-3 composites with only the TGS phase poled (Fig. 8a) and with both phases poled (Fig. 8b) as a function of temperature from room temperature to 60°C (above T_c of TGS). For comparison, the result of pure copolymer is also plotted in Fig. 8. The pyroelectric coefficient of the composites increases with increasing volume fraction of TGS. At any volume fraction of TGS, the pyroelectric coefficient of the composites increases with increasing temperature and reaches a maximum at about 50°C. Above the Curie transition of TGS, the magnitude of p decreases drastically because of the depolarization of TGS. The pyroelectric coefficient for the composites with only the TGS phase poled and the composites with both phases poled show similar temperature dependences and the magnitude of p for the samples with both phases poled is enhanced due to the contribution of the poled copolymer phase.

Using the effective-medium approach [6] (Equation 1), the effective pyroelectric coefficient p of the composites can be calculated. Fig. 9 shows the experimental results of the pyroelectric coefficients for composites with only the TGS phase poled and composites with two phases poled as a function of ϕ measured at room temperature. The theoretical curves are obtained by least square fitting of the data for the copolymer and composite samples to Equation 1. In the calculation, ε_m and p_m are measured values and ε and ε_i have been obtained using the Bruggeman model. It is seen that the experimental data and the theoretical calculation are in good agreement. The pyroelectric coefficient of the TGS powder obtained by the curve fitting is ~210 μC/m²K which is lower than that measured using the poled TGS single crystal ($p =$ 251 μC/m²K).

There are several figures of merit (FOM) which describe the contribution of the physical properties of a material to the performance of a pyroelectric sensor. The current FOM F_i, voltage FOM F_V and detectivity FOM F_D are given by:

$$F_i = \frac{p}{c}; \quad F_V = \frac{p}{c\varepsilon_r\varepsilon_o}; \quad F_D = \frac{p}{c(\varepsilon_r\varepsilon_o \tan\delta)^{1/2}} \qquad (3)$$

where c is the heat capacity per unit volume (calculated from DSC thermograms). The figures of merit for composites with TGS phase poled and composites with two phases poled are calculated using the measured p, c, ε_r (at 1 kHz) and tan δ (at 1 kHz) at room temperature. The results are plotted in Fig. 10a (with only the

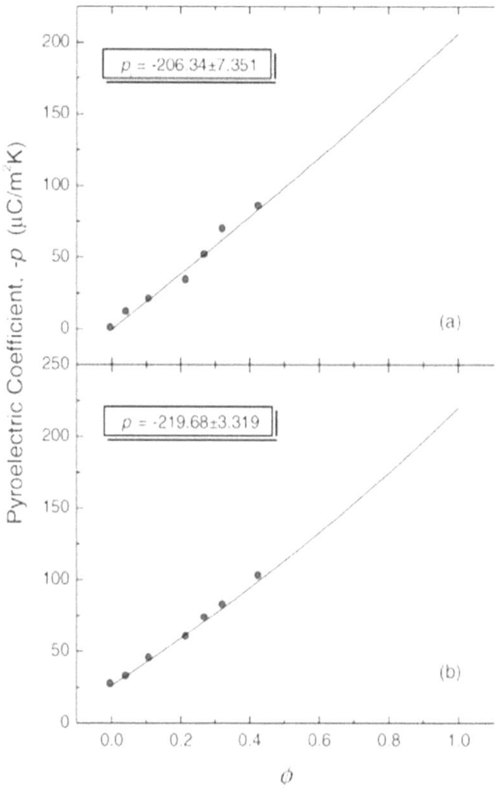

Figure 9 Pyroelectric coefficient as a function of ϕ for the composites (a) with only the TGS phase poled and (b) with both phases poled. The solid circles and lines represent the experimental data and effective medium model predictions, respectively.

TGS phase poled) and Fig. 10b (both phases poled). Compared to the values for the copolymer, all three FOM are improved in the composites, and the FOM values increase steadily with increasing ϕ. Tables I and II summarize the properties of the TGS/P(VDF-TrFE) 0-3 composites with only the TGS phase poled and the composites with both phases poled, respectively.

4. Conclusions

TGS/P(VDF-TrFE) 0-3 composites with various TGS volume fractions were fabricated by compression moulding. The composites were characterized by SEM and DSC. Three series of samples were prepared: unpoled composites, composites with only the TGS phase poled and composites with both phases poled. The dielectric, piezoelectric and pyroelectric properties of these samples were investigated. The relative permittivity and pyroelectric coefficient of the composites were found to increase with increasing TGS content. The relative permittivity and pyroelectric coefficient were compared with the theoretical predictions of the Bruggeman model and effective medium model, respectively, and good agreements have been obtained. It was also found that if the two phases were poled in the same direction, the pyroelectric contributions from TGS and P(VDF-TrFE) reinforced while piezoelectric contributions partially cancelled. Compared to the pure copolymer, the composites exhibited high pyroelectric coefficients and high pyroelectric figures of merit together with lower piezoelectric coefficients. The low piezoelectric activity is a significant advantage in pyroelectric sensor application since it will give rise to a low

TABLE I Properties of TGS/P(VDF-TrFE) 0-3 composites with the TGS phase poled

ϕ	c (MJ/m^3·K)	ε_r (1 kHz)	tan δ (1 kHZ)	d_{33} (pC/N)	$-p$ (μC/m^{2}K)	F_i (10^{-12}m/V)	F_V (m^2/C)	F_D (10^{-6}Pa$^{-1/2}$)
0	2.09	12.20	0.029	–	–	–	–	–
0.05	2.08	12.44	0.025	0.7	11.2	7.69	0.07	4.66
0.11	2.10	12.46	0.019	4.3	20.3	9.52	0.09	6.63
0.22	2.10	13.24	0.016	7.8	33.4	15.9	0.14	11.8
0.27	2.12	13.60	0.015	9.2	51.2	24.1	0.20	18.0
0.33	2.13	13.74	0.012	10.3	69.4	32.4	0.27	26.7
0.43	2.14	14.26	0.009	14.2	85.2	39.7	0.31	38.3
1[a]	2.18	18.6	0.004	40	251	115	0.70	142
1[b]	2.3	38	0.01	–	280	122	0.36	66

[a]Data obtained from the measurements of a poled TGS single crystal supplied by the Institute of Crystal Materials, Shandong University, China.
[b]From reference [10].

TABLE II Properties of TGS/P(VDF-TrFE) 0-3 composites with both phases poled

ϕ	c (MJ/m^3·K)	ε_r (1 kHz)	tan δ (1 kHZ)	d_{33} (pC/N)	$-p$ (μC/m^{2}K)	F_i (10^{-12}m/V)	F_V (m^2/C)	F_D (10^{-6}Pa$^{-1/2}$)
0	2.09	9.20	0.024	−26.2	26.4	12.6	0.16	8.96
0.05	2.08	9.66	0.021	−18.2	32.3	15.5	0.18	11.6
0.11	2.10	9.87	0.018	−14.2	44.5	21.2	0.24	16.9
0.22	2.10	10.73	0.016	−12.3	59.5	28.3	0.30	22.8
0.27	2.12	11.31	0.011	−11.0	73.0	34.4	0.34	32.5
0.33	2.13	11.55	0.010	−6.7	82.6	38.8	0.38	37.3
0.43	2.14	12.27	0.008	−0.5	102	47.7	0.44	50.0

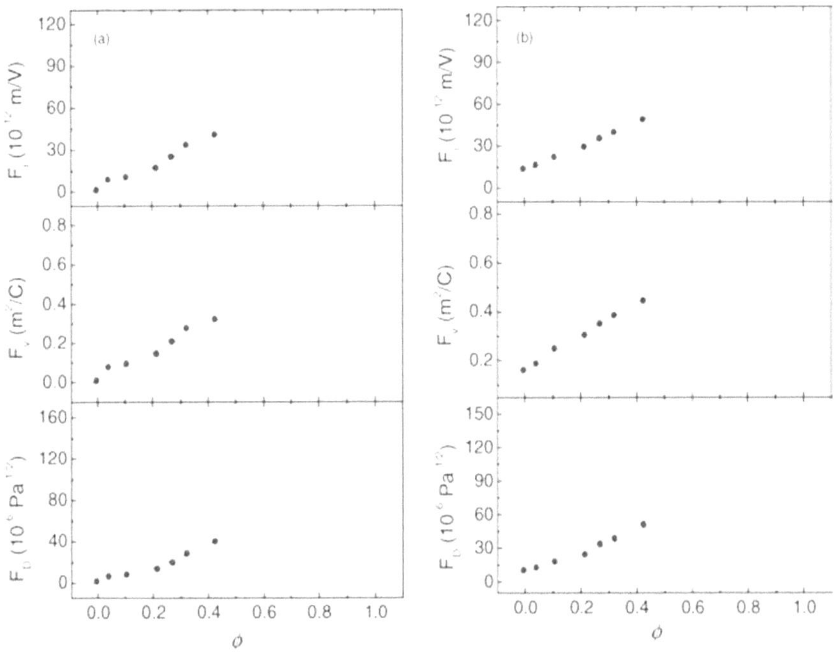

Figure 10 Room-temperature figures of merit F_i, F_V and F_D as a function of ϕ for the composites (a) with only the TGS phase poled and (b) with both phases poled.

vibration induced noise. Therefore, TGS/P(VDF-TrFE) 0-3 composite is a good candidate to be used as the sensing element for pyroelectric devices.

Acknowledgement

This work was supported by the Center for Smart Materials of the Hong Kong Polytechnic University.

References

1. C. DIAS and D. K. DAS-GUPTA, *IEEE Trans. Dielectr. Electr. Insul.* **3** (1996) 706.
2. T. YAMADA, T. UEDA and T. KITAYAMA, *J. Appl. Phys.* **53** (1982) 4328.
3. H. L. W. CHAN, W. K. CHAN, Y. ZHANG and C. L. CHOY, *IEEE Trans. Dielectr. Electr. Insul.* **5** (1998) 505.
4. H. L. W. CHAN, P. K. L. NG and C. L. CHOY, *Appl. Phys. Lett.* **74** (1999) 3029.
5. M. J. ABDULLAH and D. K. DAS-GUPTA, *IEEE Trans. Dielectr. Electr. Insul.* **25** (1990) 605.
6. K. H. CHEW, F. G. SHIN, B. PLOSS, H. L. W. CHAN and C. L. CHOY, *J. Appl. Phys.* **94** (2003) 1134.
7. Y. G. WANG, W. L. ZHONG and P. L. ZHANG, *ibid.* **74**, (1993) 521.
8. C. FANG, M. WANG and H. ZHOU, *Proc. 7 Intl. Sypm. Electrets (ISE 7)* (1991) 507.
9. B. PLOSS, B. PLOSS, F. G. SHIN, H. L. W. CHAN and C. L. CHOY, *Appl. Phys. Lett.* **76** (2000) 2776.
10. A. J. MOULSON and J. M. HERBERT, *Electroceramics: Materials, Properties, Applications* (John Wiley & Sons Press, 2003) p. 420.
11. D. A. G. BRUGGEMAN, *Ann. Phys. Lpz.* **24** (1935) 635.
12. B. PLOSS, W. M. FUNG, H. L.W. CHAN and C. L. CHOY, *Ferroelectrics* **263** (2001) 229

JOURNAL OF MATERIALS SCIENCE **41** (2006) 259–270

Near-field acoustic and piezoresponse microscopy of domain structures in ferroelectric material

Q. R. YIN*, H. R. ZENG, H. F. YU, G. R. LI
State Key Lab of High Performance Ceramics and Superfine Microstructure, Shanghai Institute of Ceramics, Chinese Academy of Sciences, 1295 Dingxi Road, Shanghai 200050, China
E-mail: qryin@sunm.shcnc.ac.cn

Three kinds of near-field microscopy imaging mode including SEAM (Scanning electron acoustic microscopy), PFM (Piezoresponse force microscopy) and SPAM (Scanning probe acoustic microscopy) have been developed to investigate domain structures of ferroelectric ceramics, crystals and thin films in our studies. The domain imaging mechanisms are presented individually in three imaging modes. Sub-surface micro-domain configuration of ferroelectric $BaTiO_3$ ceramics and single crystal and their dynamic behavior under external fields were clearly visualized by SEAM. Ferroelectric domain structures of ferroelectric PZT thin film and PMN-PT single crystal were characterized by PFM. Nanoscale switching behavior and local field-induced nanoscale displacement behavior of domain structures in ferroelectric thin film were studied by PFM. Antiparallel domain patterns in ferroelectric transparent PLZT ceramics were also characterized by SPAM. The combination of SEAM, PFM and SPAM in application to imaging domain structures undoubtedly enrich our understanding of the nature of piezoelectricity and ferroelectricity at submicro-, even nano-meter scale.

1. Introduction

Piezoelectric and ferroelectric materials have been widely used in microelectronics and micro mechanical application fields due to their piezoelectric, ferroelectric, pyroelectric effects, and so on [1, 2]. These properties and application background are closely related to the domain structure of ferroelectrics and their dynamic behavior under different conditions. Thus *in-situ* imaging and characterization of domain structure with high resolution are very important and imperative to understand local ferroelectric phenomena in ferroelectric materials. In the present study, three kinds of near-field microscopy imaging mode including SEAM (Scanning electron acoustic microscopy), PFM (Piezoresponse force microscopy) and SPAM (Scanning probe acoustic microscopy) have been implemented based on the commercial SEM and AFM, and used to perform studies of domain structures and their dynamic behavior under external fields.

2. Experimental

2.1. Scanning electron acoustic microscopy

Scanning electron acoustic microscopy (SEAM) is a technique which can produce images that show variations in an object's thermal and elastic properties with a resolution on the order of microns [3–7]. The SEAM system was shown in Fig. 1. SEAM was installed on a conventional SEM KYKY-1000B by attaching an electron beam chopping system consisting of a pair of deflection plates, an electron beam blanking power, a square-wave signal generator, a PZT and a signal preamplifier EG&G Model 5316A with a lock-in amplifier Model 5302. A computer control system was also established to allow the control of the scanning parameter, image capturing and image processing. The electron acoustic image (EAI) and secondary electron image (SEI) can be obtained simultaneously. The physical processes involved in electron acoustic microscopy are shown in Fig. 2. A periodically

*Author to whom all correspondence should be addressed.

0022-2461 © *2006 Springer Science + Business Media, Inc.*
DOI: 10.1007/s10853-005-7244-2

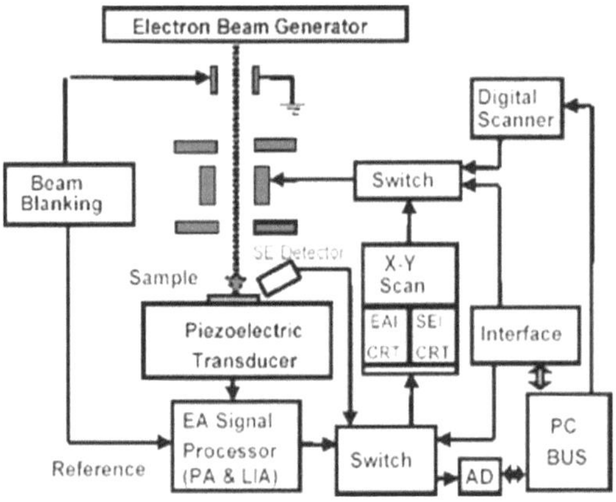

Figure 1 The schematic of scanning acoustic electron microscope.

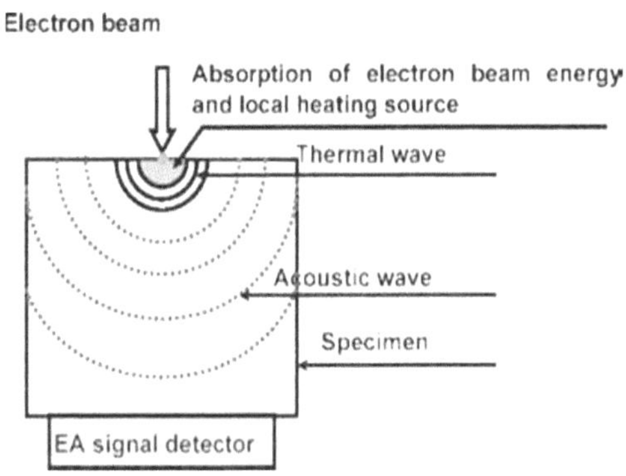

Figure 2 A diagram for physical principle of SEAM.

chopped electron beam serves as a local probe on materials being examined, causing periodic heating and local thermal expansion in the dissipation volume. The periodic expansion and contraction produce acoustic waves that propagate through the sample and are detected by a piezoelectric transducer coupled to the sample. Spatial variations or discontinuities in any of these thermal and elastic properties produce contrast in the EAI.

2.2. Piezoresponse force microscopy

Piezoresponse force microscopy (PFM) has now been used as a powerful tool for imaging and characterization of nanoscale domain structures in ferroelectrics [8–12]. PFM in our studies was developed based on the commercially available atomic force microscope instrument (Seiko SPA 300/SPI3800). The working principle for PFM in imaging domain structures are shown in Fig. 3, where an ac-voltage $V_{ac} = V_0 \sin \omega t$ is applied between the conductive tip and the bottom electrode of the sample. Due to the converse piezoelectric effect, the alternating external electric field gives rise to the piezoelectric vibration of the sample, or a change in the sample's local thickness. The sign of piezoelectric coefficient depends on the polarization direction, and thus the regions with opposite polarization states will vibrate out of phase upon the action of the ac field. The amplitude and phase of the vibration signal provide information about the magnitude of the piezoelectric coefficient and the direction of local polarization, respectively. Therefore, regions with opposite polarization directions, which vibrate in counter phase with respect to each other under the applied ac filed, will appear as regions of different contrasts in the piezoresponse image.

2.3. Scanning probe acoustic microscopy

The experimental set up of the scanning probe acoustic microscopy (SPAM) is shown in Fig. 4. The SPAM is a modified commercial atomic force microscope [13]. An ac voltage is applied between the bottom electrode of the piezoelectric sample and the conducting tip. The tip acts as a local excitation source of acoustic waves due to the converse piezoelectric effects. The induced acoustic

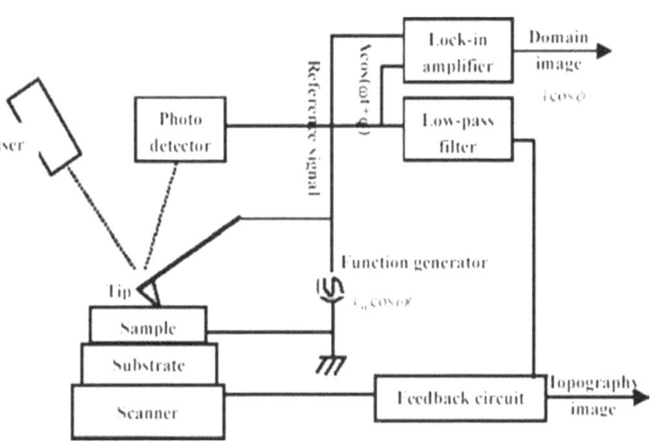

Figure 3 The schematic of piezoresponse force microscope.

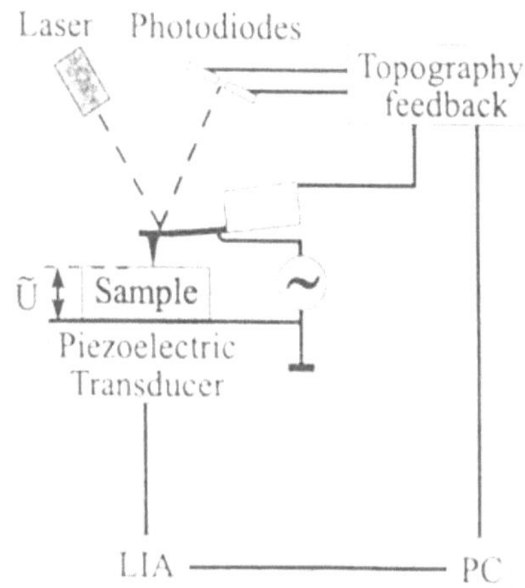

Figure 4 The set up of scanning probe acoustic microscope.

waves are detected by a piezoelectric transducer that is in intimate contact with the bottom surface of the sample. A PC unit is installed to *in-situ* process samples topography and acoustic imaging. In this work, only longitudinal acoustic waves were detected.

3. Results and discussion

3.1. Scanning electron acoustic microscopy of ferroelectric domain

3.1.1. Electron acoustic image of sub-surface micro-domain configuration in $BaTiO_3$ ceramics

Fig. 5a is the SEI of surface topography of $BaTiO_3$ ferroelectric ceramics without any pretreatment. The SEI only shows the grain and grain boundaries of $BaTiO_3$ ceramics. Fig. 5b–d are the corresponding EAI of the same area at different modulation frequencies. The electron acoustic images exhibit 90° domain configurations with different orientation occurring in the individual $BaTiO_3$ grains, as symbol "S" indicated in Fig. 5b. Some domain structures cross in the grain boundaries of the neighboring grains, as symbol "Q" showns in Fig. 5b. The domain-crossing grain boundary phenomena are also clearly seen in Fig. 5e and f. Based on the SEAM imaging principle [14], the EAIs at different modulation frequencies reflect microstructure information at different depth profile away from the sample surface. As a result, the square and circle symbol in Fig. 5b–d represent sub-surface domain configuration in $BaTiO_3$ ceramics. So SEAM provides a nondestructive

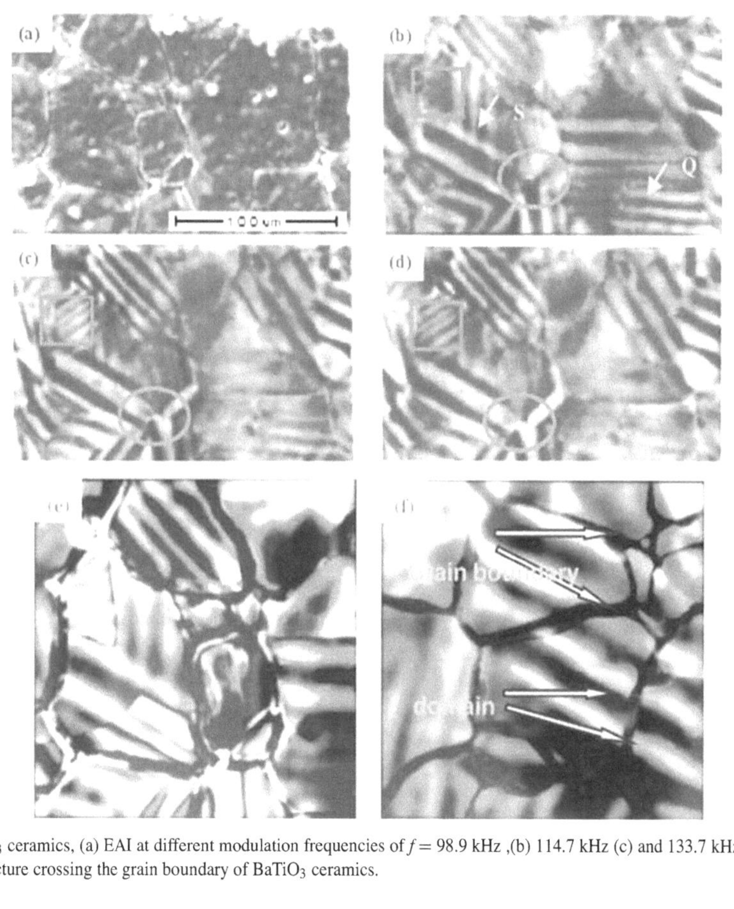

Figure 5 SEI of $BaTiO_3$ ceramics, (a) EAI at different modulation frequencies of $f = 98.9$ kHz ,(b) 114.7 kHz (c) and 133.7 kHz, (d) respectively (e) and (f) showing domain structure crossing the grain boundary of $BaTiO_3$ ceramics.

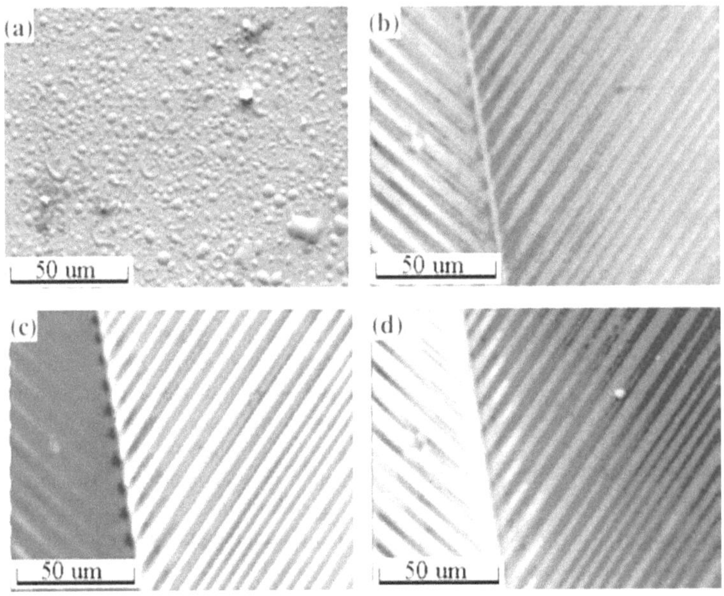

Figure 6 Modulation frequency dependence of domain structure in $BaTiO_3$ single crystal. (a) SEI, (b) EAI at 79.7 kHz, (c) EAI at 80.6 kHz, and (d) EAI at 132.7 kHz.

tool to image subsurface domain structures in ferroelectric materials through modulating imaging frequency.

3.1.2. Modulation frequency dependence of ferroelectric domain imaging contrasts

Fig. 6a is the SEI of the surface morphology of $BaTiO_3$ single crystals. Fig. 6b–d are the corresponding EAI of $BaTiO_3$ single crystals at modulation frequencies of 79.7, 80.6, and 132.7 kHz, respectively. The electron acoustic image contrasts of domain structures in Fig. 6b–d reveal an interesting frequency dependence of domain contrasts. In Fig. 6b, both domain groups contribute nearly the same degree of domain contrasts. But in Fig. 6c, strong and sharp contrasts appear in the domain group at the right side, and the left-side domain contrasts are suppressed. When the modulation frequency is increased to 132 kHz, the domain contrasts at right side of Fig. 6d are suppressed, while the left-side domain contrasts show very strong and sharp imaging contrasts. From the above results, we proposed that frequency characteristics of EAI of ferroelectric domain might be related to the frequency resonant properties of domain structures.

3.1.3. Domain structures dynamic behavior under electric fields and temperature fields

The domain evolution of 0.65PMN-0.35PT single crystal under external electric fields are shown in Fig. 7 [14]. Fig. 7a is the initial domain configuration when external field $E = 0$. There are two groups of domains with different orientations in the figure. The domains shown in Fig. 7b–d evolve gradually with increasing electric field E perpendicular to the sample surface. All domains coalesce together when electrical field reaches 7.2 kV/cm. It approaches to a single domain state. Actually some micro-domains distributing in the macrodomains might be observed if enhancing SEAM's resolution. The results indicate that the proper poling conditions can be designed for exploring performance of PMN-PT crystal.

Fig. 8a is an EAI that shows the macroscopic ferroelectric domains present in a sample at room temperature prior to annealing [15]. The macrodomains have the well-defined lamellar morphology of tetragonal 90° domains. Fig. 8b demonstrates that the tetragonal 90° macrodomain structure still exists after the annealing process. Geanerally speaking, no ferroelastic domain structure was observed in the sample prior to annealing, but this was not the case after annealing. Many complex domain structures were found on the surface of the annealed sample, as illustrated in Fig. 8c and d. The twofold symmetrical butterfly- shaped domains with alternate black and white contrasts are clearly revealed in the EAIs Fig. 8c and b. The butterfly-shaped domains pattern was supposed to be due to ferroelastic domains induced by dislocations, and might regarded as ferroelastic domains induced by the presence of dislocations in the crystal formed during the slow cool step of the cooled annealing process.

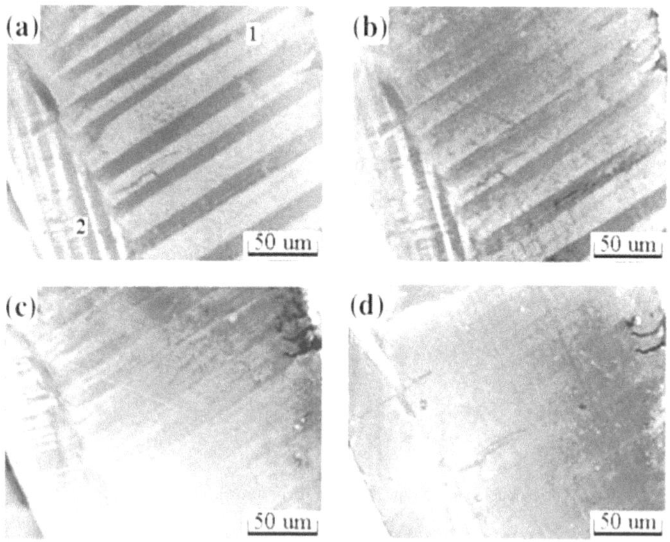

Figure 7 Domain evolution on 0.65 PMN-0.35PT crystal in poling process. (a) EAI at 133.3 kHz, $E = 0$; (b) 133.1 kHz, $E = 4.5$ kV/cm; (c) 133.4 kHz, $E = 5.4$ kV/cm; and (d) 133.4 kHz, $E = 7.2$ kV/cm.

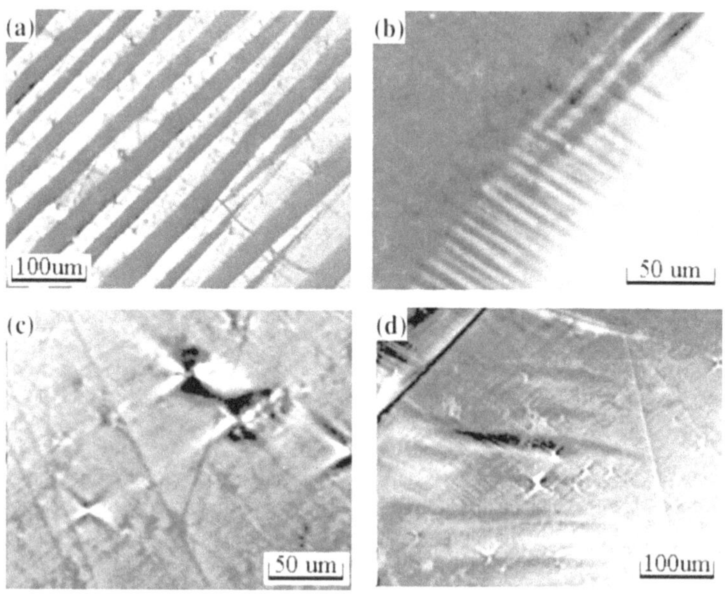

Figure 8 The comparison of ferroelectric domain structures of 0.65 PMN-0.35PT single crystal before (a) and after annealing (b) at 154.8 kHz; (c) and (d) are EAI image of the butterfly shaped domain structure in the annealed 0.65 PMN-0.35 PT single crystals at 154.8 and 153.2 kHz, respectively.

3.2. Piezoresponse force microscopy of ferroelectric domain

3.2.1. Piezoresponse image of domain structures in ferroelectric PZT thin film

Fig. 9a shows a topographic image of as-deposited $Pb_{0.40}Zr_{0.60}TiO_3$ thin film with 300 nm thickness, which reveals a polycrystalline structure with ~200 nm sized grains. A corresponding piezoresponse image is shown in Fig. 9b, where the areas appearing as bright and dark contrasts were undoubtedly 180° domain with antiparallel polarization direction, as indicated in Grain 1 and 2 of Fig. 9, which were usually visualized in many PFM studies of ferroelectric thin films. But interestingly, 90° domain configurations with 80–100 nm in width presenting stripe contrasts were clearly shown in Fig. 9b, as shown in Grain 3 of Fig. 9. The similar phenomena were remarkably

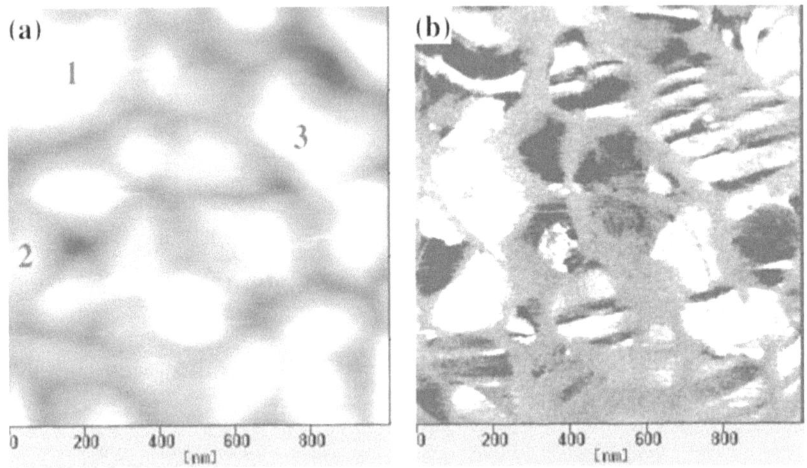

Figure 9 The topography image (a) and the corresponding piezoresponse image (b) of ferroelectric PZT thin film.

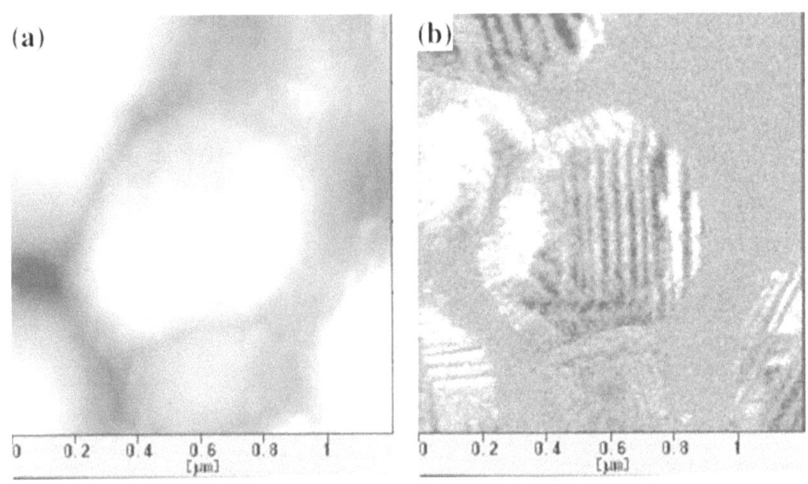

Figure 10 The topography image (a) and the corresponding piezoresponse image (b) of ferroelectric PZT thin film.

demonstrated in Fig. 10. 90° domain structures as small as 30 nm in size were clearly observed in the individual grains of Fig. 10b. Noted that the grains dominated with 90° domain walls are separated from each other in Fig. 10, so grain-calmping effect can be excluded, and nanoscale 90° domain configuration are induced by the periodic lattice constant mismatch between the substrate and the film. It is believed that the local mechanical stress is the predominant cause of the extensive equal-width domain configuration, which is compatible with the periodic stress inside the grains in the film.

3.2.2. Piezoresponse image of domain structures in PMN-PT single crystals

Fig. 11a and b show the topography and domain images of the investigated (001)-oriented PMN-30%PT single crystals. The topography image of Fig. 11a only represents the surface image of the sample without any additional information. The piezoresponse image contrast in Fig. 11b is determined by the out-of-plane component of polarization, where black and white areas correspond to the opposite polarization directions. Irregular, fingerprint-like domain patterns with average size of 200 nm were clearly observed in Fig. 11b. The remarkable contrasted domain regions are not uniform but a mixture of small domain regions with opposite contrasts. The irregular domain patterns are not the unique features of the domain structures visualized in the sample. Another type of regular domain structure was surprisingly imaged in other scanning areas of the same PMN-30%PT single crystal, as shown in Fig. 11c and d. Fig. 11c shows the regular, narrow strip-like domain patterns as compared with irregular those in Fig. 11b, while narrow strip-like domain

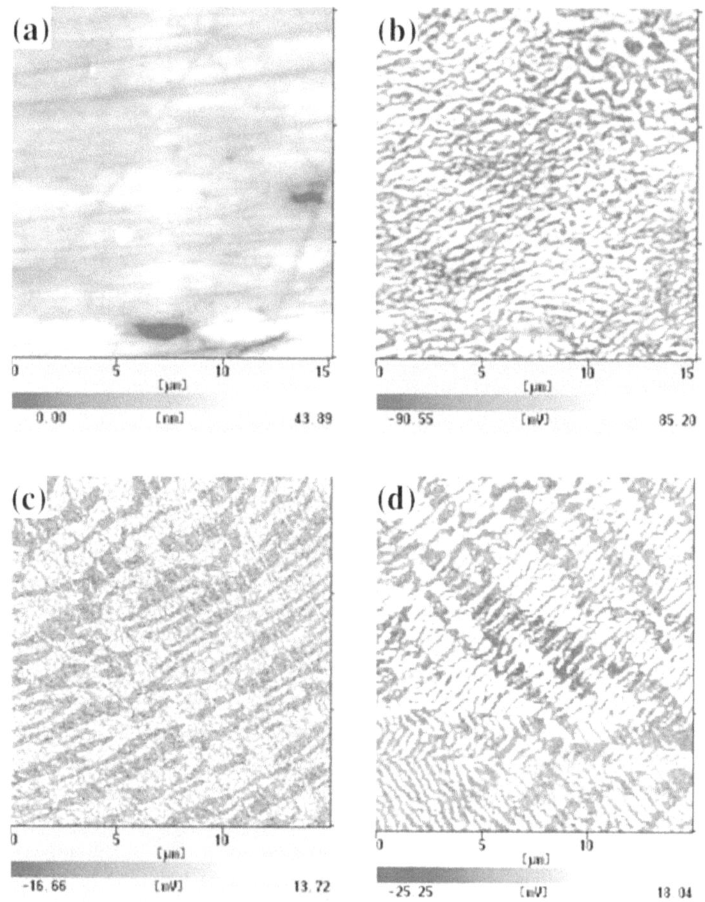

Figure 11 The spatially inhomogeneity of domain structure in PMN-PT single crystal. (a) the topography image and (b) the corresponding piezoresponse image of 15×15 μm^2 scanning areas. (c) and (d) are the piezoresponse image in different scanning areas of PMN-PT single crystal.

pattern and finger print domain configuration appear simultaneously in Fig. 11d. The spatially inhomogeneity of domain structure in PMN-PT single crystals reflect that random fields from nanoscale structure irregularity affect greatly the domain arrangement of PMN-PT single crystals [16].

Fig. 12a is the piezoresponse image of domain configurations in (110)-oriented PMN-PT single crystal. The fingerprint domain patterns and the tweed-like domain patterns are simultaneoustly visualized in the piezoresponse image. The former is the antiparalle, non-ferroelastic domain, and the latter ferroelastic domain structures. A question arises is that what is the relationship between them. Fig. 12b shows that the gradual transition regions between non-ferroelastic domain (fingerprint pattern) and ferroelastic domain (tweed pattern) appeaing in PMN-PT single crystal, which reveals that dynamic transformation behavior from non-ferroelastic to ferroelastic domain existing in the crystal due to the local stress during cooling process from high temperatures [17].

3.2.3. Nanoscale domain switching behavior in PZT thin film

The polarization reversal behavior of the PZT 40/60 thin film at the nanometer scale was studied with SFM in order to gain a better understanding domain switching mechanism in ferroelectric thin films. The 5×5 μm^2 area was successively scanned with the tip bias of -1.5 V and $+1.5$ V. Immediately after each application of the bias, the SFM piezoresponse mode was used to image the area to obtain simultaneously the topographic image and the corresponding piezoresponse image. Fig. 13a is the topographic image of the PZT 40/60 thin film, showing a dense polycrystalline structure with an average lateral grain size of $\sim$180 nm. The corresponding piezoresponse image after application of -1.5 V and $+1.5$ V are shown in Fig. 13b and c, respectively. The most significant differences in comparison of Fig. 13b with c are the formation of step structure as arrows indicated in Fig. 13c, but not in Fig. 13b. In addition, the step width is found nearly equal in a grain, and the smallest step width is $\sim$30 nm, The presence

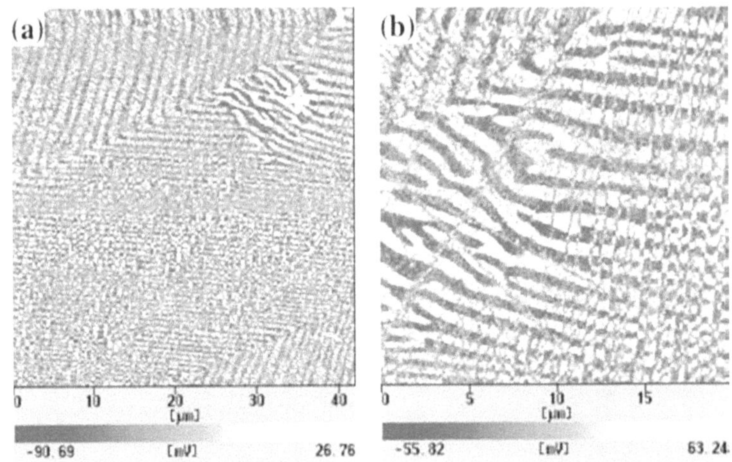

Figure 12 (a) The piezoresponse image of domain structure in PMN-PT single crystal, (b) is the zoomed piezoresponse image with scanning central point located in "+" position marked in Fig. 12a.

of the step domain structure formed during polarization reversal indicates that the domains grow sideways faster than forward in which each of the step doesn't broaden. Our present results substantiate the validity of Hong *et al.*'s claim, i.e. the forward domain growth mechanism being the rate limiting mechanism prevails in our (111) oriented PZT thin films. The presence of the step structure observed by the SFM piezoresponse mode is the reflection of real physical process of nanoscale domain switching [18].

3.2.4. Nanoscale displacement behavior of ferroelectric domains

Fig. 14a and b show the topography image and the corresponding piezoresponse image of ferroelectric PZT thin film with graded composition. The field-induced nanoscale displacement behavior of ferroelectric domain was carried out in points A, B, C and D of Fig. 14b. The related results are shown in Fig. 15a–d. Fig. 15a and b clearly exhibit nanoscale displacement-bias butterfly loops of ferroelectric domain under tips, especially typical in Fig. 15a. In contrast, asymmetry butterfly loops were clearly observed in Fig. 15c and d indicates parabola-like loops without any butterfly features.

Based on phenomenological field-induced displacement equation and the displacement-field behavior shown in Fig. 14, it can be deduced that Fig. 15d reflects the square dependence of displacement with electric field due to electrostatic effects. While the hysteresis loop in Fig. 15c implies a combination of piezoelectric and electrostatic effects. For the butterfly displacement-field loops in Fig. 15a and b, we postulate that their origins are from combined contribution of linear piezoelectric effect and domain-switching effect. These combined effects were proposed by Caspari and Merz to explain the macroscopically strain vs field butterfly loop in bulk $BaTiO_3$ ceramics [19]. So Caspari-Merz's theory is still valid at nanometer scale.

3.3. Scanning probe acoustic microscopy of ferroelectric domain

3.3.1. Acoustic image of domain structure in PLZT ceramics

Fig. 16a shows the topography of the PLZT ceramics obtained by SPAM. The main features in the topography are the surface scratches. No information associated with the domain structure was revealed. Fig. 16b is an acoustic image of the PLZT ceramics using the SFM acoustic mode, which clearly reveals fingerprint patterns related to domains structures with antiparallel polarization [14]. These stripes show a pronounced and different contrast in the different areas (A, B, C) appearing in Fig. 16b due to different crystallographic orientations of the individual grains. The bend and split of the domain marked by arrows at the grain boundary regions may be attributed to the existence of inhomogeneous lattice distortions or spatial defects which destroyed the continuity of ferroelectric domains and minimized the elastic energy and depolarization fields at the grain boundaries. The fingerprint patterns in Fig. 16b are relatively regular and are almost periodically spaced in each individual grain. The stripe-structures in Fig. 16b are 300 nm in width.

Fig. 17 illustrates the spatial distribution of the phase difference signal cos ϕ (ϕ—phase difference between the modulation voltage and the piezoresponse signal). The phase difference corresponds to the direction of the polarization vector. The phase images indicate that the bright and black domain images can be ascribed to the antiparallel polarization, i.e. −c and

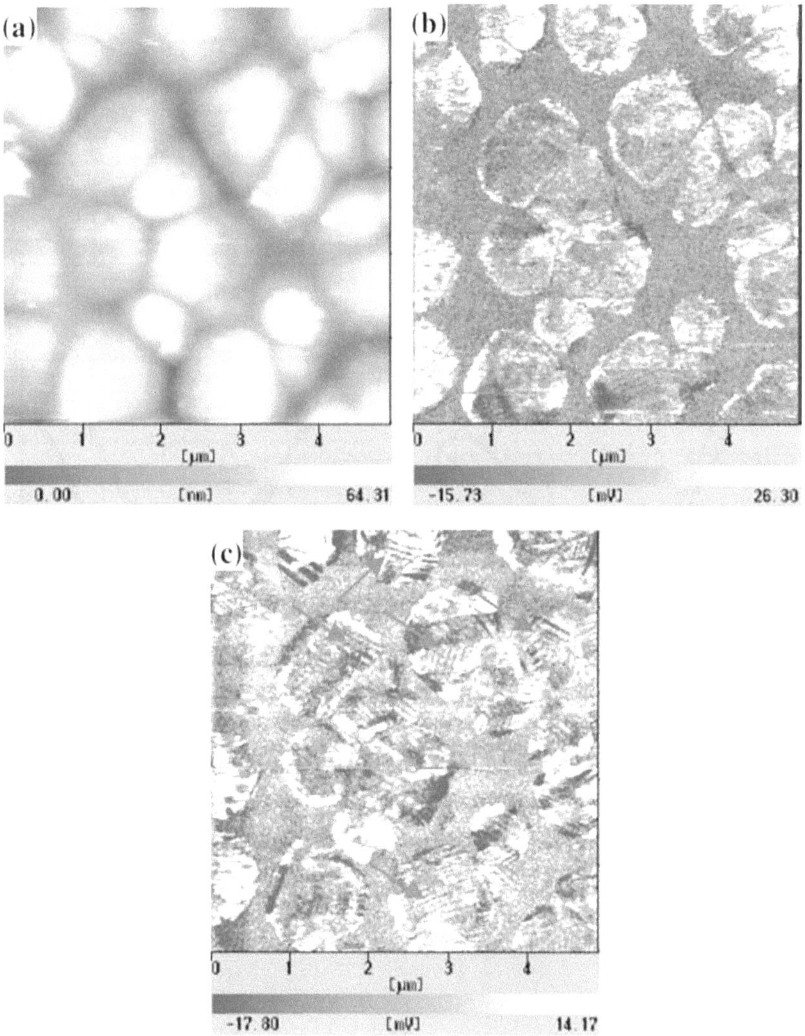

Figure 13 The topographic image (a) and piezoresponse image of 5×5 μm^2 regions scaned with tip biased at -1.5 V (b) and $+1.5$ V (c) in PZT 40/60 thin film.

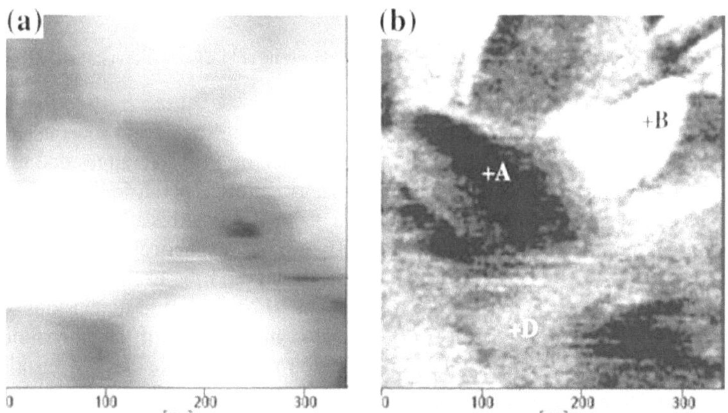

Figure 14 The topography image (a) and the corresponding piezoresponse image (b) of ferroelectric PZT thin film.

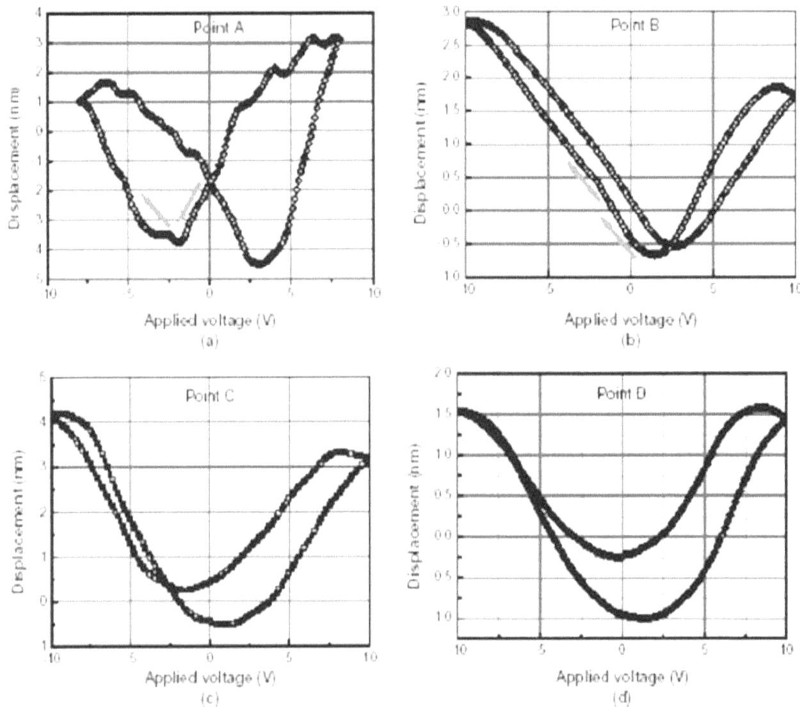

Figure 15 Nanoscale displacement vs electric field hysteresis loop of ferroelectric domains in the point A (a), point B (b), point C (c) and point D (d) shown in Fig. 14b for PZT thin film.

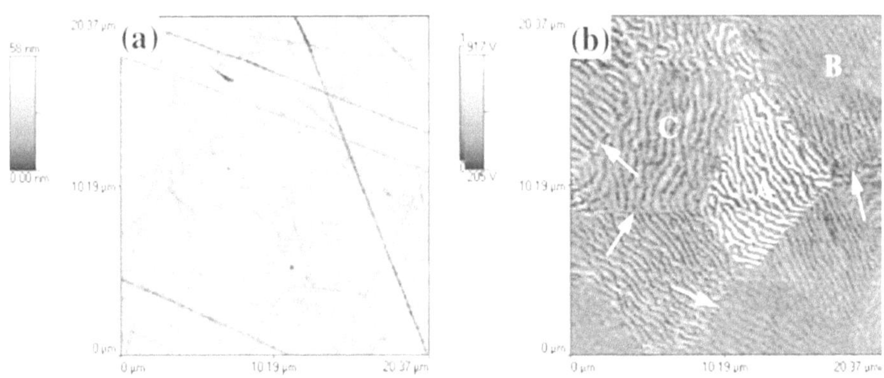

Figure 16 (a) Topography of PLZT ceramics by SFM, and (b) acoustic mode SFM imaging of PLZT ceramics at f=131.5 kHz.

+c domains, respectively. They are consistent with the former analysis, and give confirmation of the observed 180° domain structure in the acoustic image. In addition, the domains appear as bending and splitting patterns at the grain boundaries, as shown in Fig. 16b.

3.3.2. Imaging mechanism of antiparallel domains

In SPAM, When an ac voltage is applied between the conductive tip and the rear surface of the sample, the tip acts as an exciting source and generates acoustic waves in the piezoelectric PLZT ceramics due to the converse piezoelectric effect. Therefore the generation of the acoustic signal in the SPAM is attributed to piezoelectric coupling. The tip has a spherical shape with a radius of curvature of 10 nm, and the excited local region under the tip is about 50 nm in lateral dimensions, 200 nm in depth [4] and is independent of the surrounding area. Each local region of the sample can be considered as an independent piezoelectric vibrator. The vibration model of each piezoelectric vibrator is related to the interaction between the applied electrical field and the polarization vector.

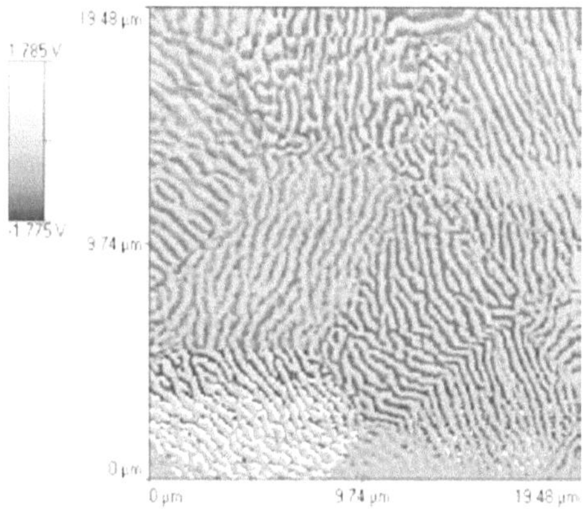

Figure 17 Phase image of acoustic mode SFM scan on PLZT ceramics.

Obviously, this kind of piezoelectric vibration is closely connected with the locally effective piezoelectric constant (d_{ij}^{eff}) of the sample. The effective d_{ij}^{eff} piezoelectric tensor is not only related to the piezoelectric coefficients d_{31}, d_{33}, d_{15}, d_{22} of the PLZT sample which has a La/Zr/Ti ratio of 6/65/35, but also related to the deviation angle of the spontaneous polarization vector from the normal direction (upward) of the sample surface. This composition is in the rhombohedral ferroelectric phase and has eight domain states with spontaneous polarization vectors oriented in the directions of cube diagonals of the cubic phase. The multi-domain state system with different orientations will give different contribution to the domain contrast observed in the amplitude image in the SFM acoustic mode, which can be considered as an equivalent function of the local effective piezoelectric constants. Different d_{ij}^{eff} will undoubtedly give rise to different domain contrast in the acoustic image. For a local area with the spontaneous polarization orientation close to the normal direction of the sample (i.e. nearly parallel to the ac-field direction), its larger d_{ij}^{eff} will lead to stronger piezoelectric vibrations and corresponding bright domain tones as shown in the a area of Fig. 16b. In contrast, the local area with larger deviating angle θ has a smaller d_{ij}^{eff}, thus causing weaker piezoelectric vibrations and showing a gray, dark domain tone, as reflected in the B and C area of Fig. 16b. So the differences in the effective piezoelectric constants of the local areas under the SPAM tip are the intrinsic characteristics resulting in the antiparallel domain contrasts of the PLZT ceramics in SPAM [13].

Fig. 18 shows a magnified acoustic imaging of the stripe-domains. A line scan (A-A′) drawn in Fig. 18a is shown in Fig. 18b. The amplitude distribution of the acoustic signal is periodically spaced, reflecting the periodicity of the domain distribution and the homogeneity of the piezoelectricity in PLZT ceramics. In addition, by observing the minimum resolvable peak separation marked by arrows in Fig. 18b, the estimated spatial resolution of acoustic mode SFM imaging in this work is about 70 nm.

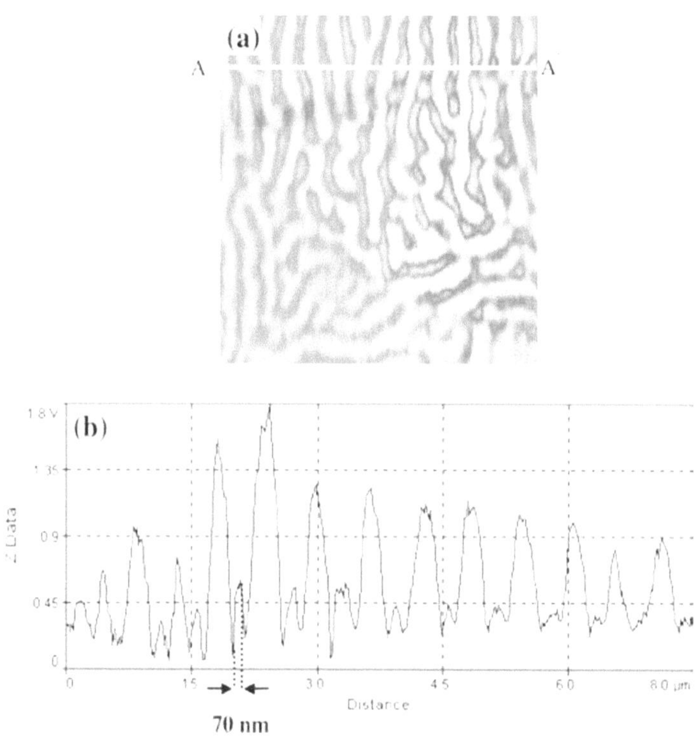

Figure 18 Acoustic mode SFM scan (a) and a line scan (C-D) on PLZT ceramics.

4. Conclusions

In summary, three kinds of near-field microscopy imaging mode including SEAM (Scanning electron acoustic microscopy), PFM (Piezoresponse force microscopy) and SPAM (Scanning probe acoustic microscopy) have been successfully developed based on the commercial SEM and AFM, and used to perform studies of domain structures and their dynamic behavior under external fields in our studies. The domain imaging contrast mechanisms were discussed in terms of their imaging principle, respectively. Sub-surface micro-domain configuration of ferroelectric ceramics and single crystal were clearly visualized by SEAM, and their evolution behavior with dependence of the modulation frequency, temperature and electric field were investigated in details. Ferroelectric domain structures of ferroelectric PZT thin film and PMN-PT single crystal were characterized by PFM. Nanoscale switching behavior and local field-induced nanoscale displacement behavior of domain structures in ferroelectric thin film were studied by PFM. SPAM provides a powerful, nondestructive, near-field imaging tool to explore subsurface domain structures at nanometer scale. Acoustic images of ferroelectric transparent PLZT clearly reveal domain arrangement associated with polarization modulations and homogeneity in the samples. The combination of SEAM, PFM and SPAM in application to imaging domain structures undoubtedly enrich our understanding the nature of piezoelectricity and ferroelectricity at submicro-, evern nano-meter scale.

References

1. D.L. POLLA and L.F. FRANCIS, *Annu. Rev. Mater. Sci.* **28** (1998) 563.
2. J.F. SCOTT, *Fer. Rev.* **1** (1998) 1.
3. E. BRANDIS and A. ROSENCWAIG, *Appl. Phys. Lett.* **37** (1980) 98.
4. Q. R. YIN, F. M. JIANG and S. X. HUI, *Ferroelectrics*, **151** (1994) 97.
5. B. Y. ZHANG, F. M. JIANG, Y. YANG and Q. R. YIN, *J. Appl. Phys.* **80** (1996) 1916.
6. Idem., *Appl. Phys. Lett.* **70** (1997) 589.
7. M. L. QIAN, X. M. WU, Q. R. YIN, B. Y. ZHANG and J. M. CANTRELL, *J. Mater. Res.* **14** (1999) 3096.
8. A. GRUVERMAN, O. AUCIELLO and H. TOKUMOTO, *Annu. Rev. Mater. Sci.* **28** (1998) 101.
9. C. S. GANPULE, V. NAGARAJAN, B. K. HILL, A. L. ROYTBURD, E. D. WILLIAMS, R. RAMESH, S. P. ALPAY, A. ROELOFS, R. WASER and L. M. ENG, *J. Appl. Phys.* **91** (2002) 1477.
10. V. NAGARAJAN, A. ROYTBURD, A. STANISHIEVSKV, S. PRASERTCHOUNG, T. ZHAO, L. CHEN, J. MELNGAILIS, O. AUCIELLO and R. RAMESH, *Nat. Mater.* **2** (2003) 43.
11. H. R. ZENG, H. F. YU, R. Q. CHU, G. R. LI, H. S. LUO and Q. R. YIN, *J. Cryst. Growth* **267** (2004) 194–198
12. S. V. KALININ and D. A. BONNELL, *Phys. Rev. B* **65** (2002) 125408.
13. Q. R. YIN, G. R. LI, H. R. ZENG, X. X. LIU, R. HEIDERHOFF and L. J. BALK, *App. Phys. A* **78** (2004) 699.
14. Q. R. YIN, H. R. ZENG, S. X. HUI and Z. K. XU, *Mater. Sci & Eng. B* **99** (2003) 2.
15. J. LIAO, X. P. JIANG, G. S. XU, H. S. LUO and Q. R. YIN, *Mater. Character* **44** (2000) 453.
16. H. R. ZENG, H. F. YU, R. Q. CHU, G. R. LI and Q. R. YIN, *Mater. Letter* **59** (2005) 2380.
17. H. F. YU, H. R. ZENG, R. Q. CHU, G. R. LI and Q. R. YIN, *J. Phys. D. App. Phys.* **21** (2004) 2914.
18. H. R. ZENG, G. R. LI, Q. R. YIN and Z. K. XU, *App. Phys. A* **76** (2003) 401.
19. M.E. CASPARI and W.J. MERZ, *Phys. Rev. B* **80** (1950) 1082.

JOURNAL OF MATERIALS SCIENCE 41 (2006) 271–280

Normal ferroelectric to ferroelectric relaxor conversion in fluorinated polymers and the relaxor dynamics

SHIHAI ZHANG*†, ROB J. KLEIN, KAILIANG REN, BAOJIN CHU, XI ZHANG, JAMES RUNT
Materials Research Institute, The Pennsylvania State University, University Park, PA 16802, USA
E-mail: zhangs@research.ge.com

Q. M. ZHANG
Materials Research Institute, The Pennsylvania State University, University Park, PA 16802, USA; Department of Materials Science and Engineering, The Pennsylvania State University, University Park, PA 16802, USA; Department of Electrical Engineering, The Pennsylvania State University, University Park, PA 16802, USA

To elucidate the molecular origin of the polarization dynamics in the ferroelectric relaxor poly(vinylidene fluoride–trifluoroethylene-chlorofluoroethylene) (P(VDF-TrFE-CFE)) terpolymer, a broadband dielectric study was carried out in the frequency range from 0.01 Hz to 10 MHz and temperatures from −150°C to 120°C for the terpolymer and a normal ferroelectric P(VDF-TrFE) copolymer. The relaxation processes were also studied using dynamic mechanical analysis. It was shown that in the terpolymer, which was completely converted to a ferroelectric relaxor, there is no sign of the relaxation process associated with the ferroelectric-paraelectric transition which occurs in the P(VDF-TrFE) copolymer. In the copolymer, three additional relaxation processes have been observed. It was found that the relaxation process β_a, which was commonly believed to be associated with the glass transition in the amorphous phase, in fact, contains significant contribution from chain segment motions such as domain boundary motions in the crystalline region. In the temperature range studied, the terpolymer exhibits the latter three relaxation processes with the one (termed β_r) near the temperature range of β_a significantly enhanced. This is consistent with the observation that in conversion from the normal ferroelectric to a ferroelectric relaxor, the macro-polar domains are replaced by nano-polar-clusters and the boundary motions as well as the reorientation of these nano-clusters generate the high dielectric response. The experimental data also reveal a broad relaxation time distribution related for the β_r process whose distribution width increases with reduced temperature, reflecting the molecular level heterogeneity in the crystalline phase due to the random introduction of the CFE monomer in the otherwise ordered macro-polar domains. The random interaction among the nano-clusters as well as the presence of the random fields produces ferroelectric relaxor behavior in the terpolymer.

*Author to whom all correspondence should be addressed.
†*Present Address*: GE Global Research, Niskayuna, NY 12304, USA.

0022-2461 © *2006 Springer Science + Business Media, Inc.*
DOI: 10.1007/s10853-006-6081-2

1. Introduction

In the last several decades, poly(vinylidene fluoride) (PVDF) and its copolymer with trifluoroethylene (P(VDF-TrFE)) have been studied extensively for their ferroelectric and electromechanical properties [1–4]. Recently, it was discovered that by proper defect modifications, the normal ferroelectric P(VDF-TrFE) copolymer can be converted to a relaxor ferroelectric polymer, which is the first and only known ferroelectric polymer relaxor [5, 6]. Interestingly, the relaxor ferroelectric polymers exhibit large electrostriction and high room temperature dielectric constant, which are attractive for a broad range of applications such as acoustic transducers, microactuators in MEMS, and electro-optic devices. These advances indicate the great potential of polymeric materials for achieving high functional responses compared with their inorganic counterparts, as well as the great need to understand the molecular origins of these responses.

For high energy electron irradiated P(VDF-TrFE) copolymers, it was established that the defects introduced by the irradiation break up the macroscopic polar-domains, leading to the conversion to the ferroelectric relaxor [5, 7, 8]. Analogously, in the terpolymer of poly(vinylidene fluoride-trifluoroethylene-1,1-chlorofluoroethylene) (P(VDF-TrFE-CFE)), the bulky termonomer CFE units serve as local defects to the polarization ordering in the crystalline regions, which convert the all-trans polar-conformation to a phase with a mixture of trans-gauche (TGTG'), all-trans and TG_3TG_3' conformations, which is macroscopically non-polar and exhibits ferroelectric relaxor behavior [9–17]. Under external electric field, both the local polar-region reorientation and the induced local conformation change result in large macroscopic electrostriction [11, 17]. For example, for a P(VDF-TrFE-CFE) terpolymer with certain composition, it was observed that a thickness strain of higher than -7% and transverse strain of 5% can be achieved [17].

While the microstructure and electromechanical properties of these polymers have been investigated in the past, the molecular origins of different dielectric relaxation processes, some of which contribute to the ferroelectric relaxor response of the polymer, are not clearly understood [18–22]. These relaxations reflect the molecular motions in both crystalline and amorphous phases and are also closely related to the macroscopic electromechanical response. In this paper, we report a detailed study on the broadband dielectric spectra of a ferroelectric relaxor P(VDF-TrFE-CFE) terpolymer with different thermal histories and electrical poling history, with the aim to explore the relaxor dynamics and their molecular origins. To gain more insight into the dynamics of the terpolymer, the polarization dynamics of the parent P(VDF-TrFE) copolymer was also studied under similar conditions.

2. Experimental

2.1. Sample preparation

A P(VDF-TrFE) copolymer with VDF content of 68 mol% was used in this study and it has a weight-average molecular weight (M_w) of 367,000 and a polydispersity of 2.7. P(VDF-TrFE-CFE) with VDF/TrFE/CFE molar ratio of 63/37/7.5 was synthesized via suspension polymerization using an oxygen-activated initiator (For easy comparison with the P(VDF/TrFE) copolymer, we use $VDF_x/TrFE_{1-x}/CFE_y$ to describe the composition of the terpolymer, where $x/(1-x)$ is the VDF/TrFE mole ratio and y is the mol% of CFE in the terpolymer) [9]. The terpolymer has a M_w of 670,000 and M_n of 238,000. Molecular weights were determined by gel permeation chromatography using narrow molecular weight polystyrene as standards and tetrahydrofuran as the mobile phase. These two compositions were selected since they have similar VDF/TrFE ratio.

Polymer films with thickness around 20 μm were prepared by solution casting with dimethylformamide (DMF) as the solvent. The samples were annealed in a vacuum oven overnight, at 120°C for terpolymer and 140°C for copolymer, to enhance their crystallinity. Quenched samples were obtained by melting the above films and then cooling to room temperature within 5 s. For electrical characterization, a 40 nm layer of gold was sputtered on both sides of the film.

2.2. Characterization

DSC measurements were carried out with a TA Q100 instrument at a heating rate of 10°C/min, with sample weights around 5 mg. IR spectra were recorded using a Nicolet FT-IR spectrometer with 64 scans averaged at a resolution of 2 cm^{-1}. Wide angle X-ray diffraction (WAXD) studies were performed using a Scintag Cu-Kα diffractometer with an X-ray wavelength of 1.54 Å. Polarization hysteresis loops at room temperature were collected using a Sawyer-Tower circuit with a frequency of 10 Hz.

Isothermal dielectric spectra $\varepsilon^*(f, T)$ were collected using a Novocontrol GmbH Concept 40 broadband spectrometer in the frequency domain (10 MHz to 0.01 Hz). The AC voltage for weak field dielectric measurement is 1.5 V. The temperature was controlled by a Novocontrol Quatro Cryosystem with stability better than ±0.1°C. Data collection did not start until the temperature had been stabilized at least 10 min. To better reveal details of the dynamics, the isothermal data at different temperatures were reorganized and the dielectric spectra at constant frequency as a function of temperature were plotted, that is, isochronal presentation.

Dynamic mechanical analysis (DMA) was performed with a TA DMA 2980 instrument in the step-scan mode at frequencies from 20 to 0.1 Hz. Samples for DMA were $\sim$200 μm thick.

3. Results and discussions

3.1. Property comparison between the normal ferroelectric P(VDF-TrFE) copolymer and the ferroelectric relaxor P(VDF-TrFE-CFE) terpolymer

To provide some background, we will first review the experimental data related to the conversion from the normal ferroelectric phase of the P(VDF-TrFE) copolymer to the ferroelectric relaxor character of the P(VDF-TrFE-CFE) terpolymer.

Presented in Figs 1 and 2 are a comparison of the polarization and dielectric (real part) responses of the copolymer and terpolymer. As a ferroelectric, the copolymer exhibits a nearly square polarization hysteresis loop due to the large nucleation barrier for domain wall motion. The remnant polarization (P_r) is 75 mC/m^2 and the cohesive electric field (E_c) is around 60 MV/m. This hysteresis is nearly eliminated in the P(VDF-TrFE-CFE) terpolymer, as expected for a relaxor ferroelectric at temperatures near the dielectric constant maximum (due to the absence of macroscopic polarization domains). P_r and E_c for the relaxor terpolymer are much smaller, i.e., 2.5 mC/m^2 and 4 MV/m, respectively. For the dielectric constant, the P(VDF-TrFE) copolymer exhibits a sharp ferroelectric-paraelectric (F-P) transition at 110°C (see Fig. 2) and the peak position does not depend on frequency. In contrast, the P(VDF-TrFE-CFE) terpolymer shows a much broad peak in the dielectric constant near room temperature and the peak position shifts progressively to higher temperature with frequency, which is a feature typical of ferroelectric relaxors [23]. It should be noted that at some temperatures below the dielectric constant maximum, the relaxor polymer exhibits large polarization hysteresis, which is another characteristic of ferroelectric relaxors [5, 24].

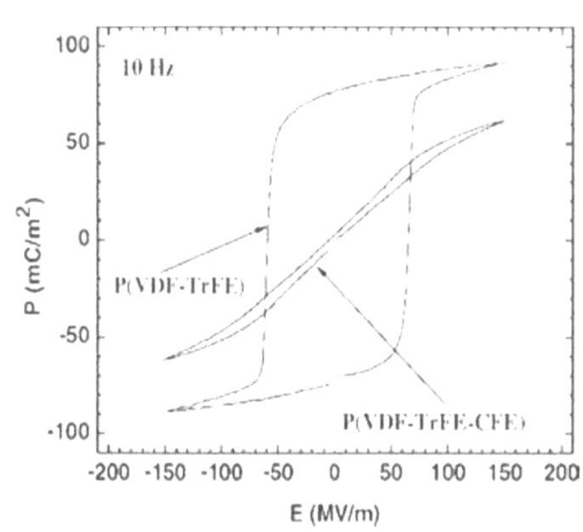

Figure 1 Polarization hysteresis loop measured at room temperature using 10 Hz triangular voltage signal.

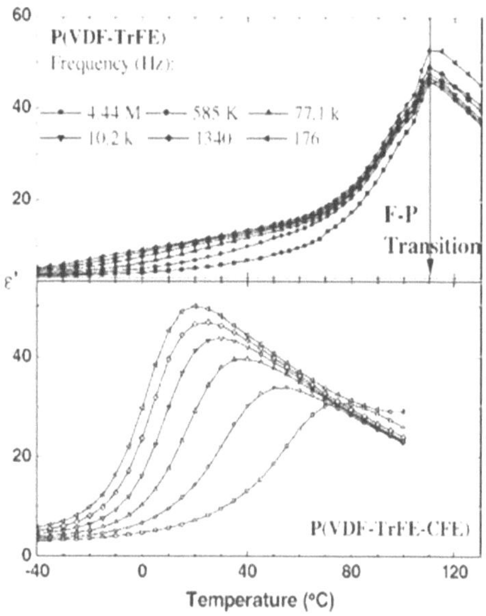

Figure 2 Comparison of the dielectric spectra of the normal ferroelectric P(VDF-TrFE) and ferroelectric relaxor P(VDF-TrFE-CFE).

The conversion from the normal ferroelectric polymer to a ferroelectric relaxor is also reflected by molecular conformation changes in the polymer. In ferroelectric copolymers, a majority of the crystallite chains ($\sim$75% for 68/32 composition) are in the polar *all trans* conformation, leading to a high effective dipole moment. In contrast, the majority of the P(VDF-TrFE-CFE) terpolymer crystallite chains adopt non-polar conformations, as a result of the defects introduced by the bulky CFE termonomer units. The different chain conformations associated with the crystalline phases can be distinguished by their characteristic bands in FTIR spectra, which are presented in Fig. 3. The bands at 1290 cm^{-1} and 850 cm^{-1}

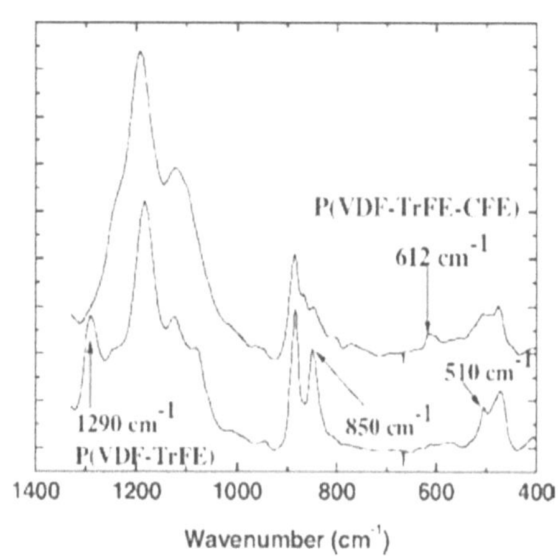

Figure 3 FTIR spectra (absorbance) at room temperature. The spectrum of the terpolymer is shifted vertically.

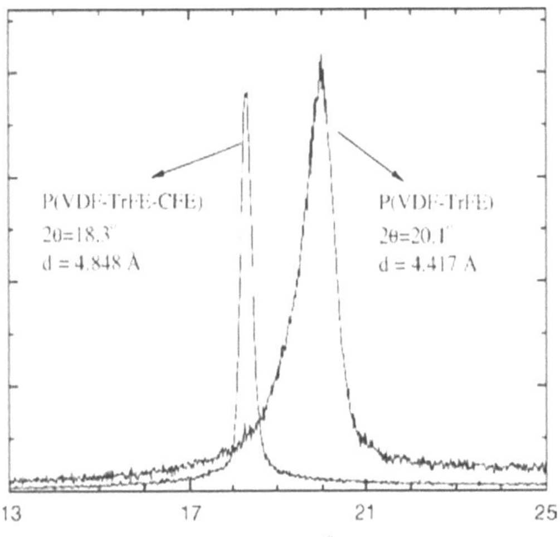

Figure 4 Wide angle X-ray diffraction data in the (110, 200) reflection region. Wavelength used is 1.54 Å (Cu-Kα).

represent the *all trans* conformation [25] and they exhibit large absorbances in the spectrum of the copolymer, but significantly reduced in the relaxor terpolymer. The band at 612 cm^{-1} corresponds to the TGTG$'$ conformation and it can only be observed in the spectrum of the terpolymer. The absorbance at 510 cm^{-1} represents the C-F bending mode in the polar T_3GT_3G' conformation and it is observed in both spectra. From the FTIR data, it can be deduced that ~55% of the terpolymer chains exhibit TGTG$'$ non-polar conformations, while only 20% of the chains are in the all-trans polar conformation.

The ferroelectric crystallites in the copolymer are transformed to nonpolar crystallites in the terpolymer after introducing the CFE defects. This is also evident in their corresponding X-ray diffraction patterns provided in Fig. 4. The ferroelectric P(VDF-TrFE) copolymer displays a

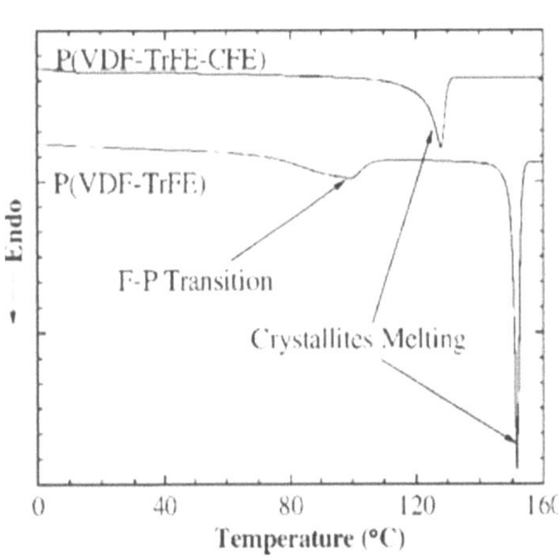

Figure 5 DSC traces of copolymer and terpolymer. Heating rate is 10°C/min.

diffraction peak at 2θ=20.1° (originating from the (110, 200) reflection and corresponding to a lattice spacing of 4.417 Åin ferroelectric crystalline phases). This peak shifts to lower diffraction angle (18.3°) for the P(VDF-TrFE-CFE) terpolymer, indicating a larger lattice spacing of 4.848 Å(from the (110, 200) reflection in nonpolar crystallites) [11, 15]. The non-polar nature of the terpolymer crystallites is also supported by the disappearance of the F-P transition in the DSC thermogram, compared with the strong ferroelectric-paraelectric transition exhibited by ferroelectric P(VDF-TrFE) copolymers (~100°C for the 68/32 composition, Fig. 5).

3.2. Dynamics of the normal ferroelectric P(VDF-TrFE) copolymer

To facilitate an understanding of the dielectric dynamics in the ferroelectric relaxor terpolymers, the dynamics of the normal ferroelectric copolymer [25–27] having a similar VDF/TrFE ratio was investigated initially. Fig. 6 presents dielectric spectra (measured during the heating run) for the P(VDF-TrFE) copolymer in the temperature range from −60 to 140°C. Dielectric data acquired at temperatures from −150 to −60°C will be presented later. In the temperature range from −150 to 140°C, four dielectric relaxation processes can be identified. As has been pointed out earlier, the peak in the dielectric constant at 110°C corresponds to the F-P transition. In the DSC thermogram, the same process is also observed as an endotherm with ΔH=25 J/g. Reflecting its first-order phase transition nature, a large thermal hysteresis is observed and the

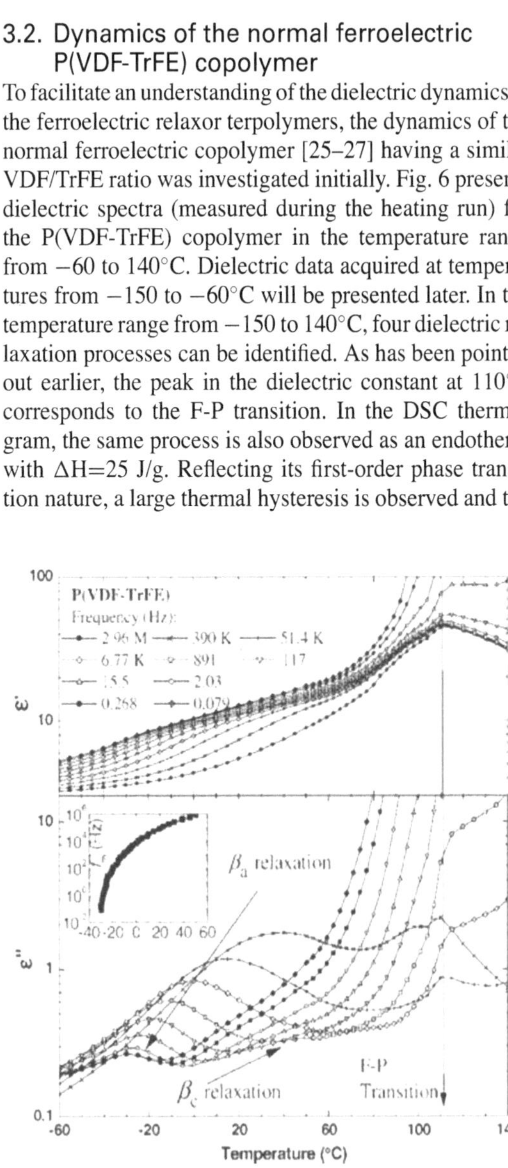

Figure 6 Dielectric spectra of P(VDF-TrFE) copolymer. Inset: peak locations of the β_a relaxation taken from the ε'' spectra.

reverse transition during cooling occurs at a temperature 50°C lower than that measured on heating.

Two frequency-dependent processes are visible (particularly in the loss spectra) around room temperature in Fig. 6. The process at lower temperatures (located at 0°C at 10.2 kHz) is relatively strong and has been traditionally assigned to the dynamic glass transition of segments in the amorphous portion of the copolymer (denoted as β_a hereafter) [28, 29]. However, a recent study by Omote and coworkers found that this process is also present in 'single crystalline' P(VDF-TrFE), although with slightly lower intensity [30]. Since there is essentially no amorphous phase in the single crystalline polymer, it was proposed that this process may also reflect chain motions due to structural defects in the crystalline phase. As demonstrated later, this dielectric relaxation includes a significant contribution from domain wall and polar defect motions in the normal ferroelectric P(VDF-TrFE) copolymer.

The β_a anomaly shifts to higher temperatures with increasing frequency, reflecting the speeding up of polymer chain motions with temperature. Its temperature-frequency relationship is usually modeled with the Vogel-Fulcher (VF) law $f_m = f_0 \times \exp\,[-U/k(T-T_0)]$ (inset in Fig. 6) [31], where f_m represents the average relaxation frequency, U is a constant, and T_0 is the Vogel temperature at which the relaxation process becomes frozen. Fitting the data in Fig. 6 leads to the following fitting parameters: f_0 = 253 MHz, U = 5.55 kJ/mol = 0.0575 eV, and T_0=209.0 K. It should be emphasized that to precisely determine the parameters in the VF equation, very reliable f_m data at temperatures close to T_0, or at least close to the glass transition temperature T_g ($T_g \approx T_0 + 30 \sim 50$°C) are required, which necessitate long time data acquisition (i.e., the dielectric constant measured at near static conditions). Fitting the VF law without sufficient experimental data can generate significantly different f_0, U, and T_0 values. This readily explains the scattered fitting results in the literatures [5, 7, 18, 21]. The dipole reorientation associated with the β_a process leads to a moderate increase in the relaxation strength $\Delta\varepsilon$ (defined as ε'[high temperature]$-\varepsilon'$[low temperature]) ~9.5 at 1 kHz, significantly smaller than that at the F-P transition. However, $\Delta\varepsilon$ is typically 3~5 for the segmental (glass) transition in other completely amorphous polymers [32], much smaller than that observed for the β_a process in P(VDF-TrFE) (particularly considering that it contains only ~25% amorphous segments). The large $\Delta\varepsilon$ for β_a, however, is consistent with the scenario that the domain wall and other polar-defect motions in the crystalline phase contribute significantly to the observed dielectric β_a relaxation process.

While the above two dielectric anomalies, i.e., the F-P transition of the crystalline phase and the complex β_a relaxation, have been reported in many studies [25–29], a third process, taking place at slightly higher temperature than the β_a process, can only be observed in low-frequency dielectric spectra. Although the nature of this process (referred to as β_c) is still under debate, most results support its origin in the crystallite/amorphous interface [33]. β_c is significantly weaker than β_a in the dielectric spectra, and they apparently merge into one process at frequencies above 10 kHz.

The β_c process is also observed in the DMA spectrum (Fig. 7) where it is more prominent than the β_a process. Furthermore, the relative f-T locations of the β_a process in DMA and dielectric spectra depend on the manner in which they are defined. In loss spectra (tan $\delta = \varepsilon''/\varepsilon'$ or E″/E′), the mechanical relaxation is about one decade slower than the comparable dielectric process (Fig. 8). This behavior is reasonable since the slowest (and

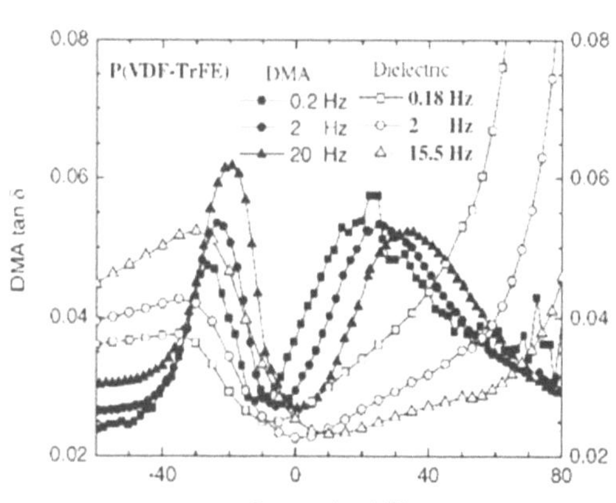

Figure 7 Comparison of the dielectric and DMA spectra of P(VDF-TrFE) copolymer. The β_c relaxation can be clearly identified in the DMA spectra.

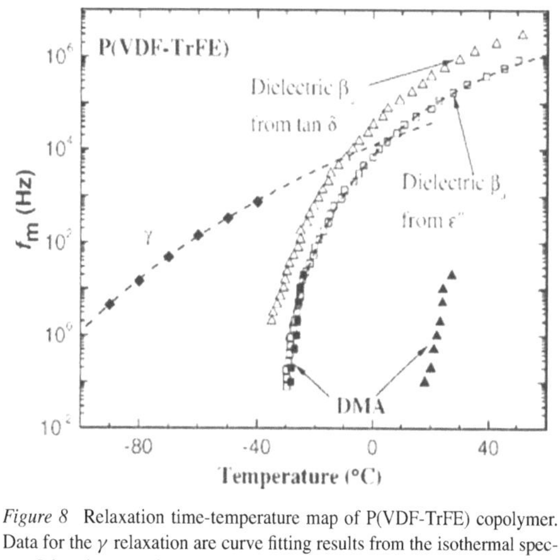

Figure 8 Relaxation time-temperature map of P(VDF-TrFE) copolymer. Data for the γ relaxation are curve fitting results from the isothermal spectra. Other data are directly read from the isochronal spectra. Dashed lines represent the VF - law and Arrhenius fitting to the experimental data.

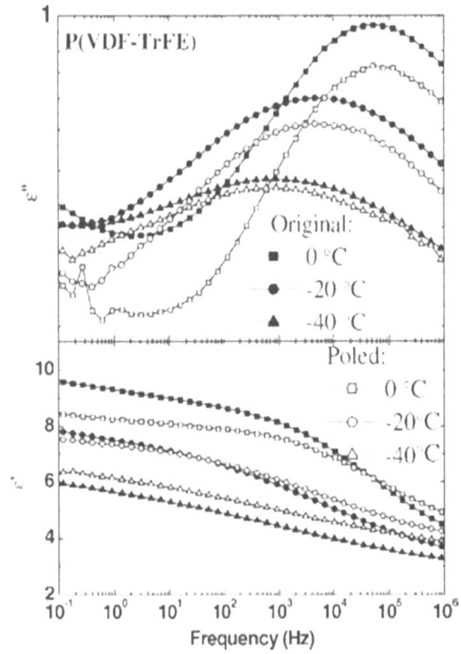

Figure 9 Poling effect on the β_a relaxation of the P(VDF-TrFE) copolymer. Samples were poled at 100 MV/m at 50°C for 10 min.

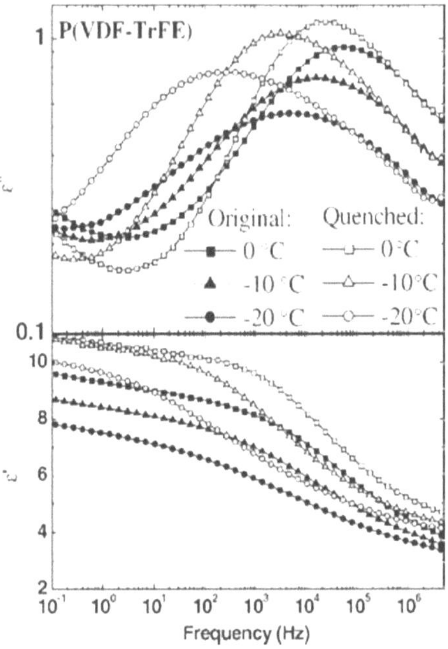

Figure 10 Isothermal dielectric spectra of quenched P(VDF-TrFE) copolymer. The β_a relaxation is weakened as compared with annealed samples.

largest) chain segments carry most mechanical stress and determine the average relaxation time in DMA, whereas the dielectric relaxation time is simply a weight-average quantity.

To further explore the molecular origin of the β_a dielectric process in P(VDF-TrFE), electric poling experiments were performed. In normal ferroelectric materials, electrical poling usually improves the polar-ordering in the crystalline region and reduces the domain wall density, and so dielectric dispersion originating from domain wall motion should be weakened [23, 24]. After poling at 100 MV/m at 50°C for 10 min, the F-P transition of the copolymer is clearly sharpened due to the formation of larger polarization domains. The β_a process, however, is weakened considerably after poling (Fig. 9), and its location shifts to slightly higher frequencies compared with the unpoled samples. The low-frequency dielectric constant in the glass transition region also decreases after poling. Since a poling field of 100 MV/m at 50°C will not have a marked effect on the crystallinity (converting amorphous chain segments into crystallites), the decrease in strength of the β_a process is attributed to reduction in the density of domain wall and other defects in the crystalline phase due to poling. The result is consistent with the scenario that there is a quite significant contribution to the β_a process from the domain walls and other defects in the crystalline phase.

A dielectric study was also performed on copolymer samples quenched from the melt. This leads to a reduction in crystallinity and hence an increase in the fraction of the amorphous phase. Consequently, the β_a process would be expected to intensify. Quenching also generates smaller crystallites and increases the concentration of domain walls and their contribution should also enhance the β_a process. Fig. 10 compares the isothermal dielectric spectra of annealed and quenched P(VDF-TrFE) in the glass transition region. The strength of the β_a relaxation in the quenched sample is about 25% larger than that in the spectra of the annealed one, and the dielectric constant ε' of the former is consistently higher than that of the latter at -10°C. Similar enhancement of the β_a process for quenched P(VDF-TrFE) is also observed at other temperatures. As will be demonstrated later, quenching the ferroelectric relaxor terpolymer leads to quite different dielectric behavior, where quenching results in a reduction of both the real and imaginary dielectric constants associated with this relaxation process.

In addition to these three dielectric relaxation processes, there is a fourth relaxation occurring at lower temperatures, which is manifested as a weak shoulder on the low temperature tail of theβ_a process in isochronal dielectric spectra [25–27]. This relaxation can be better presented in an isothermal plot (Fig. 11). At temperatures far below the T_g of P(VDF-TrFE) (~ -40°C), a clear dispersion is observed in ε''. This relaxation has been denoted as the local γ process and proposed to originate from the fast rotation of the C-F bonds or motion of one monomer unit [32]. The local γ relaxation contributes an increase of 1.7 to the dielectric

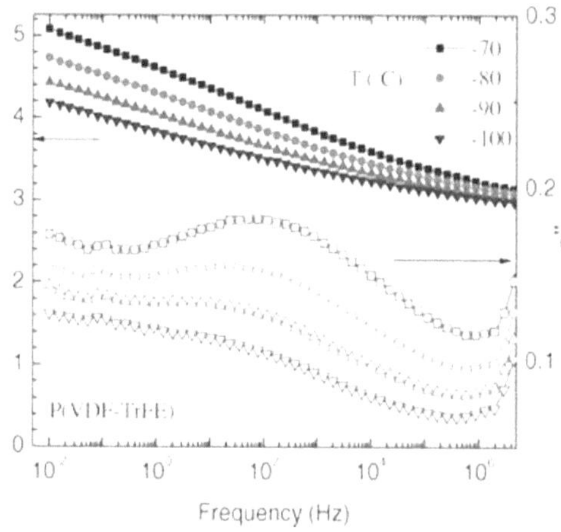

Figure 11 Isothermal dielectric spectra of P(VDF-TrFE) copolymer at low temperatures (temperatures below −70°C).

constant from 10 MHz to 0.01 Hz at −80°C. This process follows an Arrhenius law with $f_m = 1.26 \times 10^{11} \times \exp(-36.6\ \text{kJ/mol}/RT)$ (Fig. 8). The relatively low activation energy and Arrhenius behavior strongly support the non-cooperative local origin of the γ process.

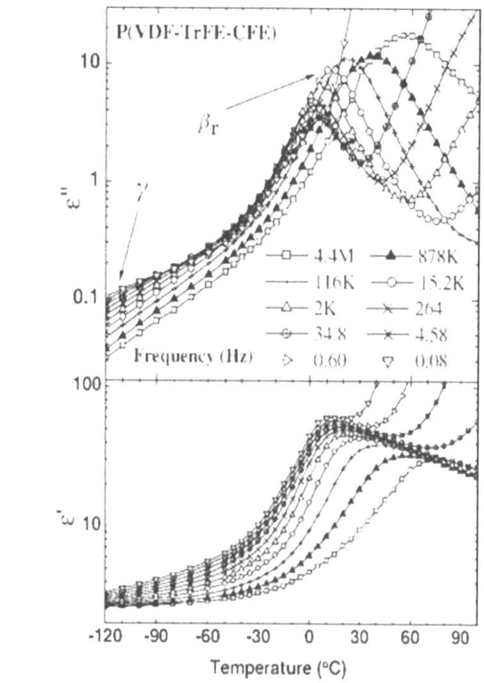

Figure 12 Dielectric spectra of P(VDF-TrFE-CFE) terpolymer at different frequencies.

3.3. Dynamics in relaxor ferroelectric P(VDF-TrFE-CFE) terpolymer

For the terpolymer investigated here, there is no sign in the dielectric spectra of an F-P transition (Fig. 12). There is no hysteresis in the dielectric data when measured in heating and in cooling. On the other hand, for terpolymers with lower CFE concentration or non-uniform CFE distribution in the terpolymer, large coherent polar-domains and a relaxor phase coexist, and a frequency-independent phase transition process is observed in both ε' and ε'' spectra, although it is shifted to lower temperatures [17]. In previous studies on high-energy electron irradiated copolymers, similar incomplete normal-relaxor conversion has also been observed in samples with insufficient irradiation treatment of the P(VDF-TrFE) copolymer [7, 8, 34]. However, increasing the irradiation dosage eliminates the ferroelectric phase in the copolymer and hence there is no sign of F-P phase transition process in the dielectric spectra in these properly treated copolymers [7].

It is interesting to note that in the same temperature range as the β_a process in the normal ferroelectric copolymer, a strong dielectric dispersion (denoted as β_r) is observed in the relaxor terpolymer, which follows the Vogel-Fulcher law Figs 12 and 13, with f_0 = 987 MHz, $U = 3.86$ kJ/mol $= 0.0400$ eV, and $T_0 = 244.3$ K. However, this relaxation process exhibits a significantly larger increase in dielectric constant ($\Delta\varepsilon = \varepsilon'$[low frequency]

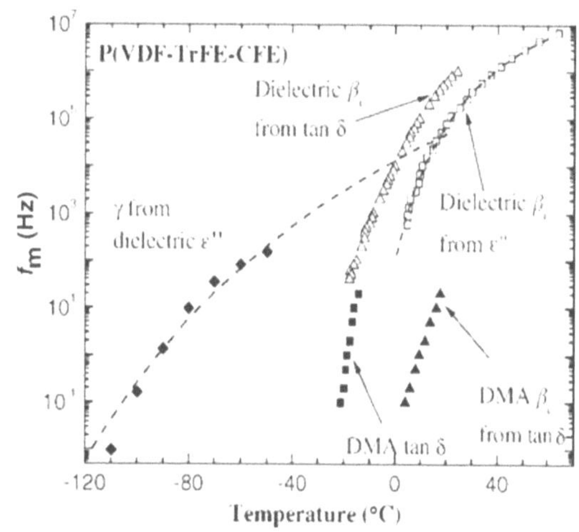

Figure 13 Relaxation times of different relaxation processes for P(VDF-TrFE-CFE) terpolymer. The γ process follows the Arrhenius equation; the linearity can be seen in the log f_m *vs.* 1/T plot. Dashed lines represent the VF - law and Arrhenius fitting to the experimental data.

$-\varepsilon\prime =$ [high frequency]$\frac{2}{\pi}\int_{-\infty}^{\infty} d(\ln\omega)\varepsilon''(\omega)$ or ε'[high temperature]$-\varepsilon'$[low temperature]) and has much larger ε''_{max} value than β_a does. Analogous to the dielectric relaxation in the normal ferroelectric P(VDF-TrFE) copolymer, the observed β_r relaxation should include contributions from the glass transition in the amorphous phase. However, the significant increase in the dielectric constant is a reflection of the large increase in the "domain wall" and other defects contributions of the crystalline phase. As the macroscopic polar-domains are replaced by the nano-clusters in the conversion from the normal ferroelectric to the ferroelectric relaxor, it is conceivable that there is a considerable increase in the contribution from the polar-cluster boundary motion (expanding and contracting of the nano-polar regions) and reorientation to the dielectric relaxation process. These nano-clusters have locally ordered structures; thereby their contribution to the dielectric relaxation is markedly larger than that from the completely disordered amorphous phase. This provides a direct link between the dielectric relaxation in the ferroelectric relaxor terpolymer and normal ferroelectric copolymer.

With decreasing temperature, the mobility of these nano-clusters decreases and this leads to their eventual freezing at the Vogel temperature T_0. The relaxation time distribution for the β_r process is very broad, with peak-width at half maximum of ~4.4 decades at 0°C, and it becomes even broader with decreasing temperature or increasing frequency, suggesting the breakdown of the time-temperature superposition principle. This broad distribution in general reflects heterogeneity at the molecular level: for example, the presence of random defects in the crystalline phase (due to CFE), the random coupling between the local nano-clusters due to random spacing and orientation among them, lead to a distribution of dynamic cooperativity. The broadening with decreasing temperature also suggests that the intermolecular cooperativity is enhanced at lower temperatures due to the reduced fractional free volume and/or chain mobility [31]. Furthermore, this heterogeneity scenario also leads to a broad distribution in T_0. That is, with decreasing temperature, the freezing-in occurs gradually from those processes with long relaxation times to those with short relaxation times. This is consistent with earlier studies employing a reduced dielectric constant scheme [18, 19]. It was found that the longest relaxation time follows a Vogel-Fulcher form and diverges at a freezing temperature near 0°C. The shortest relaxation time, however, exhibits Arrhenius behavior and remains active to temperatures well below −70°C. [18] Therefore, the fitting parameters (such as T_0) to the V-F law of the data in Fig. 12 are a measure of the averaged freezing process in the ferroelectric relaxor, rather than the freezing temperature T_f of the system at which the longest relaxation time in the system diverges.

The β_c process is also observed at low frequencies in the terpolymer at temperatures higher than the β_r relaxation, in both dielectric and DMA spectra Figs 12 and 14 . Both processes shift to higher temperatures with increasing frequency. However, dielectric loss of the β_r relaxation, associated with segmental motion and nanocluster reorientation, is enhanced much more significantly than

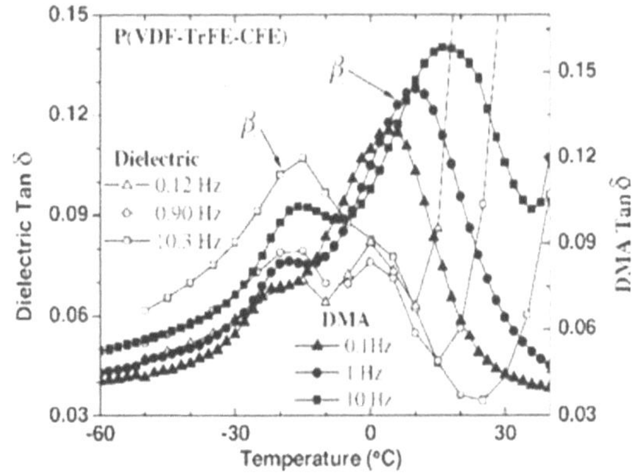

Figure 14 Comparison of the dielectric and DMA spectra of P(VDF-TrFE-CFE) terpolymer. The β_c relaxation can be clearly identified in the DMA spectra.

β_c, As a result, the latter process is completely masked by the former at higher frequencies and only one combined peak is visible in ε'' spectra at frequencies larger than 100 Hz.

The dielectric properties of the terpolymer were also measured after being subjected to an electric field of 100 MV/m at 50°C for 10 min, similar to the poling conditions applied to the copolymer. There is essentially no change in the dielectric β_r process before and after the application of the electric field, which is expected for the ferroelectric relaxor poled at temperatures where the slim polarization loop is observed. This supports complete conversion to the ferroelectric relaxor phase in this terpolymer.

In contrast to the behavior observed for the normal ferroelectric copolymer, quenching the relaxor ferroelectric terpolymer leads to a 50% decrease of the β_r relaxation strength in the ε'' spectra at 0°C and a concurrent reduction in ε' (Fig. 15). This result supports the notion that the β_r process in the terpolymer is not simply associated with segmental motion in the amorphous phase, which would otherwise have led to an increase in the relaxation strength. An earlier study by Klein *et al.* found that quenching P(VDF-TrFE-CFE) not only retards the crystallization process, but also induces the formation of polar *all trans* conformation, that is, reducing the content of relaxor nanoclusters [17]. Since the boundary motions and reorientation of nanoclusters dominate the β_r process, the reduction of ε' and ε'' in the quenched terpolymer is consistent with the proposed molecular origin.

Analogous to the copolymer, at temperatures significantly below T_g, the local γ relaxation is observed in the terpolymer (Figs12 and 16) . Unlike the β_r process, the γ transition exhibits Arrhenius behavior with $f_m = 1.88 \times 10^{12} \times \exp(-42.6\ \text{kJ/mol}/RT)$. When compared with the copolymer, the γ relaxation of the terpoly-

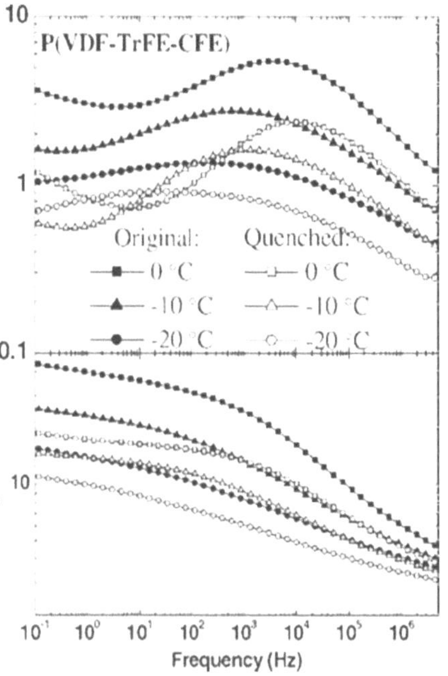

Figure 15 Isothermal dielectric spectra of quenched P(VDF-TrFE-CFE) terpolymer. The β_r relaxation is significantly weakened and sped up as compared with annealed samples.

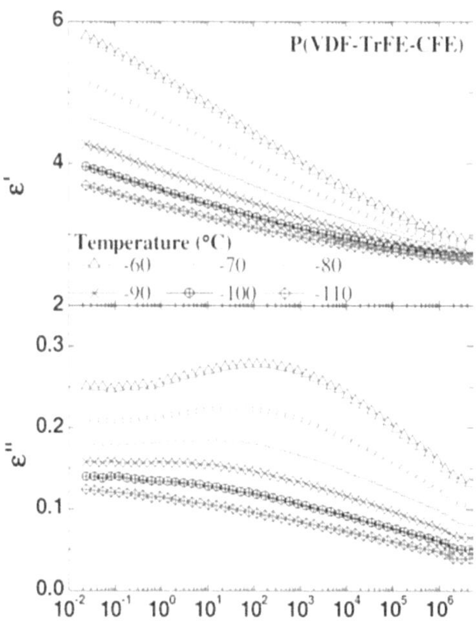

Figure 16 Isothermal dielectric spectra of P(VDF-TrFE-CFE) terpolymer at low temperatures. The dielectric constant exhibits clear step increase as a result of this relaxation.

mer is about 2 times slower and exhibits a larger activation energy. This may be due to the introduction of CFE in the polymer chain, whose bulky size may cause the local motion to become more difficult. The low temperature γ process contributes an increase of 2 to the dielectric constant at -80°C. This value much larger than that observed for the local low temperature relaxation in other common polymers with similar T_g, due to the large dipole moment of C-F [32].

4. Conclusions

In this study, we present comprehensive experimental results related to the conversion from the normal ferroelectric P(VDF-TrFE) copolymer to a relaxor ferroelectric P(VDF-TrFE-CFE) terpolymer due to the introduction of bulky CFE termonomers into the crystalline phase. Compared with the copolymer, the relaxor terpolymer exhibits high electrostrictive response and high room temperature dielectric constant. The experimental data from XRD, FTIR, DSC, and dielectric studies indicate that the macro-polar crystallites in the normal ferroelectric copolymer are transformed into a non-polar phase with local nano-clusters. There is no indication of an F-P transition in the terpolymer investigated.

The broadband dielectric spectra of both polymers exhibit rich dynamic processes, which are affected by the random introduction of the CFE in the polymer chains and by the sample treatment conditions. In the temperature range investigated (from -150 to 120°C), the copolymer exhibits at least four dielectric anomalies. The ferro-paraelectric phase transition occurs at 110°C (measured on heating). It does not depend on frequency and exhibits a large thermal hysteresis, a reflection of a first order transition process. Dynamic relaxation of polymer chains at the crystalline/amorphous interface gives rise to a weak dielectric dispersion β_c at temperatures from -20 to 40°C, depending on frequency, which is also visible in the relaxor terpolymer. The glass transition (segmental relaxation) of the amorphous phase and the domain wall relaxation in the crystalline regions, whose contributions are difficult to separate, generate a third dielectric β_a process in ε'' spectra, which follows the Vogel-Fulcher law. For the copolymer, quenching from the melt results in a stronger β_a relaxation due to increased domain wall density as well as an increase in the amorphous phase content. Poled samples exhibit a weaker β_a dispersion as a result of reduction of polar-defects and domain wall content in the crystalline regions under the high electric field. At temperatures below -60°C, both copolymer and terpolymer have a local dielectric γ relaxation process, which follows an Arrhenius law with an activation energy $\sim$40 kJ/mol. The γ process represents polymer C - F rotation and leads to an increase of $\sim$2 in dielectric constant at -80°C.

Reflecting the fact that the terpolymer P(VDF-TrFE-CFE) has been completely converted into a relaxor, no ferro-paraelectric phase transition is observed in the dielectric spectra. The terpolymer has a β_r dielectric dispersion, in addition to β_c and γ relaxations, that is located at similar temperatures to the β_a of the copolymer. However, there is a much larger dielectric constant change associated with the β_r process compared to β_a. Combining this result with those from quenched and poled samples, as well as the behavior of the copolymer, indicates that the β_r process originates primarily from the boundary motions and reorientation of nanoclusters in the crystalline phase, similar to the dynamics observed in relaxor ferroelectric ceramics. The experimental data also reveal a broad relaxation time distribution associated with the β_r process, whose distribution width increases with reduced temperature. Fitting of the dielectric data demonstrates that the averaged relaxation time (as measured by the peak of the imaginary dielectric constant) follows the V-F law.

Acknowledgements

This work was supported by the Office of Naval Research under Grant No. N00014-02-10418 and N00014-04-10292.

References

1. H. S. NALWA, (Ed). "Ferroelectric Polymers" (Marcel Dekker, Inc. NY, 1995).
2. M. E. LINES and A. M. GLASS, "Principles and Applications of Ferroelectrics and Related Materials" (Clarendon Press, Oxford, 1977).
3. Y. BAR-COHEN, (Ed). "Electroactive Polymer (EAP) Actuators as Artificial Muscles" (SPIE, Bellingham, WA, 2001).
4. A. J. LOVINGER, *Science* **220** (1983) 1115.
5. Q. M. ZHANG, V. BHARTI and X. ZHAO, *ibid.* **280** (1998) 2101.
6. A. J. LOVINGER, D. D. DAVIS, R. E. CAIS, J. M. CAIS and J. M. KOMETANI, *Polymer* **28** (1987) 617.
7. Z. Y. CHENG, Q. M. ZHANG and F. B. BATEMAN, *J. Appl. Phys.* **92** (2002) 6749.
8. V. BHARTI, H. S. XU, G. SHANTHI, Q. M. ZHANG and K. LIANG, *ibid.* **87** (2000) 452.
9. H. S. XU, Z.-Y. CHENG, D. OLSON, M. TAI, Q. M. ZHANG and G. KAVARNOS, *Appl. Phys. Lett.* **78** (2001) 2360.
10. F. BAUER, E. FOOUSSON, Q. M. ZHANG and L. M. LEE, *Proc. 11th Int. Symp. Electrets* (2002) 355.
11. F. XIA, Z. Y. CHENG, H. S. XU, H. F. LI, Q. M. ZHANG, G. J. KAVARNOS, R. Y. TING, G. ABDEL-SADEK and K. D. BELFIELD, *Adv. Mater.* **14** (2002) 1574.
12. T. C. CHUNG and A. PETCHSUK, *Macromolecules* **35** (2002) 7678.
13. C. M. ROLAND, J. T. GARRETT, R. CASALINI, D. F. ROLAND, P. G. SANTANGELO and S. B. QADRI, *Chem. Mater.* **16** (2004) 857.
14. J. T. GARRETT, C. M. ROLAND, A. PETCHSUK and T. C. CHUNG, *Appl. Phys. Lett.* **83** (2003) 1190.
15. R. J. KLEIN, J. RUNT and Q. M. ZHANG, *Macromolecules* **36** (2003) 7220.
16. R. J. KLEIN, F. XIA, Q. M. ZHANG, and F. BAUER, *J. Appl. Phys.* **97** (2005) 094105 submitted.
17. R. J. KLEIN and M. S. THESIS, The Pennsylvania State University, 2004.
18. V. BOBNAR, B. VODOPIVEC, A. LEVSTIK, M. KOSEC, B. HILCZER and Q. M. ZHANG, *Macromolecules* **36** (2003) 4436.
19. B. VODOPIVEC, V. BOBNAR, A. LEVSTIK and Q. M. ZHANG, *Ferroelectrics* **304** (2004) 857.
20. Z. YU and C. ANG, *Appl. Phys. Lett.* **84** (2004) 2145.
21. C. ANG and Z. YU, *Adv. Mater.* **16** (2004) 979.
22. Z. YU, C. ANG, L. E. CROSS, A. PETCHSUK and T. C. CHUNG, *Appl. Phys. Lett.* **84** (2004) 1737.
23. L. E. CROSS, *Ferroelectrics* **151** (1994) 305.
24. *Idem., ibid.* **76** (1987) 241.
25. T. FURUKAWA, *Phase Transitions* **18** (1989) 143.
26. T. FURUKAWA, M. OHUCHI, A. CHIBA and M. DATE, *Macromolecules* **17** (1984) 1384.
27. T. FURUKAWA, Y. TAJITSU, X. ZHANG and G. E. JOHNSON, *Ferroelectrics* **135** (1992) 401.
28. Y. ISHIDA, S. SAITO, M. ASABINA and H. KAKUTANI, *J. Polym. Sci. Part A2* **7** (1969) 1405.
29. S. YANO, *ibid.. Sci. Part A2* **8** (1970) 1057.
30. K. OMOTE, H. OHIGASHI and K. KOGA, *J. Appl. Phys.* **81** (1997) 2760.
31. J. D. FERRY, "Viscoelastic Properties of Polymers" (Wiley, New York, 1980).
32. N. G. MCCRUM, B. E. READ and G. WILLIAMS, "Anelastic and Dielectric Effects in Polymeric Solids" (Wiley, London, 1967).
33. T. YAGI, M. TATEMOTO and J. SAKO, *Polym. J.* **12** (1980) 209.
34. V. BOBNAR, B. VODOPIVEC, A. LEVSTIK, Z. Y. CHENG and Q. M. ZHANG, *Phys. Rev. B* **67** (2003) 94205.

Index

MIX
Papier aus verantwortungsvollen Quellen
Paper from responsible sources
FSC® C105338

If you have any concerns about our products,
you can contact us on
ProductSafety@springernature.com

In case Publisher is established outside the EU,
the EU authorized representative is:
Springer Nature Customer Service Center GmbH
Europaplatz 3, 69115 Heidelberg, Germany

Printed by Libri Plureos GmbH
in Hamburg, Germany